Climate Crisis: Impacts of Climate Change

Climate Crisis: Impacts of Climate Change

Editor: Vivian Moritz

Callisto Reference
www.callistoreference.com

Callisto Reference,
118-35 Queens Blvd., Suite 400,
Forest Hills, NY 11375, USA

Visit us on the World Wide Web at:
www.callistoreference.com

ISBN: 978-1-63239-925-0 (Hardback)

The publisher's policy is to use permanent paper from mills that operate a sustainable forestry policy. Furthermore, the publisher ensures that the text paper and cover boards used have met acceptable environmental accreditation standards.

Printed in China.

Cataloging-in-Publication Data

Climate crisis : impacts of climate change / edited by Vivian Moritz.
p. cm.
Includes bibliographical references and index.
ISBN 978-1-63239-925-0
1. Climatic changes. 2. Global environmental changes. 3. Environmental impact analysis. 4. Climatology.
I. Moritz, Vivian.

QC903 .C55 2018
363.738 74--dc23

Table of Contents

Preface

This book aims to highlight the current researches and provides a platform to further the scope of innovations in this area. This book is a product of the combined efforts of many researchers and scientists, after going through thorough studies and analysis from different parts of the world. The objective of this book is to provide the readers with the latest information of the field.

Global warming is a growing concern in our times. Some human activities have adversely affected the ecology of our planet. The changes in climate that have been observed are rising sea levels, changing precipitation patterns, global warming, etc. The changes can also result in the occurrence of natural hazards. This book presents the complex subject of climate crisis in the most comprehensible and easy to understand language. It includes contributions of experts and scientists which will provide innovative insights into this field.

I would like to express my sincere thanks to the authors for their dedicated efforts in the completion of this book. I acknowledge the efforts of the publisher for providing constant support. Lastly, I would like to thank my family for their support in all academic endeavors.

Editor

Increased Drought Impacts on Temperate Rainforests from Southern South America: Results of a Process-Based, Dynamic Forest Model

Alvaro G. Gutiérrez[1,2*¤], Juan J. Armesto[3,4], M. Francisca Díaz[5], Andreas Huth[1]

1 Department of Ecological Modeling, Helmholtz Centre for Environmental Research (UFZ), Leipzig, Germany, **2** Forest Ecology Group, Institute of Terrestrial Ecosystems, Department of Environmental Sciences, Swiss Federal Institute of Technology (ETH Zürich), Zürich, Switzerland, **3** Instituto de Ecología y Biodiversidad (IEB), Santiago, Chile, **4** Departamento de Ecología, Facultad de Ciencias Biológicas, Universidad Catolica de Chile, Santiago, Chile, **5** Departamento de Ciencias Biológicas, Facultad de Ciencias Biológicas, Universidad Andrés Bello, Santiago, Chile

Abstract

Increased droughts due to regional shifts in temperature and rainfall regimes are likely to affect forests in temperate regions in the coming decades. To assess their consequences for forest dynamics, we need predictive tools that couple hydrologic processes, soil moisture dynamics and plant productivity. Here, we developed and tested a dynamic forest model that predicts the hydrologic balance of North Patagonian rainforests on Chiloé Island, in temperate South America (42°S). The model incorporates the dynamic linkages between changing rainfall regimes, soil moisture and individual tree growth. Declining rainfall, as predicted for the study area, should mean up to 50% less summer rain by year 2100. We analysed forest responses to increased drought using the model proposed focusing on changes in evapotranspiration, soil moisture and forest structure (above-ground biomass and basal area). We compared the responses of a young stand (YS, ca. 60 years-old) and an old-growth forest (OG, >500 years-old) in the same area. Based on detailed field measurements of water fluxes, the model provides a reliable account of the hydrologic balance of these evergreen, broad-leaved rainforests. We found higher evapotranspiration in OG than YS under current climate. Increasing drought predicted for this century can reduce evapotranspiration by 15% in the OG compared to current values. Drier climate will alter forest structure, leading to decreases in above ground biomass by 27% of the current value in OG. The model presented here can be used to assess the potential impacts of climate change on forest hydrology and other threats of global change on future forests such as fragmentation, introduction of exotic tree species, and changes in fire regimes. Our study expands the applicability of forest dynamics models in remote and hitherto overlooked regions of the world, such as southern temperate rainforests.

Editor: Bruno Hérault, Cirad, France

Funding: This study was supported by grants from Millennium Scientific Initiative (P05-002) and CONICYT PFB-23 to the Institute of Ecology and Biodiversity. AGG was funded by a DAAD doctoral fellowship, by a Marie Curie Intra European Fellowship within the 7th European Community Framework Programme (PIEF-GA-2010-274798) and CONICYT-PAI grant (82130046). Additional funding was provided by the ERC advanced grant 233066. The funders had no role in study design, data collection and analysis, decision to publish, or preparation of the manuscript.

Competing Interests: The authors have declared that no competing interests exist.

* Email: bosqueciencia@gmail.com

¤ Current address: Instituto de Conservación Biodiversidad y Territorio, Facultad de Ciencias Forestales y Recursos Naturales, Universidad Austral de Chile, Valdivia, Chile

Introduction

Climate and forests are dynamically linked through the spatial and temporal variability of soil moisture [1], with climate system effects on ecological processes which are still poorly understood. Forest dynamics models, particularly those based on interactions among individual trees (i.e. gap models [2]), provide a simple, and general framework to assess the impacts of climate on forest dynamics. These models simulate the fate of single trees on the basis of species' life-history traits and limited resource availability (e.g. soil moisture), thereby facilitating the analysis of climate-forest interactions [3].

Forest gap models use a variety of approaches to model forest hydrology. While some gap models use a simple bucket water balance model [3,4], others include physiology-based representations of plant and soil controls on water uptake and evapotranspiration [5,6]. Regardless of the level of detail used to model forest hydrology, it seems necessary that forest gap models address water availability (i.e. soil moisture) as an integrating factor, with effects on canopy transpiration [7]. Changes in rainfall regimes, summarized by changes in the duration and frequency of periods of water stress during the year, should influence soil moisture dynamics limiting plant productivity [8]. Introducing dynamic linkages of ecological processes with soil moisture variation in gap models will contribute to predict drought-induced changes in forest dynamics. Such model improvements are increasingly relevant to understanding how forests can adapt to climate change [6,9].

Forest gap models have successfully simulated the dynamics of a variety of forest types including temperate rainforests of the southern hemisphere [10,11]. In southern South America (SSA, 37–43°S), the progressive loss, fragmentation and subsequent degradation of temperate rainforests due to unsustainable logging and fire is threatening the integrity of ecosystem functions [12,13] and modifying their hydrological balance [14,15]. Annual precipitation has decreased in the same region by about 40% in the last century (time period 1901–2005, [16]) and summer rainfall is expected to decrease up to 50% by the year 2100 [17,18]. SSA forests share similar structural characteristics with temperate rainforests of the Pacific Northwest of North America, Tasmania, and New Zealand [19]. In addition, SSA forests represent the largest area of temperate forest remaining in the southern hemisphere [20]. Floristic richness is the highest among evergreen temperate rainforests worldwide and the high concentration of endemism has given this region a unique global conservation value [20,21]. The global relevance of SSA forests and climate trends predicted for the coming decades make it urgent to expand model applications into this region, as a tool to predict temperate rainforest responses to impending declines in rainfall.

This study introduces a forest gap model specifically designed for assessing the responses of temperate rainforests in southern South America to increased drought. The model provides accurate estimates of forest water fluxes and incorporates dynamical linkages among rainfall regimes, soil moisture, and individual tree growth. We assessed model performance by comparing the results with detailed field measurements of water cycling in a stand located on northern Chiloé Island, Chile (41°50′S). We also conducted a sensitivity analysis of the response of current forests to drought, i.e. when rainfall is decreased. Model predictions of forest hydrology (evapotranspiration and soil moisture) and structure (above-ground biomass and basal area) under increased drought predicted for 2100 in the study area were compared for a young-secondary (YS) and an old-growth (OG) forest stand to analyze differences in responses to expected changes in rainfall.

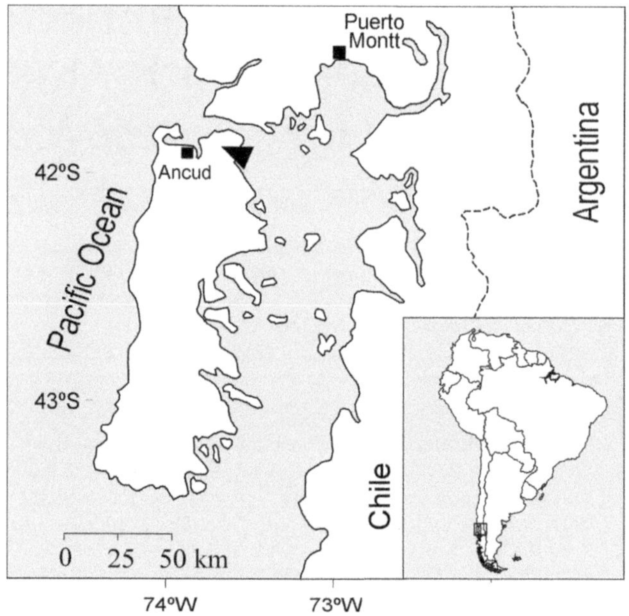

Figure 1. Location of study site (triangle) on northern Chiloé Island, Chile.

Materials and Methods

Study area

The study was conducted on northern Chiloé Island, Chile (41°50′ S, Fig. 1) at the private protected area *Estación Biológica Senda Darwin* (EBSD), with permission granted by the owner. Fragments of secondary and primary forests occur over rolling hills of low altitude (50–100 m) dispersed in a matrix of bogs, shrublands and grazing pastures. The present landscape has been shaped by a history of widespread use of fire to clear land for pastures since the late 1800s, followed by selective logging of remaining forest patches [22]. Soils are generally thin (<0.5 m), originated from moraine fields and outwash plains from the last glaciation, often with poor drainage [23]. Soils have high organic matter content, soil texture loam to silty loam, and a 2–4 mm thick iron silicate layer or hardpan (found at ca. 52 cm depth), where roots cannot penetrate [24]. The prevailing climate is wet-temperate with strong oceanic influence [25]. Rainfall occurs throughout the year, with an annual average of 2158 mm (25% occurring in summer). Mean annual temperature is 9.1°C. Maximum and minimum monthly temperatures are 13.9°C (January) and 4.2°C (July) [26].

Floristically, forests of the study area belong to the North Patagonian temperate rainforest type [27]. The canopy is dominated by evergreen trees, mainly *Podocarpus nubigena* (Podocarpaceae), *Drimys winteri* (Winteraceae) and *Nothofagus nitida* (Nothofagaceae), with the common presence of *Tepualia stipularis* (Myrtaceae) and several Myrtaceae tree species in the understory. Ferns (e.g. *Hymenophyllum* spp., *Hymenoglossum cruentum, Polypodium feullei)* and angiosperms (e.g. Gesneriaceae and Bromeliaceae) growing epiphytically are frequent. Detailed descriptions of structure and dynamics of this forest type have been previously published [26,28]. The study did not involve endangered or protected tree species.

The forest model

Here, we introduce an individual-oriented dynamic forest model (`FORMIND-CL v.1.0`) that includes calculations of hydrologic balance. The model is based on `FORMIND`, a forest model comprehensively tested to simulate the dynamics of temperate rainforests in SSA [11,13]. `FORMIND` is a generalized forest growth model that simulates the spatial and temporal dynamics of uneven-aged, mixed species forest stands [29–31]. The model simulates forest dynamics (in annual time steps, *t*) as a mosaic of interacting forest patches of 20×20 m, which is the approximate crown size of a large mature tree in the forest. Within these patches, stand dynamics is driven by competition for light and space following the gap model approach [2]. For the explicit modeling of the competition for light, each patch is vertically divided into height layers of 0.5 m, where leaf area is summed up and the light environment under the canopy is calculated via a light extinction law. The carbon balance of each individual tree is modeled explicitly, including the main physiological processes (photosynthesis and respiration [13]). Allometric functions and geometrical relations are used to calculate above-ground biomass, tree height, crown diameter and stem volume from the stem diameter at 1.3 m height of the tree (*dbh*). Tree mortality can occur either through self-thinning in densely populated stands, tree senescence, gap formation by large falling trees, slow tree grow, or external disturbances (e.g. windthrow). Gap formation links neighboring forest areas. Tree regeneration rates are formulated as maximum rates of recruitment of small trees at *dbh* threshold of 1 cm, with seed loss through predation and seedling mortality being incorporated implicitly [13]. Maximum recruitment rates are reduced by shading. Nutrient availability is considered to be homogeneous at the stand scale. A description of the core model and its equations is given elsewhere [11,13]. We focus below on the extensions added to incorporate forest hydrology.

Table 1. Structure and composition of the young-secondary North Patagonian forest stand (YS) in Senda Darwin Biological Station, compared to values reported for other secondary forest stands, elsewhere in Chiloé Island, Chile.

	Study site		Literature*			
Species	Density (trees ha^{-1})	Basal area (m^2 ha^{-1})	Density (trees ha^{-1})		Basal area (m^2 ha^{-1})	
Drimys winteri	1550	32.3	1800	4532	24.7	61.4
Eucryphia cordifolia	175	0.9	112	480	1.4	14.7
Myrtaceae†	1014	3.5	252	827	2.3	4.4
Nothofagus nitida	713	17.7	4	260	0.1	4
Podocarpaceae¥	38	0.1	80		0.5	
Weinmannia trichosperma	88	0.3				
Others	287.5	1.9				
Total	3575	54.8	2431	7950	46	86.9

*Source: [26,77–80] only forest stands <100 years old. Only listed tree species present in YS.
†*Tepualia stipularis, Myrceugenia spp.* and *Amomyrtus spp.*;
¥*Podocarpus nubigena, Saxegothaea conspicua.*

The hydrologic submodel

Soil moisture dynamics is described at a daily timescale, treating soil as a reservoir with an effective storage capacity that is intermittently filled by rainfall events. Soil water losses occur via transpiration, interception by the forest canopy, and drainage below the root zone. We neglected lateral water flow, thus the model applies mainly to flat terrains. This is a reasonable assumption in forests of the study area because during the rainy season soils tend to be saturated and accumulated water cannot infiltrate the soil.

Soil moisture s (dimensionless, $0 \leq s \leq 1$), vertically averaged over the soil depth z (mm), was considered as central state variable [8]. Thus, the water balance equation for a given point in the forest can be expressed as [1]:

$$n \cdot z \cdot \frac{ds}{dd} = Pnet_d - Tr_d - Q(s,d) \qquad (1)$$

where d is the Julian day of the year, n is the porosity (volume of voids/total volume of soil, i.e. dimensionless, vertically averaged); $Pnet_d$ is the net precipitation falling to the soil surface (mm day^{-1}); Tr_d is the transpiration rate (mm day^{-1}); and $Q(s,d)$ is the soil drainage (mm day^{-1}). Both n and z are assumed to be time-invariant parameters [1]. The volumetric water content (θ, m^3 water/m^3 soil, i.e. dimensionless) can be calculated as follows [1]:

$$\theta = s \cdot n \qquad (2)$$

The normalized version of equation (1) is used through the text where all terms are divided by $n \bullet z$. Both the local vertical and horizontal variability of soil moisture are considered negligible at the daily timescale, assuming an equal propagation of the wetting front and equal soil moisture redistribution over the rooting zone [8,32].

Net precipitation. Daily net precipitation falling to the soil surface ($Pnet_d$) is described by,

$$Pnet_d = P_d - Ec_d \qquad (3)$$

where, Ec_d is the canopy interception (mm day^{-1}), defined here as the total daily rainfall (P_d, mm day^{-1}) that is retained by the canopy and is evaporated so that it does not reach the ground. Following [33], we assumed that Ec_d asymptotically approaches the canopy retention capacity and can be modeled at daily intervals as:

$$Ec_d = S_t \cdot (1 - e^{-\alpha_h \cdot P_d}) \qquad (4)$$

where S_t is the canopy water retention capacity of the stand at year t and α_h is a parameter describing the slope of the saturation curve. The parameter α_h represents, in a simplified terms, the complex process of water partitioning into throughfall and stem flow [34]. S_t depends on leaf area index of the forest patch at simulated year t (LAI_t) and is calculated by the expression [34,35]:

$$S_t = \frac{LAI_t}{LAI_{\max}} \cdot f_h \cdot \log(1 + LAI_t) \qquad (5)$$

where, LAI_{max} is the maximum leaf area index of the forest and f_h is a shape parameter. We avoided unrealistic canopy interception values in the model by setting $Ec_d = P_d$ when $Ec_d > P_d$.

Table 2. Parameter descriptions and parameterization methods used for running simulations in `FORMIND-CL v.1.0`.

	Description	Value	Units	Method	Reference
Weather generator					
$1/\lambda$	Mean interval time between rainfall events[¥]	†	days	a	This study
η	Mean depth of rainfall events[¥]	†	mm day^{-1}	a	This study
T_μ	Mean daily temperature[¥]	†	°C	a	This study
T_σ	Standard deviation daily temperature[¥]	†	°C	a	This study
Rg_μ	Daily global radiation above canopy[¥¶]	†	µmol (photons) $m^{-2}\,s^{-1}$	a	This study
Rg_σ	Standard deviation of Rg_μ [¥¶]	†	µmol (photons) $m^{-2}\,s^{-1}$	a	This study
Hydrologic submodel					
n	Vertically averaged porosity of the soil	0.757	-	b	[53]
z	Soil depth	520	mm	b	[14]
k_{soil}	Saturated hydraulic conductivity	4	mm day^{-1}	b	[36,54]
α_h	Slope of the canopy saturation curve	0.7	-	b, c	[34]
f_h	Parameter of the relationship LAI and canopy storage capacity	3	mm day^{-1}	c	[35]
LAI_{max}	Maximum LAI of the studied forest	5.5	$m^2\,m^{-2}$	b	[81]
WUE	Water-use efficiency	9	g CO_2 $kg^{-1}H_2O$	e	[14]
γ	Psychometer constant	65	Pa K^{-1}	b	[43,44]
L	Latent heat of vaporization of water	2.56×10^6	J kg^{-1}	e	[43,44]
Δ	Rate of change of saturated vapor pressure with temperature	c	Pa K^{-1}	d	[43,44]
θ_{wp}	Wilting point of the soil	0.125	-	b	[36]
θ_{fc}	Field capacity of the soil	0.3	-	b	[36]

Method refers to a: calculated from daily meteorological data from *Senda Darwin Biological Station*, period 1998–2009, b: from literature, c: calibrated with field data, d: calculated, e: calibrated using literature.
†values indicated in Table 3.
¥per season.
¶calculated for dry ($Pd<1$ mm) and wet days ($Pd\geq1$ mm).

Soil moisture modeling. Drainage out of the root zone ($Q(s,d)$) was modeled according to [1]. When the soil is saturated ($s=1$), soil water is permitted to percolate at a rate equivalent to the saturated hydraulic conductivity of the soil (k_{soil}, mm day^{-1}, [36]). Runoff occurs when the soil is saturated and no more water can be held in place. The excess of water is assumed to leave the system, which is reasonable to assume given the large rainfall intensity in the study area. When $s<1$, soil deep percolation rate is calculated using the empirical relationship of Neilson [37],

$$Q(s,d)=k_{soil}\cdot s^2 \quad (6)$$

Transpiration. Water-use efficiency describes the proportion of water used for the assimilation of a unit of carbon in the photosynthesis [38,39]. This concept can be used to estimate transpiration of trees (Tr, mmol H_2O $m^{-2}\,s^{-1}$) from:

$$Tr=\frac{PB}{WUE} \quad (7)$$

where, PB is the gross biomass production of the tree (µmol carbon dioxide $m^{-2}\,s^{-1}$), and WUE is a parameter denoting water-use efficiency at stand level. PB is obtained from the rate of single-leaf photosynthesis following [40], which is integrated over the total LAI of the tree to account for self-shading [13]. The resulting photosynthetic rate is then multiplied by the crown area of the tree to obtain PB (see also equations in Appendix S1 and [13]). Daily transpiration (Tr_d) of trees is obtained from equation 8 and dividing Tr by the length of the active photosynthetic period per year.

The daily potential evapotranspiration (PET_d, mm day^{-1}) describes a physical limit for the amount of water that can be held and transported away from the canopy under given climatic conditions. Evaporation is neglected in the model; therefore, it is assumed that maximum water losses by vegetation are limited by the difference between PET_d and the canopy interception of the day (Ec_d), as follows:

$$PET_d-Ec_d\geq Tr_d \quad (8)$$

PET_d is calculated using a modified Penman-Monteith expression in case of aerodynamic conductance [41,42] and determined by the variation of the daily net radiation flux (Rn_d, J m^{-2} day^{-1}):

$$PET_d=(\frac{\Delta}{\Delta+\gamma})Rn_d/L \quad (9)$$

where γ is the psychometric constant (ca. 65 Pa K^{-1}, slightly depends on temperature), L is the latent heat of vaporization of

Table 3. Parameter values used to run the weather generator under different climatic scenarios.

	Current climate			Future scenarios		
Rainfall	Seasonal sum (mm, average ± sd)	1/λ (days)	η (mm day^{-1})	Seasonal sum (mm, average)	1/λ (days)	η (mm day^{-1})
DJF	284.6±132.7	0.9	8.3	238–131	0.99–1.36	4.2–7.5
MAM	543.6±138.6	0.55	12.8	813.4	0.55	12.8
JJA	813.4±197.2	0.29	16	543.6	0.29	16
SON	424.5±144.7	0.54	9.5	382–212	0.59–0.8	4.7–8.5
Annual sum (mm, average ± sd)	2094.8±353.8					
Temperature	T_μ (°C)	T_σ (°C)			T_μ (°C)	
DJF	12.5	2.6			16.5	
MAM	10.1	3.6			13.1	
JJA	8.4	4.4			10.4	
SON	9.8	3.3			11.8	
Radiation	Rg_μ (Rg_σ)	Rg_μ (Rg_σ)				
	[P_d<1 mm]	[P_d≥1 mm]				
DJF	1413.9 (317.0)	986.9 (378.2)				
MAM	701.2 (294.1)	398.8 (261.6)				
JJA	408.9 (159.1)	229.7 (139.4)				
SON	1065.5 (348.7)	640.8 (320.1)				

Current climate indicate parameters used to run the model under current climate based on instrumental records (weather station at *Senda Darwin Biological Station*, period 1998–2009). Radiation describes parameters daily global radiation Rg_μ and Rg_σ (the latter in brackets, µmol(photons) m^{-2} s^{-1}). Temperature is mean daily air temperature. *Future scenarios* are the range of climatic parameters that were varied to run the model under increased drought (36 scenarios in total, see *Methods* for details). *DJF:* December to February (austral summer, growing season); *MAM*: March to May (austral autumn); *JJA*: June to August (austral winter), *SON*: September to November (spring, growing season). sd: standard deviation.

water (ca. 2.56×10^6 J kg^{-1} slightly depends on temperature). The rate of change of saturated vapor pressure with temperature (Δ, Pa K^{-1}) is calculated as [43,44]:

$$\Delta = \frac{2.503x10^6 e^{17.269\frac{T_d}{237.3+T_d}}}{(237.3+T_d)^2} \quad (10)$$

Rn_d was calculated from latitude, day of the year, sunshine hours and daily air temperature (T_d, °C) following [12,45].

Soil moisture impact on tree biomass production. The dependence of water uptake for tree biomass production on soil moisture is described by a function representing a reduction factor due to water scarcity ($\omega(s)$, $0\leq \omega(s) \leq 1$ [46]). This factor accounts indirectly for the impact of water demand on potential photosynthetic production (i.e. possible to achieve under competition for light). $\omega(s)$ is implemented as a daily reduction factor due to water scarcity by,

$$\omega(s) = 0 \text{ if } s < \theta_{wp}$$

$$\omega(s) = \frac{s-\theta_{wp}}{\theta_{msw}-\theta_{wp}} \text{ if } \theta_{wp} < s \leq \theta_{msw}$$

$$\omega(s) = 1 \text{ if } s > \theta_{msw} \quad (11)$$

where, θ_{wp} is the wilting point, and θ_{msw} represents a threshold when enough soil moisture is available for potential tree biomass production. We calculated θ_{msw} from:

$$\theta_{msw} = \theta_{wp} + \frac{1}{3}(\theta_{fc} - \theta_{wp}) \quad (12)$$

where, θ_{fc} is the soil field capacity. θ_{wp}, θ_{msw}, and θ_{msw} are expressed as normalized soil moisture. In the model, the wilting point (θ_{wp}) determines the minimum soil moisture content necessary for tree biomass production. Thus, we assumed a linear reduction of biomass production when soil water content was between θ_{msw} and θ_{wp}. The water required for biomass production of trees is completely removed from the soil compartment when soil moisture reaches θ_{msw} (i.e., $\omega(s) = 1$), after the calculation of maximum possible transpiration of trees. Both biomass production and water supply are reduced until the water needed for biomass production corresponds with θ_{wp}. The calculated rate of biomass production influences tree respiration rate through maintenance and growth respiration, which are calculated subsequently in the model (see [13] for equations and Fig. S1 for a diagram). All calculations are performed for every tree and pooled together to calculate the stand-level values.

Weather generator. Rainfall time series, representing the frequency and depth of rainfall events, were constructed as series of random numbers generated by probability distributions. The interval between rainfall events, τ (day) can be expressed as an exponential distribution given by [47].

$$f_T(\tau) = \lambda \cdot e^{-\lambda\tau}, \text{ for } \tau \geq 0 \quad (13)$$

Table 4. Model estimates of water balance components for a young secondary forest (YS, 60 years-old) and an old-growth North Patagonian forest (OG, >500 years-old) located in northern Chiloé Island, Chile, under current climate.

Variable	This study		Literature	
	mm year^{-1}± sd	%	Value	Source*
Young secondary stand (YS)				
Canopy Interception	381.5±30.3	20.0	20–40%	1
Deep percolation	980.7±156.8	49.9	47%	1
ET	573.6±35.1	30.3	45.2%	1
Net precipitation	1617.2±434.6	80.0	60–80%	1
Runoff	722.9±397.7	33.3		
Soil moisture	-	62.5		
Transpiration	192.1±18.8	10.3	22%	1
Old-growth stand (OG)				
Canopy Interception	378.3±33.6	19.9	17.8%	2, 3
Deep percolation	907.4±142.4	46.5	66.5%	2
ET	648.0±44.1	34.3	19.9–33.3%	2
Net precipitation	1591.8±368.6	80.1		
Runoff	665.2±321.9	31.8	30–55%	4
Soil moisture	-	55.3		
Transpiration	269.7±26.1	14.4		
Potential evapotranspiration	769±4		576–724 mm year^{-1}	5

Model results are the average of 100 simulations per stand (see *Methods* for details), with annual sum of rainfall averaging 1970–2000 mm year^{-1}. Literature refers to values reported by independent studies in Chilean temperate rainforests and comparable forests elsewhere. sd: standard deviation; %: percentage of total annual rainfall; ET: Evapotranspiration (sum of canopy interception and transpiration). sd: standard deviation.

*(1) Data for other broad-leaved evergreen forests, ca. 200 years old. Annual rainfall 2500 mm year^{-1} [82,83]. (2): Mixed deciduous-broad-leaved old-growth forest. Annual rainfall 2400 mm year^{-1}[12]. (3) Mixed broad-leaved and conifer forest, ca. 200 years old in New Zealand. Annual rainfall 3400 mm year^{-1}[84]. (4) Annual rainfall 1700–4500 mm year^{-1}, data from evergreen, broad-leaved forests with 90% cover [85]. (5) Annual rainfall of 2427–3991 mm year^{-1} [86], weather stations of Castro and Punta Corona.

where $1/\lambda$ is the mean time interval between rainfall events (days). Total daily rainfall (P_d) depends on the amount of rain of each event (h, mm day^{-1}), which is also assumed to be an independent random variable, expressed by an exponential probability density function [47]:

$$P_d(h) = \frac{1}{\eta} e^{-\frac{1}{\eta}h}, \text{ for } h \geq 0 \quad (14)$$

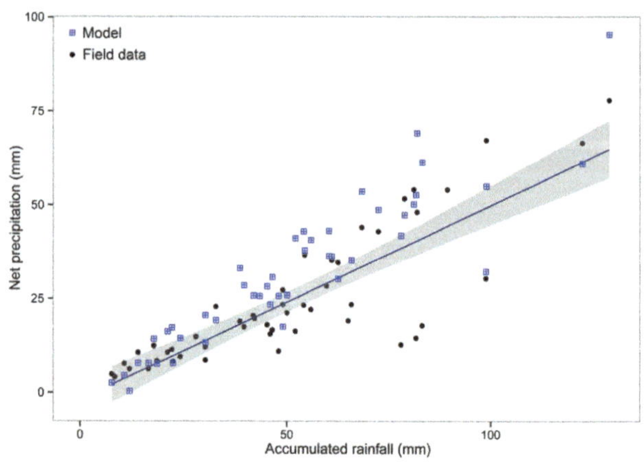

Figure 2. Comparison between measured and modeled net precipitation for 50 rain events recorded in a young secondary stand in northern Chiloé Island, Chile, for the time period 2007–2010. The line represents a linear regression between field measured net precipitation and accumulated rainfall during each event. The gray area represents 0.95 confidence intervals. Model results are for a forest patch are of 400 m^2, with a LAI = 5.

where η is the mean depth of rainfall events (mm day^{-1}). Both $1/\lambda$ and η parameters are calculated for each season of the year. We obtained daily global radiation (Rg_d) from EBSD instrumental records (period from May 1998 to February 2009). Rg_d varied among seasons in relation to daily rainfall (T-test, $p<0.001$). Therefore, in the model, Rg_d was distributed as a Gaussian variable whose mean and standard deviation depended on P_d (mean, Rg_μ, and standard deviation, Rg_σ). Values of Rg_μ and Rg_σ were obtained from instrumental records and varied depending on a threshold value of 1 mm of P_d and the season of the year. Daily temperature (°C, T_d) was simulated by a Gaussian random variable with parameters (mean, T_μ, and standard deviation, T_σ) that varied according to season of the year.

Field data

Stand structure. We characterized stand structure in terms of tree species dominance (basal area, m^2 ha^{-1}) and size (*dbh*) distribution in a young secondary and an old-growth North Patagonian forest stand (hereafter YS and OG respectively) found in a flat forested area at EBSD (Fig. 1). The YS stand was initiated from a stand-replacing fire about 60 years ago and currently presented no evidence of logging. In 2007, we set up two 20×20 m

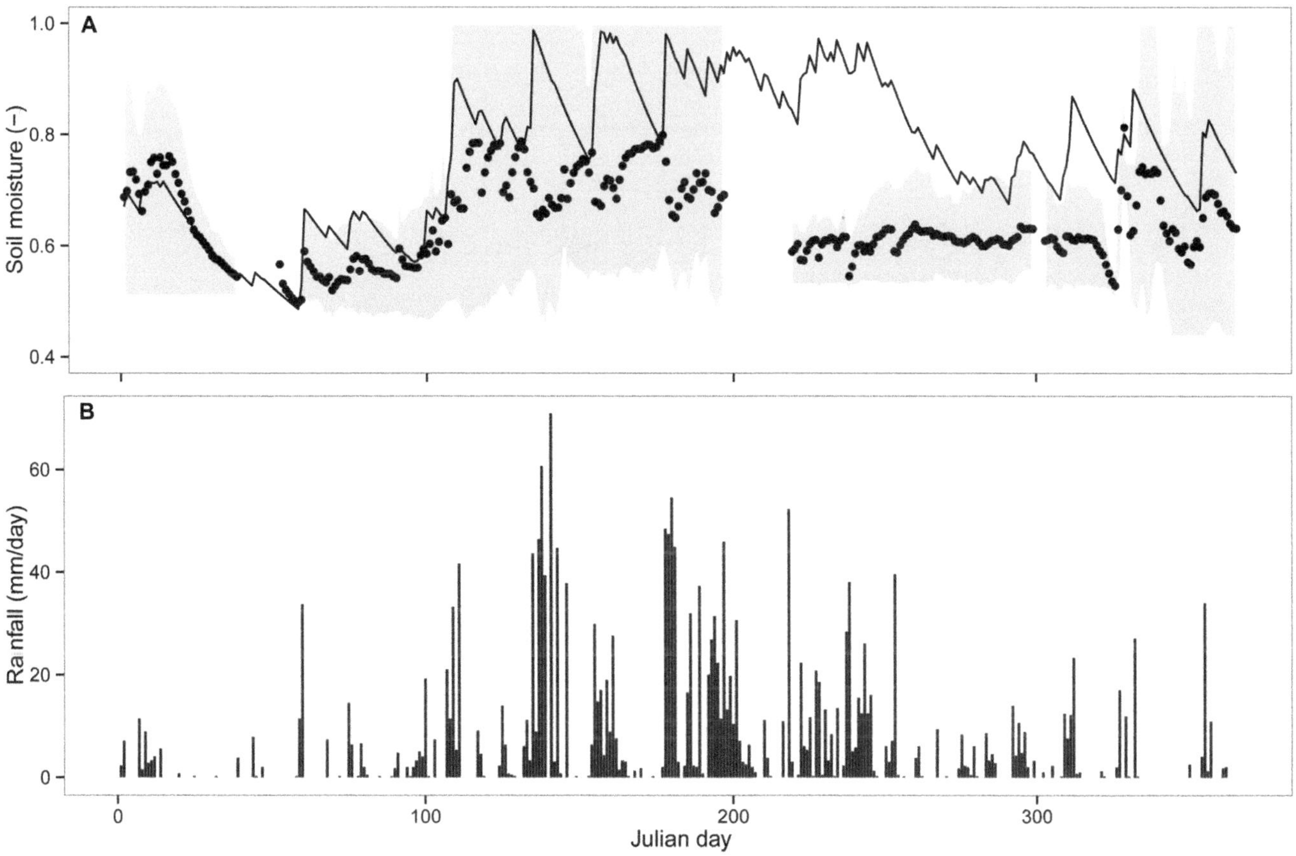

Figure 3. Variation in soil moisture during 2008 in a young secondary, North Patagonian forest stand (YS) in northern Chiloé Island, Chile. a) Comparison between observed and simulated soil moisture (normalized data, dimensionless). Observed data were obtained from five sensors randomly placed inside a 400-m^2 plot in YS (daily means denoted by filled dots, gray area showing the range of data). Simulated soil moisture (line) is for one-hectare forest with a successional age of 60 years (LAI = 4.5). Soil parameters are the same as described in Table 2 (see also *Methods*). b) Rain events recorded instrumentally during 2008.

plots to measure the hydrologic balance and stand structure. All trees rooted within each plot with stems >1.3 m height and > 5 cm *dbh* were permanently marked with numbered aluminum tags, identified to species, and their *dbh* measured to the nearest cm. Structure and composition of YS is comparable to that described for young-secondary stands elsewhere on northern Chiloé Island (Table 1). OG was an unmanaged forest stand >590 years old, without evidence of recent human disturbance, and representative of old-growth North Patagonian forest on Chiloé Island and elsewhere in the region [26]. Sampling methods, stand history, species composition and structure of OG is described by [28]. OG had total basal area of 72 m^2 ha^{-1} and density of 2610 trees ha^{-1}, with a mixed dominance of Podocarpaceae species, *D. winteri, N. nitida* and Myrtaceous tree species in the understory [28].

Hydrologic measurements. We estimated net precipitation using throughfall measurements taken in two YS forest plots (i.e. rainfall falling through canopy gaps plus canopy drip), adding stemflow (water running down the stems). We conducted these measurements of volume accumulated during rainfall events occurred between June 2007 and December 2010. During this time period, we also analyzed hourly records of rainfall from the meteorological station at EBSD to obtain daily incident rainfall above the canopy. Rainfall events considered in the analysis occurred with a separation of at least two hours without rain to allow for full drip from the forest canopy. Stemflow collectors consisted of a 2 mm thick smooth polycarbonate sheet molded around the stem to form a funnel. A hose led from the lowest point of the funnel to a 25 l polythene container, where the stemflow volume was collected after each rain event. Stemflow collectors were placed in 10 randomly selected trees of the two main canopy species, *D. winteri* and *N. nitida* (*dbh* >10 cm), in each plot. We eliminated two trees from our stemflow estimates that died during the study period. We converted the volume of collected water to millimeters of rain assuming that the surface of the collectors equals the projected tree crown area. Crown area was approximated by the area of an ellipse. Throughfall collectors, 0.12×2 m long (0.7 m^2 total area per plot) gutters were held, with a slight inclination, 0.5 m above the ground at three different locations within each plot. Collectors were connected with a funnel to a 25 l polythene container. Soil matric potential was measured every 30 minutes with four sensors per plot (*WatchDog Data Loggers* 450 and 800) placed approximately in every quarter of each plot, beneath the canopy and at ca. 15 cm soil depth. Continuous soil moisture measurements were obtained for the period January 2007 to March 2009.

Model parameterization

We used a previous model parameterization for North Patagonian forests including all main canopy tree species (11 tree species) occurring in the studied forests. The calibration, validation and robustness of this parameter set to reproduce forest stand

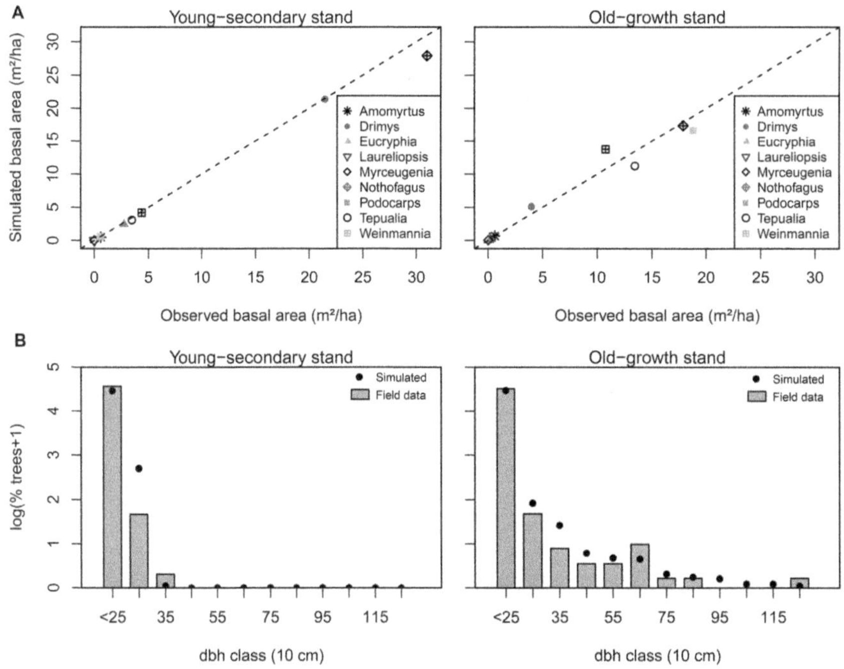

Figure 4. Comparison of forest structure between (a) observed (field data) and simulated basal area of tree species (Spearman's r^2>0.9, p<0.01 in both cases) and (b) *dbh* distributions for the young secondary (YS) and old-growth (OG) North Patagonian forest stands studied in Chiloé Island, Chile. Simulated OG structure was obtained initializing the model with inventory data.

structure is discussed in detail by [11]. Here, we describe calibration of parameters related to the inclusion of hydrologic balance into the model. New parameters needed to run `FORMIND-CL v.1.0` and their values are shown in Table 2.

The parameter f_h describing the relationship between leaf area index and canopy water storage capacity was calibrated following [34] and assuming that storage capacity reaches 4.9 mm day^{-1} at a leaf area index of 5.0 as measured by [14]. LAI_{max} was set to 5.5 following the maximum value observed in other Chilean temperate rainforests [48]. The slope of the saturation curve of the canopy rain retention capacity (α_h) was set according to common values for broad-leaved temperate trees [34]. To the best of our knowledge no estimation exists for water-use efficiency at stand scale in forests of the study area. Therefore we calibrated *WUE* using transpiration estimates of Díaz *et al.* [14] in Chiloé Island and the potential canopy photosynthetic rate estimated by the model for the study area under current climate ($Tr = 296$ mm year^{-1} and $PB = 32.9$ tC ha^{-1}. The selected *WUE* was then confirmed by comparison with reported values from other temperate rainforests [49–52]. Soil characteristics (porosity and depth) followed field descriptions from Chiloé Island [53]. We set water-retention and percolation properties of the soil (parameters θ_{wp}, θ_{fc} and k_{soil}) to average values [36,54] using texture classes (loam to silty loam) described for soils in the study area [53]. Daily records of rainfall from the EBSD weather station (60 m a.s.l, period from May 1998 to February 2009) were directly used to calculate rainfall parameters for the current climate simulations (Table 3). EBSD is the nearest and most representative weather station for the climate at the study site. We calculated the mean time interval between rainfall events from the duration (days) of rain events occurring in each season. The mean depth of rainfall events was calculated by dividing seasonal rainfall sum by the amount of wet days (P_d >0). These calculations were done only for seasons with >85 daily records. We avoided the potential overestimation of annual rainfall by normalizing predicted seasonal rainfall sum by prescribed seasonal rainfall averages (Table 3). The weather generator reproduced well the seasonal fluctuations in rainfall, temperature and radiation during the year with no significant departures from observed climatic records (Fig. S2, S3). Calibrated climatic parameters were assumed representative for growing conditions of North Patagonian forests on Chiloé Island and neighboring regions in the mainland.

Analyses

Model verification. We compared field measurements of net precipitation with model predictions at daily temporal scale. For this analysis we selected 50 rain events, representing field measurement intervals <20 days long, for which accumulated rainfall during the event (i.e. sum of daily rainfall from first to last day of the event) was correctly measured by EBSD weather station, and for which <50% of containers were filled completely during the rain event. Hydrologic parameters used for model estimation of net precipitation are indicated in Table 2. LAI measurements were unavailable for YS, thus we set LAI to comparable, averaged values reported by Diaz *et al.* [14] for the same forest type. We qualitatively compared daily variation in soil moisture produced by the model with field observations. For this analysis soil matric potential obtained for an entire year was transformed to soil moisture contents using a water retention curve [55]. We set parameters of the water retention curve following [56] for loam to silty loam texture classes and particle density of the soil type under study (ca. 2.0 g/cm^3, [53]). Model simulations were run using rainfall data for the same period of field measurements conducted over a whole year, i.e. 2008. Note that for both analyses spatial scale of model results (>1 ha) differed from the scale of field measurements (400 m^2 plots). We also compared model results with field measurements of hydrologic

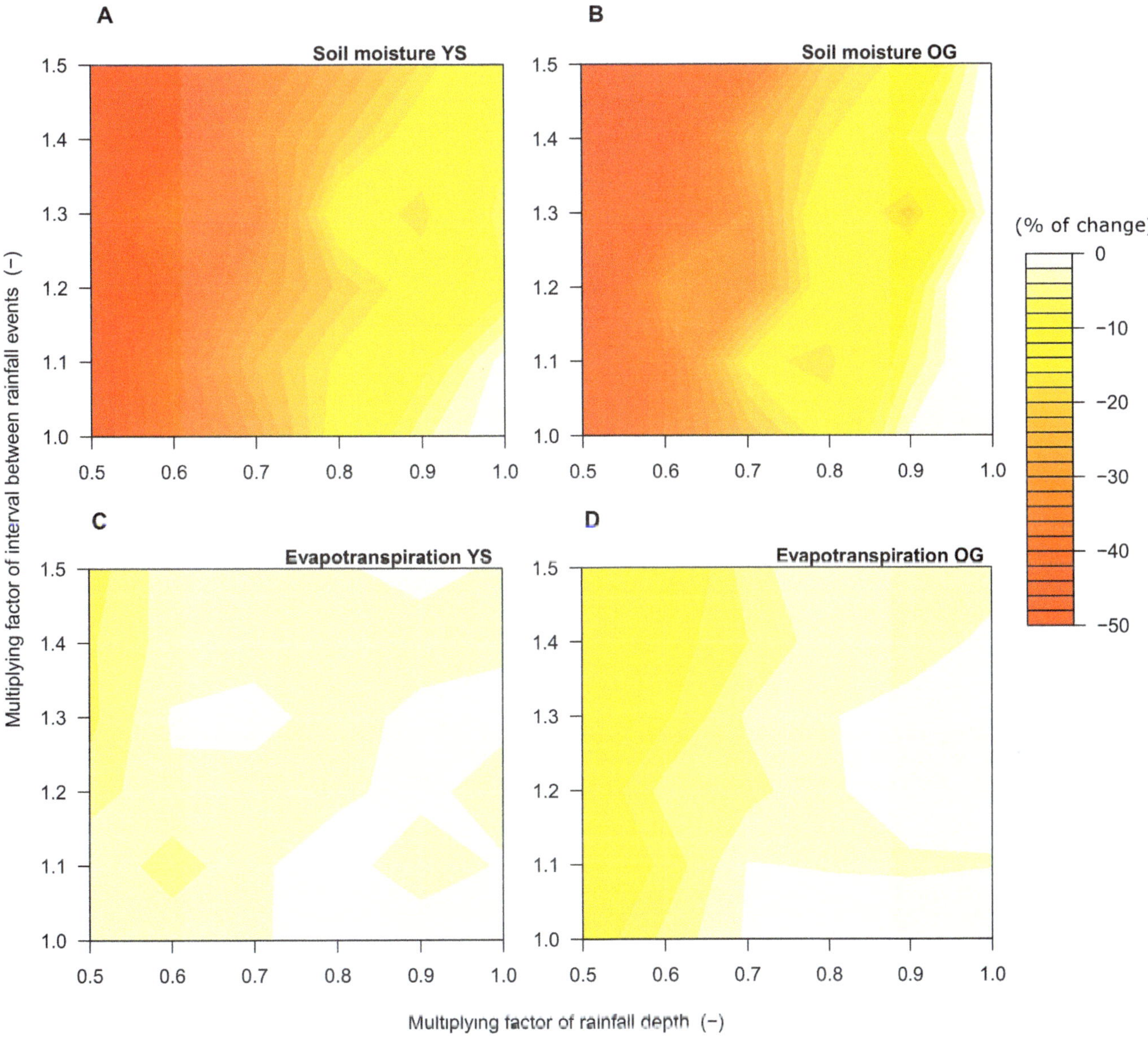

Figure 5. Sensitivity of soil moisture (a, b) and evapotranspiration (c, d) predicted by the model under increased drought in a young-secondary (YS) and an old-growth (OG) North Patagonian forest. Results represent the difference between the average under current climate (indicated in Table 4) and the average under future scenarios. A value of 0% indicates no change. The y-axis represents the variation of mean interval time between rainfall events (parameter $1/\lambda$) and the x-axis represents the variation in the mean depth of rainfall events (η) under increased drought. To represent increased drought scenarios, the parameter $1/\lambda$ was multiplied by a factor ranging from 1 (current climate) to 1.5 whereas the parameter η was multiplied by a factor ranging from 1 (current climate) to 0.5. The axes of the figures correspond to these multiplying factors of rainfall parameters. Results are the averages of 30 simulations per scenario for YS and OG.

balance at yearly temporal scales in temperate rainforests elsewhere in Chile (i.e. independent studies).

We tested model performance to reproduce forest structure under current climate. For this analysis, we compared tree basal areas and stand *dbh* distributions predicted by the model with measured structure of both YS and OG North Patagonian stands. Model comparison for the YS was performed after 60 years of succession, with succession initiated from a treeless state. To compare OG forest structure, we initiated simulations with stand inventory data and run the model for 1000 years to allow the simulated stand to reach dynamic equilibrium. We compared data at the end of the simulations with the known OG structure [28]. For each stands, we ran 100 simulations of 1 ha (i.e. 25×0.04 ha forest patches, 2500 patches in total) using current climate parameters and model parameters listed in Table 2 and 3. Demographic and species parameters were taken from [11] (site Tepual). To assess the consequences of the hydrologic submodel for the simulated forest composition, we compared species-specific basal areas reported by [11] with data from the simulations using the model version described in this study. Simulations were conducted following methods outlined by [11] and results were compared at the corresponding successional ages of YS and OG.

Simulations under increased drought. We tested the sensitivity of model predictions for forest structure and hydrological balance to changes in rainfall regimes. Climatic scenarios were selected to represent the potential range of expected climate

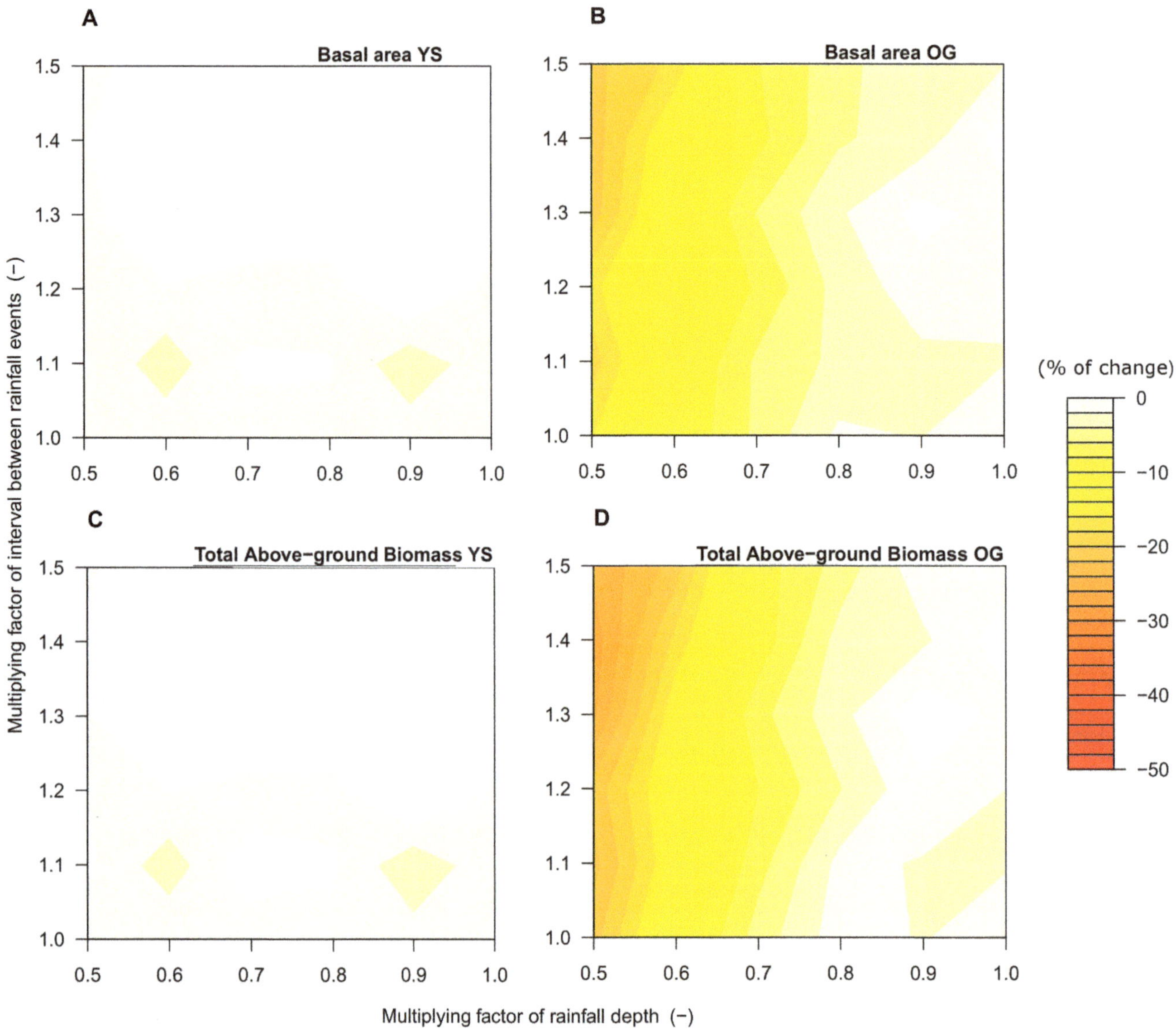

Figure 6. Sensitivity of total basal area (a, b) and above-ground biomass (c, d) predicted by the model under increased drought in a young-secondary (YS) and an old-growth (OG) North Patagonian forest. Results are the percentage of change between the average under current climate (indicated in Table 4) and the average under future scenarios. Results are the average of 30 simulations per scenario for YS and OG. The axes of the figures are as in Figure 5.

change (here increased drought) predicted for this century in the study area. To this end, we used a regional climate model downscaled for Chilean landscapes (PRECIS-DGF model, [17]). The business-as-usual scenario provided by PRECIS-DGF suggests a 50% decrease in rainfall during the growing season by year 2100. We used the daily scale output of the PRECIS-DGF model for year 2100 to calculate seasonal climatic parameters for year 2100. We linearly interpolated this seasonal rate of change between 1998–2009 and 2100. We used this scenario to set the limit of change in rainfall by year 2100 and developed climatic scenarios covering the change from current climate to the business-as-usual scenario. Climate scenarios were developed by gradually changing current climate parameters $1/\lambda$ and η until they reached the estimated value for year 2100, i.e. 50% of the current value. First, we reduced η (mean depth of rainfall events) multiplying current values by 0.9 to 0.5 in steps of 0.1. Then, we increased the parameter $1/\lambda$ (interval between events) multiplying current values by 1.1 to 1.5 in steps of 0.1. These two parameters were first varied separately (keeping the second parameter constant) and then both together. Parameter variations produced a total of 36 climate scenarios, i.e. six levels of each of the two rainfall parameters, including the current climate scenario (Table 3). In addition, we assessed the impact of warming trends on potential evapotranspiration, and its influence on other hydrologic components and forest structure. To this end, we ran an additional scenario including temperature changes expected for year 2100, with parameters $1/\lambda$ and η kept at their current values (Table 3). We considered four output variables computed at yearly temporal scale: total basal area, above-ground biomass, evapotranspiration (computed as the sum of canopy transpiration and interception) and soil moisture. The latter is dynamically linked to forest processes such as annual gross biomass production of each tree (*PB*, cf. Eq. 8, see also Fig. S1). We described changes in forest structure at different successional stages based on simulations for

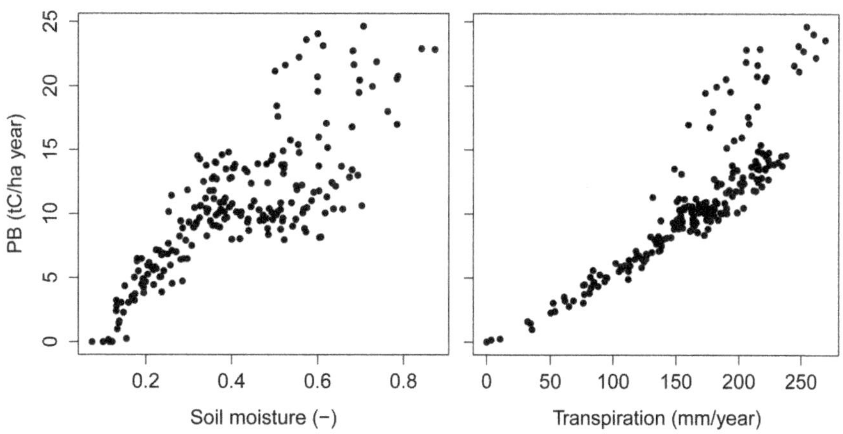

Figure 7. Shifts in annual gross biomass production (*PB*) of the YS North Patagonian forest stand in response to soil moisture and canopy transpiration changes in northern Chiloé Island, Chile. Results are simulations for one hectare of forest with a successional age of 60 years.

both YS and OG forest stands. Simulations were initialized and ran under the same conditions described in section *Model verification*. We ran 30 simulations per scenario and for each forest stand. To assess the impacts of increased drought on forests we compared the differences between means of the studied variable under current climate and means of the same variable under climate change scenarios. Finally, we checked soil moisture influence on *PB* and *Tr* by running simulations under different soil moisture conditions. We ran 216 simulations, at a scale of one ha, by monotonically varying current rainfall depth parameter ($1/\lambda$) from 0 to 100% of the current value (Table 2). All other parameters were kept constant. We assessed results at the corresponding successional ages of YS and OG, with succession initiated from a treeless state. All statistical analyses were done in R [57].

Results

Model results under current climate

Model predictions for the major components of forest hydrologic balance at a yearly temporal scale were similar to reported values for broad-leaved temperate rainforests in Chile and elsewhere (Table 4, independent studies). At a daily scale, the hydrologic model captured a large portion of net precipitation variability measured in the field (N = 50, $r^2 = 0.8$, $P<0.001$, Fig 2). However, for rain events accumulating >100 mm, the model predicted higher net precipitation than recorded in the field (Fig. 2). Modeled daily variation in soil moisture resembled the daily pattern of soil matric potential measured in the field during the year (Fig. 3, $r^2 = -0.65$, $p<0.001$). Soil moisture increased during austral fall and winter (Fig. 3a, 59< julian day <242) due to a higher frequency and depth of rainfall events and reduced transpiration of trees. In contrast, during the growing season, soil moisture gradually decreased according to the model and field data (Fig. 3). Forest structure simulated by the model with the inclusion of the hydrologic submodel was qualitatively similar to field data (Fig. 4a), but with some departures from observed data for small dbh-classes in YS (<35 cm, Fig. 4b). Total basal area of YS simulated by the model was similar to basal area observed in the field but total basal area of OG was underestimated (Fig. S4). However, model predictions for forest structure including the hydrologic submodel varied in the same manner across species as in the original model version and resembled field data for both stands ($r^2 > 0.97$, $P<0.001$, Fig. S4).

Net precipitation and runoff predictions under current climate did not differ between YS and OG stands ($P>0.12$, two-sample Wilcoxon test, Table 4). Water loss through deep percolation was lower in OG than in YS ($P<0.001$, two-sample Wilcoxon test, Table 4). Evapotranspiration was higher in OG than YS, mainly due to higher canopy transpiration rates in OG (Table 4, $P<0.001$, two-sample Wilcoxon test). Modeled soil moisture was lower in OG than in YS (0.55 and 0.63 respectively, $P<0.001$, two-sample Wilcoxon test, Table 4).

Simulations under increased drought

The model predicted changes in hydrological components and forest structure when simulations were run under expected trends of increasing drought (Fig. 5, 6). In both forest stands, YS and OG, we found similar responses of soil moisture to changes in rainfall parameters, with declines of soil moisture up to 50% (Fig. 5a, b). Reducing η to values <80% of the current value (Table 2, e.g. multiplying factor of 0.8) consistently reduced soil moisture by 20% (Fig. 5a, b). The influence of $1/\lambda$ (mean interval time between rainfall events) was negligible when η was kept constant at its current value (Fig. 5a, b). The model consistently predicted less evapotranspiration (hereafter ET) relative to current values (Fig. 5c, d). In YS, decreases in ET under drier climate were < 50 mm year^{-1} (<8% reduction relative to current values, Fig. 5c). In OG forest, ET was reduced up to 94.4 mm year^{-1} from the current value (15% reduction relative to current values Fig. 5d). Such decreases in ET occurred when current value of η was multiplied by 0.6 and $1/\lambda$ was multiplied by 1.2 (Fig. 5d). An increase in PET was predicted by the model when simulations included warming trends ($P<0.001$, two-sample Wilcoxon test, Fig. S5). Only transpiration and ET in YS changed in response to warming trends and increased drought ($P<0.05$, two-sample Wilcoxon test, Fig. S5). Moreover, in both forest stands, changes in PET due to warming trends did not transfer to changes in forest structure ($P>0.05$, two-sample Wilcoxon test, see also Fig. S5).

We did not find distinct differences in basal area and aboveground biomass (AGB) in YS attributable to changes in rainfall parameters (Fig. 6). In contrast, in OG forest increased drought produced decreases in basal area by 21% of the current value (Fig. 6b, current basal area of 63.8 m^2 ha^{-1}) and decreases in AGB by 27% of the current value (Fig. 6d, current AGB of

309.5 tC ha^{-1}). Main changes in basal area and AGB of OG stand were predicted when current η and $1/\lambda$ parameters were multiplied by 0.6 and 1.3, respectively (Fig. 6b, d).

Discussion

Model performance

We developed and evaluated the performance of an individual- and process-based dynamic forest model that incorporates detailed calculations of water cycling for temperate rainforests of southern South America. The model allows for the investigation of dynamic linkages between rainfall trends and forest processes at stand scale. Parameters selected for this study were taken from the literature or calibrated using our own field data (Table 2), thus they can be considered as empirically based. The model incorporates the main known hydrological controls on forest processes in temperate rainforests of southern South America (Table 2, 4).

The accuracy of model predictions regarding forest composition and structure were comparable to previous results (Fig. S4), which were performed using a previous model version without hydrologic balance calculations [11]. Model prediction of total basal area of forests decreased with the inclusion of the hydrologic module, mainly due to fewer large trees (>100 cm dbh) predicted by the model in OG stands (Fig. 4b). However, model predictions of forest structure were considered realistic at small spatial scales (< 0.2 ha) used for field sampling, compared to model results (25 ha, see also discussion in [11]). Some departures between observed and predicted daily values of net precipitation could also be attributed to canopy heterogeneity operating at different spatial scales of model results and empirical data. These discrepancies may appear when comparing forest hydrologic balance estimated for 1 ha against net precipitation measured within considerably smaller areas (<1 m^2). Further, model predictions of daily net precipitation were based on average estimated LAI values for YS because of the lack of field measurements. LAI is a relevant variable to understanding biogeochemical cycles [58], and it should therefore be incorporated in future hydrologic analyses. Additionally, sampling errors during field measurements cannot be disregarded. Despite these limitations, we considered model predictions of forest hydrology acceptable at daily temporal scale for YS (Fig. 2, 3).

At yearly temporal scale, the model reproduced the main hydrologic components reported for similar evergreen, broad-leaved temperate rainforests in southern South America and elsewhere (Table 4, independent studies). Differences in deep percolation predicted by the present model compared to values reported in Table 4 might be accounted for specific physical characteristics of glacially originated soils on northern Chiloé Island. Deep percolation in the model is mainly depending on k_{soil}, which was calibrated specifically for soils from northern Chiloé Island. Applying this model to forests developing on other soil types in SSA (e.g. volcanic originated soils) will require a site-specific calibration of soil parameters (Table 2). Moreover, we compared model results against a different forest type (Table 4), with shared dominance by evergreen and deciduous tree species, and differing in LAI and annual net precipitation. Despite these broad differences, Table 4 suggests that selected hydrologic parameters (e.g. α_{soil}, f_h, and LAI_{max}, see Table 1) yielded reasonable values for canopy water retention capacity of broad-leaved temperate rainforests in SSA.

Forest responses to increased drought

According to predictions from our model, North Patagonian forests are likely to be altered by increased drought predicted for this century by climate change models. In our modeling study, the simulation of 50% reduction in summer rainfall predicted for the study area (business-as-usual scenario, [17]) can induce changes in both hydrological balance (up to ca. 100 mm $year^{-1}$ decrease in ET, Fig. 5) and forest structure (up to 83 tC ha^{-1} decrease in current AGB, Fig. 6), even without considering the potential ecological effects of concomitant global warming. A direct interpretation of changes in rainfall regimes is possible because the model accounts explicitly for changes in frequency and depth of rainfall events [8,32,47]. Decreasing in the depth of rainfall alone can induce some structural changes in the studied forest type (Fig. 6), but simultaneous changes in the frequency and depth of rainfall produced the strongest changes in hydrology and structure of stands (Fig. 6). These results highlight the impact on forest structure and growth of the duration and frequency of water limitation periods.

Reductions of basal area due to increased drought (Fig. 6) conform to empirical findings that drought increases the likelihood of mortality of large trees [59–61]. Our model formulation implies that trees assimilating greater biomass will have increasing demands of water for growth (Eq. 7, see also [62]). For example, annual gross biomass production increased with greater soil moisture availability during the year, a mechanism triggered by increased canopy transpiration (Fig. 7). Consequently, big trees that occur primarily in old-growth forests (Fig. 4b) experience increased stress-induced mortality due to greater hydrologic limitations during dry years. Trees that die under increased drought produce a decrease in both stand basal area and above-ground biomass (AGB, Fig. 6). Moreover, the model predicted that the OG forest has a higher *PB* than the young secondary stand (31.6±2.3 vs. 18.2±1.2 tC ha^{-1} $year^{-1}$, both obtained under current climate), and consequently a higher water demand for biomass production. Under increased drought, water demand for biomass production in OG forest is not fully covered by soil moisture supply, which causes the predicted decline of AGB and basal area (Fig. 6). We propose that the contrasting ET and structural patterns between YS and OG predicted for the coming decades under increasing drought are mainly due to significant limitation of available soil moisture for biomass production in OG, with lower impact on YS.

Research needs and model application

Here, we focused on developing an accurate model for assessing the influence of hydrologic processes involved in forest dynamics. Using the model, we quantitatively demonstrated the relative importance of soil moisture on forest structure (Fig. 6). We excluded other hydrologic processes to keep model formulation simple and results tractable with empirical information available. Processes such as water table dynamics, root dynamics, increased run off in slopes, and soil moisture dynamics across multiple soil layers can be incorporated as more empirical data becomes available for model calibration and validation. However, we strongly suggest that future model applications prioritize processes known to have an influence on the system under study [63]. In our study area, variations in the height of the water table may interfere with ecological processes such as tree establishment and mortality [14,24]. Our model provides a convenient starting point to incorporate water table dynamics into the analysis of climate change impacts, and to explore its effects on long-term forest dynamics. Our results also highlight the need of further fieldwork and experimental research on less known, mechanistic soil parameters (e.g. k_{soil}).

The rise of atmospheric carbon dioxide concentration and ensuing climate change are influencing water-use efficiency of forests [64,65]. WUE under future climate is likely to differ among forests [66], local scales [67], and species [68]. Forest water-use efficiency is a sensitive parameter in our model (e.g. Fig. S6) and illustrates the need for detailed studies of the expected variations in this parameter under drought, and their connection to gas exchange capacity of trees in different topographic settings and for a large set of species in SSA. It was beyond the scope of the present study to discuss model behavior along drought-to-moist gradients operating at regional scales because such analyses require an accurate and quantitative assessment of species-specific water-use efficiency [4]. In SSA, WUE variations have been experimentally tested in few study sites and for only three tree species included in our study [49,69]. Additional experimental research should address the complex interaction between photosynthetic carbon assimilation and water loss via transpiration with declining water supply [70]. Future research can address this question by studying changes in photosynthetic parameters along climatic gradients. Incorporating WUE as a species-specific parameter or a state dependent variable in our model is straightforward based on further empirical information.

Here, we deliberately excluded the influence of expected regional changes in temperature on forest processes (e.g. tree growth) to rather emphasize direct impacts of drought on model results. However, the model suggests that changes in PET due to expected warming trends in the study region are negligible compared to the strong impacts of increased drought (Fig. S5). Our simulated scenarios have been done using the common assumption of vegetation dynamic models that climate-forest interactions under inter-annual variation of climatic conditions can be used as proxy also for impacts of long-term climatic variations. In the later case it would be possible that some tree species would show adaptations effects. However, tree species adaptation to climate variability is still poorly understood in SSA. Moreover, warming can modify individual tree growth by affecting photosynthesis and both plant and soil respiration [38] [71], nutrient dynamics [72] and tree establishment [73]. An undergoing study analyzes the combined effect of temperature and rainfall changes on tree demography in the study area [74].

To the best of our knowledge, this is the first application of a forest gap model in temperate rainforests of SSA that integrates dynamic calculations of forest hydrology. The present work uses the best information available to ensure that climatic patterns were directly comparable to hydrologic field measurements used for model calibration. However, our results on climate change impacts should be interpreted with caution because our baseline climate is constrained to a short period (1998–2009) within a long-term trend of rainfall (time period 1901–2005). Long-term monitoring of forest hydrology and dynamics can corroborate our results. To date, long term monitoring (>10 years) of forest hydrology is lacking in forests of SSA. As more empirical data becomes available, the model can be revised and updated. The model developed here allows for the analysis of multiple environmental factors driving forest dynamics. For example, our model can help us understand the long-term responses of regional forest to drought events induced by El Niño Southern Oscillation that amplify background tree mortality rates in *Nothofagus* forest of SSA [75]. Moreover, increased regional drought is likely to interact with other drivers of global change such as changes in fire regimes, massive introduction of exotic forestry species, and forest fragmentation. To date, the implications of these drivers in the context of a changing climate remain poorly understood in SSA. These interacting growing threats demand from ecologists to understand and integrate multiple dimensions of global change on forest functioning. The model presented here is a particularly suitable tool for analyzing broad global change questions in forests of SSA because it also includes logging and fragmentation submodels [13,76]. Model-based experiments can also contribute to develop sound management strategies that anticipate forest responses to increasing drought and other drivers of climate change.

Conclusions

We developed and applied a forest dynamic model to analyze the impact of climate-driven increased drought on ecological and hydrological processes. The developed model was accurate for depicting forest hydrology at stand scales (i.e. <100 ha) and allowed the analysis of the dynamical linkages among rainfall regimes, soil moisture variation, and individual tree growth. Using the model we demonstrated that evergreen, broad-leaved temperate rainforests in southern South America are expected to be highly sensitive to future climate change, particularly increases in drought during parts of the year. Increased summer drought predicted for this century will likely impair biomass carbon accumulation, and amplify background tree mortality rates in this region. The developed model expands the range of applicability of gap models to assess climate change impacts in remote and understudied regions of the world, such as temperate forests of the southern hemisphere. It also represents an advance in the development of simple, general models to account for complex and dynamical processes operating at multiple spatial scales in forests.

Supporting Information

Figure S1 A diagram of the hydrologic submodel of FORMIND-CL v.**1.0.** Interaction between processes and variables in the hydrologic submodel, and their respective time scales of calculations. Arrows indicate whether the results of a model calculation influence the calculations of another submodel. Blue boxes represent analyzed variables of this study. All calculations are done in yearly time steps in the model, excepting the ones indicated in the dashed box. Variable notations follow the text. AGB: Above-ground biomass, LAI: leaf area index.

Figure S2 Weather generator results. Density functions of daily rainfall and daily mean temperature predicted by the weather generator compared to observed weather records from EBSD weather station.

Figure S3 Weather generator results. Comparison between simulated and observed climatic patterns during the year. Simulations were run for 100 years using parameters in Table 3. Daily data were averaged by seasons (mean daily temperature and daily radiation). Rainfall is the amount of rainfall during each season. Observed weather data are from EBSD weather station and seasons according to table 3.

Figure S4 Forest composition predictions. Model results for forest composition using different model versions compared to field data. Simulations run under the same conditions detailed in Methods section.

Figure S5 Drought induced simulations with warming included. Changes in hydrologic components and forest structure when warming and increased drought was considered. PET: Potential evapotranspiration (mm $year^{-1}$), T: transpiration (mm year^{-1}), Ec: canopy interception (mm year^{-1}), ET: evapotranspiration (mm year^{-1}), BAT: total basal area (m^2 ha^{-1}), BT: Total biomass (tC ha^{-1}). Result of a two-sample Wilcoxon test is shown on the upper right of each panel. Pink lines, drought induced simulations with warming included, blue lines drought induced simulations without warming, circles represent the values of simulation results. Note different scales for the axes.

Figure S6 Sensitivity of evaporatranspiration. Changes in evapotranspiration (ET) of the old-growth stand under current climate when using different water-use efficiency values (WUE). Simulations run under the same conditions detailed in Methods section.

Appendix S1 Calculation of canopy photosynthetic rate in FORMIND-CL **v.1.0.**

Acknowledgments

We appreciate the useful comments offered by Hans Pretzsch, and anonymous reviewers in previous versions of the manuscript. This is a contribution to the Research Program of the Chilean LTSER network.

Author Contributions

Conceived and designed the experiments: AGG AH JJA MFD. Performed the experiments: AGG JJA MFD. Analyzed the data: AGG AH MFD. Contributed reagents/materials/analysis tools: AGG JJA AH. Wrote the paper: AGG.

References

1. Rodriguez-Iturbe I, Porporato A, Ridolfi L, Isham V, Cox DR (1999) Probabilistic modelling of water balance at a point: the role of climate, soil and vegetation. Proceedings of the Royal Society a-Mathematical Physical and Engineering Sciences 455: 3789–3805.
2. Botkin DB, Wallis JR, Janak JF (1972) Some Ecological Consequences of a Computer Model of Forest Growth. Journal of Ecology 60: 849–872.
3. Bugmann H (2001) A review of forest gap models. Climatic Change 51: 259–305.
4. Bugmann H, Cramer W (1998) Improving the behaviour of forest gap models along drought gradients. Forest Ecology and Management 103: 247–263.
5. Bugmann HKM, Wullschleger SD, Price DT, Ogle K, Clark DF, et al. (2001) Comparing the performance of forest gap models in North America. Climatic Change 51: 349–388.
6. Reynolds JF, Bugmann H, Pitelka LF (2001) How much physiology is needed in forest gap models for simulating long-term vegetation response to global change? Challenges, limitations, and potentials. Climatic Change 51: 541–557.
7. Asbjornsen H, Goldsmith GR, Alvarado-Barrientos MS, Rebel K, Van Osch FP, et al. (2011) Ecohydrological advances and applications in plant-water relations research: a review. Journal of Plant Ecology-Uk 4: 3–22.
8. Porporato A, Daly E, Rodriguez-Iturbe I (2004) Soil water balance and ecosystem response to climate change. American Naturalist 164: 625–632.
9. Vose JV, Sun G, Ford CR, Bredemeier M, Otsuki K, et al. (2011) Forest ecohydrological research in the 21st century: what are the critical needs? Ecohydrology 4: 146–158.
10. Hall GMJ, Hollinger DY (2000) Simulating New Zealand forest dynamics with a generalized temperate forest gap model. Ecological Applications 10: 115–130.
11. Gutierrez AG, Huth A (2012) Successional stages of primary temperate rainforests of Chiloe Island, Chile. Perspectives in Plant Ecology Evolution and Systematics 14: 243–256.
12. Echeverría C, Newton AC, Lara A, Benayas JMR, Coomes DA (2007) Impacts of forest fragmentation on species composition and forest structure in the temperate landscape of southern Chile. Global Ecology and Biogeography 16: 426–439.
13. Rüger N, Gutiérrez AG, Kissling WD, Armesto JJ, Huth A (2007) Ecological impacts of different harvesting scenarios for temperate evergreen rain forest in southern Chile - A simulation experiment. Forest Ecology and Management 252: 52–66.
14. Díaz MF, Bigelow S, Armesto JJ (2007) Alteration of the hydrologic cycle due to forest clearing and its consequences for rainforest succession. Forest Ecology and Management 244: 32–40.
15. Little C, Lara A, McPhee J, Urrutia R (2009) Revealing the impact of forest exotic plantations on water yield in large scale watersheds in South-Central Chile. Journal of Hydrology 374: 162–170.
16. Trenberth KE, Jones PD, Ambenje P, Bojariu R, Easterling D, et al. (2007) Observations: Surface and Atmospheric Climate Change. In: Solomon S, Qin D, Manning M, Chen Z, Marquis M et al., editors. Climate Change 2007: The Physical Science Basis Contribution of Working Group I to the Fourth Assessment Report of the Intergovernmental Panel on Climate Change. Cambridge, United Kingdom and New York, NY, USA.: Cambridge University Press. 336
17. DGF CONAMA (2006) Estudio de la variabilidad climática en Chile para el siglo XXI. Santiago: Departamento de Geofisica. Universidad de Chile. 63 p.
18. Christensen JH, Hewitson B, Busuioc A, Chen A, Gao X, et al. (2007) Regional Climate Projection. In: Solomon S, Qin D, Manning M, Chen Z, Marquis M et al., editors. Climate Change 2007: The Physical Science Basis Contribution of Working Group I to the Fourth Assessment Report of the Intergovernmental Panel on Climate Change. Cambridge, United Kingdom and New York, NY, USA: Cambridge University Press. 94
19. Alaback PB (1991) Comparative Ecology of Temperate Rain-Forests of the America Along Analogous Climatic Gradients. Revista Chilena de Historia Natural 64: 399–412.
20. Armesto JJ, Smith-Ramírez C, Carmona MR, Celis-Diez JL, Díaz I, et al. (2009) Old-growth temperate rain forests of South America: Conservation, plant-animal interactions, and baseline biogeochemical processes. In: Wirth C, Gleixner G, Heimann M, editors. Old-growth forests: Function, fate and value. Berlin, Heidelberg: Springer New York. 367–390
21. Armesto JJ, Rozzi R, Smith-Ramírez C, Arroyo MTK (1998) Conservation targets in South American temperate forests. Science 282: 1271–1272.
22. Willson MF, Armesto JJ (1996) The natural history of Chiloé: on Darwin's trail. Revista Chilena de Historia Natural 69: 149–161.
23. Holdgate MW (1961) Vegetation and soils in the South Chilean Islands. Journal of Ecology 49: 559–580.
24. Díaz MF, Armesto JJ (2007) Physical and biotic constraints on tree regeneration in secondary shrublands of Chiloe Island, Chile. Revista Chilena De Historia Natural 80: 13–26.
25. di Castri F, Hajek E (1976) Bioclimatología de Chile. Santiago: Universidad Católica de Chile.
26. Gutiérrez AG, Armesto JJ, Aravena JC, Carmona M, Carrasco NV, et al. (2009) Structural and environmental characterization of old-growth temperate rain-forests of northern Chiloe Island, Chile: Regional and global relevance. Forest Ecology and Management 258: 376–388.
27. Veblen TT, Schlegel FM, Oltremari JV (1983) Temperate broad-leaved evergreen forest of South America. In: Ovington JD, editor. Temperate Broad-Leaved Evergreen Forest. Amsterdam: Elsevier Science Publishers. 5–31
28. Gutiérrez AG, Armesto JJ, Aravena JC (2004) Disturbance and regeneration dynamics of an old-growth North Patagonian rain forest in Chiloe Island, Chile. Journal of Ecology 92: 598–608.
29. Köhler P (2000) Modelling anthropogenic impacts on the growth of tropical rain forests-using an individual-oriented forest growth model for the analyses of logging and fragmentation in three case studies [PhD thesis http://hdl.handle.net/10013/epic.15101].Osnabrück, Germany: Center for Environmental Systems Research and Department of Physics. University of Kassel. Der Andere Verlag.
30. Köhler P, Chave J, Riera B, Huth A (2003) Simulating the long-term response of tropical wet forests to fragmentation. Ecosystems 6: 114–128.
31. Köhler P, Huth A (1998) The effects of tree species grouping in tropical rainforest modelling: Simulations with the individual-based model FORMIND. Ecological Modelling 109: 301–321.
32. Kumagai T, Katul GG, Saitoh TM, Sato Y, Manfroi OJ, et al. (2004) Water cycling in a Bornean tropical rain forest under current and projected precipitation scenarios. Water Resources Research 40: W01104.
33. Rutter AJ, Morton AJ (1977) Predictive Model of Rainfall Interception in Forests.3. Sensitivity of Model to Stand Parameters and Meteorological Variables. Journal of Applied Ecology 14: 567–588.
34. Wattenbach M, Hattermann F, Weng R, Wechsung F, Krysanova V, et al. (2005) A simplified approach to implement forest eco-hydrological properties in regional hydrological modelling. Ecological Modelling 187: 40–59.
35. Rey JM (1999) Modelling potential evapotranspiration of potential vegetation. Ecological Modelling 123: 141–159.
36. Maidment DR (1993) Handbook of Hydrology; McGraw-Hill, editor: McGraw-Hill.
37. Neilson RP (1995) A Model for Predicting Continental-Scale Vegetation Distribution and Water-Balance. Ecological Applications 5: 362–385.
38. Lambers H, Chapin FS, Pons TL (1998) Plant physiological ecology. New York: Springer. 540 p.
39. Bazzaz FA (1979) Physiological ecology of plant succession. Annual Review of Ecology and Systematics 10: 351–371.

40. Thornley HMJ, Johnson IR (1990) Plant and Crop Modelling – A mathematical approach to plant and crop physiology. Oxford, UK.: Clarendon Press.
41. Gerten D, Schaphoff S, Haberlandt U, Lucht W, Sitch S (2004) Terrestrial vegetation and water balance - hydrological evaluation of a dynamic global vegetation model. Journal of Hydrology 286: 249–270.
42. Venevsky S, Maksyutov S (2007) SEVER: A modification of the LPJ global dynamic vegetation model for daily time step and parallel computation. Environmental Modelling & Software 22: 104–109.
43. Prentice IC, Sykes MT, Cramer W (1993) A Simulation-Model for the Transient Effects of Climate Change on Forest Landscapes. Ecological Modelling 65: 51–70.
44. Haxeltine A, Prentice IC (1996) BIOME3: An equilibrium terrestrial biosphere model based on ecophysiological constraints, resource availability, and competition among plant functional types. Global Biogeochemical Cycles 10: 693–709.
45. Prentice IC, Cramer W, Harrison SP, Leemans R, Monserud RA, et al. (1992) A Global Biome Model Based on Plant Physiology and Dominance, Soil Properties and Climate. Journal of Biogeography 19: 117–134.
46. Dingman SL (2002) Physical hydrology. New Jersey: Prentice Hall. 646 p.
47. Laio F, Porporato A, Ridolfi L, Rodriguez-Iturbe I (2001) Plants in water-controlled ecosystems: active role in hydrologic processes and response to water stress - II. Probabilistic soil moisture dynamics. Advances in Water Resources 24: 707–723.
48. Lusk CH (2001) When is a gap not a gap? Light levels and leaf area index in bamboo-filled gaps in a Chilean rain forest. Gayana Botanica 58: 25–30.
49. Piper FI, Corcuera LJ, Alberdi M, Lusk C (2007) Differential photosynthetic and survival responses to soil drought in two evergreen Nothofagus species. Annals of Forest Science 64: 447–452.
50. Zuñiga R, Alberdi M, Reyes-Diaz M, Olivares E, Hess S, et al. (2006) Seasonal changes in the photosynthetic performance of two evergreen Nothofagus species in south central Chile. Revista Chilena de Historia Natural 79. 489–504.
51. Cunningham SC (2005) Photosynthetic responses to vapour pressure deficit in temperate and tropical evergreen rainforest trees of Australia. Oecologia 142: 521–528.
52. Brodribb T, Hill RS (1998) The photosynthetic drought physiology of a diverse group of southern hemisphere conifer species is correlated with minimum seasonal rainfall. Functional Ecology 12: 465–471.
53. Janssen I, Kruemmelbein J, Horn R, Ellies AS (2004) Physical and hydraulic properties of the ñadi soils in south Chile - Comparison between untilled and tilled soil. Revista de la Ciencia del Suelo y Nutrición Vegetal 4: 14–28.
54. Sitch S, Smith B, Prentice IC, Arneth A, Bondeau A, et al. (2003) Evaluation of ecosystem dynamics, plant geography and terrestrial carbon cycling in the LPJ dynamic global vegetation model. Global Change Biology 9: 161–185.
55. van Genuchten MT (1980) A closed-form equation for predicting the hydraulic conductivity of unsaturated soils. Soil Science Society of America Journal 44: 892–898.
56. Seki K (2007) SWRC fit-a nonlinear fitting program with a water retention curve for soils having unimodal and bimodal pore structure. Hydrology and Earth System Sciences Discussions 4: 407–437.
57. R-Development-Core-Team (2012) R: A language and environment for statistical computing. Vienna, Austria.: R Foundation for Statistical Computing.
58. Breda NJJ (2003) Ground-based measurements of leaf area index: a review of methods, instruments and current controversies. Journal of Experimental Botany 54: 2403–2417.
59. Condit R, Hubbell SP, Foster RB (1995) Mortality-Rates of 205 Neotropical Tree and Shrub Species and the Impact of a Severe Drought. Ecological Monographs 65: 419–439.
60. Allen CD, Macalady AK, Chenchouni H, Bachelet D, McDowell N, et al. (2010) A global overview of drought and heat-induced tree mortality reveals emerging climate change risks for forests. Forest Ecology and Management 259: 660–684.
61. Phillips OL, van der Heijden G, Lewis SL, Lopez-Gonzalez G, Aragao LEOC, et al. (2010) Drought-mortality relationships for tropical forests. New Phytologist 187: 631–646.
62. Larcher W (2001) Ökophysiologie der Pflanzen. Stuttgart: 6th edition, Verlag Eugen Ullmer.
63. Grimm V, Railsback SF (2005) Individual-based modeling and ecology. New Jersey: Princeton University Press.
64. Penuelas J, Canadell JG, Ogaya R (2011) Increased water-use efficiency during the 20th century did not translate into enhanced tree growth. Global Ecology and Biogeography 20: 597–608.
65. Keenan TF, Hollinger DY, Bohrer G, Dragoni D, Munger JW, et al. (2013) Increase in forest water-use efficiency as atmospheric carbon dioxide concentrations rise. Nature 499: 324–+.
66. Silva LCR, Anand M (2013) Probing for the influence of atmospheric CO_2 and climate change on forest ecosystems across biomes. Global Ecology and Biogeography 22: 83–92.
67. Penuelas J, Hunt JM, Ogaya R, Jump AS (2008) Twentieth century changes of tree-ring delta C-13 at the southern range-edge of Fagus sylvatica: increasing water-use efficiency does not avoid the growth decline induced by warming at low altitudes. Global Change Biology 14: 1076–1088.
68. Levesque M, Saurer M, Siegwolf R, Eilmann B, Brang P, et al. (2013) Drought response of five conifer species under contrasting water availability suggests high vulnerability of Norway spruce and European larch. Global Change Biology 19: 3184–3199.
69. Figueroa JA, Cabrera HM, Queirolo C, Hinojosa LF (2010) Variability of water relations and photosynthesis in *Eucryphia cordifolia* Cav. (Cunoniaceae) over the range of its latitudinal and altitudinal distribution in Chile. Tree Physiology 30: 574–585.
70. Chaves MM, Pereira JS, Maroco J, Rodrigues ML, Ricardo CPP, et al. (2002) How plants cope with water stress in the field. Photosynthesis and growth. Annals of Botany 89: 907–916.
71. Ryan MG, Law BE (2005) Interpreting, measuring, and modeling soil respiration. Biogeochemistry 73: 3–27.
72. Aerts R, Chapin FS (2000) The mineral nutrition of wild plants revisited: A re-evaluation of processes and patterns. Advances in Ecological Research, Vol 30 30: 1–67.
73. Hobbie SE, Chapin FS (1998) An experimental test of limits to tree establishment in Arctic tundra. Journal of Ecology 86: 449–461.
74. Gutiérrez AG (2010) Long-term dynamics and the response of temperate rainforests of Chiloé Island (Chile) to climate change [Phd Thesis]. Freising, Germany: Technische Universität München. 170 p.
75. Suarez ML, Ghermandi L, Kitzberger T (2004) Factors predisposing episodic drought-induced tree mortality in Nothofagus-site, climatic sensitivity and growth trends. Journal of Ecology 92: 954–966.
76. Groeneveld J, Alves LF, Bernacci LC, Catharino ELM, Knogge C, et al. (2009) The impact of fragmentation and density regulation on forest succession in the Atlantic rain forest. Ecological Modelling 220: 2450–2459.
77. Donoso PJ, Soto DP, Bertin RA (2007) Size-density relationships in *Drimys winteri* secondary forests of the Chiloe Island, Chile: Effects of physiography and species composition. Forest Ecology and Management 239: 120–127.
78. Navarro C, Donoso C, Sandoval V (1999) Los renovales de Canelo. In: Donoso C, Lara A, editors. Silvicultura de los bosques nativos de Chile. Santiago, Chile: Editorial Universitaria.
79. Naulin P (2002) Estimacion de la biomasa en un renoval de canelo (Drimys winteri J.R. et Foster) en la Comuna de Ancud. Santiago: Universidad de Chile. 45 p.
80. Aravena JC, Carmona MR, Perez CA, Armesto JJ (2002) Changes in tree species richness, stand structure and soil properties in a successional chronosequence in northern Chiloe Island, Chile. Revista Chilena De Historia Natural 75: 339–360.
81. Lusk CH (2001) Leaf life spans of some conifers of the temperate forests of South America. Revista Chilena De Historia Natural 74: 711–718.
82. Huber A, Iroume A (2001) Variability of annual rainfall partitioning for different sites and forest covers in Chile. Journal of Hydrology 248: 78–92.
83. Oyarzun CE, Godoy R, Staelens J, Donoso PJ, Verhoest NEC (2011) Seasonal and annual throughfall and stemflow in Andean temperate rainforests. Hydrological Processes 25: 623–633.
84. Barbour MM, Hunt JE, Walcroft AS, Rogers GND, McSeveny TM, et al. (2005) Components of ecosystem evaporation in a temperate coniferous rainforest, with canopy transpiration scaled using sapwood density. New Phytologist 165: 549–558.
85. Lara A, Little C, Urrutia R, McPhee J, Alvarez-Garreton C, et al. (2009) Assessment of ecosystem services as an opportunity for the conservation and management of native forests in Chile. Forest Ecology and Management 258: 415–424.
86. CIREN (1994) Delimitación y descripción de microregiones para la transferencia tecnológica de INDAP. Santiago: CORFO. 99 p.

2

Pushing the Pace of Tree Species Migration

Eli D. Lazarus[1]*, Brian J. McGill[2]

1 Environmental Dynamics Lab, Earth Surface Processes Group, School of Earth and Ocean Sciences, Cardiff University, Cardiff, Wales, United Kingdom, 2 School of Biology and Ecology, University of Maine, Orono, Maine, United States of America

Abstract

Plants and animals have responded to past climate changes by migrating with habitable environments, sometimes shifting the boundaries of their geographic ranges by tens of kilometers per year or more. Species migrating in response to present climate conditions, however, must contend with landscapes fragmented by anthropogenic disturbance. We consider this problem in the context of wind-dispersed tree species. Mechanisms of long-distance seed dispersal make these species capable of rapid migration rates. Models of species-front migration suggest that even tree species with the capacity for long-distance dispersal will be unable to keep pace with future spatial changes in temperature gradients, exclusive of habitat fragmentation effects. Here we present a numerical model that captures the salient dynamics of migration by long-distance dispersal for a generic tree species. We then use the model to explore the possible effects of assisted colonization within a fragmented landscape under a simulated tree-planting scheme. Our results suggest that an assisted-colonization program could accelerate species-front migration rates enough to match the speed of climate change, but such a program would involve an environmental-sustainability intervention at a massive scale.

Editor: Gil Bohrer, The Ohio State University, United States of America

Funding: This research was supported by the U.S. National Science Foundation award EPS-0904155. The funders had no role in study design, data collection and analysis, decision to publish, or preparation of the manuscript.

Competing Interests: The authors have declared that no competing interests exist.

* Email: LazarusED@cf.ac.uk

Introduction

Plants and animals have responded to past climatic changes by shifting the spatial boundaries of their geographic ranges. Such migratory responses to climate change have proved remarkably successful for many species. For example, in North America, no species of small mammal [1] and only one species of tree [2] went extinct in the dozen-plus glacial cycles over the Quaternary. With the approximately 7°C warming that occurred 20–10 kya (punctuated by rapid rates of temperature change over shorter intervals within that period), plant and animal ranges in North America shifted up to the order of 1000 km [3–5]. Pollen records show evidence that temperate tree species migrated at rapid rates (100–1000 m yr^{-1}) following the last Ice Age [6]. Warming of approximately 3.5°C is expected over the next 100 years [7]. As a first approximation, linearly extrapolating the ratio of warming (7°C) to range adjustment in the early Holocene (1000 km) suggests that ranges could shift an average of 500 km in the coming century (5 km yr^{-1}) [8]. Reconciling paleo-rates with observed spreading patterns of plant taxa has motivated a wealth of ecology research into mechanisms for rapid migration, such as long-distance dispersal [9,10]. However, paleo-range shifts occurred in the absence of anthropogenically dominated landscapes. How human uses of otherwise suitable habitat will affect climatically driven range migration across landscapes is an open, fundamental question in environmental sustainability. Here we explore this question in terms of temperate-forest tree species.

Field measurements of typical seed dispersal distances would suggest that tree fronts migrate across a landscape by a process of local diffusion, at rates significantly slower than the velocities reflected in pollen data [9]. But seeds are occasionally carried long distances from their source by wind or by animals [11]. If those seeds mature into trees that in turn dispense seeds, the plant species may migrate at rates that far exceed diffusive propagation [6,9]. Existing models of tree migration by long-distance dispersal produce migration rates between approximately 100–200 m yr^{-1} [9,10,12]. A global analysis of temperature-change rates across geographic gradients and biomes finds that temperate broad-leaf and mixed forests, which includes the North American taxa that spread by wind-blown dispersal [10], will need to shift at a mean velocity of 350 m yr^{-1} [13]. These required migration rates appear to exceed the fastest modeled rates, but may fall within the ranges empirically observed in the last deglaciation. Migration rates barely sufficient to track with climate, combined with the well-documented effect that landscape fragmentation further impedes migration [14–19], points to the apparently unequivocal conclusion that climatic change will outpace the migration of wind-dispersed tree species through human-dominated landscapes. One of the few known cases in which climate-driven species migration was impeded comes from Europe, where east-west mountain ranges and the Mediterranean Sea prevented trees and plants from advancing far enough south during Pleistocene glaciation, resulting in a high proportion of extinction [20]. By extension, understanding how human fragmented landscapes interfere with migration rates might mean the difference between minimal extinction rates and massive extinction rates in next few hundred years.

Assisted migration, assisted colonization, and species translocation are already common conservation practices applied to variety of plant and animal species [21,22]. Unwitting or unintended human introduction of alien, invasive species to new environs via long-distance transport pathways is arguably a version of the same idea. Assuming disparity between rates of climate change and temperate-forest migration is an inherent condition of the future sets up a new question: how much human intervention might windblown-dispersal taxa need to keep pace with climate change? Although assisted migration remains controversial [23], here we neither advocate for nor reject it. We simply ask: if humans were to deliberately augment the migration rates of temperate-forest tree species, what would the scale of that intervention need to be relative to the natural dynamics driving the migration front?

Here we present results from a probabilistic spatial model that simulates the passage of a generic, wind-dispersed tree species through a two-state, cellular landscape in which cells are either suitable or unsuitable for trees (Fig. 1 and fig. S1; additional model explanation, parameters, and source code are provided in File S1; data shown in figures are provided in File S2.). Carefully abstracted models that capture the salient dynamics of well-defined systems are useful tools for insight into how natural systems may respond to changes in forcing, and by design are more exploratory than explicitly predictive [24,25]. Cellular architecture has been used in several idealized spatial models of migration patterns through heterogeneous habitats and fragmented landscapes [16,26–28]. We use our model to test how augmenting the number of successful seedlings accelerates the rate of tree front migration. Our results suggest that relatively small changes in the annual number of successful seedlings could have a comparatively large effect on biome migration rates.

Model Design

Spatial domain

Our model operates on a gridded domain with a width-to-length ratio of 1:200 km and a periodic (i.e. wrap-around) boundary in the width dimension (fig. S1). The periodic boundary condition makes the domain infinitely wide; the extended length dimension allows a fixed aerial view (200 km) large enough to frame potentially rapid incoming migration fronts (> km yr $^{-1}$). Each grid cell is 10×10 m, and can accommodate a single tree. Although these cell dimensions are coarse for small, young trees, they are consistent with crown widths of large trees measured via remote sensing in a North American temperate mixed forest [29]. We run the model for 1000 years so that initial conditions dissipate and migration dynamics may play out in full (e.g. until available area is fully populated, or all trees die out).

Seed dispersal

Accounting for infrequent long-distance dispersal events requires a statistical function with a "heavy tail," so that the maximum dispersal distance may be very large even if the average dispersal distance is short. Other models have used various heavy-tailed probability distribution functions to drive dispersal distance [9,10,12,16,30–32]. Seed dispersal distance in our model is generated by a *t*-distribution with a degrees-of-freedom variable that changes the "heaviness" of the distribution's tail: the minimum degrees of freedom (DOF = 1) allows the longest dispersal distances; by comparison, a high degree of freedom (e.g. DOF = 100) approximates a Gaussian distribution, with a dispersal pattern limited to local diffusion (Fig. 2). We use the *t*-distribution because of the straightforward means by which the distribution can be shifted from Gaussian to heavy tailed for comparative scenarios. In the results presented here, the initial model domain (Fig. 1) is populated with trees according to a *t*-distribution (with DOF = 1) centered on the far left edge.

Tree characteristics

Each tree in our model is assigned biological characteristics for tree fecundity (seeds tree $^{-1}$ yr $^{-1}$, as a function of tree age), seed dispersal distance, and seedling survival according to probability-distribution functions constrained by empirically derived parameters for North American wind-dispersed trees (see Table S1) [10]. Maturation age (yr) and seed-crop interval (yr) are sampled from normal probability distributions. Longevity (yr) is sampled from a gamma distribution, which allows a few long-lived trees to persist among a community of trees whose average age is younger. Each model tree broadcasts some number of "successful" seeds (seeds that will germinate into seedlings) every few years as a function of its maturation age, seed-crop interval, and fecundity. We treat tree fecundity as an increasing parabolic function of age; the number of viable seedlings per tree is then a percentage of that fecundity, determined by a probability of seedling survival for a given tree's seed crop in a given model year. Each seed is assigned a randomly selected compass direction and a dispersal distance that corresponds to a grid cell in the domain, which, if unoccupied, becomes a new tree with parametric values resampled from the probability-distribution functions for biological characteristics.

Landscape disturbance

We create disturbance patterns by first generating a domain-scale layer of random values between 0–1 that are either highly spatially autocorrelated (e.g. high values occur near other high values, and likewise for low values) or are spatially uncorrelated (spatially random), with the degree of correlation controlled by the Hurst exponent [17,19,33]. We use Hurst exponents of $H = 0.99$ and $H = 0$ for our correlated and uncorrelated landscapes, respectively. We control the percentage of landscape disturbance by isolating cells in the grid that exceed a given threshold value between 0–1; cells whose values exceed the threshold are rendered "disturbed" and assigned a value = 1; all other cells (undisturbed) are set = 0. For comparability across scenarios, the underlying Hurst-model landscape (for a given value of *H*) is kept constant. The threshold changes, and therefore the overall pattern of disturbance, but retaining the same underlying Hurst model means that a patch disturbed at a high threshold (yielding a low percentage of disturbance) will persist at lower thresholds. Furthermore, the disturbance pattern is held constant through time within a model run. Trees may occupy any undisturbed cells in the domain; seedlings distributed to disturbed cells are removed.

Assisted colonization

Assisted colonization occurs in the model when new trees are "planted" at the end of a model year, after all existing, active trees in the domain have broadcast their seeds (fig. S1). Grid cells in which new trees are planted are randomly selected from among all undisturbed cells not already occupied by an existing tree. We do not account for ways in which proximal areas of disturbance might impinge upon tree health, or disturbance "edge effects" [34,35]: planted trees are governed by the same parameters as trees growing "naturally" in the model.

The number of trees planted is based on the average number of new trees that grow in the model when (a) there is no landscape disturbance and (b) dispersal distances are generated by a heavy-tailed *t*-distribution function (DOF = 1). In the results shown here, that average number is 2000 trees yr $^{-1}$. These conditions constitute the model's the ecological baseline. Assistance is thus a function of

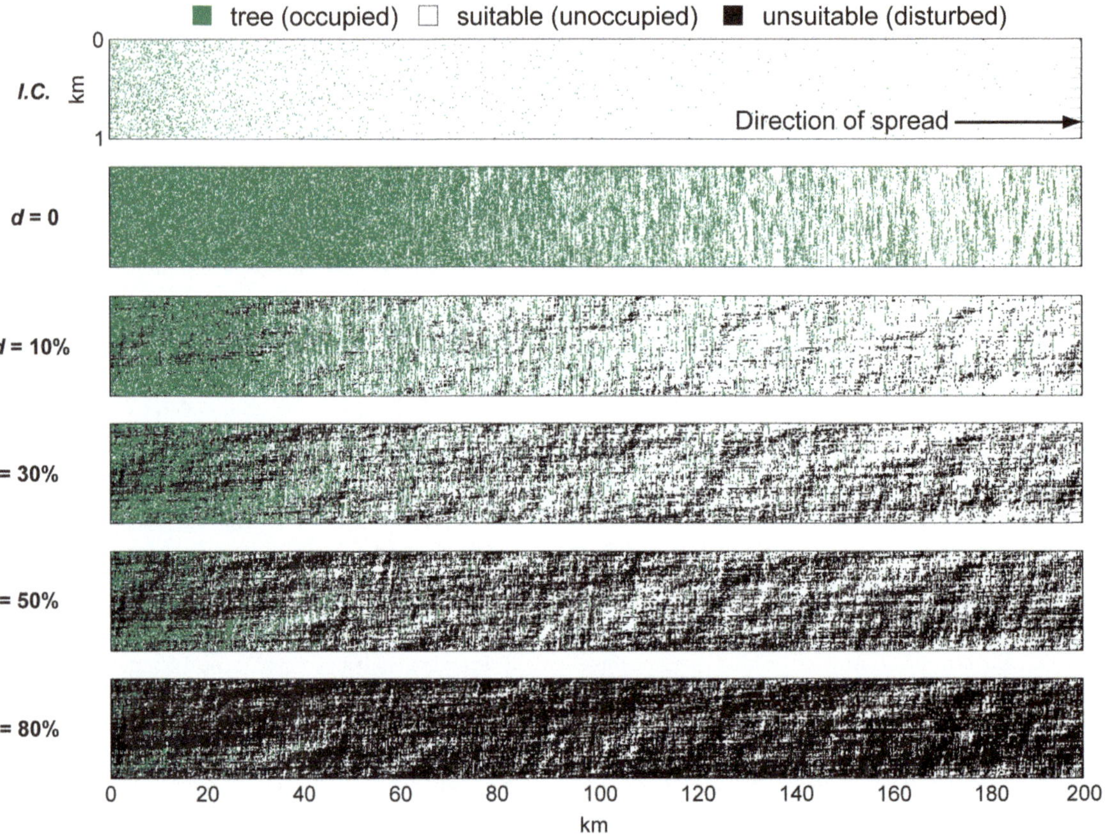

Figure 1. Representative model domains. Top panel shows the model's initial condition (*I.C.*) for tree positions in an undisturbed landscape. Subsequent panels show the resulting domains after 1000 years given different extents of disturbance (*d*), where disturbance is spatially autocorrelated (Hurst exponent $H = 0.99$). Apparent lineations in the ordinate dimension are a distortion in the aspect ratio, not an artifact of the model.

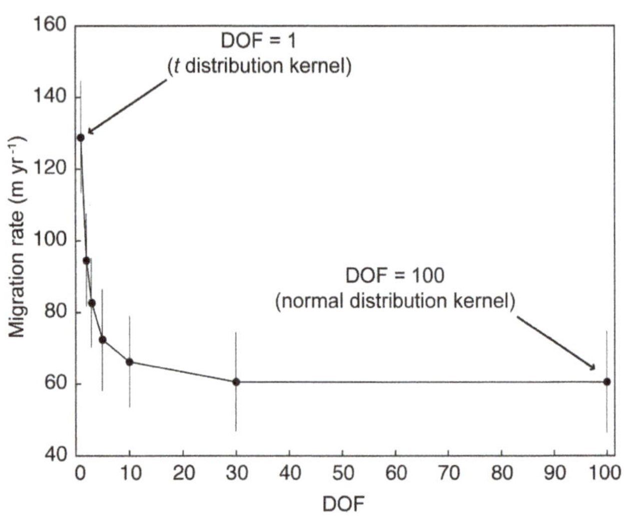

Figure 2. Mean migration rates with increasing degrees of freedom (DOF) in a *t*-distribution kernel for long-distance dispersal. Bars indicate envelope of one standard deviation. When DOF = 1, the dispersal kernel is "heavy tailed." When DOF is >>1 (e.g. DOF = 100), the dispersal kernel has the comparatively limited range of a Gaussian distribution, and resulting mean migration rates represent the rate of spreading by local diffusion.

increasing the fixed number of new trees per year by a percentage *K*.

Planting in the model is not explicitly organized in the direction of migration: available sites at the distal edge of the domain are not prioritized over other available sites. However, because the distal edge of the initial model domain is comparatively empty, distal planting sites are more numerous and therefore more likely to be selected from among the sites available, especially when the rate of assistance is high.

Results

When the entire model domain is hospitable to seedling growth, heavy-tailed dispersal distances drive front migration rates of approximately 130 m yr^{-1} that are nearly double those driven by diffusive dispersal (Fig. 2). Front migration rates in this model, measured as the change in mean tree position across the domain, are comparable to rates reported for other, analogous long-distance dispersal models [9,12], including the study from which our ecological parameters derive [10]. This agreement in rates is not a foregone conclusion. Although the independent biological variables in the model are constrained by parametric distributions that are similar to (but not the same as) those used in other studies, migration rate is an implicit variable that arises from the collective state of the full model domain over time. The agreement we find provides a good validation of our modeling approach.

Fragmenting the model domain with landscape disturbance shows that migration rate and disturbance are negatively

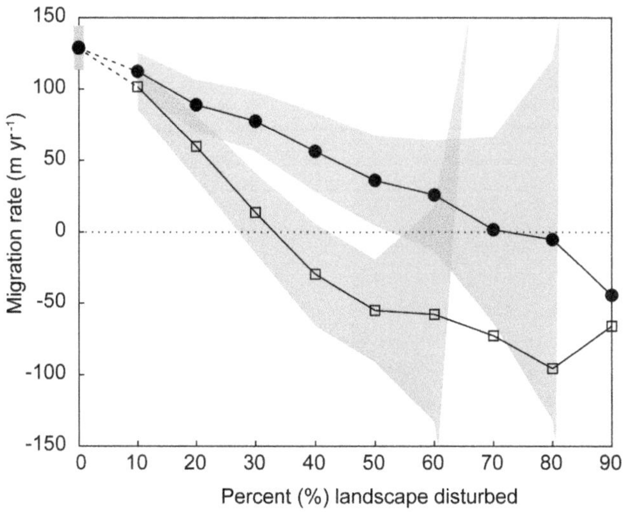

Figure 3. Mean migration rates (via long-distance dispersal, given *t*-distribution kernel with DOF = 1) through increasing proportions of landscape disturbance when disturbance is spatially autocorrelated (black dots; *H* = 0.99) and spatially random (open squares; *H* = 0). Gray regions denote envelope of one standard deviation. Negative values indicate conditions under which the tree front receded from the domain (died back) rather than advanced across it.

correlated (Fig. 3), as other disturbance models have also shown [16,17,19]. Furthermore, the presence or absence of pattern in the disturbance regime affects the migration rate. When disturbance across the model landscape is spatially random ($H = 0$), the migration front ceases to advance once the total area disturbed reaches 40%. This propagation result grants some comparison to percolation models for square lattices. The percolation threshold for a square lattice is approximately 59% [36], meaning a fully connected route through the lattice exists as long as disturbance to the lattice is less than 41%. Long-distance dispersal in model does not necessarily succeed where random walks cannot: even a species capable of dispersal between non-adjacent sites does not advance through a randomly fragmented domain ($H = 0$) in which disturbance exceeds the percolation threshold. When disturbed areas are highly autocorrelated in the model ($H = 0.99$), leaving contiguous reaches of potential habitat (corridors) relatively intact, migration rates decrease more slowly, tolerating up to 70% total disturbance (Fig. 3). Most importantly, none of these migration rates—including the maximum rates in our idealized, undisturbed scenario (Fig. 3)—come close to matching the 350 m yr^{-1} "velocity of climate change" modeled for temperate broad-leaf and mixed forests [13].

That apparent lag motivates our simplified version of assisted colonization to accelerate migration rates. The key assumption we make is that all of the trees deliberately introduced to the domain can grow to be viable adults, subject to the probabilistic limits of tree longevity and fecundity imposed in the model. To consider the effect of assisted colonization on rates of tree front migration through an undisturbed landscape (Fig. 4), we systematically increase the number of planted trees by $K = 10$–200% in increments of 10% (e.g. when $K = 10\%$, 200 trees are planted annually). Planting begins in model year = 200 and continues until year = 1000. For these experiments, we calculate mean migration rate over the first hundred years of assisted colonization (model years 201–300); after 100 years of planting, the migration front

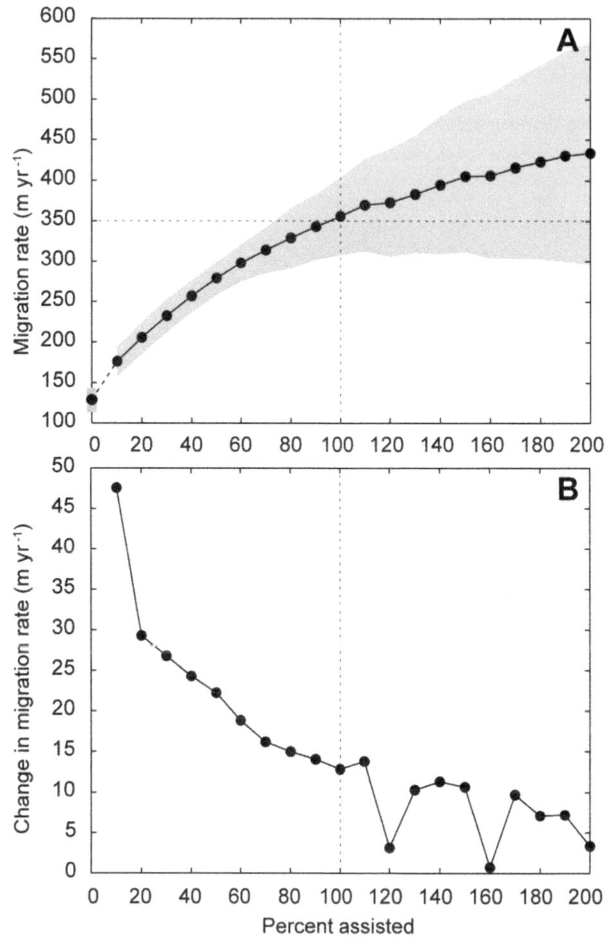

Figure 4. Mean migration rates with assistance. (A) Mean migration rate through an undisturbed landscape with increasing levels of assisted colonization. Gray regions denote envelope of one standard deviation. When the number of naturally successful seedlings is doubled ($K = 100\%$), migration rates become comparable to the measured mean velocity of range shift for temperate deciduous forests [13]. (B) First derivative of migration rates plotted in (A), showing that increasing assistance has a diminishing return on migration-rate acceleration, especially for $K > 100\%$.

effectively halts as available spaces in the model domain fill up (fig. S2).

These simulations suggest that doubling the number of annually successful trees ($K = 100\%$) in an undisturbed landscape would result in migration rates equal to the necessary mean velocity of range shift of 350 m yr^{-1}. We also find that even the weakly organized strategy of assisted colonization in this model has a pronounced, accelerating effect on migration rates in fragmented landscapes where disturbance is highly autocorrelated ($H = 0.99$) (Fig. 5). This result is not necessarily surprising. Unlike stochastic seed dispersal as a means of colonizing undisturbed areas, planting ensures that trees occupy available space. In a highly disturbed landscape, planting makes colonization a comparatively targeted process, which only makes migration rates faster. However, a caveat associated with determining the migration rate within a disturbed domain is that the calculation is based on fewer trees overall than are present in an undisturbed domain. A high migration rate can appear to be maintained through large extents

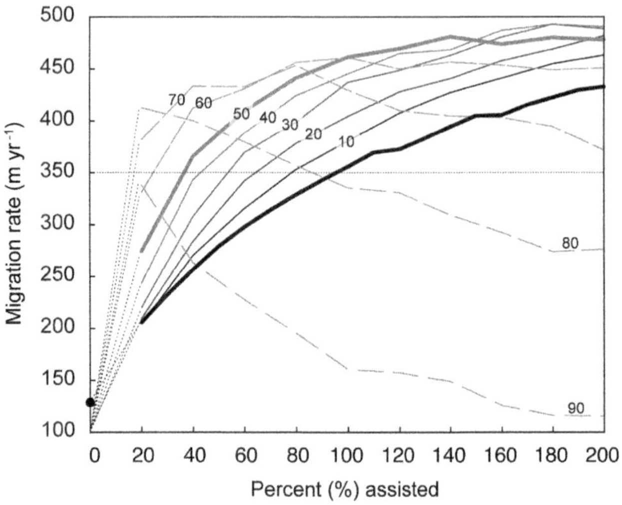

Figure 5. Assisted migration rates through landscapes with increasing levels of disturbance (lines labeled 10–90%), where disturbance is spatially autocorrelated (*H*= 0.99). Bold line shows assisted migration rates through an undisturbed landscape. Even small amounts of assisted colonization have a strong effect on migration rates as landscape disturbance increases.

of disturbance with less intervening assistance (Fig. 5), but the total population of trees is smaller.

Discussion

Model uncertainty

Migration rates in the model depend principally on the presence or absence of a heavy tail in the distribution of dispersal distances (Fig. 2), and on the empirical constraints on the probability-distribution functions [10]. We ran our model using other heavy-tailed probability-distribution functions for dispersal distances (e.g. power-law functions) constrained by the same statistical ranges and found that, although the resulting landscapes differed in detail, the model produced comparable migration rates with similar migration-front dynamics.

Given that our interest is in the migration rate of a generic pattern at the domain scale, our model is more phenomenological than mechanistic and shares the same parametric sensitivities as others of its kind [32]. The spatial distribution of surviving seedlings, not just the distribution of dispersal distances, is an important component of migration, but here we do not account for potential differences in seedling success (e.g., germination, competition, survivorship, seed production) with distance from the parent tree [32,37].

Furthermore, although climate change motivates our interest in modeling tree-species migration through fragmented landscapes, we do not attempt to model climate-change effects on ecological traits [38] or on habitat traits such as soil conditions [39]. To keep our model focused on dispersal, we treat climate change here as the exogenous driver of the minimally required rate of migration needed to avoid extinction. Assuming that the trees in this model migrate into and across a landscape already ecologically suitable for them to inhabit emphasizes the relative effects of disturbance and assistance on migration dynamics. We thus treat climate change as an implicit migration-rate threshold: even in this "fastest-case" scenario, for a given pair of disturbance and assistance conditions, are migration rates faster or slower than the projected rate they must achieve to survive?

Of course, the inherent uncertainty in our model derives far less from tree ecology than from patterns of landscape disturbance and the application of assisted migration, the abstracted human activities we represent. As Higgins et al. [32] remark in their review of plant-migration forecasting, "inherent uncertainty cannot be reduced by improving model or parameter uncertainty." By design, our model is a suitable tool for reasoning by analysis rather than by analogy [40]: a means of exploring inherent uncertainty through the encapsulation of relevant dynamics rather than in a scaled simulacrum of the real world.

Landscape corridors and assisted migration

Protecting contiguous reaches of undisturbed habitat by concentrating new disturbance near existing disturbed areas is the premise behind ecological-corridor conservation practices [23,41,42]. Some researchers have discussed the extent to which the mechanism of long-distance dispersal could enable a tree species, in the absence of an uninterrupted habitat corridor, to migrate along a series of proximal habitat "islands" [16,17]. Our results suggest that even when available habitat is strongly autocorrelated, once half the landscape domain has been disturbed, long-distance dispersal advances the front no faster than diffusive passage through an undisturbed landscape (Figs. 2 and 3). Moreover, beyond approximately 70% disturbance, the migration front fails to advance even with long-distance dispersal and in the presence of autocorrelated habitat islands (Fig. 3).

Indeed, if migration fronts effectively halt with disturbances exceeding 70% (Fig. 3), then migration by assisted colonization might be the only means by which a front of trees could move through a landscape so heavily disturbed. Moreover, such planting and could accelerate migration rates within decades (fig. S2), commensurate with time scales relevant to climate-change projections and mitigation actions [7]. We find that migration rates in our model increase with increased assistance, but augmenting the number of annually successful trees by more than 60–80% has a diminishing effect on acceleration (Fig. 4). Within a fixed domain (e.g. a spatial domain that fills as the migration front advances, rather than tracking with the front) an economy of scale therefore may govern the rate-accelerating effects of assisted colonization. Assuming tree planting has an associated fixed cost (i.e. each tree requires time and labor to be planted; trees for planting need to be raised in nurseries, etc.), at least in the context of a coordinated, top-down climate-adaptation program [43,44], this diminishing advantage in the relative effect of planting more trees could manifest in both ecological and practical terms.

Representing a tree-planting strategy as a stochastic, opportunistic process is at least as reasonable as our stylized treatment of landscape fragmentation, and likewise is as reasonable as assuming that in a real tree-planting campaign every site available ecologically would be available in practice. Targeting planting efforts at the far edge of a species' geographic range is one way to accelerate migration, but trees planted beyond the advancing front are not the only ones capable of affecting the migration rate. A consequence of long-distance dispersal is that even a tree planted behind the migrating front is capable of broadcasting a large number of viable seeds toward the distal edges of the geographic range. Perhaps the primary benefit of assisting tree-species colonization and migration, as opposed to managing animal species that roam, is the ability to selectively and specifically establish sensitive taxa in undisturbed environs irrespective of fragmentation intensity.

Implications

The lag between tree migration rates and the velocities of their preferred environments raises interesting implications for ecological dynamics at ecotonal boundaries. For example, temperate-broadleaf and mixed-forest biome environments in North America will need to move to higher latitudes at an average rate of 350 m yr^{-1} to track with climate, but boreal forest and taiga environments, also moving northward, are shifting at a faster average of 430 m yr^{-1} [13]. Ecotones between biomes demarcate zones where individuals are living at the fringe of their environmental tolerances and are arguably the most sensitive to relative differences in surface-temperature velocities [45]. Habitat fragmentation in a shifting ecotonal zone is therefore especially complicated and problematic [45,46]. Conceivably, habitat fragmentation across an ecotone could result in one principal biome departing before the next can establish, making the liminal landscape a kind of anthropogenic biome with its own ecological signature [46].

An intentional planting program therefore represents a compromise: a direct intervention that enables an erstwhile natural biome to persist under anthropogenic pressure. Our model results suggest that this would have to occur on a massive scale to be effective for a continental ecotone. Targeted plantings at specific locations assumes a top-down approach to climate-change adaptation, but such adaptation is at least as likely to occur—and is perhaps more realistic—through a bottom-up approach in which people act locally (i.e., plant trees in their neighborhood) [43,44]. An environmental conservation campaign resembling the simplified one in our model—a tree-planting campaign without explicit spatial targets but enacted at a large enough scale to potentially accelerate the migration rates of a forest biome—is not without precedent. The Green Belt Movement, founded by the late Nobel Laureate Prof. Wangari Maathai, is credited with having planted more than 51 million trees in Kenya since 1977 through grass-roots community organization and mobilization [47]. The Billion Tree Campaign, a United Nations Environment Programme initiative that Maathai also championed, has registered over 12 billion trees planted worldwide since 2006 [48]. Given that landscape fragmentation and land-use changes will only further hinder slow-moving tree fronts as they follow fast-moving climate change, adding various common tree types to the list of migrating species that large-scale and local conservation organizations actively assist is both a necessary action and an environmental sustainability problem with, it would seem, a viable solution.

Supporting Information

Figure S1 Schematics of our tree migration model: (A) representative initial condition without disturbance (arrows indicate how the periodic boundary operates on seed dispersal); final landscapes (B) without disturbance and (C) with disturbance; (D) sequence in a model year in which (D2) new trees are added by natural dispersal and (D3) by assisted colonization, resulting in (D4) the final landscape as shown. Actual model outputs are shown in Fig. 1.

Figure S2 Plot of mean front position versus time for increasing levels of assisted migration in the absence of landscape disturbance. Black line shows the natural, background migration rate (~130 m yr^{-1}), given a *t*-distribution dispersal kernel with DOF = 1. In this baseline scenario, approximately 2000 new trees grow in the domain per year. Blue lines show the effects of assisted migration regimes in which an additional K% trees per year are "planted", for $K = 10$–200%. Bold blue lines denote 50% increments; $K = 200$% is shown in red.

File S1 Document file containing model source code and explanatory text.

File S2 Spreadsheet file containing data shown in the published figures.

Acknowledgments

We thank A. Payson and D. Hall for helpful discussions, the Sustainability Solutions Initiative at the University of Maine, and the anonymous reviewers who contributed their comments.

Author Contributions

Conceived and designed the experiments: EDL BJM. Performed the experiments: EDL. Analyzed the data: EDL BJM. Contributed reagents/materials/analysis tools: EDL BJM. Contributed to the writing of the manuscript: EDL BJM.

References

1. McGill BJ, Hadly EA, Maurer BA (2005) Community inertia of quaternary small mammal assemblages in North America. P Natl Acad Sci USA 102: 16701–16706.
2. Jackson ST, Weng CY (1999) Late quaternary extinction of a tree species in eastern North America. P Natl Acad Sci USA 96: 13847–13852.
3. Davis MB, Shaw RG (2001) Range shifts and adaptive responses to Quaternary climate change. Science 292: 673–679.
4. Lyons SK (2003) A quantitative assessment of the range shifts of Pleistocene mammals. J Mammal 84: 385–402.
5. Williams JW, Shuman BN, Webb T, Bartlein PJ, Leduc PL (2004) Late-quaternary vegetation dynamics in North America: scaling from taxa to biomes. Ecol Monogr 74: 309–334.
6. Huntley B, Birks HJB (1983) An atlas of past and present pollen maps for Europe 0–13000 years ago. Cambridge: Cambridge University Press. 667 p.
7. IPCC (2007) Contribution of Working Groups I, II and III to the Fourth Assessment Report of the Intergovernmental Panel on Climate Change, Core Writing Team. Fourth Assessment Report of the Intergovernmental Panel on Climate Change. Geneva, Switzerland: IPCC.
8. Parmesan C, Yohe G (2003) A globally coherent fingerprint of climate change impacts across natural systems. Nature 421: 37–42.
9. Clark JS, Fastie C, Hurtt G, Jackson ST, Johnson C, et al. (1998) Reid's paradox of rapid plant migration: dispersal theory and interpretation of paleoecological records. Bioscience 48: 13–24.
10. Nathan R, Horvitz N, He Y, Kuparinen A, Schurr FM, et al. (2011) Spread of North American wind-dispersed trees in future environments. Ecol Lett 14: 211–219.
11. Nathan R (2006) Long-distance dispersal of plants. Science 313: 786–788.
12. Higgins SI, Richardson DM (1999) Predicting plant migration rates in a changing world: the role of long-distance dispersal. Am Nat 153: 464–475.
13. Loarie SR, Duffy PB, Hamilton H, Asner GP, Field CB, et al. (2009) The velocity of climate change. Nature 462: 1052–1055.
14. Schwartz MW (1993) Modeling effects of habitati fragmentation on the ability of trees to respond to climatic warming. Biodivers Conserv 2: 51–61.
15. Hill MF, Caswell H (1999) Habitat fragmentation and extinction thresholds on fractal landscapes. Ecol Lett 2: 121–127.
16. Collingham YC, Huntley B (2000) Impacts of habitat fragmentation and patch size upon migration rates. Ecol Appl 10: 131–144.
17. Pearson RG, Dawson TP (2005) Long-distance plant dispersal and habitat fragmentation: identifying conservation targets for spatial landscape planning under climate change. Biol Conserv 123: 389–401.
18. Pearson RG (2006) Climate change and the migration capacity of species. Trends Ecol Evol 21: 111–113.
19. McInerny G, Travis JMJ, Dytham C (2007) Range shifting on a fragmented landscape. Ecol Inform 2: 1–8.
20. Latham RE, Ricklefs RE (1993) Continental comparisons of temperate-zone tree species diversity. In: Ricklefs RE, Schluter D (eds), Species diversity in ecological

communities: historical and geographical perspectives. Chicago: University of Chicago Press. 294–314 p.
21. Hoegh-Guldberg O, Hughes L, McIntyre S, Lindenmayer DB, Parmesan C, et al. (2008) Assisted colonization and rapid climate change. Science 321: 345–346.
22. Vitt P, Havens K, Kramer AT, Sollenberger D, Yates E (2010) Assisted migration of plants: changes in latitudes, changes in attitudes. Biol Conserv 143: 18–27.
23. Simberloff D, Farr JA, Cox J, Mehlman DW (1992) Movement corridors: conservation bargains or poor investments? Conserv Biol 6: 493–504.
24. Murray AB (2007) Reducing model complexity for explanation and prediction. Geomorphology 90: 178–191.
25. Guisan A, Thuiller W (2005) Predicting species distribution: offering more than simple habitat models. Ecol Lett 8: 993–1009.
26. Travis JMJ (2003) Climate change and habitat destruction: a deadly anthropogenic cocktail. P Roy Soc B-Biol Sci 270: 467–473.
27. Carey PD (1996) DISPERSE: A cellular automaton for predicting the distribution of species in a changed climate. Global Ecol Biogeogr 5: 217–226.
28. Iverson LR, Prasad A, Schwartz MW (1999) Modeling potential future individual tree-species distributions in the eastern United States under a climate change scenario: a case study with Pinus virginiana. Ecol Model 115: 77–93.
29. Popescu SC, Wynne RH, Nelson RF (2003) Measuring individual tree crown diameter with lidar and assessing its influence on estimating forest volume and biomass. Can J Remote Sens 29: 564–577.
30. Johst K, Brandl R, Eber S (2002) Metapopulation persistence in dynamic landscapes: the role of dispersal distance. Oikos 98: 263–270.
31. Iverson LR, Schwartz MW, Prasad AM (2004) How fast and far might tree species migrate in the eastern United States due to climate change? Global Ecol Biogeogr 13: 209–219.
32. Higgins SI, Clark JS, Nathan R, Hovestadt T, Schurr F, et al. (2003) Forecasting plant migration rates: managing uncertainty for risk assessment. J Ecol 91: 341–347.
33. Hastings HM, Sugihara G (1993) Fractals: a user's guide for the natural sciences. New York: Oxford University Press 235 p.
34. Alston KP, Richardson DM (2006) The roles of habitat features, disturbance, and distance from putative source populations in structuring alien plant invasions at the urban/wildland interface on the Cape Peninsula, South Africa. Biol Conserv 132: 183–198.
35. Saunders DA, Hobbs RJ, Margules CR (1991) Biological consequences of ecosystem fragmentation: a review. Conserv Biol 5: 18–32.
36. Ziff RM, Sapoval B (1986) The efficient determination of the percolation threshold by a frontier-generating walk in a gradient. J Phys A Math Gen 19: L1169–1172.
37. Clark DA, Clark DB (1984) Spacing dynamics of a tropical rainforest tree: evaluation of the Janzen-Connell model. Am Nat 124: 769–788.
38. Morin X, Viner D, Chuine I (2008) Tree species range shifts at a continental scale: new predictive insights from a process-based model. J Ecol 96: 784–794.
39. Pastor J, Post WM (1988) Response of northern forests to CO_2-induced climate change. Nature 334: 55–58.
40. Paola C, Straub K, Mohrig D, Reinhardt L (2009) The "unreasonable effectiveness" of stratigraphic and geomorphic experiments. Earth-Science Reviews 97: 1–43.
41. Earn DJD, Levin SA, Rohani P (2000) Coherence and conservation. Science 290: 1360–1364.
42. Damschen EI, Haddad NM, Orrock JL, Tewksbury JJ, Levey DJ (2006) Corridors increase plant species richness at large scales. Science 313: 1284–1286.
43. Rayner S (2010) How to eat an elephant: a bottom-up approach to climate policy. Clim Policy 10: 615–621.
44. Wilbanks TJ, Kates RW (1999) Global change in local places: How scale matters. Climatic Change 43: 601–628.
45. Neilson RP (1993) Transient ecotone response to climatic change: some conceptual and modeling approaches. Ecol Appl 3: 385–395.
46. Lloyd KM, McQueen AAM, Lee BJ, Wilson RCB, Walker S, et al. (2000) Evidence on ecotone concepts from switch, environmental and anthropogenic ecotones. J Veg Sci 11: 903–910.
47. The Greenbelt Movement. Available: http://www.greenbeltmovement.org. Accessed February 2013.
48. United Nations Environment Programme (UNEP) Billion Tree Campaign. Available: http://www.plant-for-the-planet-billiontreecampaign.org/. Accessed February 2013.

3

The Reduced Effectiveness of Protected Areas under Climate Change Threatens Atlantic Forest Tiger Moths

Viviane G. Ferro[1], Priscila Lemes[2], Adriano S. Melo[1], Rafael Loyola[1]*

1 Departamento de Ecologia, Universidade Federal de Goiás, Goiânia, Goiás, Brazil, 2 Programa de Pós-Graduação em Ecologia e Evolução, Universidade Federal de Goiás, Goiânia, Goiás, Brazil

Abstract

Climate change leads to species' range shifts, which may end up reducing the effectiveness of protected areas. These deleterious changes in biodiversity may become amplified if they include functionally important species, such as herbivores or pollinators. We evaluated how effective protected areas in the Brazilian Atlantic Forest are in maintaining the diversity of tiger moths (Arctiinae) under climate change. Specifically, we assessed whether protected areas will gain or lose species under climate change and mapped their locations in the Atlantic Forest, in order to assess potential spatial patterns of protected areas that will gain or lose species richness. Comparisons were completed using modeled species occurrence data based on the current and projected climate in 2080. We also built a null model for random allocation of protected areas to identify where reductions in species richness will be more severe than expected. We employed several modern techniques for modeling species' distributions and summarized results using ensembles of models. Our models indicate areas of high species richness in the central and southern regions of the Atlantic Forest both for now and the future. However, we estimate that in 2080 these regions should become climatically unsuitable, decreasing the species' distribution area. Around 4% of species were predicted to become extinct, some of them being endemic to the biome. Estimates of species turnover from current to future climate tended to be high, but these findings are dependent on modeling methods. Our most important results show that only a few protected areas in the southern region of the biome would gain species. Protected areas in semideciduous forests in the western region of the biome would lose more species than expected by the null model employed. Hence, current protected areas are worse off, than just randomly selected areas, at protecting species in the future.

Editor: Brock Fenton, University of Western Ontario, Canada

Funding: VGF work is funded by CNPq (grant #563332/2010-7). PL received a PhD studentship from CNPq. RDL and ASM received productivity fellowships granted by CNPq (grants #304703/2011-7 and #307479/2011-0, respectively). RDL work is funded by CAPES-FCT Program, by the Brazilian Research Network on Global Climate Change (Rede CLIMA), and Conservation International Brazil. This study benefited from resources provided by Site 13 (Parque Nacional das Emas) of the Brazilian Long Term Ecological Research Network (CNPq, grant #558187/2009-9). The funders had no role in study design, data collection and analysis, decision to publish, or preparation of the manuscript.

Competing Interests: The authors have declared that no competing interests exist.

* Email: rdiasloyola@gmail.com

Introduction

The implementation and maintenance of protected areas is still the cornerstone of conservation actions [1]. However, due to broad-scale environmental changes which will potentially shift the distribution of suitable habitats for many species across the geographic space [2,3], scientists have expressed concern that existing networks of protected areas might not be able to guarantee the long-term persistence of the species they are supposed to protect [4,5].

Climate change poses a new challenge to the traditionally static way conservation planning is usually done, by forcing planning to become more dynamic [6]. Several species have already shifted their ranges to cooler regions, both in temperate regions and in the tropics, as a response to a warming climate [7–9]. Most solutions ofereded by conservation scientists and practitioners to deal with spcies' range shifts focus on the establishment of new protected areas that should cope with the effects of climate change on species distribution [6,10–16]. However, the effectiveness of protected areas may decrease as they become climatically unsuitable for most species [5] and more suitable to invasive species [17,18]. Climate-driven modifications in species composition within protected areas may disrupt species interactions [19] by altering ecosystem functioning [20]. Therefore, it is necessary to take into account forecasted changes in species distributions to evaluate the future effectiveness of protected areas [5,6,21].

The effects of climate change on species distributions have been generally inferred through ecological niche models (ENM) [22], also referred to as "bioclimatic envelope models" (BEM) [23] or "species distribution models" (SDM) [24]. Araújo and Peterson [25], Peterson and Soberón [26], and Rangel and Loyola [27] have all provided recent clarifications on their conceptual differences. However, different modeling methods and climate models may produce very different outputs, increasing uncertainties in projected distributions and their applicability to conservation efforts [3,28]. In the last decade, new modeling techniques have been developed that take into consideration such impediments (e.g. model ensemble forecasting). Nonetheless, few studies have applied modern techniques to predict the distribution of invertebrates (but see Diniz-Filho et al., 2010a, b), despite an urgent need to evaluate the consequences of climate change to this hyperdiverse group in order to plan for its conservation [29].

Lepidoptera is the second richest order of insects, with 150,000 species recorded in the world [30]. Butterflies and moths are exclusive pollinators of many plant species [31,32]. The vast majority of Lepidoptera larvae are herbivores, consuming almost all orders of gymnosperms and angiosperms as well as mosses and ferns [33]. Herbivory can influence the fitness [34], distribution, composition and abundance of plant species [35], as well as the rate of litter decomposition [36]. Lepidopterans are also important food items for arthropods (in particular spiders and other insects) and vertebrates (especially birds and bats). Changes in Lepidoptera diversity, abundance, phenology, distribution and assemblage composition driven by climate changes may affect ecosystem functions and services, species interactions, as well as the structure of plant communities and economic losses due insect infestation and pestilence [37].

Many studies have found that climate change will alter patterns of phenology, horizontal or vertical range (distribution and area), and the abundance of Lepidoptera species (reviewed in [38]). For example, Conrad et al. [39] studied the population dynamics of the tiger moth *Arctia caja* (Arctiinae) between 1968 and 1999 in Great Britain and observed a decrease of about 30% of abundance and proportion of occupied sites after 1984. Arctiinae (Erebidae) (classification following [40]), in particular, comprises almost 11,000 species of moths, of which about 6,000 are found in the Neotropics [41]. In Brazil, there are records for 1,391 species [42]. The Atlantic Forest has the richest Arctiinae fauna among all Brazilian biomes (1193 species) and approximately 40% of these species are endemic to this biome [43]. Arctiinae larvae feed on angiosperms and gymnosperms, as well as algae, lichens, and mosses [44]. Although Arctiinae are among the most polyphagous lepidopterans [44], the proportion of generalist species decreases toward the tropics. Many Arctiinae larvae and adults have conspicuous coloration, are diurnal and many adults form mimetic rings with Hemiptera, Hymenoptera, Coleoptera and unpalatable butterflies [45]. As well as exhibiting warning coloration, most of these moths are also toxic or unpalatable. Several secondary compounds were found in all stages of Arctiinae (eggs, larvae, pupae and adults), including pyrrolizidine alkaloids, which are generally sequestered from their host plants during the larval stage [45]. We choose this group because of its important roles in ecosystems and also because of their extremely high diversity in the region, so that we can run models using a great amount of data and keeping the taxonomic group narrow.

Here, we evaluated the current and future climatic suitability of protected areas located in the Atlantic Forest Biodiversity Hotspot (in Brazil) based on species' ecological niche models and diversity patterns of tiger moths (Lepidoptera: Erebidae: Arctiinae). More specifically, we addressed the following questions: (1) how will climate change affect the geographical pattern of Arctiinae species richness in the region? And (2) how does the spatial location of a given protected area determine if it will gain or lose species under different climate change scenarios?

We selected tiger moths as our case-study because they comprise a species-rich subfamily, are well represented in Brazilian collections (especially in the Atlantic Forest, which is the geographical area covered in this study) [42] and because there are several active researchers currently working with Neotropical Arctiinae. These researchers helped with the identification of some species and allowed for a nomenclatural update, incorporating recent occurrence records in our dataset.

Methods

Study region

We focused our analyses on the Atlantic Forest Biodiversity Hotspot [46]. The Atlantic Forest originally covered around 150 million ha (Fig. 1) with heterogeneous environmental conditions. Its latitudinal range extends into tropical and subtropical regions, and its wide longitudinal range harbors differences in forest composition due to a diminishing gradient in rainfall from coast to interior [47]. Although the Atlantic Forest has high diversity and endemism (with more than 20,000 plant species, 261 mammal species, 688 bird species, 200 reptile species, 280 amphibian species, to name well-studied taxonomic groups), currently, only *ca.* 1% of the original forests are legally protected [47]. An effective reserve network, taking into account climate change to ensure the species persistence in the long-term, is therefore imperative to address conservation investment in appropriate sites.

Ecological niche models

We obtained occurrence records for 703 tiger moth species inhabiting the Atlantic Forest from field surveys and museum records. Tiger moth records included the period of 1920 to2008. We overlaid these point-locality records for each species into an equal-area grid (10 km×10 km of spatial resolution) that covered the full extent of the Atlantic Forest. Then, we built a species by grid cell matrix, considering presences of species inside grid cells. Species with less than five occurrences were excluded to avoid model bias, and therefore, a total of 507 species were studied.

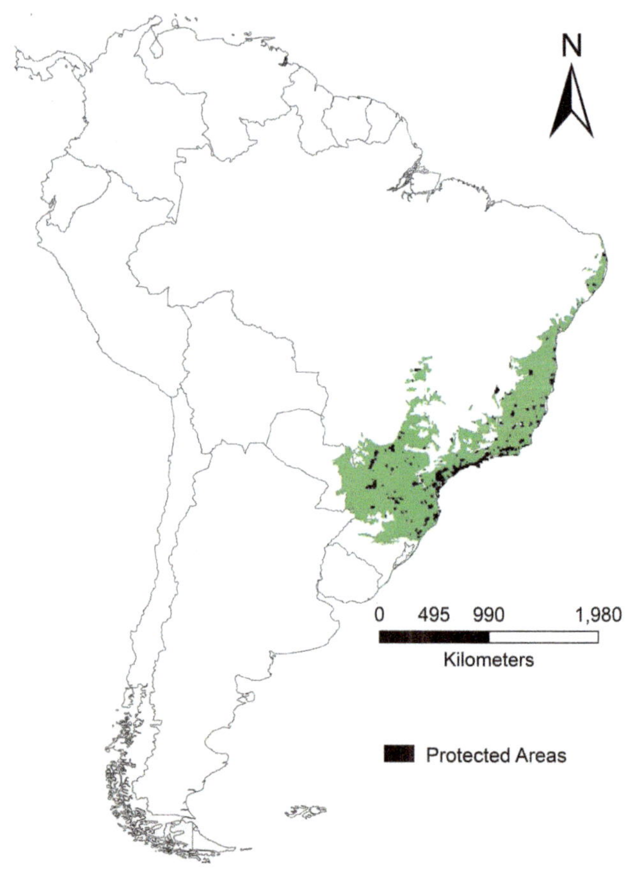

Figure 1. Original extent of the Atlantic Forest Biodiversity Hotspot in Brazil.

We obtained current climatic data from the WorldClim database (http://www.worldclim.org/current) and future climatic scenarios from CIAT (http://ccafs-climate.org). These future scenarios were developed by the Intergovernmental Panel on Climate Change (IPCC) Fourth Assessment Report (AR4). For each species, we modeled its distribution as a function of four climatic variables: annual mean temperature, temperature seasonality (standard deviation * 100), annual precipitation and precipitation seasonality (coefficient of variation). These variables represent interpolated climate data from 1950 to2000 [48]. For future climatic conditions we used the same climate variables for the year 2080, obtained from three Atmosphere-Ocean General Circulation Models (AOGCMs) of the A2 emission scenario (CCCMA-CGCM2, CSIRO-MK2.0 and UKMO-HadCM3), that were generated by application of delta downscaling method on the original data from IPCC's report. Data original resolution was 30 arc-seconds and both current and future climate variables were re-scaled to our grid resolution.

We used presence derived from species' occurrences and climatic variables to model species' ecological niche and project their distributions. As reliable absence data were not available, we fitted six presence-only modeling methods (which differ both conceptually and statistically [27]), grouped them into two separate sets (distance methods and machine-learning methods) and applied the ensemble forecasting approach within each set (see text below). Distance methods were Euclidian and Gower distances [49] and Ecological Niche Factor Analysis (ENFA) [50]. Machine learning methods were Maximum Entropy (MaxEnt) [51], Genetic Algorithm for Rule Set Production (GARP) [52] and Artificial Neural Networks (ANN) [53]. These presence-only methods for modeling species' ecological niches can be grouped into three types of presence-only methods [54]. Firstly, into methods based solely on presence records (e.g. Euclidian and Gower distances), the prediction is made without reference to other samples from the study area. Secondly, into methods using "background" climatic data for the whole study area (e.g. ENFA, MaxEnt), which evaluate how the climatic conditions where species are known to occur relates to the climate across the rest of the study area (the 'background'). Thirdly, into methods that generate (sample) "pseudo-absences" from the study area (e.g. GARP, ANN), assessing differences between occurrence sites and a set of sites chosen from the study area which are used instead of real absence data. In this case, the set of "pseudo-absences" were selected randomly [52].

For all models, we randomly partitioned presence and pseudo-absence data of each species in 75% for calibration (or training) and 25% for validation (or test); repeating this process 10 times (i.e. a cross-validation) and maintaining the observed prevalence of each species. We converted continuous predictions in presence and pseudo-absences finding the threshold with maximum sensitivity and specificity values in the receiver operating characteristic (ROC) curve and calculated the True Skill Statistics (TSS) to evaluate model performance [55]. The ROC Curve is created by plotting the fraction of true positives out of the positives vs. the fraction of false positives out of the negatives, at various threshold settings. The TSS range from −1 to +1, where values equal +1 is a perfect prediction and values equal to or less than zero is a prediction no better than random [55]. Although the area under the receiver operating characteristic curve (AUC) is the most common method to evaluate the accuracy of predictive distribution models, we decided to use TSS. There are several reasons why AUC should not be used for this purpose [56]. In particular, AUC weighs omission and commission errors equally and the total geographic extent of the study highly influences its scores [56].

We did an ensemble of forecasts to produce consensual predictions of species distributions [3,10,23,28,57–59]. We projected distributions to current climatic conditions and obtained 30 projections per species within each set of methods (3 modeling methods ×10 randomly partitioned data). We also projected distributions to future climate, obtaining 90 projections per species (3 modeling methods ×3 climate models ×10 randomly partitioned data). This allowed us to generate a frequency of projections in the ensemble. We then generated the frequency of projections weighted by the TSS statistics for each species and timeframe within each set of methods, i.e. best models have more weight in our consensus projections. We considered the presence of a species only in cells with 50% or more of frequency of projections, but a continuous value was held when this occurred. Then, we defined species richness as the sum of the ranges which overlapped (predicted by ENMs) for each cell. So, for example, if twenty different species were projected in a given cell the species richness for that cell was considered as twenty. Finally, we calculated species turnover between contemporary and future species distributions in each cell (G+L/S)/S+G, where "G" was the number of species gained, "L" the number of species lost and "S" is the contemporary species richness found in the cell.

Evaluation of protected area effectiveness under climate change

The locations of the 187 protected areas (IUCN Categories I–IV) currently established in the Atlantic Forest were obtained from the United Nations Environmental Programme, World Conservation Monitoring Cemntre (UNEP-WCMC) [60]. These protected areas comprised 820 cells distributed in 11,461 Atlantic Forest cells. We overlaid protected area polygons onto our grid considering a grid cell as "protected" even if only a portion of it is protected, assuming that all species occurring in that cell could potentially benefit from the occurrence of a protected area in that cell.

Our aim was to evaluate whether current locations of protected areas are better than random allocations in protecting tiger moth diversity in the face of climate change. For this, we generated a null model that maintained size, form and orientation of protected areas but removed other intrinsic effects that likely will affect their suitability in the face of climate change (i.e. latitude, altitude). The null model allocated the protected areas randomly in the Atlantic Forest and obtained species richness in the present and the future based on the projections of species distribution models. Since there are many distinct possibilities to randomly allocate protected areas in the Atlantic Forest, this procedure was repeated 1,000 times and the average species richness obtained.

Results

For most species, TSS values were relatively high (mean TSS ± SD = 0.55±0.05 for distance methods; and 0.64±0.14, for machine-learning methods) indicating good model fit (Table S1). Different modeling methods projected similar patterns of tiger moth species richness both for current and future climates, except for ENFA and ANN (Fig. 2). For all modeling methods, ensemble projections indicated areas of high species richness in the central and southern regions of the Atlantic Forest both for now and for 2080 (Fig. 3). However, these regions should become climatically unsuitable in 2080 for many species, decreasing species' distribution areas (Fig. 3, Table S2 and S3). We also found high species temporal turnover (up to 100% for all methods) across the

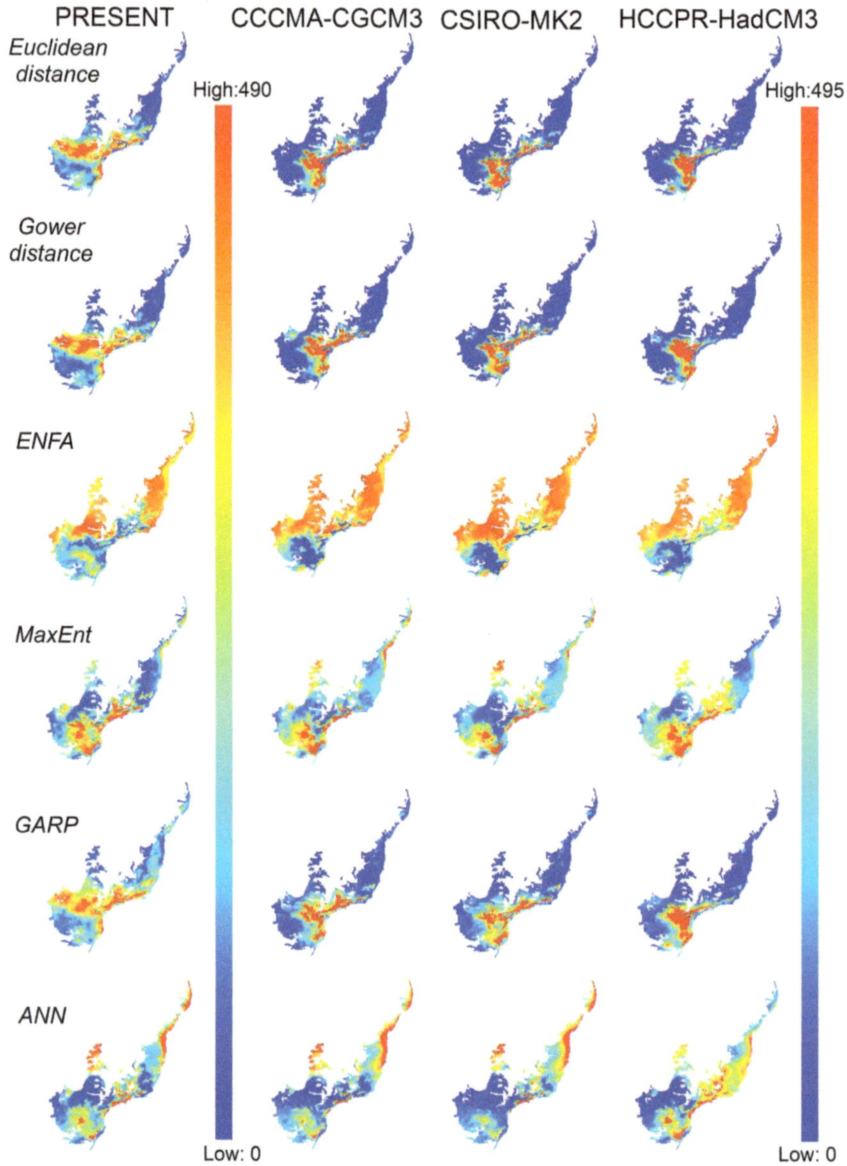

Figure 2. Species richness patterns for tiger moths in the Atlantic Forest Biodiversity Hotspot, Brazil. Tiger moth species richness patterns in the Atlantic Forest, Brazil (present and future, 2080, climate models CCCMA-CGCM3, CSIRO-MK2, and HCCPR-HadCM3) forecasted by ecological niche models generated by different distance modeling methods (Euclidian and Gower distances, Ecological Niche Factor Analysis, ENFA) and machine learning methods (Maximum Entropy, MaxEnt; Genetic algorithm for Rule set Production, GARP; Artificial Neural Networks, ANN).

projections between current and future climates (Fig. 4). This means that future scenarios showed dramatic changes not only in species richness but also in turnover of species composition (Table S4).

Central and northern sections of the Atlantic Forest should face higher temperatures, with lower seasonality, whereas the south should receive more rainfall, yet also have lower seasonality (Fig. 5). Most species had a significant range contraction (Table S3). Consensus of machine-learning methods projected up to 4.3% of species that had their ranges reduced by 100%, whereas consensus of distance methods projected 0.4% of species in the same situation (Table S3).

Future species richness should be lower than contemporary richness for most of the protected areas (Fig. 6). Differences between future and contemporary species richness were more varied for Distance than Machine Learning methods (Fig. 6A1 and 6B1). Distance models show a more optimistic result, with some protected areas holding more species (green dots in Fig. 6A1 and protected areas in the Fig. 6A2) in the future than in current time. These protected areas predicted to gain species richness are mostly located in the mountainous southeastern portion of the biome.

Machine-learning methods predicted that almost all protected areas will lose species in the future (Figs. 6B1 and 6B2). The main expectation of the null model of random location of protected areas is the loss of species in the future, both for Distance (Fig. 6A1) and Machine Learning (Fig. 6B1) methods. Among the protected areas predicted to lose species (orange and red dots and protected areas in Fig. 6), around half of them should experience losses worse than those predicted by the null model (red dots and protected areas in Fig. 6). These protected areas are concentrated in the southwestern and northern sections of the biome.

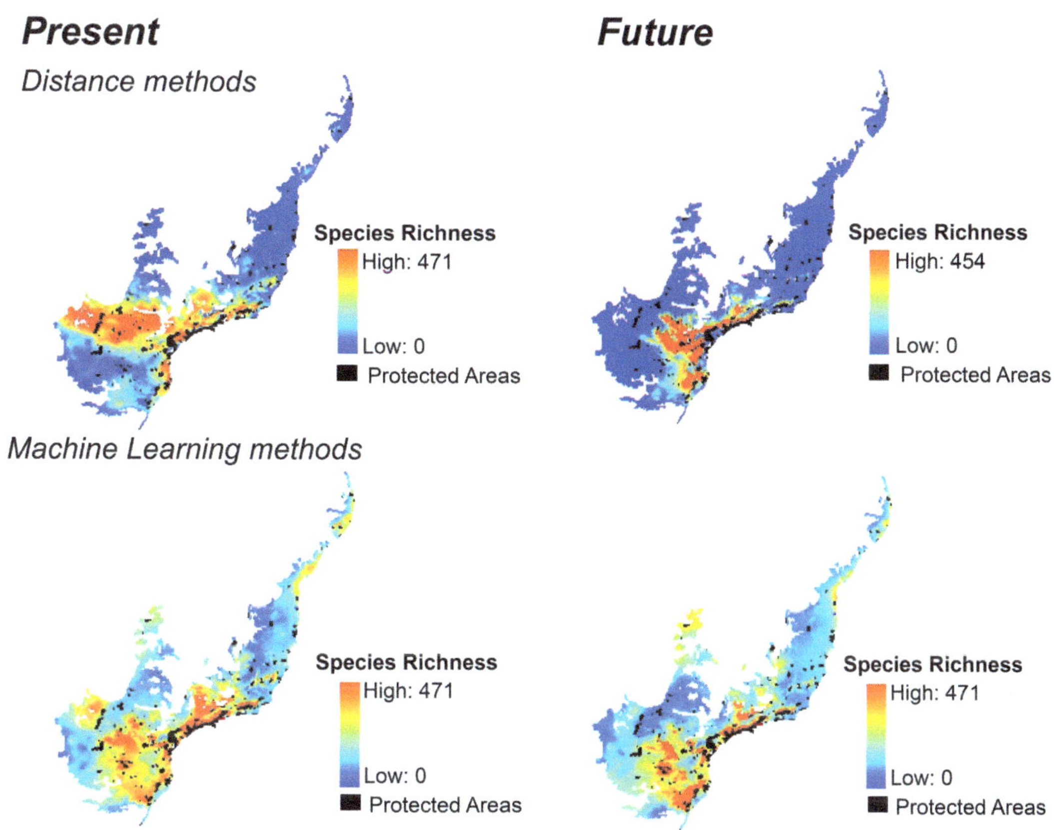

Figure 3. Consensus maps for tiger moth species richness in the Atlantic Forest Biodiversity Hotspot, Brazil. Maps of modeled tiger moth species richness based on consensus projections of 507 species predicted to occur in the Atlantic Forest Biodiversity Hotspot, Brazil, for current time (1950–2000) and 2080 (2051–2080) according to two different types of modeling methods and climate models. Models from distance and machine-learning methods were combined through an ensemble of forecasts to generate these consensus maps.

In addition to the general trend of species loss, the baseline comparison provided by the null model of random allocation of protected areas (blue dots) indicated great variation between how many species protected areas would lose (and also gain in case of distance methods). For machine-learning methods, species richness for current and future climates for the real locations of protected areas was generally lower (red dots and protected areas in Fig. 6) than those obtained from the random locations of protected areas (blue dots in Fig. 6). Nearly half of currently established protected areas should become climatically unsuitable at rates higher than expected by a random distribution of protected areas (red dots and protected areas in Fig. 6).

Protected areas predicted to gain species (particularly for distance methods) are mostly located in the cooler southern region of the Atlantic Forest (Fig. 6). Protected areas along the coast or in adjacent mountainous areas should lose species, although those in the north would do so at rates higher than those expected by our null model (protected areas indicated in red). The western region of the biome, which includes semideciduous forests, will experience a severe reduction in species richness due to climate change (Fig. 6).

Discussion

We forecasted species' range shifts and range contractions for most tiger moth species inhabiting the Atlantic Forest to evaluate the effectiveness of existing protected areas [61,62]. Our findings indicate that most protected areas should become climatically unsuitable for sustaining their current number of species under climate change. As Atlantic Forest protected areas will become less effective in safeguarding moths, it is important to anticipate how climatic changes will lead to a decreasing species representation across the entire network of protected areas.

Our results agree with Pearson and colleagues [63] which detected similar accuracy among species distribution models, although the spatial pattern in the predictions was different. Distance methods are simple methods that do not consider complex relationships between species occurrence and predictors, use presence-only data and tend to underestimate the distribution in novel conditions like those expected to occur under climate change [24,63]. Alternatively, machine-learning methods are very complex, assume different relationships and can underestimate or overestimate distributions in novel conditions [21,63]. These features help explain the differences in our results obtained by distance methods when compared with machine-learning methods, especially regarding the pattern of species richness and turnover rates.

The choice of a perfect set of modeling methods, however, is not an easy task. When predicting climate change effects on species distributions, commission errors lead to the overestimation of range expansions whereas omission errors produce overestimates of range contractions. If one is employing such models to predict regions of climatic stability to guide conservation actions, both omission and commission errors are of particular interest, as these errors are likely to produce huge bias in the results of gap-analysis as well. Distance models tend to inflate commission errors and

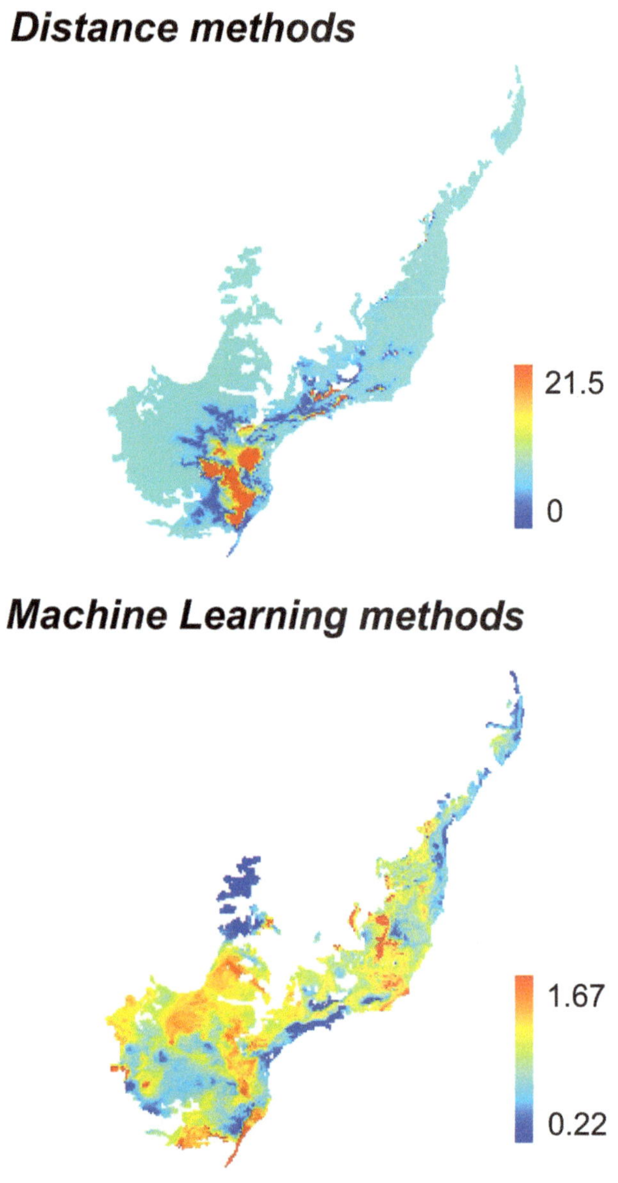

Figure 4. Species turnover pattern for tiger moths in the Atlantic Forest Biodiversity Hotspot, Brazil. Maps of modeled tiger moth species turnover based on consensus projections of 507 species predicted to occur in the Atlantic Forest Biodiversity Hotspot, Brazil, for current time (1950–2000) and 2080 (2051–2080) according to two different types of modeling methods and climate models. Models from distance and machine-learning methods were combined through an ensemble of forecasts to generate these consensus maps.

individual species distribution, which are usually larger than predicted by other models. Machine-learning models usually result in smaller distributions when compared to other models. Thus, if conservation practitioners want to reduce commission errors, machine-learning models are perhaps the best option. Although, in this case, one will probably face the inconvenience of a lack of clarity and in turn, a difficulty of explaining results to stakeholders and decision makers involved in the conservation process [27].

Our study predicts that up to 4.3% of tiger moth species should face 100% of range contraction in the Atlantic Forest's future. Among these species that would disappear from the region, three of them are endemic (*Clemensia marmorata*, *Cosmosoma cingla*, and Arctiini NI92 morphospecies). The majority (n = 18) of species predicted to have a 100% of range contraction belong to the Arctiini tribe (in general robust, medium-sized moths and polyphagous larvae) and the Lithosiini tribe (in general slender, small-sized moths, and lichen-feeding larvae). Unfortunately, there islittle, scattered information on the natural history of and the appropriate climatic conditions for these species available (as for the vast majority of Neotropical species). It is therefore hard to make recomendations for conservation strategies for these species.

In general, the consequences of these extinctions can spread through networks of interaction causing extinction cascades and consequent disrupted ecological functions in ecosystems [19,64]. In addition to causing local and regional extinctions, climate change can alter the size and location of the range and patterns of abundance and phenology of species [38]. Many lepidopterans, including Arctiinae species [65], are agricultural pests and changes in range, abundance and phenology of these species may increase the rate of invasion and increase the damage intensity they can cause [66,67]. Furthermore, climate change can modify patterns of synchronization between larvae and host plants, among adults and flowers, and among the two life stages and their natural enemies [68]. Some tiger moth species are important pollinators of Atlantic Forest plants, including orchids [69], and the lack of synchronization between flowering and pollinating may cause more extinctions.

We also found an increase in species richness in the future in a few southern areas of the biome, due to the expansion of the range toward higher latitudes. This pattern has been observed in studies on Lepidoptera [9,70] and other groups [8]. It is known, however, that species with wide latitudinal ranges could be preadapted to cope with climate change because they already find considerable temperature variation across their habitats. However, low dispersal capability and difficulties to cross open habiats may prevent or at least slow down range shifts of Arctiinae [38]. For instance, several studies have found that Arctiinae species disperse over short distances in natural conditions [71,72]. The average distance traveled by *Dysauxes ancilla* L. moths in two consecutive days (at mark-released-recapture study), for example, was only 43 m [72]. The same author observed that moths rarely leave their breeding area and did not cross very open areas or dense forest. In fact, two other studies in Brazilian Biodiversity Hotspots, the Atlantic Forest [73] and the Cerrado [74] found that Arctiinae fauna of open vegetation (grassland) differed from the denser vegetation (forest) and a very small number of species co-occurred in these vegetation types. More importantly, these differences were evident even when the sample sites were close to each other (about 100 m). Accordingly, the low dispersal capability of the species of Arctiinae compounded with the low ability of the species to cross open areas in fragmentated landscapes of the Atlantic Forest should further complicate the future of these species. This is particularly concerning in the Atlantic Forest, given that human activity has degraded more than 85% of the biome [47].

One strategy to mitigate the loss of species in the Atlantic Forest protected areas in the future would be the creation of protected areas in southern regions and at higher altitudes of the Atlantic Forest. Klorvuttimontara and colleagues [75], for example, despite having recorded a reduction of approximately 30% of butterfly richness in the protected areas of Thailand in the future (A2 scenario), observed that the effectiveness of protected areas remained almost the same in the future. The explanation for this was that most protected areas in Thailand are located at high altitudes, allowing the butterflies to migrate from lowland areas to higher altitudes as the climate becomes unsuitable in the future. Of course, critical questions should be addressed to actually imple-

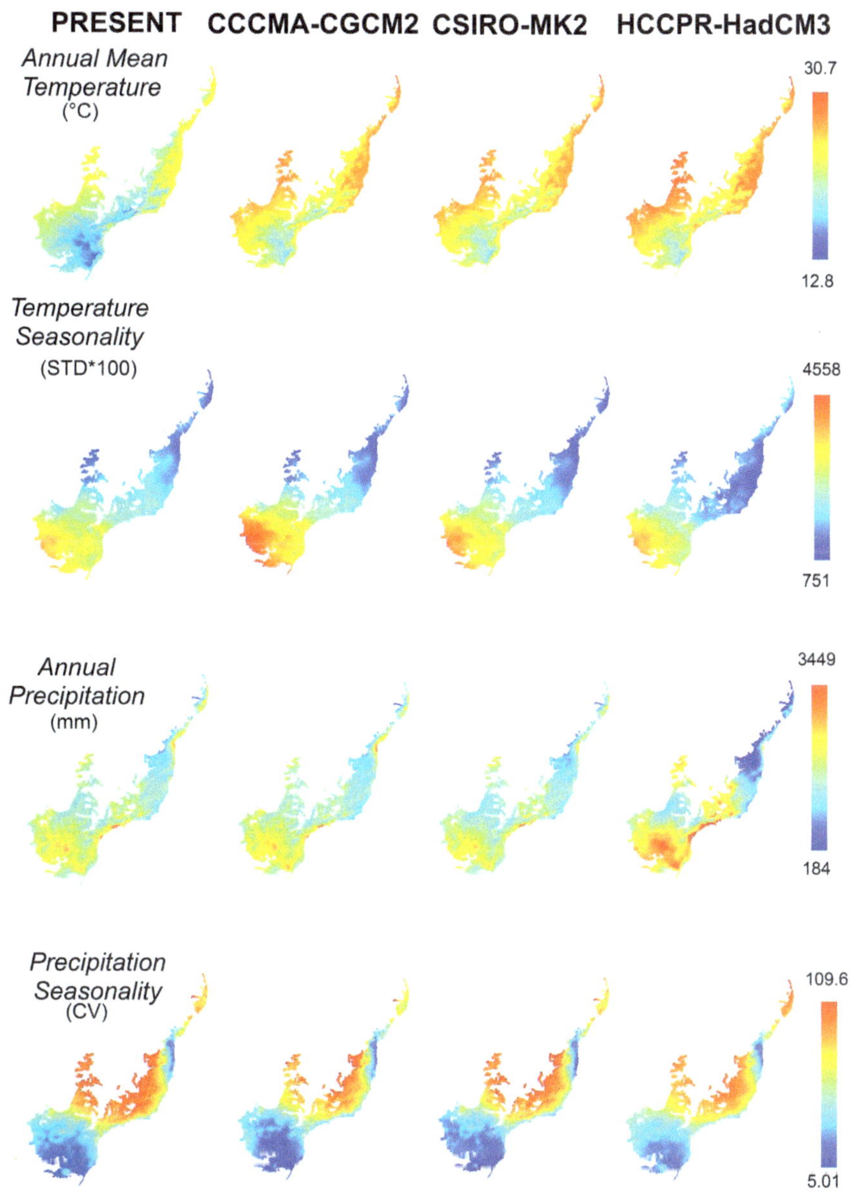

Figure 5. Expected changes for the four climatic variables used to model species' ecological niches in the Atlantic Forest Biodiversity Hotspot, Brazil. Maps show present conditions and values for future climate models CCCMA-CGCM3, CSIRO-MK2, and HCCPR-HadCM3, in 2080).

ment a decision like this, such as infostering the implementation of protected areas in high-altitude regions. Some questions would be, for example: what are the indications that these regions would protect other taxa than moths? Weel, it seems defensible to predict that they likely will be, because many species and communities are likely to migrate south and to higher elevations in response to climate change [7,8,10,17]. Should protected areas be placed to span altitudinal ranges within the protected area itself, so that species can start low and move steadily higher as the climate changes? These questions are still difficult to answer, but we hope our study is a step towards scientifically driven decisions in future protected area establishment and management.

The establishment of protected areas is still one of our best conservation actions to protect biodiversity. The Atlantic Forest has been indicated as aconservation priority by several world studies using different criteria [76,77]. Our approach is one of the first to incorporate climate change threats on the long-term assessment of protected areas (see [5,17,21,77,78]). In fact, climatic changes are increasingly driving species out of protected areas due to species range shifts [5,79,80].

It is important to highlight some of the caveats of our study. Firstly, our database has a large number of records, which were gathered from field studies and museum records. Although the data is dense and trustworthy, there could be some bias towards easily assessed sampling regions or toward large, colorful species, as in all kinds of samplings. Further, some records were from as early as 1920 and climate has likely shifted already since then. Secondly, our models assume tiger moths are in equilibrium with the current climate and have unlimited dispersal to tackle suitable climates as they move in the geographic space. These are simple assumptions allowing us to model all species distribution at a time. Thirdly, we predicted future species distribution assuming that the

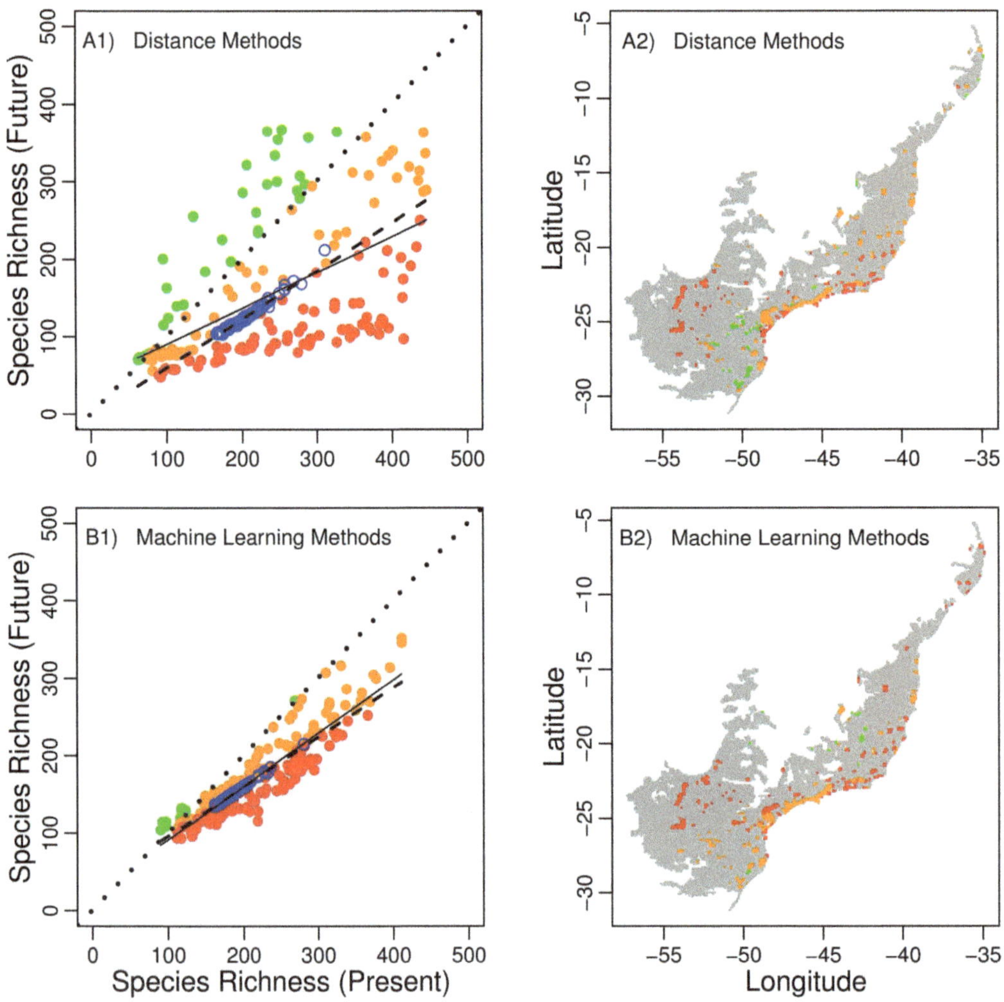

Figure 6. Relationship between present and future tiger moth species richness in the Atlantic Forest Biodiversity Hotspot, Brazil. Modeled present and future species richness of protected areas (filled circles) in the Atlantic Forest biome (A1 and B1). Open blue circles indicate expected species richness according to a null model of random location of protected areas in the biome. Dashed lines are the extrapolated regressions of the expected species richness according to the null model (A1: y~−1.983+0.623*x; B1: y ~32.178+0.640*x). Filled red circles indicate protected areas predicted to have severe species richness losses, defined as those in which future species richness will be lower than the predicted by a null model of random location of protected areas (below dashed regression line). Orange filled circles indicate protected areas predicted to have mild species richness losses, defined as those in which future species richness will be higher than the predicted by a null model of random location of protected areas. Green filled circles indicate protected areas predicted to gain species richness. Solid lines indicates the regression of modeled species richness in the future against modeled species richness in the present (A1: y~43.817+0.464*x, $R^2=0.362$, $F_{1,185}=105.2$, $P<0.001$; B1: y~22.736+ 0.688*x, $R^2=0.850$, $F_{1,185}=1051.0$, $P<0.001$). Maps of protected areas predicted to gain (green) or lose (orange, red) species in future changing climate (A2 and B2).

vegetation types in the Atlantic Forest will remain in the same regions of the current distribution. This last assumption can affect our species distribution model predictions. Species' range shifts outside of the current limits of the biome or their preferential habitat cannot be measured by our methods. Fourthly, the effectiveness of protected areas tends to be overestimated given that species were considered to be protected if any part of their range overlapped with protected areas polygons [81]. However, simple presence within a protected area is insufficient to ensure the long-term persistence of many species. Most existing protected areas in the Atlantic Forest were created without following any ecological criterion and species representation in this network is highly variable. Clearly, there is a need to update the current protected areas to improve species representation and climate change is an important factor to be considered.

As a final message, it is important to remember that species will generally alter their distributions independently. How do we prioritize which species to try to track through these changes? In this paper we recognize this conundrum and selected Tiger moths as our model group because of available data and diversity of species. It is now cear that we simply can't model or predict the response of all species on Earth to climate change. However, there are some generalities, such as the general tendency of species to mode southward or uphill in the southern hemisphere that could be helpful, if combined with models like the one we presented here, to guiding decisions about how to better allocate finite resources for biodiversity conservation.

Supporting Information

Table S1 Tiger moth tribes, species, True Skill Statistics (TSS) for each ecological niche modeling method and the ensembles among distance methods (Euclidian and Gower distance and ENFA) and machine-learning methods (MaxEnt, GARP, and Artificial Neural Networks).

Table S2 Tiger moth tribes, species, and number of grid cells predicted to be occupied during baseline climate (1950–2000) and future climate (2051–2080) under different climate models and ecological niche modeling methods.

Table S3 Tiger moth tribes, species, and percent of range contraction predicted to during baseline climate (1950–2000) and future climate (2051–2080) under different climate models and ecological niche modeling methods.

Table S4 Turnover measures for each of 11,464 grid cells overlaying the Atlantic Forest, Brazil, in this study. Turnover is shown for each modeling method (Euclidian and Gower distance, ENFA, MaxEnt, GARP, and Artificial Neural Networks), and climate model (CCCMA-CGCM, CSIRO-MK2, and HCCPR-HadCM3). Turnover found across the ensemble of distance and machine learning methods are also shown.

Author Contributions

Conceived and designed the experiments: VF RL. Analyzed the data: PL ASM. Wrote the paper: RL VF ASM PL.

References

1. Loucks C, Ricketts TH, Naidoo R, Lamoreux J, Hoekstra J (2008) Explaining the global pattern of protected area coverage: relative importance of vertebrate biodiversity, human activities and agricultural suitability. J Biogeogr 35: 1337–1348. doi:10.1111/j.1365-2699.2008.01899.x.
2. Parmesan C (2006) Ecological and Evolutionary Responses to Recent Climate Change. Ann Rev Ecol Evol Sys 37: 637–669. doi:10.1146/annurev.ecolsys.37.091305.110100.
3. Diniz-Filho JAF, Bini LM, Fernando Rangel T, Loyola RD, Hof C, et al. (2009) Partitioning and mapping uncertainties in ensembles of forecasts of species turnover under climate change. Ecography 32: 897–906. doi:10.1111/j.1600-0587.2009.06196.x.
4. Scott JM, Davis FW, McGhie RG, Wright RG, Groves C, et al. (2001) Nature reserves: do they capture the full range of america's biological diversity? Ecol Appl 11: 999–1007.
5. Araujo MB, Cabeza M, Thuiller W, Hannah L, Williams PH (2004) Would climate change drive species out of reserves? An assessment of existing reserve-selection methods. Global Change Biol 10: 1618–1626. doi:10.1111/j.1365-2486.2004.00828.x.
6. Hannah L (2010) A global conservation system for climate-change adaptation. Conserv Biol 24: 70–77. doi:10.1111/j.1523-1739.2009.01405.x.
7. Colwell RK, Brehm G, Cardelús CL, Gilman AC, Longino JT (2008) Global warming, elevational range shifts, and lowland biotic attrition in the wet tropics. Science 322: 258–261.
8. Chen I, Hill JK, Ohlemüller R, Roy DB, Thomas CD (2011) Rapid range shifts of species associated with high levels of climate warming. Science 333: 1024–1026. doi:10.1126/science.1206432.
9. Parmesan C, Ryrholm N, Stefanescu C, Hill JK, Thomas CD, et al. (1999) Poleward shifts in geographical ranges of butterfly species associated with regional warming. Nature 399: 579–583. doi:10.1038/21181.
10. Lemes P, Loyola RD (2013) Accommodating Species Climate-Forced Dispersal and Uncertainties in Spatial Conservation Planning. PLoS ONE 8: e54323. doi:10.1371/journal.pone.0054323.
11. Faleiro F V., Machado RB, Loyola RD (2013) Defining spatial conservation priorities in the face of land-use and climate change. Biol Conserv 158: 248–257. doi:10.1016/j.biocon.2012.09.020
12. Loyola RD, Lemes P, Nabout JC, Trindade-Filho J, Sagnori MD, et al. (2013) A straightforward conceptual approach for evaluating spatial conservation priorities under climate change. Biodivers Conserv 22: 483–485. doi:10.1007/s10531-012-0424-x.
13. Dawson TP, Jackson ST, House JI, Prentice IC, Mace GM (2011) Beyond predictions: biodiversity conservation in a changing climate. Science 332: 53–58. doi:10.1126/science.1200303.
14. Williams P, Hannah L, Andelman S, Midgley G, Araújo M, et al. (2005) Planning for Climate Change: Identifying Minimum-Dispersal Corridors for the Cape Proteaceae. Conserv Biol 19: 1063–1074. doi:10.1111/j.1523-1739.2005.00080.x.
15. Mawdsley JR, O'Malley R, Ojima DS (2009) A review of climate-change adaptation strategies for wildlife management and biodiversity conservation. Conserv Biol 23: 1080–1089. doi:10.1111/j.1523-1739.2009.01264.x.
16. Mawdsley J (2011) Design of conservation strategies for climate adaptation. Wiley Interdisciplinary Reviews: Climate Change 2: 498–515. doi:10.1002/wcc.127.
17. Nori J, Urbina-Cardona JN, Loyola RD, Lescano JN, Leynaud GC (2011) Climate change and American Bullfrog invasion: what could we expect in South America? PloS ONE 6: e25718. doi:10.1371/journal.pone.0025718.
18. Loyola RD, Lemes P, Faleiro FV, Trindade-Filho J, Machado RB (2012) Severe loss of suitable climatic conditions for marsupial species in Brazil: challenges and opportunities for conservation. PLoS ONE 7: e46257. doi:10.1371/journal.pone.0046257.
19. Tylianakis JM, Didham RK, Bascompte J, Wardle DA (2008) Global change and species interactions in terrestrial ecosystems. Ecol Lett 11: 1351–1363. doi:10.1111/j.1461-0248.2008.01250.x.
20. Terborgh J, Lopez L, Nuñez P, Rao M, Shahabuddin G, et al. (2001) Ecological meltdown in predator-free forest fragments. Science 294: 1923–1926. doi:10.1126/science.1064397.
21. Hannah L, Midgley G, Andelman S, Araújo M, Hughes G, et al. (2007) Protected area needs in a changing climate. Frontiers Ecol Env 5: 131–138.
22. Peterson AT, Soberón J, Pearson RG, Anderson RP, Martínez-Meyer E, et al. (2011) Ecological Niches and Geographic Distributions. Levin SA, Horn HS, editors Princeton University Press. p.
23. Garcia RA, Burgess ND, Cabeza M, Rahbek C, Araújo MB (2012) Exploring consensus in 21st century projections of climatically suitable areas for African vertebrates. Global Change Biol 18: 1253–1269. doi:10.1111/j.1365-2486.2011.02605.x.
24. Franklin J (2009) Mapping species distributions: spatial inference and prediction. Cambridge: Cambridge University Press.
25. Araújo MB, Peterson AT (2012) Uses and misuses of bioclimatic envelope modeling. Ecology 93: 1527–1539.
26. Peterson AT, Soberón J (2012) Species Distribution Modeling and Ecological Niche Modeling: Getting the Concepts Right. Nat Conservação 10: 102–107. doi:10.4322/natcon.2012.019.
27. Rangel TF, Loyola RD (2012) Labeling Ecological Niche Models. Nat Conservação 10: 119–126. doi:10.4322/natcon.2012.030.
28. Araújo MB, New M (2007) Ensemble forecasting of species distributions. Trends Ecol Evol 22: 42–47.
29. Moilanen A, Possingham HP, Wilson KA (2009) Spatial conservation prioritization-past, present and future. In: Moilanen A, Wilson KA, Possingham HP, editors. Spatial Conservation Prioritization: Quantitative Methods and Computational Tools. Oxford, UK: Oxford University Press. pp. 260–268.
30. Grimaldi D, Engel MS (2005) Evolution of the Insects. Grimaldi D, Engel MS, editors Cambridge University Press.
31. Wasserthal LT (1997) The pollinators of the Malagasy star orchids Angraecum sesquipedale, *A. sororium* and *A. compactum* and the evolution of extremely long spurs by pollinator shift. Acta Botanica 110: 343–359.
32. Pellmyr O (2003) Yuccas, yucca moths, and coevolution: A review. Ann Mo Bot Gard 90: 35–55. doi:10.2307/3298524.
33. Duarte M, Marconato G, Specht A, Casagrande M (2012) Lepidoptera. In: Rafael J, Melo G, Carvalho C, Casari S, Constantino R, editors. Insetos do Brasil Diversidade e Taxonomia. Ribeirão Preto: Holos Editora. pp. 625–682.
34. Marquis RJ (1984) Leaf herbivores decrease fitness of a tropical plant. Science 226. 537–539.
35. Bouchard M, Kneeshaw D, Bergeron Y (2006) Forest dynamics after successive spruce budworm outbreaks in mixedwood forests. Ecology 87: 2319–2329.
36. Chapman SK, Hart SC, Cobb NS, Whitham TG, Koch GW (2003) Insect Herbivory Increases Litter Quality and Decomposition: an Extension of the Acceleration Hypothesis. Ecology 84: 2867–2876. doi:10.1890/02-0046.
37. Speight M, Hunter M, Watt A (2008) Ecology of Insects: Concepts and Applications. Oxford, UK: Wiley Blackwell.
38. Kocsis M, Hufnagel L (2011) Impacts of climate change on lepidoptera species and communities. Applied Ecology and Environmental Reserach 9: 43–72.
39. Conrad KF, Woiwod IP, Perry JN (2002) Long-term decline in abundance and distribution of the garden tiger moth (*Arctia caja*) in Great Britain. Biol Conserv 106: 329–337. doi:10.1016/S0006-3207(01)00258-0.
40. Zahiri R, Holloway JD, Kitching IJ, Lafontaine JD, Mutanen M, et al. (2012) Molecular phylogenetics of Erebidae (Lepidoptera, Noctuoidea). Syst Entomol 37: 102–124. doi:10.1111/j.1365-3113.2011.00607.x.
41. Heppner JB (1991) Faunal regions and the diversity of Lepidoptera. Tropical lepidoptera 2.

42. Ferro V, Diniz I (2010) Riqueza e composição das mariposas Arctiidae (Lepidoptera) no Cerrado. In: Diniz I, Marinho-Filho J, Machado R, Cavalcanti R, editors. Cerrado: conhecimento científico quantitativo como subsídio para ações de conservação. Brasília: Theraurus Editora. pp. 255–313.
43. Ferro VG, Melo AS (2010) Diversity of tiger moths in a Neotropical hotspot: determinants of species composition and identification of biogeographic units. J Insect Conserv 15: 643–651. doi:10.1007/s10841-010-9363-6.
44. Wagner DL, Conner WE (2008) The immature stages: structure, function, behavior, and ecology. In: Conner WE, editor. Tiger moths and wolly bears: Behavior, Ecology, and Evolution of the Arctiidae. Oxford: Oxford University Press. pp. 31–55.
45. Weller SJ, Jacobson NL, Conner WE (1999) The evolution of chemical defences and mating systems in tiger moths (Lepidoptera: Arctiidae). Biol J Linn Soc 68: 557–578. doi:10.1111/j.1095-8312.1999.tb01188.x.
46. Myers N, Mittermeier Ra, Mittermeier CG, da Fonseca GA, Kent J (2000) Biodiversity hotspots for conservation priorities. Nature 403: 853–858. doi:10.1038/35002501.
47. Ribeiro MC, Metzger JP, Martensen AC, Ponzoni FJ, Hirota MM (2009) The Brazilian Atlantic Forest: How much is left, and how is the remaining forest distributed? Implications for conservation. Biol Conserv 142: 1141–1153. doi:10.1016/j.biocon.2009.02.021.
48. Hijmans RJ, Cameron SE, Parra JL, Jones PG, Jarvis A (2005) Very high resolution interpolated climate surfaces for global land areas. International Journal of Climatology 25: 1965–1978. doi:10.1002/joc.1276.
49. Carpenter G, Gillison AN, Winter J (1993) DOMAIN: a flexible modelling procedure for mapping potential distributions of plants and animals. Biodivers Conserv 2: 667–680. doi:10.1007/BF00051966.
50. Hirzel AH, Hausser J, Chessel D, Perrin N (2002) Ecological-niche factor analysis: hw to compute habitat-suitability maps without absence data? Ecology 83: 2027–2036.
51. Phillips S, Anderson R, Schapire R (2006) Maximum entropy modeling of species geographic distributions. Ecol Model 190: 231–259. doi:10.1016/j.ecolmodel.2005.03.026.
52. Stockwell DRB, Noble IR (1992) Induction of sets of rules from animal distribution data: a robust and informative method of data analysis. Mathematics and Computers in Simulation 33: 385–390. doi:10.1016/0378-4754(92)90126-2.
53. Manel S, Dias JM, Buckton ST, Ormerod SJ (1999) Alternative methods for predicting species distribution: an illustration with Himalayan river birds. J Appl Ecol 36: 734–747. doi:10.1046/j.1365-2664.1999.00440.x.
54. Pearson RG (2007) Species' Distribution Modeling for Conservation Educators and Practitioners: 1–50.
55. Allouche O, Tsoar A, Kadmon R (2006) Assessing the accuracy of species distribution models: prevalence, kappa and the true skill statistic (TSS). J Appl Ecol 43: 1223–1232. doi:10.1111/j.1365-2664.2006.01214.x.
56. Lobo JM, Jiménez-Valverde A, Real R (2008) AUC: a misleading measure of the performance of predictive distribution models. Glob Ecol Biogeogr 17: 145–151. doi:10.1111/j.1466-8238.2007.00358.x.
57. Marmion M, Parviainen M, Luoto M, Heikkinen RK, Thuiller W (2009) Evaluation of consensus methods in predictive species distribution modelling. Divers Distrib 15: 59–69. doi:10.1111/j.1472-4642.2008.00491.x.
58. Diniz-Filho JAF, Nabout JC, Bini LM, Loyola RD, Rangel TF, et al. (2010) Ensemble forecasting shifts in climatically suitable areas for *Tropidacris cristata* (Orthoptera: Acridoidea: Romaleidae). Insect Conserv Div 3: 213–221. doi:10.1111/j.1752-4598.2010.00090.x.
59. Faleiro FV., Machado RB, Loyola RD (2013) Defining spatial conservation priorities in the face of land-use and climate change. Biol Conserv 158: 248–257. doi:10.1016/j.biocon.2012.09.020.
60. IUCN, UNEP (2010) World Database on Protected Areas (WDPA). Annual Release: 1–9.
61. Griffith B, Scott JM, Adamcik R, Ashe D, Czech B, et al. (2009) Climate change adaptation for the US National Wildlife Refuge System. Environ Manag 44: 1043–1052.
62. Araújo MB, Alagador D, Cabeza M, Nogués-Bravo D, Thuiller W (2011) Climate change threatens European conservation areas. Ecol Lett 14: 484–492.
63. Pearson RG, Thuiller W, Araújo MB, Martinez-Meyer E, Brotons L, et al. (2006) Model-based uncertainty in species range prediction. J Biogeogr 33: 1704–1711. doi:10.1111/j.1365-2699.2006.01460.x.
64. Cardinale BJ, Srivastava DS, Duffy JE, Wright JP, Downing AL, et al. (2006) Effects of biodiversity on the functioning of trophic groups and ecosystems. Nature 443: 989–992. doi:10.1038/nature05202.
65. Zanuncio JC, Mezzomo JA, Guedes RNC, Oliveira AC (1998) Influence of strips of native vegetation on Lepidoptera associated with Eucalyptus cloeziana in Brazil. For Ecol Manag 108: 85–90. doi:10.1016/S0378-1127(98)00215-1.
66. Hódar J, Zamora R (2004) Herbivory and climatic warming: a Mediterranean outbreaking caterpillar attacks a relict, boreal pine species. Biodivers Conserv 13: 493–500.
67. Gutierrez AP, D'Oultremont T, Ellis CK, Ponti L (2006) Climatic limits of pink bollworm in Arizona and California: effects of climate warming. Acta Oecol 30: 353–364. doi:10.1016/j.actao.2006.06.003.
68. Logan JA, Régnière J, Powell JA (2003) Assessing the impacts of global warming on forest pest dynamics. Frontiers Ecol Env 1: 130–137.
69. Singer RB, Breier TB, Flach A, Farias-Singer R (2007) The Pollination Mechanism of ***Habenaria pleiophylla*** Hoehne & Schlechter (Orchidaceae: Orchidinae). Functional Ecosystems and Communities 1: 10–14.
70. Diniz Filho JAF, Ferro VG, Santos T, Nabout JC, Dobrovolski R, et al. (2010) The three phases of the ensemble forecasting of niche models: geographic range and shifts in climatically suitable areas of Utetheisa ornatrix (Lepidoptera, Arctiidae). Rev Bras Entomol 54: 339–349. doi:10.1590/S0085-56262010000300001.
71. Yamanaka T, Tatsuki S., Shimada M (2001) Flight characteristics and dispersal patterns of fall webworm (Lepidoptera: Arctiidae) males. Env Entomol 30: 1150–1157.
72. Betzholtz PÉ (2002) Population structure and movement patterns within an isolated and endangered population of the moth *Dysauxes ancilla* L. (Lepidoptera, Ctenuchidae): implications for conservation. J Insect Conserv 6: 57–66.
73. Ferro VG, Romanowski HP (2012) Diversity and composition of tiger moths (Lepidoptera: Arctiidae) in an area of Atlantic Forest in southern Brazil: is the fauna more diverse in the grassland or in the forest? ZOOLOGIA 29: 7–18. DOI: 10.1590/S1984-46702012000100002
74. Ferro VG, Diniz IR (2007) Composição de espécies de Arctiidae (Insecta, Lepidoptera) em áreas de Cerrado. Revta Bras Zool 24: 635–646.
75. Klorvuttimontara S, McClean CJ, Hill JK (2011) Evaluating the effectiveness of Protected Areas for conserving tropical forest butterflies of Thailand. Biol Conserv 144: 2534–2540. doi:10.1016/j.biocon.2011.07.012.
76. Brooks T, Mittermeier R, Fonseca Gda (2006) Global Biodiversity Conservation Priorities. Science 313: 58–61.
77. Loyola RD, Kubota U, da Fonseca GAB, Lewinsohn TM (2009) Key Neotropical ecoregions for conservation of terrestrial vertebrates. Biodivers Conserv 18: 2017–2031. doi:10.1007/s10531-008-9570-6.
78. Loyola RD, Nabout JC, Trindade-filho J, Lemes P, Urbina-Cardona JN, et al. (2012) Climate change might drive species into reserves: a case study of the American bullfrog in the Atlantic Forest Biodiversity Hotspot. Alytes 29: 61–74.
79. Monzón J, Moyer-Horner L, Palamar MB (2011) Climate Change and Species Range Dynamics in Protected Areas. BioScience 61: 752–761. doi:10.1525/bio.2011.61.10.5.
80. Wiens JA, Seavy NE, Jongsomjit D (2011) Protected areas in climate space: What will the future bring? Biol Conserv 144: 2119–2125. doi:10.1016/j.biocon.2011.05.002.
81. Rodrigues A, Andelman S, Bakarr M, Boitani L (2004) Effectiveness of the global protected area network in representing species diversity. Nature 428. doi:10.1038/nature02459.1.

4

Climate Warming May Facilitate Invasion of the Exotic Shrub *Lantana camara*

Qiaoying Zhang[1,2]*, Yunchun Zhang[2], Shaolin Peng[3]*, Kristjan Zobel[1]

1 Department of Botany, Institute of Ecology and Earth Sciences, University of Tartu, Tartu, Estonia, **2** Department of Environmental Science and Engineering, Qilu University of Technology, Jinan, Shandong, China, **3** State Key Laboratory of Biocontrol, School of Life Sciences, Sun Yat-sen University, Guangzhou, Guangdong, China

Abstract

Plant species show different responses to the elevated temperatures that are resulting from global climate change, depending on their ecological and physiological characteristics. The highly invasive shrub *Lantana camara* occurs between the latitudes of 35°N and 35°S. According to current and future climate scenarios predicted by the CLIMEX model, climatically suitable areas for *L. camara* are projected to contract globally, despite expansions in some areas. The objective of this study was to test those predictions, using a pot experiment in which branch cuttings were grown at three different temperatures (22°C, 26°C and 30°C). We hypothesized that warming would facilitate the invasiveness of *L. camara*. In response to rising temperatures, the total biomass of *L. camara* did increase. Plants allocated more biomass to stems and enlarged their leaves more at 26°C and 30°C, which promoted light capture and assimilation. They did not appear to be stressed by higher temperatures, in fact photosynthesis and assimilation were enhanced. Using lettuce (*Lactuca sativa*) as a receptor plant in a bioassay experiment, we also tested the phytotoxicity of *L. camara* leachate at different temperatures. All aqueous extracts from fresh leaves significantly inhibited the germination and seedling growth of lettuce, and the allelopathic effects became stronger with increasing temperature. Our results provide key evidence that elevated temperature led to significant increases in growth along with physiological and allelopathic effects, which together indicate that global warming facilitates the invasion of *L. camara*.

Editor: Lalit Kumar, University of New England, Australia

Funding: The study was supported by MOBILITAS Postdoctoral Research Grant (MJD305) from the European Social Fund and the Key Program of National Natural Science Foundation of China (Grant No. 2010330004103989). The funders had no role in study design, data collection and analysis, decision to publish, or preparation of the manuscript.

Competing Interests: The authors have declared that no competing interests exist.

* Email: qiaoyingzhang@163.com (QZ); lsspsl@mail.sysu.edu.cn (SP)

Introduction

Global average temperatures are increasing and are predicted to do so further in the future [1]. Changes in temperature and precipitation associated with rising concentrations of CO_2 are altering local environmental conditions, which may inhibit native species [2,3]. At the same time, this may provide some non-native species with emerging opportunities for population growth and expansions [4]. The successful invasion of new areas by non-native species can have serious ecological consequences for species interactions and ecosystem structure and functioning [5]. Therefore, it is essential to better understand the risk of invasion under climate change scenarios for effective management of invasive plants in the 21st century [6].

The abundance and distribution of plant species are tightly regulated by both climatic factors [7] and biotic interactions [8], so changes in climatic conditions are likely to cause major shifts in their population dynamics and geographic ranges [2,3,9]. Apart from changes in the potential distributions of native species, climate change may also affect the spatial distribution of invasive species [5,6,9]. Previous studies have shown that global warming has enabled alien plants to expand into regions where they previously could not survive and reproduce [9]. Any alterations of plant community structure that are caused by climate change result from underlying changes in the population dynamics of species that make up the community [10]. Thus, understanding responses to climate change at the species level is important to the prediction of future ecosystem functioning [10].

The focal species of this study is *Lantana camara* (Verbenaceae), a small perennial shrub which can grow to around 2 m in height and forms dense thickets in a variety of environments [11]. Its native range is Central America, the northern part of South America and the Caribbean [11]. *L. camara* has been identified as one of the 100 World's Worst Invasive Alien Species [12]. Since the sixteenth century, it has been subject to intense horticultural improvement in Europe, and now it exists in many different forms and varieties around the world [13]. Its global distribution includes about 60 countries and islands between the latitudes of 35°N and 35°S [14]. *L. camara* has become a major problem in many of these areas, causing reductions in native species diversity, declines in soil fertility, allelopathic alteration of soil properties, and alteration of ecosystem processes [11,14].

The model CLIMEX has been widely used to illustrate the potential distribution of species under future climate scenarios [15]. Based on CLIMEX simulations, the potential distribution of *L. camara* will expand in some areas under current and future climate scenarios [14,16,17,18], and in China specifically its distribution could potentially expand further inland [11]. This is consistent with field investigations in southern China, where *L. camara* has recently become more prevalent [19]. Climate studies have shown that winter minimum temperatures in the region (i.e., Guangdong province) started to rise in the middle and later periods of the 1960s, and it has become warmer since the 1980s [20]. The observed increase in abundance is thus likely related to elevated temperature. We hypothesized that warming leads to positive effects on the fitness parameters of the invasive shrub *L. camara*, further facilitating its invasiveness.

We planted branch cuttings of *L. camara* in different temperature treatments (22, 26 and 30°C) in three experiments: a growth experiment, a physiological experiment and a bioassay experiment designed to assess the allelopathic potential of the species. Our goal was to describe and compare the morphological, physiological and biochemical responses of the species to future climate scenarios. Specifically, we sought to address the following questions: (1) How does the growth of *L. camara* respond to elevated temperatures? (2) How are gas exchange rates and photosynthesis affected by elevated temperatures? (3) Does the allelopathic potential of *L. camara* change with increasing temperature, as has been observed in some other plant species [21]?

Methods

Growth and morphology experiment

We collected three-year-old branches of *L. camara* on March 10, 2008 in Guangzhou, China. The sampling site (23°02′–23°04′N, 113°23′–113°24′E) was neither located on farmland nor in a protected area. No specific permissions were required for these locations/activities. No endangered or protected species were involved in the sampling. The tops of the branches were cut to keep them at least 20 cm long. The cuttings were planted at the experimental field of Sun Yat-sen University, Guangzhou, Guangdong province. After three weeks, we selected uniform branch cuttings and transplanted them into plastic pots (20-cm diameter, 15-cm height, with three branches per pot). The pots were filled with equal proportions of nutrient-rich soil and vermiculite for water retention. The plants were grown in different greenhouses (14/10 h day/night cycle, 75%±2% relative humidity, photosynthetically active radiation (PAR) 400 mmol m^{-2} s^{-1}) at three constant temperatures (22, 26 and 30°C). There were 15 replicate pots per temperature treatment. All pots were randomly placed once a week to avoid internal effects. They were watered with diluted Hoagland solution (25% v/v) once a week, for a total of 18 weeks.

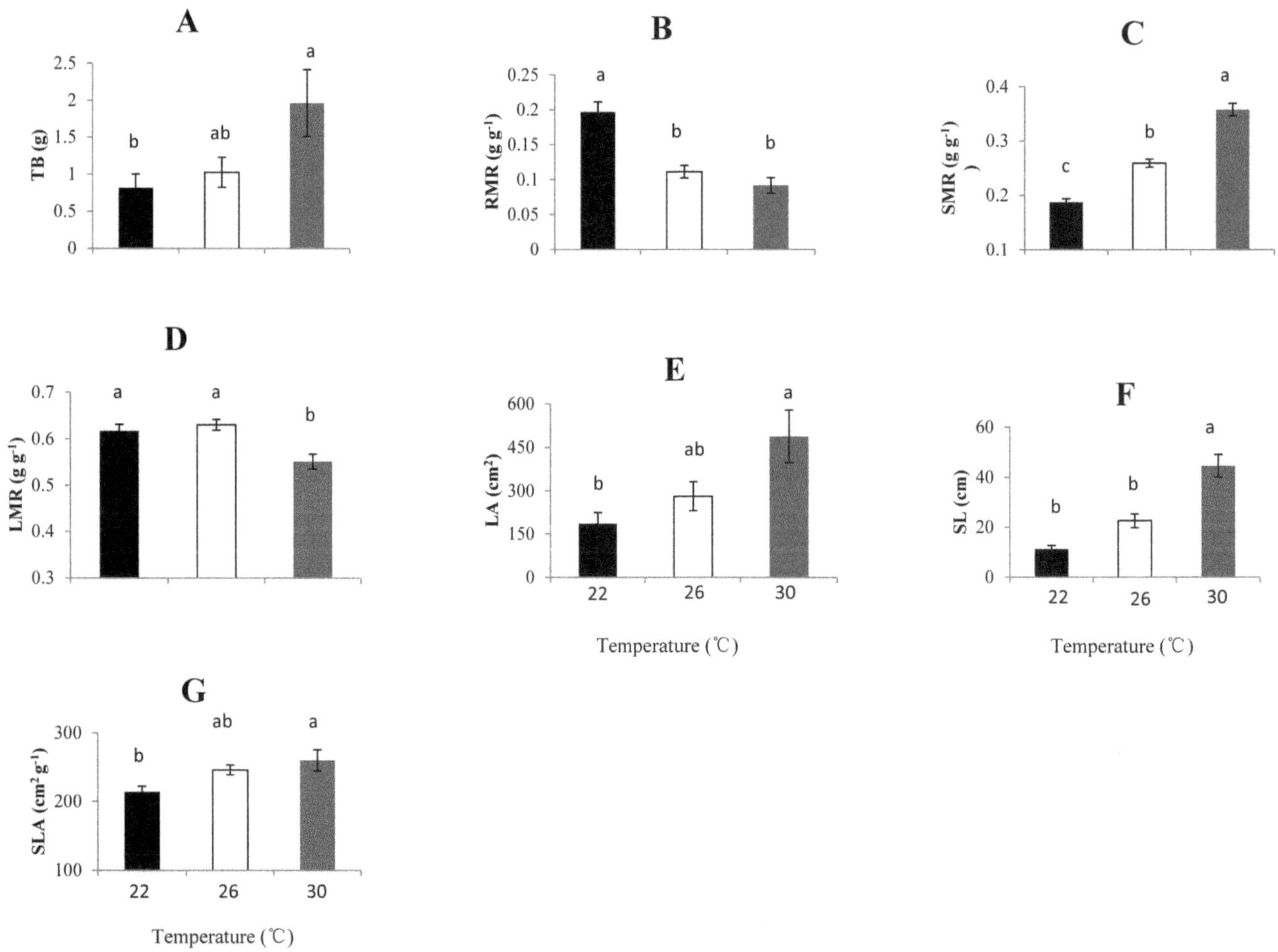

Figure 1. Responses of *Lantana camara* plants to temperature treatments (mean ± SE, *n* = 15). Different letters indicate significant differences (P<0.05) between means according to Tukey's HSD tests (The same below). (A) TB, total biomass; (B) RMR, root mass ratio; (C) SMR, support organ mass ratio; (D) LMR, leaf mass ratio; (E) LA, leaf area; (F) SL, stem length; (G) SLA, specific leaf area.

Table 1. F-values of one-way ANOVA which was used to test the effects of different temperatures on growth and morphology of *L. camara*.

	TB	RMR	SMR	LMR	LA	SL	SLA
d.f.	2, 16	2, 16	2, 16	2, 16	2, 16	2, 16	2, 16
F	4.04	22.47	91.03	8.92	5.76	27.00	4.70*
P	0.038*	<0.001***	<0.001***	0.003**	0.013*	<0.001***	0.025*

*** $P<0.001$,
** $P<0.01$,
* $P<0.05$.
Parameters of growth and morphology:
TB: total biomass; RMR: root mass ratio; SMR: support organ mass ratio; LMR: leaf mass ratio; LA: leaf area; SL: stem length; SLA: specific leaf area.

Then, at the end of the experiment, plants were harvested and divided into leaf blades, petioles, stems and roots, and were dried separately to a constant mass at 70°C. The total stem length was measured. Leaf area was determined using an LA meter (CI-203 Area-meter, CID, USA). The raw data were used to calculate the following growth parameters [22]: leaf mass ratio (leaves without petioles, LMR), root mass ratio (RMR), support organ mass ratio (stems and petioles, SMR), and specific leaf area (SLA). The mass ratio was calculated by dividing the dry mass by total plant dry mass.

Physiological experiment

Gas exchange. After potted plants had been allowed to grow for 18 weeks at different temperatures, we measured the gas exchange of *L. camara* on fully expanded leaves under controlled optimal conditions, using an open system with a portable photosynthesis measurement system (LI-6400, LI-COR, USA). The measurements were made on 15 plants per treatment. We found out that under greenhouse conditions, the net photosynthetic rate (Pnet) of *L. camara* was greatest at 800–1200 μmol m^{-2} s^{-1}, so we maintained PAR at 1200 μmol m^{-2} s^{-1}. We used an LI-6400 artificial light source and maintained the temperature at 22, 26 and 30°C for each treatment. In order to avoid the effect of midday photosynthetic depression [23], we completed the measurements on two sunny days from 08:00 to 11:30 and from 15:00 to 17:30. Pnet and stomatal conductance (Cond) were also measured, while intrinsic water use efficiency (WUE) was calculated by dividing Pnet by Cond [23].

Chlorophyll (Chl) fluorescence. We used the saturation pulse method [24] to measure the Chl fluorescence. Measurements were taken from the upper surface of the same leaves used in the previously described measurements, with a pulse-amplitude-modulated fluorometer (PAM 2100, Walz, Effeltrich, Germany) [25]. Before measurement, the leaves were placed in dark for at least 30 min. The intensity and duration of the saturation pulses was 4,000 μmol m^{-2} s^{-1} and 0.8 s, respectively. The "actinic light" was 600 μmol m^{-2} s^{-1}. We recorded the fluorescence parameters Fv/Fm and ΦPSII. Fv/Fm is the maximum quantum yield of photosystem II (PSII), which is assessed as (Fm - Fo)/Fm [26], where Fo and Fm are the minimal and maximal fluorescence values of a dark-adapted sample, respectively, with all of the PSII reaction centers fully open. It was measured at predawn, when plants were in the dark, to make sure that all the PSII reaction centers were open. ΦPSII is the effective quantum yield of PSII. It was calculated as $\Phi PSII = (Fm' - Ft)/Fm'$, where Fm′ is the maximal fluorescence value reached in a pulse of saturating light with an illuminated sample, and Ft is the fluorescence value of the leaf at a given photosynthetically active radiation [27].

Bioassay experiment on allelopathic potential

After the potted *L. camara* plants had been growing for 18 weeks at three different temperatures (22, 26 and 30°C), we collected fresh leaves (10 g) randomly from plants in each greenhouse and soaked them in distilled water (100 mL) for 24 h in darkness at 22, 26 and 30°C, respectively. We then made aqueous leachates with a concentration of 0.1 g mL^{-1} from each treatment. The pH value of all leachates was adjusted to 6.8 using 1 M NaOH or HCl, and distilled water was used as a control.

Twenty uniform lettuce (*Lactuca sativa*) seeds were selected, surface-sterilised with 0.5% $KMnO_4$ for 15 min, and then washed with sterile water. The seeds were put on top of two layers of filter paper (9-cm diameter) in a glass Petri dish (9-cm diameter). Each dish contained 5 mL of aqueous leachate obtained from *L. camara*

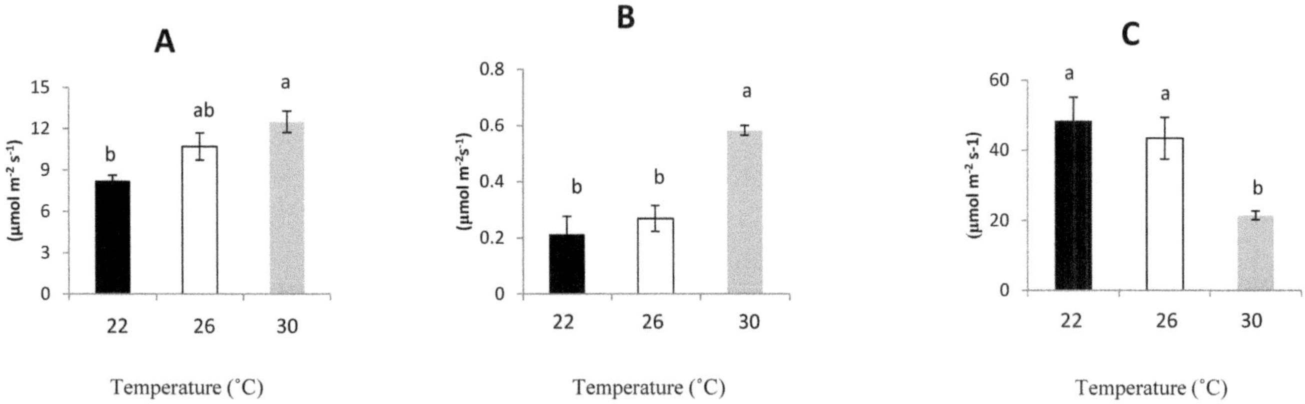

Figure 2. Gas exchange of *L. camara* seedlings growing at different temperatures. Data are means ± SE (n = 15). (A) Pnet, Net photosynthetic rate; (B) Cond, stomatal conductance; (C) WUE, water use efficiency.

plants grown at different temperatures, while the controls contained 5 mL distilled water. The Petri dishes were kept in dark conditions at room temperature (~22°C). All of the treatments were conducted with four independent replicates. The germinated seeds (once radicle length was about 1–2 mm) were counted every 12 h for the first day, and every 24 h thereafter. Germination of *L. sativa* was recorded up to 5 days, and seedling growth (root length and shoot length) was recorded at the end of the experiment, on the seventh day. The effect of the leachate on lettuce growth was evaluated using a response index (RI) [28] as follows:

$$RI(\%) = (T/C\text{-}1) \times 100 \qquad (1)$$

where C was the control value and T was the treatment value. RI>0 indicates a stimulatory effect, while RI<0 indicates an inhibitory effect.

Statistical analysis

The effect of different temperatures on growth, physiology and allelopathy of *L. camara* was assessed by one-way ANOVA and means were compared by Tukey tests. All statistical analyses were performed using the software R 3.0.1 [29].

Results

Growth and morphology

The overall biomass of *L. camara* plants, and their allocation of biomass to support organs (stem and petiole), were significantly higher in high-temperature treatments than in the control (22°C), while biomass allocation to roots (RMR) and leaves (LMR) displayed the opposite pattern (Table 1, Fig. 1). The leaf area, stem length and SLA of *L. camara* were also significantly higher at the elevated temperatures (Table 1, Fig. 1). The stem length of plants growing at 30°C was four times that of those at 22°C, while their leaf area was double. SLA at the elevated temperature of 30°C was 22.1% higher than at 22°C.

Photosynthesis and chlorophyll fluorescence

Gas exchange. Plants growing at 30°C showed significantly higher Pnet and Cond, yet lower WUE than those growing at 22°C (Table 2, Fig. 2 A–C). No significant differences in any gas exchange parameters were found between seedlings growing at 22°C and 26°C, while seedlings growing at 26°C exhibited significantly lower Cond and higher WUE than those growing at 30°C (Table 2, Fig. 2 B, C).

Chlorophyll fluorescence. No significant differences in Fv/Fm and ΦPSII were found among seedlings of *L. camara* growing at the three different temperatures (Table 2, Fig. 3 A, B).

Allelopathic potential

The allelopathic effects of aqueous leachate from *L. camara* leaves on seed germination and seedling development of lettuce at different temperatures were evaluated (Table 2, Fig. 4). The shoot and root length of lettuce significantly decreased with an increase in temperature, but there were no significant differences in germination between different temperatures. The allelopathic effects on shoot and root length were significantly greater at the higher temperatures compared to at 22°C, with the highest effect occurring at 26°C.

Discussion

Over the course of human history, people have intentionally or unintentionally moved innumerable plant species outside of their native ranges, and many of those alien plants become invasive [6]. Human activities are also partly responsible for the increase of global surface temperatures [21,30,31]. A number of recent studies on invasive plants and climate change have shown that increasing temperatures and changing precipitation might either "help" or "hinder" invasive plants, depending on the species, location and dominant forces causing changes in climate conditions [6,9,21,32]. Such variation makes it challenging to assess and understand the mechanisms that might facilitate or constrain the success of invasive species in the context of climate change [5].

As our findings illustrate, climate warming affects many aspects of the invasive species *L. camara*'s biology and ecology. Firstly, elevated temperature caused changes in the biomass allocation and morphology of plants. Plant growth is directly influenced by biomass allocation between leaves, stems, and other plant parts [33]. With rising temperature, individuals exhibited a significant increase in stem length, and biomass allocation to stems and petioles at the expense of leaves and roots. These changes may ensure greater structural support and an increased ability to capture light. Although biomass allocation to leaves decreased as in temperature increased, SLA and LA were greatest in the high-temperature treatments. SLA is a plant trait that is important for the regulation and control of functions such as carbon assimilation and carbon allocation [34,35]. Generally, the combination of

Table 2. F-values of one-way ANOVA which was used to test the effects of different temperatures on gas exchange and chlorophyll fluorescence, and allelopathic potential of *L. camara*.

	Pnet	Cond	WUE	Fv/Fm	ΦPSII	Germ	SH	RL
d.f.	2, 16	2, 16	2, 16	2, 16	2, 16	2, 16	2, 16	2, 16
F	9.047	15.61	6.63	0.627	0.452	0.756	7.893	9.608
P	0.0035*	<0.001***	0.010*	0.547	0.645	0.497	0.001**	<0.001***

*** $P<0.001$,
** $P<0.01$,
* $P<0.05$.
Parameters of gas exchange and chlorophyll fluorescence
Pnet: net photosynthetic rate; Cond: stomatal conductance; WUE: intrinsic water use efficiency; Fv/Fm; ΦPSII.
Parameters of allelopathic potential: Germ: germination; SH: shoot length; RL: root length.

increased SLA and LA results in increased light absorption, and shading of other species [36]. This may be a light utilization strategy that could enhance the competitive ability of *L. camara*, because this species cannot survive under the dense, continuous canopies of taller native forest species due to the lack of light [14]. *L. camara* usually flowers in the first growing season after its establishment and, if adequate moisture and light are available, it can flower in all seasons [37]. If the mean global temperature will rise by 1.4–5.8°C over the period of 1990–2100 as predicted [30], *L. camara* may increase in height more rapidly than its neighbours, and then suppress their growth by shading them. It may also flower more often and for longer periods of time, enabling it to produce more offspring, which could cause substantial damage to other species and their ecosystems. The responses to the increase in temperature observed in our study suggest that warming may help *L. camara* to reach further into the upper layer of the plant community, and expand its leaves as much as possible for better light capture and assimilation so as to facilitate its invasion.

Secondly, elevated temperature induced changes in the physiological parameters of *L. camara*. We found that plants exhibited a significantly higher photosynthetic capacity at higher temperatures (26 and 30°C) than at 22°C, which may be ascribed to a higher Pnet. High temperatures tended to increase stomatal conductance (Fig. 2 B, [38]), which can augment water loss. Consequently, the instantaneous water-use efficiency (WUE) of plants decreased with increasing temperature (Fig. 2 C). This response may explain why *L. camara* mostly invades wetter habitats [14]. High temperatures can influence photosynthesis in different ways, such as enhancing membrane fluidity and oxidative stress [38], or by changing the activity of the Calvin cycle and photorespiration [39]. High temperature may also inhibit the repair of PSII [40]. In this study, there were no significant differences in Fv/Fm and ΦPSII among different temperature conditions, and *L. camara* showed optimal functioning of its PSII with very low photoinhibition levels (Fv/Fm from ~0.750 to 0.870) after exposure to higher temperatures. This suggested that higher temperature did not lead to stress in *L. camara*. Higher photosynthesis can increase invasive plants' growth rates and biomass accumulation, which may enable invasive species to out-compete slower growing species and hence facilitate their colonization [41,42]. The responses we observed to elevated temperatures can be viewed as positive effects of warming on the physiological parameters of *L. camara*, i.e., increased rates of photosynthesis at higher temperatures could facilitate its invasive success. Of course, this enhancement of the plant's growth is often a "negative" effect at the ecosystem level.

Thirdly, elevated temperature induced changes in the allelopathic effects of *L. camara*. Those effects have been well-documented to cause severely reduced seedling recruitment in almost all species exposed to *L. camara*, and a reduction in the DBH growth of mature trees and shrubs [43,44]. The allelopathic effects of *L. camara* may explain why it can survive secondary succession and form monospecific thickets [14]. In our experiment, we found that its phytotoxicity increased with temperature, with respect to both seed germination and seedling growth of the receptor plants (Table 2, Fig. 4). This result is consistent with other research about how elevated temperature influences the allelopathic effects of invasive plants [21,45]. Allelopathic biochemicals produced by invasive plants function as their "novel weapons" since they can inhibit the growth of native plants in the invaded communities [46]. The increased phytotoxicity of *L. camara* in higher temperatures may be a result of the plant producing more allelochemicals, or of its allelochemicals becoming

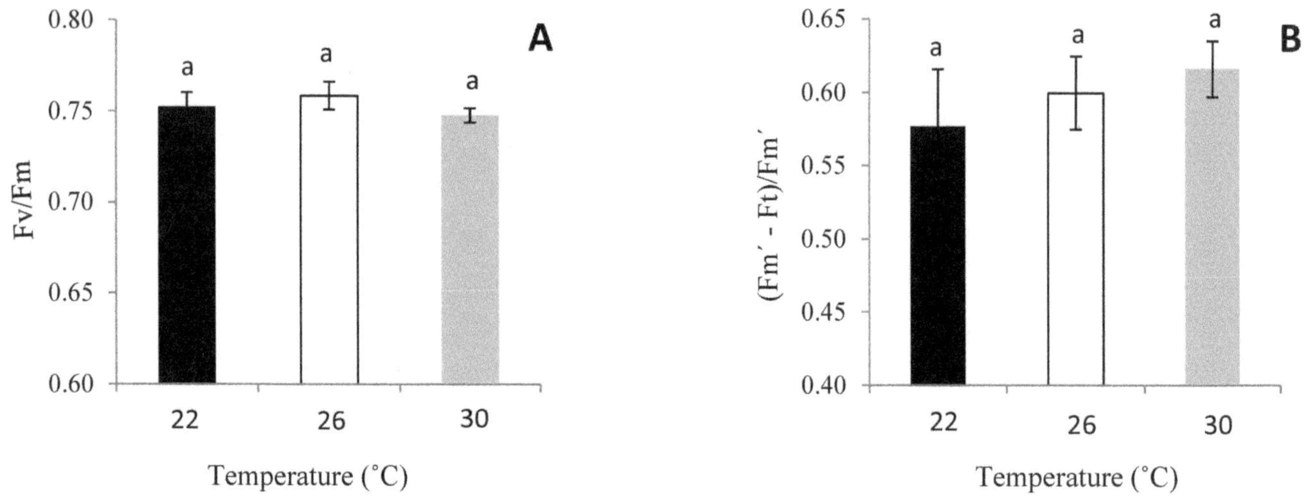

Figure 3. Chlorophyll fluorescence of *L. camara* seedlings growing at different temperatures. Data is means ±SE (n = 15). (A) Fv/Fm; (B) ΦPSII.

more phytotoxic under elevated temperatures [21]. As such, we can conclude that warming also enhanced the allelopathic potential of *L. camara*.

Biological invasions and climate change are key factors that are currently affecting global biodiversity [9] and the relationship between them is very complex [6]. In this study, we chose to study temperature, one of the most important elements of climate change, to understand its effects on plant invasion. The results showed that elevated temperature resulted in significant increases in biomass allocation and beneficial changes in morphology, photosynthesis and allelopathic effects of *L. camara*, indicating that global warming could facilitate the invasion of this plant. Based on the predictions of climate models, a 1°C increase in mean annual temperature could result in a pole-ward shift of each of the world's vegetation zones by approximately 200 km [2]. If

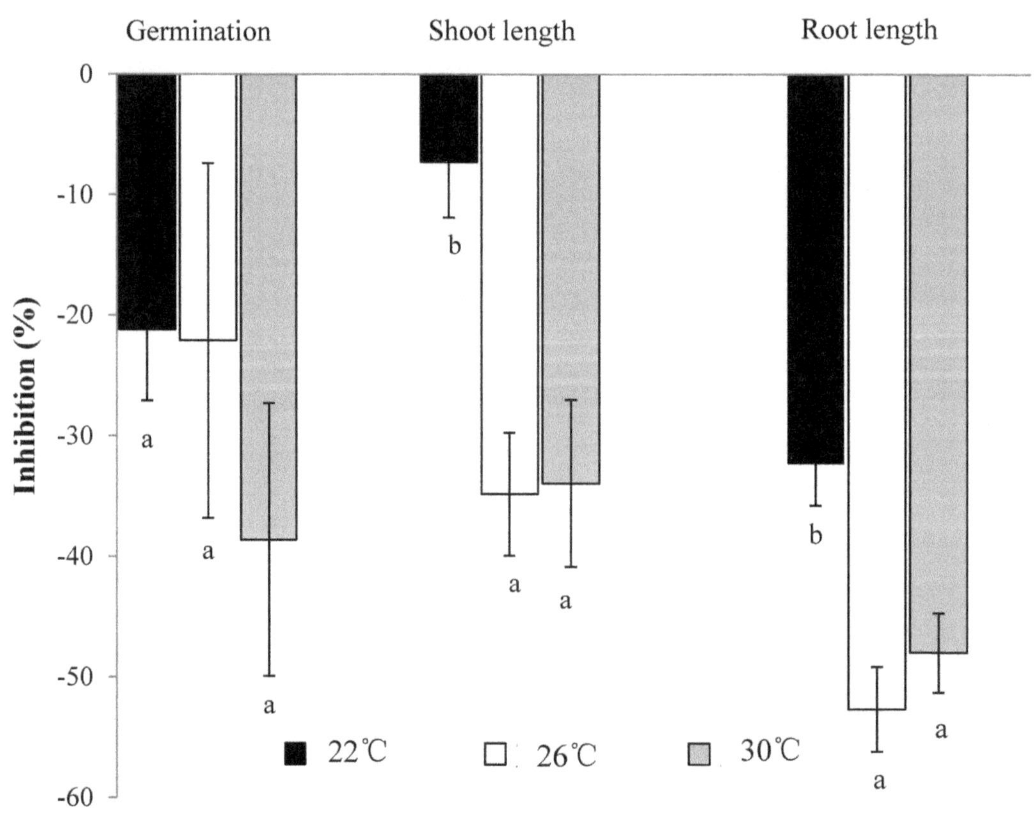

Figure 4. Allelopathic effects of aqueous leachate from fresh leaves of *L. camara* on seed germination, shoot length and root length of *Lactuca sativa*, expressed as a response index at different temperatures. Each bar represents a mean ± SE.

the global temperature increases by 1.4–5.8°C as predicted [30], *L. camara* will likely migrate toward higher latitudes.

Our experiment was conducted in the absence of competition from surrounding plants. In fact, competition with surrounding native species is one of the most important factors that influence the outcomes of invasion by alien plants. Under future climate scenarios, both native and invasive species are likely to grow more vigorously [42], which could affect competitive interactions in the invaded habitats. Future studies should also address the biotic factors that affect the invasiveness of *L. camara*.

Acknowledgments

We thank Marina Semchenko and John Davison for their helpful comments and linguistic corrections. Furong Li and Na Zhao kindly helped with the experiments.

Author Contributions

Conceived and designed the experiments: QZ YZ SP. Performed the experiments: QZ YZ. Analyzed the data: YZ QZ. Wrote the paper: QZ YZ KZ.

References

1. IPCC (2007) Climate Change 2007: Synthesis Report. Summary for Policymakers. Intergovernmental Panel on Climate Change, Cambridge University Press.
2. Walther GR, Post E, Convey P, Menzel A, Parmesan C, et al. (2002) Ecological responses to recent climate change. Nature 416: 389–395. doi:10.1038/416389a.
3. Parmesan C (2006) Ecological and evolutionary responses to recent climate change. Annual Review of Ecology, Evolution, and Systematics 37, 637–669.
4. Sorte CJB, Ibáñez I, Blumenthal DM, Molinari NA, Miller LP, et al. (2013) Poised to prosper? A cross-system comparison of climate change effects on native and non-native species performance. Ecology Letters 16(2): 261–270. doi: 10.1111/ele.12017
5. Weitere M, Vohmann A, Schulz N, Linn C, Dietrich D, et al. (2009) Linking environmental warming to the fitness of the invasive clam *Corbicula fluminea*. Global Change Biology 15, 2838–2851. doi: 10.1111/j.1365-2486.2009.01925.x
6. Bradley B A, Blumenthal DM, Wilcove DS, Ziska LH (2010) Predicting plant invasions in an era of global change. Trends in Ecology & Evolution 25(5): 310–318.
7. Woodward FI (1987) Climate and Plant Distribution. Cambridge University Press.
8. Araújo MB, Luoto M (2007) The importance of biotic interactions for modelling species distributions under climate change. Global Ecology and Biogeography 16(6): 743–753.
9. Walther GR, Roques A, Hulme PE, Sykes MT, Pysek P, et al. (2009) Alien species in a warmer world: risks and opportunities. Trends in Ecology and Evolution 24(12): 686–693.
10. Williams AL, Wills KE, Janes JK, Vander Schoor JK, Newton PC, et al. (2007) Warming and free-air CO_2 enrichment alter demographics in four co-occurring grassland species. New Phytologist 176: 365–374.
11. Taylor S, Kumar L, Reid N, Kriticos DJ (2012) Climate Change and the Potential Distribution of an Invasive Shrub, *Lantana camara* L. PLoS ONE 7(4): e35565. doi:10.1371/journal.pone.0035565
12. Invasive Species Specialist Group, IUCN (2001) 100 World's Worst Invasive Alien Species. Available: http://www.issg.org/database/species/reference_files/100 English.pdf
13. Thomas SE, Ellison CA (2000) A Century of Classical Biological Control of *Lantana camara*: Can pathogens make a significant difference? pp. 97–104. In: Spencer, N.R. [ed.], Proceedings of the X International Symposium on Biological Control of Weeds, Bozeman, USA, 1999.
14. Day MD, Wiley CJ, Playford J, Zalucki MP (2003) Lantana: Current management Status and Future Prospects. ACIAR Monograph series, Canberra.
15. Shabani F, Kumar L, Taylor S (2012) Climate change impacts on the future distribution of Date Palms: A modelling exercise using CLIMEX. PLoS ONE 7(10): e48021. doi:10.1371/journal.pone.0048021.
16. Taylor S, Kumar L (2013) Potential distribution of an invasive species under climate change scenarios using CLIMEX and soil drainage: A case study of *Lantana camara* L. in Queensland, Australia. Journal of Environmental Management 114: 414–422.
17. Taylor S, Kumar L (2012) Sensitivity analysis of CLIMEX parameters in modelling potential distribution of *Lantana camara* L. PLoS ONE 7(7): e40969. doi:10.1371/journal.pone.0040969.
18. Taylor S, Kumar L, Reid N (2012) Impacts of climate change and land-use on the potential distribution of an invasive weed: a case study of *Lantana camara* in Australia. Weed Research 52: 391–401.
19. Shan JL (2003) Preliminary studies on exotic plant communities in Hainan. Chinese Journal of tropical agriculture 23(3):1–4;51.
20. Liang JY, Wu SS (2000) Climatological diagnosis of winter temperature variations in Guangdong. Journal of Tropical Meteorology 6: 37–45.
21. Wang RL, Zeng RS, Peng SL, Chen BM, Liang XT, et al. (2011) Elevated temperature may accelerate invasive expansion of the liana plant *Ipomoea cairica*. Weed research 51: 574–580.
22. Poorter L (1999) Growth responses of 15 rain-forest tree species to a light gradient: the relative importance of morphological and physiological traits. Functional Ecology 13(3): 396–410. doi:10.1046/j.1365-2435.1999.00332.x.
23. Zhang YC, Zhang QY, Luo P, Wu N (2009) Photosynthetic response of *Fragaria orientalis* in different water contrast clonal integration. Ecological Research 24(3): 617–625.
24. Schreiber U, Bilger W, Hormann H, Neubauer C (1998) Chlorophyll fluorescence as a diagnostic tool: basics and some aspects of practical relevance. In: Photosynthesis: A Comprehensive Treatise (ed. A. S. . Raghavendra) pp. 320–336. Cambridge University Press, Cambridge.
25. Brugnoli E, Björkman O (1992) Chloroplast movements in leaves - influence on chlorophyll fluorescence and measurements of light-induced absorbency changes related to Delta-Ph and Zeaxanthin formation. Photosynthesis Research 32: 23–35.
26. Bolhar-Nordenkampf HR, Long SP, Baker NR, Oquist G, Schreiber U, et al. (1989) Chlorophyll fluorescence as a probe of the photosynthetic competence of leaves in the field. A review of current instrumentation. Functional Ecology 3: 497–514.
27. Genty B, Briantais JM, Baker NR (1989) The relationship between the quantum yield of photosynthetic electron transport and quenching of chlorophyll fluorescence. Biochimica et Biophysica Acta 990: 87–92.
28. Williamson GB, Richardson D (1988) Bioassays for allelopathy: measuring treatment responses within dependent control. Journal of Chemical Ecology 14(1): 181–187.
29. R Development Core Team (2013) R: A Language and Environment for Statistical Computing. R Foundation for Statistical Computing, Vienna, Austria. ISBN 3-900051-07-0, URL http://www.R-project.org
30. Houghton JT, Ding Y, Griggs DJ, Noguer M, van der Linden PJ, et al. (2001) The projections of the earth's future climate. In: Climate Change 2001: The Scientific Basis (eds JT . Houghton, Y . Ding, DJ . Griggs, M . Noguer, PJ . Van Der Linden et al), 62–77. Cambridge University Press, Cambridge, UK.
31. Li MH, Kräuchi N, Gao SP (2006) Global warming: can existing reserves really preserve current levels of biological diversity? Journal of Integrative Plant Biology 48: 255–259.
32. Bradley BA, Oppenheimer M, Wilcove DS (2009) Climate change and plant invasion: restoration opportunities ahead? Global Change Biology 15: 1511–152.
33. Reich PB, Tjoelker MG, Walters MB, Vanderklein DW, Buschena C (1998) Close association of RGR, leaf and root morphology, seed mass and shade tolerance in seedlings of nine boreal tree species grown in high and low light. Functional Ecology 12(3): 327–338. doi:10.1046/j.1365-2435.1998.00208.x.
34. Reich PB, Walters MB, Ellsworth DS (1997) From tropics to tundra: global convergence in plant functioning. Proc Natl Acad Sci USA 94(25): 13730–13734. doi:10.1073/pnas.94.25.13730. PMID:9391094.
35. Lambers H, Poorter H (2004) Inherent variation in growth rate between higher plants: a search for physiological causes and ecological consequences. Academic Press, London, New York.
36. Qin Z, Mao DJ, Quan GM, Zhang JE, Xie JF et al (2012) Physiological and morphological responses of invasive *Ambrosia artemisiifolia* (common ragweed) to different irradiances. Botany 90: 1284–1294.
37. Sharma GP, Raghubanshi AS, Singh JS (2005) Lantana invasion: an overview. Weed Biology and Management 5: 157–167
38. Carrion-Tacuri J, Rubio-Casal AE, de Cires A, Figueroa ME, Castillo JM (2013) Effect of low and high temperatures on the photosynthetic performance of *Lantana camara* L. leaves in darkness. Russian Journal of Plant Physiology 60(3): 322–329.
39. Yordanov I (1992) Response of Photosynthetic Apparatus to Temperature Stress and Molecular Mechanisms of Its Adaptation. Photosynthetica 26: 517–531.
40. Allakhverdiev SI, Kreslavskii VD, Klimov VV, Los DA, Carpentier R, et al. (2008) Heat stress: An overview of molecular responses in photosynthesis. Photosynthesis Research 98: 541–550.
41. Lambers H, Poorter H (1992) Inherent variation in growth rate between higher plants: A search for physiological causes and ecological consequences. Adv Ecol Res 23: 187–261.
42. Anderson LJ, Cipollini D (2013) Gas exchange, growth, and defense responses of invasive *Alliaria petiolata* (Brassicaceae) and native *Geum vernum* (Rosaceae) to elevated atmospheric CO_2 and warm spring temperatures. American Journal of Botany 100(8): 1544–1554.

43. Gentle CB, Duggin JD (1997) Allelopathy as a competitive strategy in persistent thickets of *Lantana camara* L. in three Australian forest communities. Plant Ecology 132: 85–95.
44. Zhang QY, Peng SL, Zhang YC (2009) Allelopathic potential of reproductive organs of exotic weed *Lantana camara*. Allelopathy Journal 23:213–220.
45. Lobón NC, Gallego JC, Daz TS, Garcia JC (2002) Allelopathic potential of Cistus ladanifer chemicals in response to variations of light and temperature. Chemoecology 12: 139–145.
46. Callaway RM, Aschehoug ET (2000). Invasive plants versus their new and old neighbors: A mechanism for exotic invasion. Science 290(5491): 521–523.

5

Non-Destructive Lichen Biomass Estimation in Northwestern Alaska: A Comparison of Methods

Abbey Rosso[1], Peter Neitlich[1]*, Robert J. Smith[2]

1 National Park Service, Winthrop, Washington, United States of America, **2** Department of Botany and Plant Pathology, Oregon State University, Corvallis, Oregon, United States of America

Abstract

Terrestrial lichen biomass is an important indicator of forage availability for caribou in northern regions, and can indicate vegetation shifts due to climate change, air pollution or changes in vascular plant community structure. Techniques for estimating lichen biomass have traditionally required destructive harvesting that is painstaking and impractical, so we developed models to estimate biomass from relatively simple cover and height measurements. We measured cover and height of forage lichens (including single-taxon and multi-taxa "community" samples, $n = 144$) at 73 sites on the Seward Peninsula of northwestern Alaska, and harvested lichen biomass from the same plots. We assessed biomass-to-volume relationships using zero-intercept regressions, and compared differences among two non-destructive cover estimation methods (ocular vs. point count), among four landcover types in two ecoregions, and among single-taxon vs. multi-taxa samples. Additionally, we explored the feasibility of using lichen height (instead of volume) as a predictor of stand-level biomass. Although lichen taxa exhibited unique biomass and bulk density responses that varied significantly by growth form, we found that single-taxon sampling consistently under-estimated true biomass and was constrained by the need for taxonomic experts. We also found that the point count method provided little to no improvement over ocular methods, despite increased effort. Estimated biomass of lichen-dominated communities (mean lichen cover: $84.9 \pm 1.4\%$) using multi-taxa, ocular methods differed only nominally among landcover types within ecoregions (range: 822 to 1418 g m^{-2}). Height alone was a poor predictor of lichen biomass and should always be weighted by cover abundance. We conclude that the multi-taxa (whole-community) approach, when paired with ocular estimates, is the most reasonable and practical method for estimating lichen biomass at landscape scales in northwest Alaska.

Editor: Helge Thorsten Lumbsch, Field Museum of Natural History, United States of America

Funding: This work was funded by the US National Park Service-Western Arctic National Parklands and the US National Park Service-Arctic Network. The corresponding author secured governmental funding for this project via the US National Park Service. The funders had no role in study design, data collection and analysis, decision to publish, or preparation of the manuscript.

Competing Interests: The authors have declared that no competing interests exist.

* Email: peter_neitlich@nps.gov

Introduction

Lichen biomass is an important indicator of grazing impact and the availability of winter forage for caribou, reindeer, muskox, and other animals in northern regions [1], [2]. In northwestern Alaska, these animals rely on forage lichens including many species of *Cladonia* ("reindeer lichen", previously *Cladina*), *Alectoria*, *Bryocaulon*, *Bryoria*, and *Cetraria* [3]. Lichens provide the main winter sustenance for the Western Arctic Caribou Herd, which is one of the largest caribou herds in North America and is important in the subsistence economy of native Alaskans [1]. Aside from utility as wildlife forage, terrestrial lichen biomass can also be used as a vegetation monitoring metric to assess the impact of disturbances such as fire [3]–[5], climate change [6], [7] and air pollution [8], [9]. However, estimating lichen biomass by destructive sampling is very time consuming and does not allow for assessment of the same area over time. Researchers studying epiphytic lichens have previously approached this problem in places such as Norway [10], China [11], British Columbia [12] and the U.S. Pacific Northwest [13] by developing regression equations, yet very few have estimated biomass for ground-dwelling, terrestrial lichens.

Biomass estimation requires accurate volumetric measurements for lichen height and area cover. The method by which cover is measured represents a trade-off between the potential bias of estimates (if estimated subjectively) and the potential to under-represent rare or patchily distributed taxa (if measured quantitatively) [14]. Point-count estimates have the potential to be less biased given a high enough density of points [15], while ocular cover estimates can assess larger areas rapidly and can integrate patchiness. Therefore, ocular and point count methods must be comparatively assessed for accuracy and for their influence on biomass estimates.

Climate and vegetation type are important predictors of lichen biomass [16] and should be accounted for in landscape-level assessments. The landscapes of northwestern Alaska span several climatic ecoregions and a wide variety of vegetation and landcover types including wetlands, spruce woodlands, and tundra dominated by lichens, low shrubs or graminoids [17], [18]. Within these landcover types, there are strong gradients of substrate pH and

Table 1. Number of lichen samples among habitats and ecoregions of northwestern Alaska for all sample types.

Land cover type	Ecoregion	Sample type	*Alectoria*	*Bryocaulon*	*Cetraria*	*Cladonia*	*C. stellaris*	Total
L	N	Single-taxon	4	7	1	3	4	19
		Multi-taxa	2	2	1	12	1	18
	S	Single-taxon	0	0	0	0	0	0
		Multi-taxa	0	0	0	0	0	0
M	N	Single-taxon	0	0	5	2	0	7
		Multi-taxa	0	0	0	6	1	7
	S	Single-taxon	0	0	2	3	0	5
		Multi-taxa	0	0	0	2	0	2
P	N	Single-taxon	0	0	7	2	0	9
		Multi-taxa	0	0	0	11	0	11
	S	Single-taxon	3	0	6	3	0	12
		Multi-taxa	0	0	2	14	0	16
S	N	Single-taxon	7	7	0	3	2	19
		Multi-taxa	1	5	0	8	1	15
	S	Single-taxon	0	2	0	0	0	2
		Multi-taxa	1	0	0	1	0	2
Total:			18	23	24	70	9	144

Habitat type codes [17]: L = Dwarf Shrub-Lichen Dominated; M = Mesic/Dry Herbaceous; P = Open Low Shrub Dwarf Birch/Ericaceous; S = Sparse Vegetation. "*Cladonia*" includes all species except *C. stellaris*. Ecoregion codes [18]: N = Northern Seward Peninsula/Kotzebue Sound Lowlands (Kotzebue Lowlands) and S = Southern Seward Peninsula/Bering Sea Coast (Bering Coast).

vascular vegetation physiognomy that influence lichen community composition [19]. Because species differ in growth forms and levels of dominance, it could be expected that variability in lichen biomass and bulk density would be associated with distinctive landcover types.

Lichens in northwestern Alaska occur in very diverse, highly mixed assemblages typified by high species richness [3], [19]. Yet, most estimates of terrestrial lichen biomass completed elsewhere have been limited to only a single [5] or a very restricted number of species [2], [20], mainly focusing on the critical forage lichen genus *Cladonia*. Some workers [20] have suggested that in diverse lichen communities, the most rapid and effective way to estimate biomass would be to integrate estimates over entire communities rather than parsing out unique biomass relationships among individual taxa. Whole-community surveys [21] are also the most realistic way to monitor biomass across landscapes that have heterogeneous or patchy distributions of a large number of species, or for landscapes that have large species turnover (beta-diversity) among sites.

This study explored the feasibility of using height and cover measurements to estimate terrestrial lichen biomass in northwestern Alaska. Our inferences focused mainly on forage lichens (including *Alectoria*, *Bryocaulon*, *Cetraria* and *Cladonia* spp.) considered both as individual taxa and as the dominant members of aggregate, multi-taxa lichen communities. Here, we extended the range of species to include multi-taxa "community" samples that integrated all observed species, and we presented regression slope estimates to use as conversion factors applicable to northwestern Alaska. Our primary objectives in this study were: 1) to estimate the slope of zero-intercept linear regressions for biomass vs. volume; 2) to examine model fit between single-taxon vs. multi-taxa sampling techniques; 3) to examine model fit between ocular vs. point count cover estimation techniques; 4) to determine whether observed and predicted biomass in multi-taxa communities differed significantly among landcover types within ecoregions; and 5) to examine whether height or cover by themselves could each be good predictors of lichen biomass. We also place our results in the context of previous studies and make recommendations for best survey practices.

Methods

Ethics Statement

Field sampling was conducted on public lands administered by the US Bureau of Land Management by permission of that agency. We complied with all national and international rules regarding ethics; the research did not involve measurements on humans or animals. Plant material collected for this study was sampled on a very limited scale and therefore had negligible effects on landscapes. We declare no commercial interests or conflicts of interest.

Field Sampling

Field data are available as Dataset S1. We measured lichen height, cover and biomass (both for single taxa and for all taxa aggregated) among two different ecoregions [22] and four landcover types on the Seward Peninsula, northwestern Alaska (Table 1). Although both ecoregions converge on the Seward Peninsula, they differ sharply in rainfall and climate [23]. The first ecoregion, the Northern Seward Peninsula/Kotzebue Sound Lowlands (hereafter, "Kotzebue Lowlands"), has a Chukchi Sea-influenced and more continental climate with mean annual precipitation of 250–500 mm. The second ecoregion, the Southern Seward Peninsula/Bering Sea Coast (hereafter "Bering Sea Coast") is moist with mean annual precipitation ranging from 400–800 mm. Within each ecoregion we sampled four landcover types [17] which formed the basis of long-term lichen monitoring in the Bering Land Bridge National Preserve [24]; these were: Dwarf-Shrub and Lichen Dominated (L), Mesic/Dry Herbaceous (M), Open Low Shrub – Dwarf Birch/Ericaceous (P), and Sparse Vegetation (S) [17]. While there was a newer landcover classification available at the time of sampling, we chose an older data set compatible with NPS's Arctic Network's long-term lichen monitoring studies [24].

In the study area, we assigned plot locations non-randomly. For single-taxon sampling, we targeted sites with continuous coverage of a dominant forage lichen species. For multi-taxa sampling, we targeted sites with continuous coverage of lichens in general, avoiding non-lichen vegetation. Because multi-taxa communities were often dominated by high proportions of forage lichens (*Alectoria* spp., *Bryocaulon divergens, Cetraria* spp., or branched-fruticose *Cladonia*), we recorded the dominant lichen taxon within

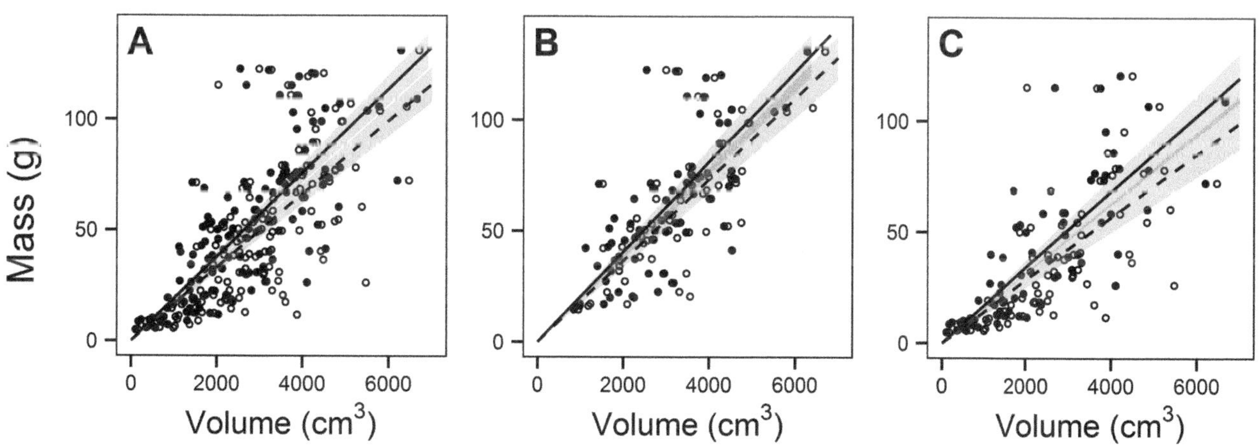

Figure 1. Biomass-volume relations using multiple methods. Shown are the ocular cover method (filled circles and solid lines) and the point count method (open circles and dotted lines), for all sample types combined (Fig. 1A), multi-taxa samples (Fig. 1B), and single-taxon samples (Fig. 1C). Lines indicate fitted model slopes (see Table 2 for estimates), while grey polygons indicate 95% confidence intervals.

Table 2. Estimated parameters and goodness-of-fit statistics for fitted models of forage lichen biomass in northwestern Alaska.

Method	Sample type	Taxon	Slope coefficient (g cm^{-3})			Bayesian Information Criterion (*BIC*)			Log-likelihood and *LLR*		
			Ocular	Pt count	Difference	Ocular	Pt count	Difference	Ocular	Pt count	*LLR*
Combined	All	All	0.0176	-	-	2607.35	-	-	−1298.01	-	-
Either	All	All	0.0189	0.0165	−12.8%	1285.43	1319.60	2.7%	−637.75	−654.83	−34.17
Either	Multi-taxa	All	0.0203	0.0183	−9.8%	641.14	647.12	0.9%	−316.31	−319.30	−5.98
Either	Single-taxa	All	0.0171	0.0143	−16.4%	642.61	666.68	3.7%	−317.01	−329.05	−24.08
Either	Multi-taxa	*Alectoria*	0.0106	0.0098	−7.2%	35.27	33.63	−4.7%	−16.25	−15.43	1.64
Either	Multi-taxa	*Bryocaulon*	0.0160	0.0154	−3.8%	54.72	53.66	−1.9%	−25.41	−24.88	1.06
Either	Multi-taxa	*Cetraria*	0.0177	0.0135	−23.4%	19.87	16.89	−15.0%	−8.84	−7.34	2.98
Either	Multi-taxa	*Cladonia*	0.0207	0.0186	−10.1%	481.10	486.65	1.2%	−236.56	−239.34	−5.56
Either	Multi-taxa	*C. stellaris*	0.0299	0.0288	−3.6%	21.56	16.35	−24.2%	−9.68	−7.08	5.21
Either	Single-taxa	*Alectoria*	0.0083	0.0056	−33.3%	93.63	95.56	2.1%	−44.18	−45.14	−1.93
Either	Single-taxa	*Bryocaulon*	0.0140	0.0120	−14.6%	128.78	123.45	−4.1%	−61.62	−58.95	5.33
Either	Single-taxa	*Cetraria*	0.0141	0.0105	−25.5%	157.81	153.10	−3.0%	−75.86	−73.51	4.71
Either	Single-taxa	*Cladonia*	0.0197	0.0184	−6.6%	155.42	158.15	1.8%	−74.94	−76.30	−2.73
Either	Single-taxa	*C. stellaris*	0.0221	0.0208	−6.1%	55.66	56.34	1.2%	−26.04	−26.38	−0.68
Height only	Multi-taxa	All	-	11.05 (g cm^{-1})	-	-	653.31	-	-	−322.4	-
Either, cover only	Multi-taxa	All	0.1188 (g cm^{-2})	0.1049 (g cm^{-2})	11.7%	672.70	682.46	1.5%	−332.1	−337.0	−9.76

"Method" column summarizes either ocular or point count methods of estimating volume, both of these, or height only. "Slope coefficient" is in units of g cm^{-3} unless otherwise noted, and is a measure of lichen mat bulk density; uncertainty of each estimate is presented graphically as 95% confidence intervals in Fig. 3. "Difference" between the ocular and point count method is expressed as a proportion of the ocular method. Model goodness-of-fit was determined by likelihood-based methods: better models have values closer to zero for Bayesian Information Criterion (*BIC*) values and greater log-likelihood values. Log-likelihood ratios (*LLR*) were calculated holding ocular method as the null hypothesis, therefore, more strongly negative values indicate better fit for models using the ocular method. Each *LLR* test was significant ($p<0.0001$).

Table 3. Observed and predicted lichen biomass (g m^{-2}) for each sample type and landcover type in northwestern Alaska.

Sample type	Land cover type	Observed	Volume (ocular)		Volume (pt ct)		Height only		Cover only (ocular)		Cover only (pt ct)	
		Mean	Mean	Deviation	Mean	Deviation	Mean	Deviation	Mean	Deviation	Mean	Deviation
Multi-taxa	L	911.9 (102.1)	943.9 (69.0)	3.5%	916.2 (61.3)	0.5%	907.3 (56.9)	−0.5%	1068.1 (25.0)	17.1%	1021.3 (14.4)	12.0%
Multi-taxa	M	1508.6 (113.1)	1418.5 (104.5)	−6.0%	1379.7 (92.7)	−8.5%	1357.7 (93.6)	−10.0%	1080 (21.9)	−28.4%	1030.7 (10.2)	−31.7%
Multi-taxa	P	789.6 (77.0)	822 (75.8)	4.1%	899.2 (77.5)	13.9%	932.7 (71.4)	18.1%	893.4 (22.3)	13.1%	960.7 (17.3)	21.7%
Multi-taxa	S	1124.1 (118.5)	1021.4 (76.0)	−9.1%	978.6 (59.1)	−12.9%	957.1 (55.6)	−14.9%	1089.7 (28.7)	−3.1%	1034.5 (10.1)	−8.0%
Multi-taxa	All	991.9 (56.8)	976.3 (45.5)	1.6%	983.4 (41.7)	0.9%	986.0 (38.8)	0.6%	1008.3 (16.7)	−1.7%	1002.6 (8.8)	−1.1%
Single taxon	L	671.3 (110.5)	653.7 (86.9)	−2.6%	710.1 (68.9)	5.8%	-	-	-	-	-	-
Single taxon	M	809.9 (192.5)	665.2 (126.7)	−17.9%	651.9 (104.7)	−19.5%	-	-	-	-	-	-
Single taxon	P	435.0 (88.4)	435.0 (89.5)	0.0%	447.0 (76.8)	2.8%	-	-	-	-	-	-
Single taxon	S	676.2 (99.9)	754.5 (74.6)	13.1%	739.3 (62.7)	9.3%	-	-	-	-	-	-
Single-taxon	All	627.5 (58.2)	624.5 (47.0)	0.5%	633.2 (40.0)	−0.9%	-	-	-	-	-	-

Mean lichen biomass, estimated using the "All Taxa" regression slope coefficients from Table 2 and parceled out by landcover type, is reported with standard errors (in parentheses) and mean deviation from observed biomass. Mean deviation is interpreted as measurement error. Landcover type codes: L = Dwarf Shrub-Lichen Dominated; M = Mesic/Dry Herbaceous; P = Open Low Shrub Dwarf Birch/Ericaceous; S = Sparse Vegetation. Caution: these biomass values were based on average quadrat cover >80% and should *not* be considered an inference about areas >1 m^2 (which could have erratic cover).

each community to facilitate later comparison with single-taxa samples, and also recorded the dominant vascular vegetation type for landcover classification according to [17].

On the ground at each sampling location, we placed a quadrat frame made of PVC pipe (internal area 25×25 cm) divided into twenty-five cells (5×5 cm). In the center of each cell, we measured the height of lichens (if present) using a narrow, ruled metal rod (3 mm diameter) lowered to the base of the lichen until firm resistance at the duff/soil layer was encountered, but not pushing into layers below. We calculated lichen height as the average of the twenty-five points within each plot (including "no-hit" zero values) because we wanted to facilitate adaptation to other techniques that might include many lichen absences (e.g., other point count or line intercept surveys). For single-taxon samples, we recorded only the height of the target lichen, while for multi-taxa samples we also recorded the identity of the lichen species and visually estimated its cover to the nearest one percent. Based on ground surface similarity of growth form (i.e., without dislodging the lichen) and bulk density, species found in single-taxon plots were aggregated to the genus level if several were present in the same sample. For example, *Alectoria ochroleuca* was grouped with *A. nigricans*, and *Cetraria laevigata* was grouped with *C. islandica*. Morphologically similar *Cladonia* species (*C. arbuscula and C. mitis*; *C. rangiferina* and *C. stygia*) were also grouped, although we measured *C. stellaris* separately because it had apparently different bulk density and a morphology that was readily recognizable in the field.

After recording height and cover, lichens were misted with water, then manually harvested by cutting vertically along the edge of each quadrat and gently separating intact mats from the ground. We removed non-lichen plant parts, non-target species, obviously dead or decaying lichen parts, and any dirt, gravel or extraneous debris, then placed samples in dry paper bags for transport to the lab. Each sample was completely oven-dried at 80°C until reaching a stable, unchanging mass (minimum 8 hours, no more than 12 hours). Note that this is considerably less drying time than the 76 hours used in a similar study [20]. We weighed each sample with a digital scale (precision ± 0.01 g) within just a few minutes of removal from the drying oven to prevent moisture uptake that might bias dry mass estimates. This also differs from [20], in which lichens were allowed to cool to room temperature before weighing.

Calculations

Our goal was to assess the relationship between biomass and lichen volume (calculated as the product of cover×height). We used two methods to calculate volume; the first was based on the ocular (visual) method of estimating cover described above, and the second was based on the point count estimation of cover. For the point count method, we estimated percentage cover by dividing the number of lichen presences by the total number of points (twenty-five) in each plot. For both methods, we multiplied percentage cover by the area of the plot to convert all measurements to cubic centimeters.

We regressed lichen biomass on volume (or height or cover) using ordinary least squares regression with forced zero intercepts implemented in R version 3.0.1 (R Core Development Team 2013) and compared model slope parameters. Although weighted regressions have been used elsewhere [20] to estimate forage lichen biomass, examination of model residuals (not reported here) suggested this was not necessary given our data. Because goodness-of-fit R^2 statistics are artificially inflated and are essentially meaningless for zero-intercept models, we instead report confidence intervals for each slope, and compare models using three metrics: percent difference in slope, Bayesian Information

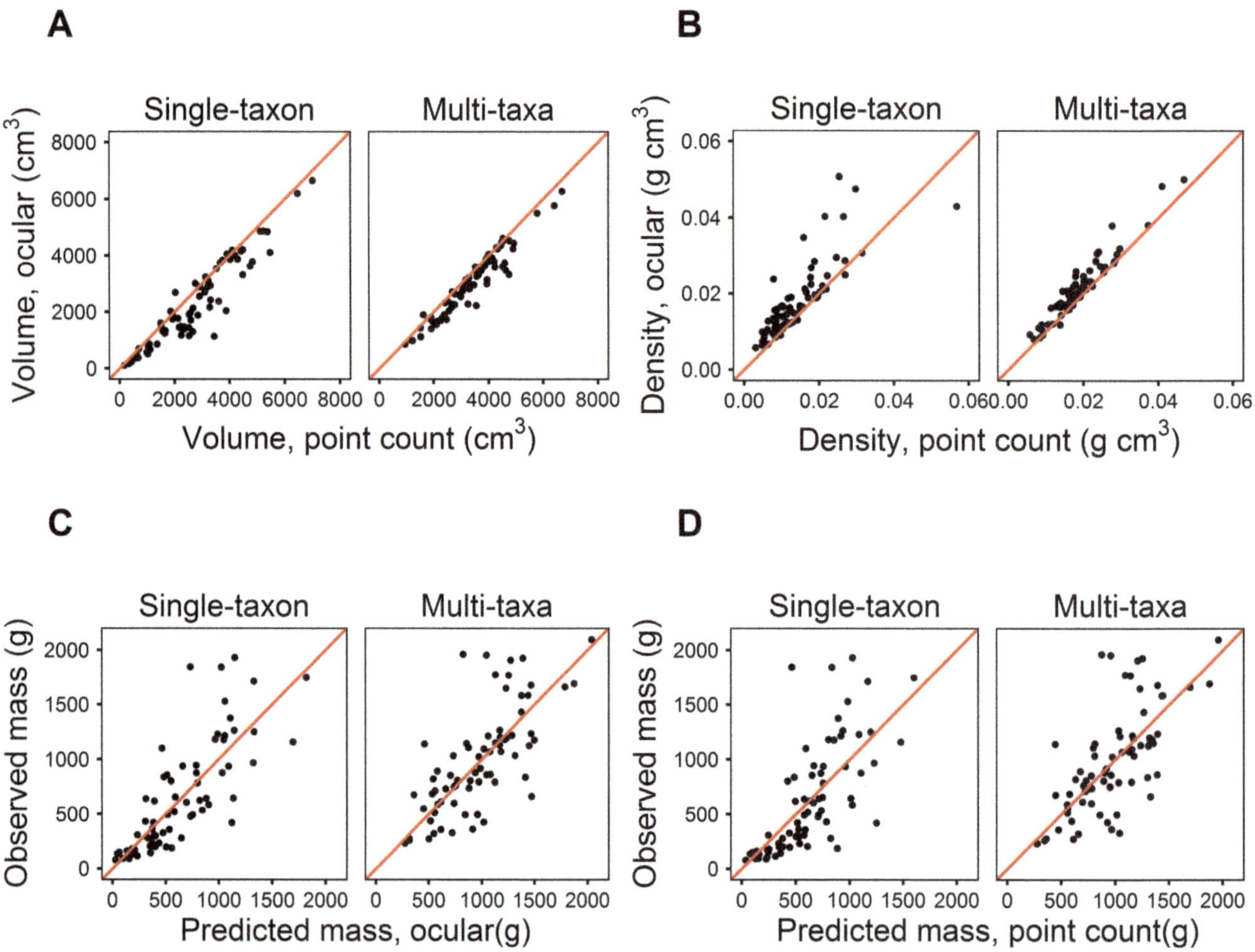

Figure 2. Relative comparisons of bias among two methods. Comparisons are between ocular vs. point count methods for estimates of volume (A) and estimates of bulk density (B); relative comparisons for observed vs. predicted lichen biomass for the ocular method (C) and the point count method (D). Each red line is a hypothetical 1:1 isoline, where deviations from this line indicate magnitude of differences.

Criterion (*BIC*) [25] and log-likelihood ratio (*LLR*) tests [26]. The latter two are likelihood-based methods that do not attempt to assess whether two model parameters differ more than expected by chance (as would a probability-based *F*-test, for example), but instead they provide a relative measure of how much better or worse each model fits, given the data at hand. For our application, this is more useful than *F*-tests because the sign and magnitude of difference between two methods provides more information than simply whether or not any such difference occurs.

To determine whether mean observed and predicted biomass in multi-taxa "communities" differed significantly among landcover types within ecoregions, we used nested analysis-of-variance *F*-tests (landcover type nested within ecoregions), accompanied by Tukey's Honest Significant Difference test for pairwise comparisons; this procedure used only the multi-taxa data to simulate realistic conditions.

Results

Biomass vs. Volume Slope

From the regression of biomass on volume, estimates of the slope ranged from 0.0143 to 0.0203 g cm^{-3} for lichen communities, depending on the method used (Table 2 and Fig. 1A). Individual taxa ranged from 0.0056 g cm^{-3} (*Alectoria*) to 0.0221 g cm^{-3} (*Cladonia stellaris*). All slopes differed significantly from zero ($p<0.05$ from *t*-test of linear slope coefficients). For both observed mass and bulk density, mean values differed among all taxon groups (*F*-tests $p<0.0001$) except the pairwise comparisons among *Bryocaulon –Cetraria* and *Cladonia–C. stellaris* (for all others, $p<0.05$ from Tukey test).

Single-taxon vs. Multi-taxa Sampling

The single-taxon method consistently underestimated biomass relative to the multi-taxa method (Fig. 1B vs. 1C). Model fit was always better for the multi-taxa method (Table 2), which in all cases also had less variation in biomass than the single-taxon method (Table 3).

Ocular vs. Point Count

The point count method had lower biomass estimates relative to ocular methods (Fig. 1, filled vs. open symbols). This is directly related to the point count method estimating higher volume (Fig. 2A) and lower bulk density (Fig. 2B) relative to the ocular method. Model fit was usually but not always better with the ocular method compared to the point count method (Table 2). Among individual taxa, differences in slope between methods (Fig. 3) were more pronounced for samples dominated by taxa

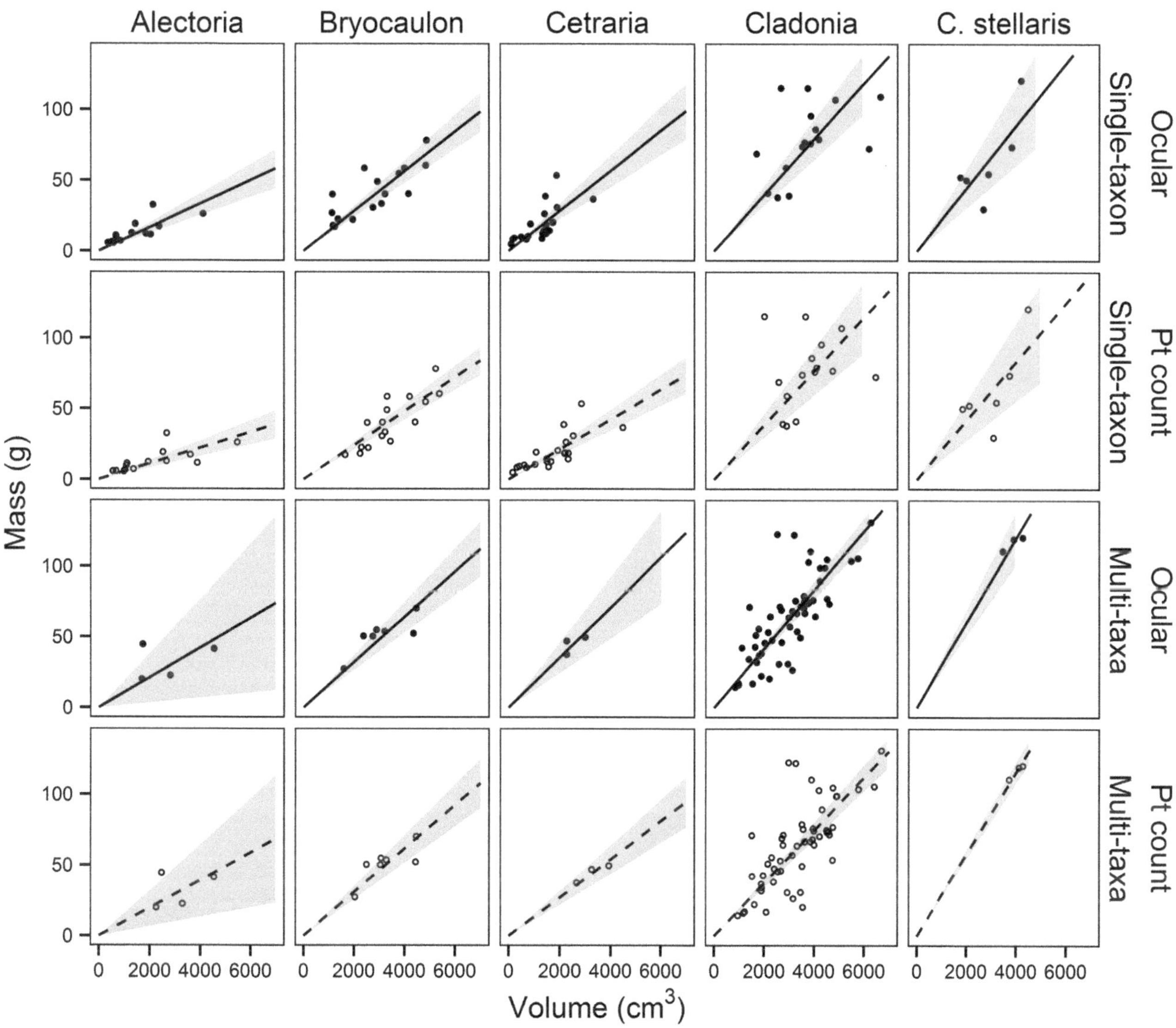
Alectoria
Bryocaulon
Cetraria
Cladonia
C. stellaris
Ocular Single-taxon
Pt count Single-taxon
Ocular Multi-taxa
Pt count Multi-taxa
Mass (g)
0
50
100
0 2000 4000 6000
Volume (cm3)

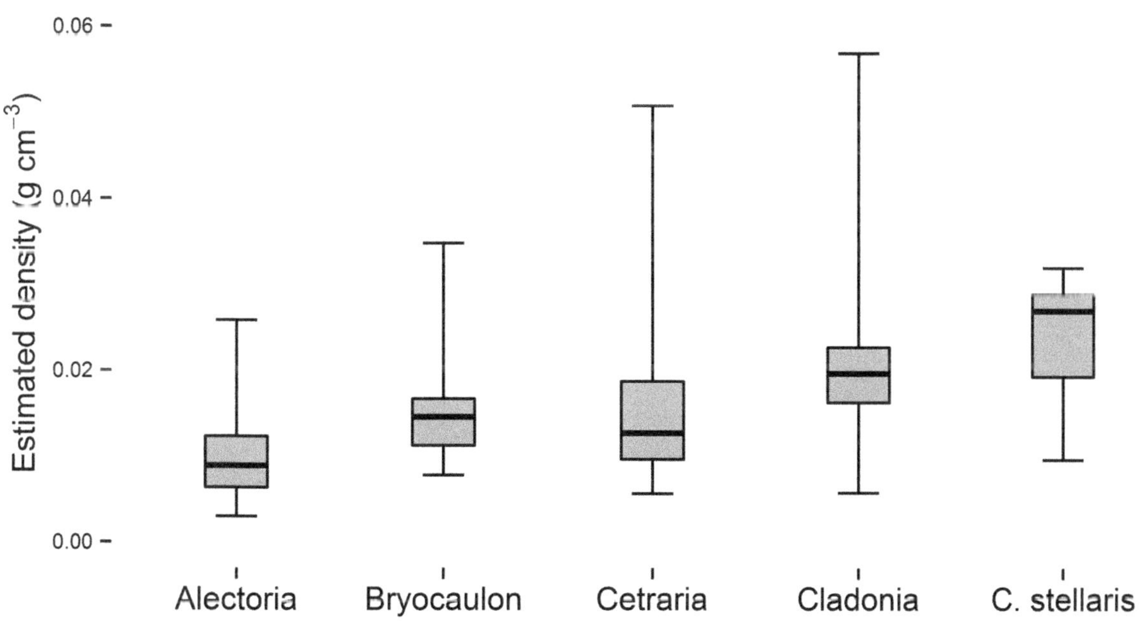
Estimated density (g cm-3)
0.00
0.02
0.04
0.06
Alectoria
Bryocaulon
Cetraria
Cladonia
C. stellaris

Figure 3. Biomass-volume relations and bulk density distributions for dominant forage lichen taxa in northwestern Alaska. Regressions were fitted using either single-taxon sampling (top two rows) or multi-taxa samples (bottom two rows). Lines indicate fitted model slopes (see Table 2 for estimates), while grey polygons indicate 95% confidence intervals. In the bottom panel, the distribution of estimated bulk density (all methods, grey boxes) for 144 lichen samples from northwestern Alaska is shown, where dark bars in boxes are median values, boxes represent the interquartile range of data values, and whiskers are the maxima/minima within each group. Mean bulk density differs among all species groups (*F*-test $p<0.0001$) except the pairwise comparisons among *Bryoria – Cetraria* and *Cladonia – C. stellaris* (for all others, $p<0.05$ or less from Tukey HSD test).

with diffuse, filamentous growth forms like *Alectoria*, rather than for denser taxa like *Cladonia* that tended to fill entire sample frames.

Ecoregions and Landcover Types

When slope estimates were used to predict standing biomass for each of the four landcover types, the point count method had generally greater deviation from observed values than the ocular method (Table 3), though this was not true in all cases; magnitudes of deviation ranged from 0.01–19.5%. The biomass values reported here for the four landcover classes studied represents the high end of the lichen biomass distribution for each cover type because sites were selected based on high lichen cover to allow for better biomass-to-volume estimates. Of the seven combinations of landcover types nested within ecoregions (Table 1), there were 21 possible pairwise comparisons. Of the 21 comparisons, Tukey tests revealed very few significant pairwise differences for observed mass (3 significant comparisons), mass predicted using the ocular method (3 comparisons), mass predicted using the point count method (2 comparisons), and mass predicted using only height (0 comparisons). Each of these nominally "significant" pairwise comparisons had *p*-values that were close to the $p = 0.05$ significance threshold, so we do not report them here.

Biomass vs. Height or Cover

Deviation of estimated values from observed values (Table 3) was similar but of slightly greater magnitude when using height as the sole predictor of biomass in multi-taxa communities than when using volume; magnitudes of deviation ranged from 0.5–18.1% for different landcover types. When using cover as the sole predictor of biomass, the deviation of estimated from observed values was far greater than when using volume (about 3–5 times greater: Table 3); this was true of the ocular method (absolute deviation: 3.1–28.4%) as well as the point count method (absolute deviation: 8.0–31.7%).

Discussion

Biomass as a Function of Volume

When lichen biomass is regressed on volume, the resulting linear slope is equivalent to a mass-to-volume ratio or a measure of the lichen mat's bulk density. Our findings indicate that taxa have different bulk densities as a result of their different growth forms, a trend that is reflected in multi-taxa communities dominated by each growth form. For individual taxa, *Cladonia* was most dense, while *Bryocaulon*, *Cetraria*, and *Alectoria* were successively less dense. There were congruent patterns in multi-taxa communities dominated by each of those genera, with *Cladonia*-dominated communities being densest.

The study most comparable with our methods [20] used volumetric estimation of biomass for four species (three *Cladonia* spp. and *Cetraria islandica*) in Sweden using 50×50 cm plots. Their mean *Cladonia* bulk densities agree closely with our estimates (within 12% of each other for ocular methods), but were slightly lower, perhaps because they manually pulled lichen mats off the ground (rather than cut them cleanly as we did), which could leave lichen material attached to the ground. By contrast, *Cetraria* spp. bulk densities from our study were nearly 50% less than that of *C. islandica* in Sweden [20]. We assume that unmeasured factors (e.g., nutrient availability, moisture, growing season, fire regime, successional status, herbivory patterns) allow *Cetraria* in Sweden to occur in much denser colonies than on the Seward Peninsula of Alaska.

In Finland, Kumpula et al. [2], reported a much lower bulk density than either our results or those from Sweden [20], possibly resulting from differing definitions of what constitutes a "living" portion of a lichen thallus. Dunford et al. [5] did not report a bulk density because they relied solely on lichen cover (not height or volume) for the single species *Cladonia mitis*. Though their scope of inference extended to 25 sites, Dunford et al. [5] also had a drastically smaller sample size than ours ($n = 8$ vs.144).

To the extent that we had outlier observations, these were likely due to the unintentional inclusion of non-target vegetation in some plots, especially where lichen thalli were layered over top of unseen moss or shrub tissues beneath. While our volumetric calculations probably included portions of non-target vegetation, we note that this is consistent with all situations in which non-destructive estimates of cover and height are used for applying conversion factors to larger study areas.

Single-taxon vs. Multi-taxa Sampling

Individual genera had generally lower bulk density than the multi-taxa samples. This is probably because lichens on the Seward Peninsula grew in mixed assemblages with intertwined and overlapping layers composed of different growth forms and bulk densities. Thus what superficially appeared to be a continuous cover of one lichen from the surface was in fact usually a tangled mat of several different lichen taxa below. This may be why we found it difficult to locate monotypic mats for every target species, even using a plot size of just 25×25 cm.

Estimating biomass–volume ratios for every species found in northwestern Alaska would be a prohibitive task. Given that multi-taxa surveys require only general taxonomic training and that they capture variation in biomass at levels equivalent to or better than individualistic single-taxon assessments, we conclude that a multi-taxa focus is preferable to single-taxon surveys when lichens across entire landscapes must be faithfully represented.

Ocular vs. Point Count

The use of point-count estimations of cover gave consistently higher cover (and volume) estimates than ocular methods, leading to lower bulk density estimates; this is consistent with other findings [20]. Despite its initial appeal for reducing observer bias, we noted that the point count method often seemed to overestimate cover by "catching" or touching disproportionately small pieces of lichen thallus in areas of the plot that lacked appreciable coverage, especially for taxa that were filamentous (e.g., *Alectoria*) or patchily distributed. Though the ocular method seemed initially less objective, it had the benefit that it allowed observers to visually integrate patchy or diffuse spatial coverage of lichens in a more realistic manner. Point count methods overestimate cover of vascular plants, while visual estimates are more accurate [27], and counting enough points to be useful would be logistically limiting in remote Arctic settings. We note that point count methods may still have utility for plots larger than our 25×25 cm size because ocular cover becomes less precise in larger plots [15]. Point count methods may also be preferable for long-term monitoring where repeatability is paramount [28] and where results may differ among observers [14]. Future work is needed to resolve variation among observers and identify sources of error.

Ecoregions and Landcover Types

The two ecoregions we examined on the Seward Peninsula appeared to have similar bulk densities (g cm^{-3}) of lichens. Bulk densities also differed far less than expected among the four landcover types. We expected that differences in substrate availability, substrate chemistry, moisture and vascular vegetation would be expressed in lichen community biomass because growth forms differ according to those conditions. For example, filamentous groups such as *Bryocaulon* and *Alectoria* dominate the Sparse Vegetation ("S") cover type on the Seward Peninsula, whereas the low-elevation L, P, and M landcover types tend to be dominated by denser *Cladonia* spp. We suspect that ecoregion and landcover would have been more influential had we sampled across a broader latitudinal gradient (see [29] for one example of an Arctic gradient spanning 1800-km north-south). Note that while the biomass values in Table 3 are a starting point for landscape modeling, they portray only one possible scenario (i.e., one with lichen cover averaging ~80% due to our preferential plot assignment). Future landscape estimates must account for patchy or sparse lichen cover, and therefore require a truly random sample of sites from within each landcover type.

Biomass vs. Height or Cover

In some instances, a direct measure of biomass as a function of average height of the lichen can be useful, avoiding the need to estimate cover. For example, line intercept methods can provide copious amounts of height data in a short time frame and may be more applicable than closed-frame methods in the open, unforested landcover types that are commonly found in northwestern Alaska. However, we found a large amount of variation in the estimates that relied on height as the sole predictor, and we advise that height-based estimates must be always weighted by some measure of cover to prevent overestimation. Given that caveat, transect-based line intercept methods could provide biomass estimates if the total area of the study site is known.

If height alone appeared to be an inadequate predictor of lichen biomass, then cover alone was even worse. This makes logical sense – two different lichen patches could cover an equal surface area yet differ by an order of magnitude in height; observers would have no way of knowing their true volume. Because we found that deviations from true values were 3 to 5 times greater when using cover alone as compared to the other methods, we do not recommend using cover as the sole predictor of lichen biomass.

Extending Spatial Coverage

Remote sensing is one way that estimates can be scaled up to landscape-level inferences. Field surveys can be coupled with remote sensing imagery to construct biomass estimates via several algorithms [4]. Other workers [30] have similarly used Landsat imagery to derive leaf area index and biomass for multiple vegetation groups including fruticose, foliose and crustose lichens (in aggregate). Spectral decomposition of remote imagery is a further refinement that can distinguish among taxonomic or functional groups of lichens [31]. Though location-specific correction factors must always be derived and validated on the ground, remote sensing is a promising technology that should allow rapid estimation of biomass for very large, remote or difficult-to-access locations in northern ecosystems.

Management Applications

The biomass functions presented here promise to have utility in several different areas of applied ecology in arctic Alaska. Studies of ungulate biology should be able to use these equations to help scale plot-level vegetation data (where both lichen cover and height measurements are available) to remote sensing imagery, as in [32]. Long-term studies of ungulate grazing exclosures are underway on Alaska's Seward Peninsula [33], [34], and conversion of lichen cover/height data to forage biomass will be an essential component of these studies. The National Park Service has other long-term lichen/vegetation studies in which both lichen cover and height are measured [33]; these may similarly be coupled with remote sensing modeling using our conversion factors for applications including changes in vegetation community structure with climate change effects [35], vegetation response to changing fire regimes [36], and carbon accounting [37]. The National Park Service's Arctic Network has sufficient density of ground-based measurements to model lichen biomass at a landscape scale [24]. Repeat measurements would allow for temporal and spatial depictions of lichen biomass on the landscape. Improved burn severity geospatial layers and historic fire perimeters would increase the resolution of these layers by allowing for the modeling of successional status, and hence lichen height and cover. Lichen winter range for the Western Arctic Caribou Herd is projected to suffer some decline with climate-driven shrub increase [1], [35], more frequent wildfire [36], and increasing inputs of nitrogen and sulfur from regional development [38]. Continued monitoring of lichen biomass will be critical for detecting and addressing ongoing changes.

Conclusions

To summarize, we found that multi-taxa ("bulk") sampling, when coupled with ocular estimation, was the preferred method for biomass estimation because it yielded the most accurate estimates and was logistically most efficient in the field. This method required no more than general taxonomic knowledge in the field and was tractable for statistical purposes. The zero-intercept regression equation (biomass = 0.0203 g cm^{-3}) which we derived by this method is readily applied to lichen communities in northwestern and arctic Alaska, though we caution that this does not necessarily apply to other geographical regions where location-specific equations will be required. Height alone was not the preferred predictor of biomass, and cover alone was a poor predictor; the best estimates should ideally include both height and cover. Because bulk densities varied among the forage lichen species we measured, we suggest that in cases where there is only a single species of interest, investigators may benefit from using separate regression equations for target species. Lichen biomass estimation has a wealth of applications that will help managers estimate wildlife forage, understand successional trends, detect climate and air quality signals, and account for landscape-level carbon. Therefore, continued monitoring will be vital for understanding how ongoing changes in lichen biomass and distribution affect other elements of Arctic ecosystems.

Acknowledgments

We thank Marci Johnson, Gina Hernandez, Matthew Jenkins and Mason London for help with field work and with cleaning samples. We also wish to thank Nichole Andler for providing field crew, and Bruce McCune and Jody Deland for the use of drying ovens.

Author Contributions

Conceived and designed the experiments: AR PN. Performed the experiments: AR PN. Analyzed the data: AR PN RJS. Contributed reagents/materials/analysis tools: AR PN. Contributed to the writing of the manuscript: AR PN RJS.

References

1. Joly K, Chapin FS, Klein DR (2010) Winter habitat selection by caribou in relation to lichen abundance, wildfires, grazing, and landscape characteristics in Northwest Alaska. Ecoscience 17: 321–333. doi:10.2980/17-3-3337.
2. Kumpula J, Colpaert A, Nieminen M (2000) Condition, potential recovery rate, and productivity of lichen (*Cladonia* spp.) ranges in the Finnish reindeer management area. Arctic 53: 152–160.
3. Holt EA, McCune B, Neitlich P (2008) Grazing and fire impacts on macrolichen communities of the Seward Peninsula, Alaska, USA. Bryologist 111: 68–83. doi:10.1639/0007-2745(2008)111[68:GAFIOM]2.0.CO;2.
4. Arseneault D, Villeneuve N, Boismenu C, Leblanc Y, Deshaye J (1997) Estimating lichen biomass and caribou grazing on the wintering grounds of northern Quebec: an application of fire history and Landsat data. J Appl Ecol 34: 65–78.
5. Dunford JS, McLoughlin PD, Dalerum F, Boutin S (2006) Lichen abundance in the peatlands of northern Alberta: implications for boreal caribou. Ecoscience 13: 469–474. doi:10.2980/1195-6860.
6. Heggberget TM, Gaare E, Ball JP (2002) Reindeer (*Rangifer tarandus*) and climate change: importance of winter forage. Rangifer 22: 13–31. doi:10.7557/2.22.1.388.
7. Lang SI, Cornelissen JHC, Shaver GR, Ahrens M, Callaghan TV, et al. (2012) Arctic warming on two continents has consistent negative effects on lichen diversity and mixed effects on bryophyte diversity. Glob Chang Biol 18: 1096–1107. doi:10.1111
8. Geiser LH, Neitlich PN (2007) Air pollution and climate gradients in western Oregon and Washington indicated by epiphytic macrolichens. Environ Poll 145: 203–218. doi:10.1016/j.envpol.2006.03.024.
9. Pardo LH, Fenn ME, Goodale CL, Geiser LH, Driscoll CT, et al. (2011) Effects of nitrogen deposition and empirical nitrogen critical loads for ecoregions of the United States. Ecol Appl 21: 3049–3082. doi:10.1890/10-2341.1.
10. Gauslaa Y, Lie M, Ohlson M (2008) Epiphytic lichen biomass in a boreal Norway spruce forest. Lichenologist 40: 257–266.
11. Li S, Liu W, Wang L, Ma W, Song L (2011) Biomass, diversity and composition of epiphytic macrolichens in primary and secondary forests in the subtropical Ailao Mountains, SW China. For Ecol Manage 261: 1760–1770. doi:10.1016/j.foreco.2011.01.037.
12. Price K, Hochachka G (2001) Epiphytic lichen abundance: effects of stand age and composition in coastal British Columbia. Ecol Appl 11: 904–913. doi:10.1890/1051-0761(2001)011[0904:ELAEOS]2.0.CO;2.
13. Rhoades FM (1981) Biomass of epiphytic lichens and bryophytes on *Abies lasiocarpa* on a Mt. Baker lava flow, Washington. Bryologist 84: 39–47. doi:10.2307/3242976.
14. McCune B, Lesica P (1992) The trade-off between species capture and quantitative accuracy in ecological inventory of lichens and bryophytes in forests in Montana. Bryologist 95: 296–304.
15. Godínez-Alvarez H, Herrick JE, Mattocks M, Toledo D, Van Zee J (2009) Comparison of three vegetation monitoring methods: Their relative utility for ecological assessment and monitoring. Ecol Indic 9: 1001–1008. doi:10.1016/j.ecolind.2008.11.011.
16. Joly K, Jandt RR, Klein DR (2009) Decrease of lichens in Arctic ecosystems: the role of wildfire, caribou, reindeer, competition and climate in north-western Alaska. Polar Res 28: 433–442.
17. Markon CJ, Wesser S (1997) The Bering Land Bridge National Preserve Land Cover Map and Its Comparability with 1995 Field Conditions. USGS Open File Report No 97-103. Reston, Virginia: US Department of the Interior, Geological Survey.
18. Jorgenson MT, Roth JE, Miller PE, Macander MJ, Duffy MS, Wells AF, Frost GV, Pullman ER (2009) An Ecological Land Survey and Landcover Map of the Arctic Network. Natural Resource Technical Report NPS/ARCN/NRTR—2009/270. Fort Collins, Colorado: US Department of the Interior, National Park Service.
19. Holt EA, McCune B, Neitlich P (2009) Macrolichen communities in relation to soils and vegetation in the Noatak National Preserve, Alaska. Botany 87: 241–252. doi:10.1139/B08-142.
20. Moen J, Danell Ö, Holt R (2007) Non-destructive estimation of lichen biomass. Rangifer 27: 41–46.
21. Bergerud AT (1971) Abundance of forage on the winter range of Newfoundland caribou. Can Field Nat 85: 39–52.
22. Jorgenson MT, Roth JE, Emers M, Davis WA, Schlentner SF, et al. (2004) Landcover Mapping for Bering Land Bridge National Preserve and Cape Krusenstern National Monument, Northwestern Alaska. Fort Collins, Colorado: US Department of the Interior, National Park Service.
23. Daly C (2009) Annual Mean Average Temperature for Alaska 1971–2000. PRISM Climate Group at Oregon State University, Corvallis, Oregon, USA. Available: http://www.prism.oregonstate.edu. Accessed 11 January 2011.
24. Holt EA, Neitlich PN (2010) Lichen Inventory Synthesis: Western Arctic National Parklands and Arctic Network, Alaska. Natural Resource Technical Report NPS/AKR/ARCN/NRTR—2010/385. Fort Collins, Colorado: US Department of the Interior, National Park Service.
25. Schwarz G (1978) Estimating the dimension of a model. Ann Stat 6: 461–464. doi:10.1214/aos/1176344136.
26. Engle RF (1984) Wald, likelihood ratio, and Lagrange multiplier tests in econometrics. In: Griliches Z, Intriligator MD, editors. Handbook of Econometrics. Elsevier, Vol. 2. pp. 775–826.
27. Bråkenhielm S, Qinghong L (1995) Comparison of field methods in vegetation monitoring. Water Air Soil Pollut 79: 75–87. doi:10.1007/BF01100431.
28. Fancy SG, Gross JE, Carter SL (2009) Monitoring the condition of natural resources in US national parks. Environ Monit Assess 151: 161–174. doi:10.1007/s10661-008-0257-y.
29. Epstein HE, Walker DA, Raynolds MK, Jia GJ, Kelley AM (2008) Phytomass patterns across a temperature gradient of the North American arctic tundra. J Geophys Res Biogeosci 113: 1–11. doi:10.1029/2007JG000555.
30. Chen W, Li J, Zhang Y, Zhou F, Koehler K, et al. (2009) Relating biomass and leaf area index to non-destructive measurements in order to monitor changes in Arctic vegetation. Arctic 62: 281–294. doi:10.14430/arctic148.
31. Nelson PR, Roland C, Macander MJ, McCune B (2013) Detecting continuous lichen abundance for mapping winter caribou forage at landscape spatial scales. Remote Sens Environ 137: 43–54. doi:10.1016/j.rse.2013.05.026.
32. Swanson DK, Neitlich PN (2014) Terrestrial Vegetation Monitoring Protocol for the Arctic Alaska Network: Establishment, Sampling, and Analysis of Permanent Monitoring Plots. Natural Resource Technical Report NPS/AKR/ARCN/NRTR. Fort Collins, Colorado: US Department of the Interior, National Park Service. In Preparation.
33. Moore KE (2013) Collaborative Monitoring of Seward Peninsula, Alaska Reindeer (*Rangifer tarandus tarandus*) Grazing Lands: A Monitoring Protocol. M.S. thesis. Fairbanks, AK: University of Alaska, Fairbanks. 55 p.
34. Swanson DK (2013) Three Decades of Landscape Change in Alaska's Arctic National Parks: Analysis of Aerial Photographs, c. 1980–2010. Natural Resource Technical Report NPS/ARCN/NRTR—2013/668. Fort Collins, Colorado: US Department of the Interior, National Park Service.
35. Tape K, Sturm M, Racine C (2006) The evidence for shrub expansion in Northern Alaska and the Pan-Arctic. Glob Chang Biol 12: 686–702.
36. Higuera PE, Chipman ML, Barnes JL, Urban MA, Hu FS (2011) Variability of tundra fire regimes in Arctic Alaska: millennial scale patterns and ecological implications. Ecol Appl 21: 3211–3226. doi:10.1890/11-0387.1.
37. Porada P, Weber B, Elbert W, Pöschl U, Kleidon A (2013) Estimating global carbon uptake by lichens and bryophytes with a process-based model. Biogeosciences 10: 6989–7033. doi:10.5194/bg-10-6989-2013.
38. Linder G, Brumbaugh W, Neitlich P, Little E (2013) Atmospheric deposition and critical loads for nitrogen and metals in Arctic Alaska: review and current status. Open J Air Poll 2(4): 76–99. doi: 10.4236/ojap.2013.24010.

6

Mean Annual Precipitation Explains Spatiotemporal Patterns of Cenozoic Mammal Beta Diversity and Latitudinal Diversity Gradients in North America

Danielle Fraser[1,2]*, Christopher Hassall[1,3], Root Gorelick[1,4,5], Natalia Rybczynski[1,2]

1 Department of Biology, Carleton University, Ottawa, Ontario, Canada, **2** Palaeobiology, Canadian Museum of Nature, Ottawa, Ontario, Canada, **3** School of Biology, University of Leeds, Leeds, United Kingdom, **4** Department of Mathematics and Statistics, Carleton University, Ottawa, Ontario, Canada, **5** Institute of Interdisciplinary Studies, Carleton University, Ottawa, Ontario Canada

Abstract

Spatial diversity patterns are thought to be driven by climate-mediated processes. However, temporal patterns of community composition remain poorly studied. We provide two complementary analyses of North American mammal diversity, using (i) a paleontological dataset (2077 localities with 2493 taxon occurrences) spanning 21 discrete subdivisions of the Cenozoic based on North American Land Mammal Ages (36 Ma – present), and (ii) climate space model predictions for 744 extant mammals under eight scenarios of future climate change. Spatial variation in fossil mammal community structure (β diversity) is highest at intermediate values of continental mean annual precipitation (MAP) estimated from paleosols (~450 mm/year) and declines under both wetter and drier conditions, reflecting diversity patterns of modern mammals. Latitudinal gradients in community change (latitudinal turnover gradients, aka LTGs) increase in strength through the Cenozoic, but also show a cyclical pattern that is significantly explained by MAP. In general, LTGs are weakest when continental MAP is highest, similar to modern tropical ecosystems in which latitudinal diversity gradients are weak or undetectable. Projections under modeled climate change show no substantial change in β diversity or LTG strength for North American mammals. Our results suggest that similar climate-mediated mechanisms might drive spatial and temporal patterns of community composition in both fossil and extant mammals. We also provide empirical evidence that the ecological processes on which climate space models are based are insufficient for accurately forecasting long-term mammalian response to anthropogenic climate change and inclusion of historical parameters may be essential.

Editor: Alistair Robert Evans, Monash University, Australia

Funding: D. Fraser was supported by a Natural Science and Engineering Research Council of Canada (NSERC) postgraduate scholarship, a Fulbright Traditional Student Award, a Mary Dawson Pre-Doctoral Fellowship grant, an Ontario Graduate Scholarship (OGS), and a Koningstein Scholarship for Excellence in Science and Engineering. C. Hassall was supported by an Ontario Ministry of Research and Innovation Postdoctoral Fellowship. R. Gorelick was supported by an NSERC Discovery Grant (#341399). N. Rybczynski was supported by an NSERC Discovery Grant (#312193). The funders had no role in study design, data collection and analysis, decision to publish, or preparation of the manuscript.

Competing Interests: The authors have declared that no competing interests exist.

* Email: danielle_fraser@carleton.ca

Introduction

Terrestrial species from all major taxonomic groups show dramatic changes in richness and diversity across the landscape [1]. One of the fundamental goals in ecology is therefore to ascertain why there are more species in some places than in others. A satisfactory answer would identify and disentangle the drivers of biodiversity at all spatial scales, from the microhabitat to the globe, as well as explain changes through time. Attempts to provide such an answer have produced many studies of species richness patterns and community composition in extant organisms [1–8]. Prime examples are the numerous studies of latitudinal richness gradients (LRGs), which have been observed in many terrestrial groups including angiosperms, birds, mammals, insects and other invertebrates. The best supported hypotheses show that richness declines toward the poles in correlation with reductions in precipitation, temperature, and net primary productivity [9]. Correlation of global climate with animal richness over the past 65 Ma, specifically a decline in richness as climates cooled, similarly supports a link between diversity and climate [10–12]. However, of the spatial and temporal dimensions of diversity, spatial patterns of community differences ("β diversity") are infrequently studied despite considerable variation on both local and regional scales [2,13,14] and their influential role in the structuring of continental-scale richness patterns including LRGs [3,4].

β diversity has been defined most broadly as the differentiation in community composition (i.e. the species that make up the community) among regions or along environmental gradients [15]. Similar to LRGs, β diversity generally declines from the tropics to the poles in correlation with climate [2]. However, temporal changes in β diversity remain poorly studied despite their potential power for illuminating the drivers of past and present richness patterns and importance in modern conservation [16–18]. This study therefore tests the hypothesis that climatic influences on mammalian β diversity apply equally to temporal patterns, i.e. that the underlying ecological processes are "ergodic" (dynamic processes that are the same in both time and space).

The mid to late Cenozoic (36 Ma to present) has been a time of dramatic mammalian diversity change, shaped in part by the transition from the productive ice-free ecosystems of the early to mid Cenozoic to the more temperate glaciated ecosystems of the late Cenozoic. Under these changing climatic conditions, mammalian communities show dramatic reductions in richness, changes in community composition, and morphology [10,19–24]. The most dramatic changes occurred at high latitudes, where ecosystems transitioned from *Metasequoia* forests during the early to mid Cenozoic [25,26] to boreal-type forests during the later Cenozoic and to modern tundra [27]. Associated with Cenozoic climate change, were changes in latitudinal climate gradients; overall, the intensity of latitudinal climate gradients increased toward the present, reflecting disproportionate polar cooling due to the formation of permanent Arctic glaciation [28,29]. We therefore predict that latitudinal diversity gradients increased in strength under cooler, less productive environmental conditions just as modern LRGs are steeper in temperate than in tropical regions. Further, we predict that β diversity declined under cooler, less productive environmental conditions just as modern β diversity declines toward the poles [2,7].

Quaternary (2.6 Ma to present) climates have been cool relative to the majority of the late Cenozoic. Recently, however, high latitudes have experienced disproportionate increases in annual temperature (up to 2°C to date), increases in plant primary productivity, and loss of large areas of perennial ice under anthropogenic global warming [30]. Flora and fauna have responded through shifts in phenology [31], *in situ* evolution [32], and, in some cases, extinction [33]. However, perhaps the most often recorded response is the climatically-correlated pattern of extirpations and colonization that manifest as shifts in the location of a species' geographic range. Distributional studies over ecological timescales (<100 yrs) have recorded dramatic poleward range shifts and expansions for a wide range of terrestrial taxa in response to northern warming [34,35]. Projections (i.e. Special Report on Emissions Scenarios) for the next 100 years predict levels of global warming similar to the middle Miocene (+ 6°C) – a time of reduced or absent perennial Arctic glaciation [36,37] – or warmer (+11°C for the most extreme case; Table S1). We therefore expect continued range expansion, extinction, evolution, and community level changes among North American animals and plants.

A common approach to predicting the long-term outcomes of climate change for terrestrial organisms is climate space modeling (CSM). CSMs use distributional information and climate data to project species ranges into the future, usually under the assumption of no evolution and without adjustment for dispersal differences among species [38–40]. Rapid evolutionary changes on very short timescales and high degrees of variation in dispersal ability under climate change have been observed across a wide range of organisms [34,39,41], therefore CSMs are unlikely to generate accurate forecasts of climate change response. The fossil record, which encompasses many disparate environments and climates, might serve as record of a natural experiment by which ecological hypotheses can be tested in the temporal dimension. Fossil collections are a rich historical record of response to various climatic events that can be incorporated into predictive models, and mammals, in particular, are an excellent group for testing the generality of ecological hypotheses because they have an extensive Cenozoic fossil record. However, studies of extinct organisms have focused largely on richness [12,22,23,42,43] or morphology [44], with limited focus on community composition [20,22]. Because changes in biological communities are not always associated with changes in richness, spatiotemporal patterns of community composition may be better indicators of climate change response [13,18].

We propose that integrating the study of fossil, modern, and projected spatiotemporal patterns of community composition i) allows for the testing of ecological principles in the temporal dimension, ii) provides the most complete picture of diversity responses to climate change, and iii) enables evaluation of the performance of commonly employed CSMs. Our approach of combining the study of fossil, modern, and projected diversity patterns provides novel insights into the ecological and evolutionary processes that drive continental patterns of biodiversity in space and time.

Methods

Data collection and preparation

We downloaded occurrences for modern North American mammals from NatureServe Canada. The extant mammal dataset included 744 species after the exclusion of a small number of unreadable or corrupted files [45]. We restricted our study of fossil mammals to the late Eocene through Pleistocene, thus avoiding the confounding effects of the early Paleogene mammal radiation. We partitioned the fossil mammal occurrence data by North American Land Mammal Age (NALMA) subdivisions because they delineate relatively temporally stable community assemblages and allowed us to obtain a nearly continuous sequence of mammal community change without large intervening gaps. Using NALMA subdivisions leads to time averaging of mammal communities and to differences in sampling (i.e. intensity, geographic coverage etc.) among time periods. However, we use a statistical approach to reduce these biases, described below. We based the dates for all NALMA subdivisions on Woodburne (2004). Further, we combined data for the entire Clarendonian and excluded for the Whitneyan, late Late Hemphillian, and early Chadronian due to poor sampling (Table 1).

We downloaded fossil mammal occurrence data for the Eocene, Oligocene, Pliocene, and Pleistocene from the the Paleobiology Database using the Fossilworks Gateway (fossilworks.org) in July and August, 2012, using the group name 'mammalia' and the following parameters: time intervals = Cenozoic, region = North America, paleoenvironment = terrestrial (primary contributor: John Alroy; literature sources summarized in Appendix S1). We downloaded Miocene mammal occurrence data from the Miocene Mammal Mapping Project in March 2011 [46] using the NALMA subdivision as our search criterion. For all analyses, with the exception of the Miocene, we used paleolatitudes and paleolongitudes. We chose to use MIOMAP for the Miocene data because it is the most complete Miocene dataset. However, MIOMAP does not provide paleo-coordinates. Fortunately, there are only small differences between modern and Miocene latitudes for the downloaded localities. We removed all taxa with equivocal species identifications (e.g. *Equus* sp.) unless they were the only occurrence for a genus. We assumed all occurrences of open nomenclature (e.g. *Equus* cf. *simplicidens*) were correct identifications.

We did not use latitudinal grids for fossil or extant mammals as in previous studies of latitudinal richness gradients [1,47] because our study is focused on community composition. We therefore do not need to clump localities by spatial proximity to employ rarefaction methods. In addition, the uneven spatial distribution of fossil localities makes the use of a grid method impractical. Instead,

Table 1. Summary of sampled North American Land Mammal Age (NALMA) subdivisions.

Epoch	NALMA subdivision	Age Range (Ma)	Midpoint Age (M)	Number of species	Number of fossil localities	Area (km^2)
Pleistocene	Rancholabrean	0.25–0.011	0.1305	222	180	176615.9
Pliocene	Irvingtonian II	0.85–0.25	0.55	189	94	144745.5
Pliocene	Irvingtonian I	1.72–0.85	1.285	102	37	60361.4
Pliocene	Blancan V	2.5–1.72	2.11	165	130	125042.6
Pliocene	Blancan III	4.1–2.5	3.3	183	163	122839.5
Pliocene	Blancan I	4.9–4.1	4.5	85	66	140433.4
Miocene	Early late Hemphillian	6.7–5.9	6.3	68	46	20108.2
Miocene	Late early Hemphillian	7.5–6.7	7.1	63	55	29446.7
Miocene	Early early Hemphillian	9–7.5	8.25	65	47	31455.8
Miocene	Clarendonian	12.5–9	10.75	104	90	36139.8
Miocene	Late Barstovian	14.8–12.5	13.6	195	194	33789.1
Miocene	Early Barstovian	15.9–14.8	15.5	150	168	51753.3
Miocene	Late Hemingfordian	17.5–15.9	16.7	100	83	25478.4
Miocene	Early Hemingfordian	18.8–17.5	18.15	107	105	45531.3
Miocene	Late late Arikareean	19.5–18.8	19.15	108	123	38307.2
Oligocene/Miocene	Early late Arikareean	23.8–19.5	21.65	71	67	37892.2
Oligocene	Late early Arikareean	27.9–23.8	25.85	95	65	20927.8
Oligocene	Early early Arikareean	30–27.9	28.95	116	124	15382.3
Oligocene	Late Orellan	33.1–32	32.55	38	36	17725.7
Oligocene	Early Orellan	33.7–33.1	33.4	88	130	5579.8
Eocene	Middle Chadronian	35.7–34.7	35.3	88	37	10349.7

we created taxon-by-locality occurrence matrices for extant and fossil mammals at the species taxonomic level excluding *Homo sapiens* [20,22]. In all cases, taxa and localities with fewer than two occurrences were removed from the dataset. Final numbers of localities and species are summarized in Table 1.

To make direct comparisons with modern mammals, we created occurrence matrices for extant mammals by pseudo fossil localities, which were generated using an iterative procedure in R with the maptools, sp, gpclib, ggplot2, rgeos, and MASS packages [48–54] (contact corresponding author for R code). To generate pseudo fossil localities and to ensure that we created pseudo fossil localities with the same spatial distributions as the fossil localities, we fit frequency distributions (normal, gamma, or β) to fossil localities for each NALMA subdivision (Fig. S1). We then generated point samples based on the frequency distributions and the number of fossil localities from which we created occurrence matrices (taxon-by-pseudo locality), repeating the procedure 100 times for each NALMA sub-age for a total of 2100 occurrence matrices. Fossil localities do not record the entire community and so show reduced richness compared to the actual communities (however, note that time averaging also increases richness at fossil localities). Further, most fossil localities, unless intensively screen washed, are biased against small species. Therefore, we also intentionally tested for the effects of sampling bias by removing 25%, 50%, and 75% of species from the extant mammal occurrence matrices for a total of 6300 occurrence matrices. Further, we tested for the effects of body mass bias by 25%, 50%, and 75% of species smaller than 5 kg for a total of 6300 occurrence matrices.

Climate space models

To create climate space models, we sampled the ranges of extant North and South American mammals at a series of 5066 points corresponding to a 1° grid (which we only used to project mammal occurrences under climate change models, but not to calculate biodiversity). Due to the focus on North America, we omitted any species with southern hemisphere ranges that did not cross the equator (n = 602; Table S2). We also excluded rare species (present in <20 cells) for which accurate species distribution models could not be generated (n = 361), leaving 706 species for the climate change projections. We extracted mean annual and winter (December, January, February) temperature and mean annual precipitation data from Climate Wizard (www.climatewizard.org) for the period of 1951–2006 and the following SRES scenarios and time periods: B1 2050s, A1b 2050s, A1b 2080s, A2 2050s, and A2 2080s [55] (Table S1). Each of these projections is based on an ensemble of 16 global circulation models [56]. However, to ensure that we sampled a range of potential warming, we also extracted the ensemble lowest B1 2050s projection (hereafter "B1 2050s low") and the ensemble highest A2 2080s projection (hereafter "A2 2080s high"). This gave a range of warming in North America from 1.49°C (B1 2050s low) to 6.78°C (A2 2080s high, see Table S1 for the full range).

We modeled species' ranges with the BIOMOD package in R using generalized linear models, generalized boosted models, classification tree analysis, artificial neural networks, surface range envelopes, flexible discriminant analysis, multiple adaptive regression splines, and random forests [57] (contact corresponding author for R code). We then used these models to make consensus forecasts for each of the projections described above, as well as current climate to evaluate the performance of the models. We tested model performance using area under the receiver operating curve (AUC), true skill statistic (TSS), and proportion correct classification (PCC, Fig. S2). Species and generic presences were determined across the 1° latitude-longitude grid to give presence or absence in each location at each time and SRES scenario.

Using the projections described above, we created pseudo localities, as before. From this, we created occurrence matrices as described above. We repeated this process 100 times for each projection for a total of 16,800 occurrence matrices.

Latitudinal turnover gradients (LTGs) and β diversity

We calculated β diversity as the change in mammalian communities across the North American landscape using multivariate dispersion and the Jaccard index for each NALMA subage, for modern mammals, and for the climate projections [58]. We calculated Euclidean distances from the centroid for localities using the R package vegan [59]. Larger distances from the centroid indicate greater spatial community turnover and thus higher β diversity. We did not regress the Jaccard index values against distance, as has been used for modern species [2] because we have found such an approach to be highly influenced by species-area relationships.

To estimate ancient, modern, and projected LTG strength for North American mammals, we calculated the amount of community change with latitude using detrended correspondence analysis (DCA; an ordination technique) in the vegan R package [59]. We used explained variance (R^2; how much of the variation in community change is explained by latitude) as a measure of LTG strength [13]. High values of explained variance indicate strong LTGs [60]. We did not compute latitudinal richness gradients because sampling bias (e.g. loss of taxa, body mass bias) is too great (Fraser, D. unpub.).

Sampling bias control

Although we have chosen methods that minimize the effects of sampling bias, we still used multiple methods to control for the non-independence of β diversity from the number of localities, the geographic area sampled, and the number of sampled taxa. We used three approaches. Firstly, we used a re-sampling approach wherein we sub-sampled (without replacement) each NALMA 100× using a standardized number of localities (thirty) and limited to localities occurring between 30°and 50° North latitude. We also re-sampled the extant mammal ranges under various conditions of bias (taxonomic bias through the removal of 25%, 50%, 75% of taxa and body mass bias where we removed 25%, 50%, and 75% of species with a body mass lower than 5 kg) as above to test for direct causality of sampling bias. We also used a method of detrending whereby we regressed LTG strength and β diversity against statistically significant sampling bias metrics and further analyzed the residuals from the model. Finally, we used multivariate linear models to simultaneously account for the model variance explained by sampling and biological phenomena. The last multivariate method is similar to [61] and [62] (also addressed in [63]) who combine the predictive properties of models of biodiversity change and taphonomic bias.

Correlation with climate

We tested for correlations of β diversity and LTG strength with stable oxygen isotopes from benthic foraminifera ($\delta^{18}O$ ‰) [64,65], mean annual precipitation estimated from paleosols [66], number of localities, sampling area (km^2), number of species, latitudinal range (degrees), and length of the sampled interval (Ma) of the fossil localities using generalized least squares and using an autocorrelation structure of order one (corAR1) to account for temporal autocorrelation in R [67,68]. Best fit models were selected using automated model selection in the MuMIn R package [69] and the Akaike Information Criterion (ΔAIC).

Results

Fossil mammal β diversity showed considerable variation with the warmest intervals (late Eocene, mid-late Oligocene, mid Miocene, and mid Pliocene), but showing generally higher β diversity than with cooler intervals (early Oligocene, late Miocene) (Fig. 1C). The best fit model includes mean annual precipitation (MAP squared), length of the NALMA subdivision, and number of taxa, which together accounts for 67% of model variance (Table 2). β diversity is statistically significant for all three predictors ($p<0.05$). Residual β diversity is significantly explained by MAP only (Table 2; Fig. 2B). Re-sampling did not alleviate the effects of sampling bias; re-sampled β diversity is significantly explained by MAP-squared, number of taxa, and NALMA subdivision length (Table 2). The remainder of the manuscript will discuss the results from the analyses of raw and residual β diversity only.

Mammalian latitudinal turnover gradients (LTGs) are weak prior to the late Miocene (Fig. 1D). Raw LTG strength (i.e. not detrended) peaks during late Miocene (Hemphillian) and late Pleistocene (Rancholabrean) (Fig. 1D). The best fit model includes mean annual precipitation (MAP) [66], number of taxa, area (km^2) and an the interaction of area and the number of taxa, which explains 47% of the model variance (Table 2; Fig. 2C). LTG strength of late Cenozoic mammal species is statistically significantly explained by all four metrics ($p<0.001$; Table 2). Residual LTG strength is significantly explained only by MAP ($p<0.05$; Table 2; Fig. 2D). As above, re-sampling did not alleviate the effects of sampling bias on LTG strength (Table 2). In other words, even accounting for variables that describe potential sources of bias, a climatic variable (MAP) still explains a significant proportion of the variance.

β diversity is much lower for extant mammals than for extinct mammals (Fig. 3A). LTG strength for extant mammals is also greater than for early to mid Cenozoic fossil mammals, but similar to the values for the late Miocene and Pleistocene (Fig. 3B). Extant mammal β diversity shows a slight decrease under incomplete sampling and a slight increase under body-mass–bias sampling (Fig. 3A), but the change is much smaller than observed for fossil mammals. LTG strength does not appear to be significantly affected by the sample size reduction.

Our forecast models (which showed a strong fit to modern mammalian distributions, see Fig. S2A–C) show a slight increase in β diversity for extant mammals (Fig. 3C), but no substantial change in LTG strength compared to the present (Fig. 3D).

Discussion

Spatiotemporal patterns of β diversity remain poorly studied despite being potentially very useful in conservation biology [17,18,70] and linkage to well-studied biogeographic phenomena such as latitudinal richness gradients [4]. Using an extensive analysis of past and present mammalian communities, we demonstrate that, over the past 36 Ma, spatiotemporal patterns of mammal community composition have varied by orders of magnitude in North America. Specifically, Cenozoic spatial turnover of mammal communities is explained by continental mean annual precipitation (MAP) (Fig. 2A–B), broadly supporting predictions drawn from published studies of modern terrestrial organisms [2,70,71] and our predictions outlined above.

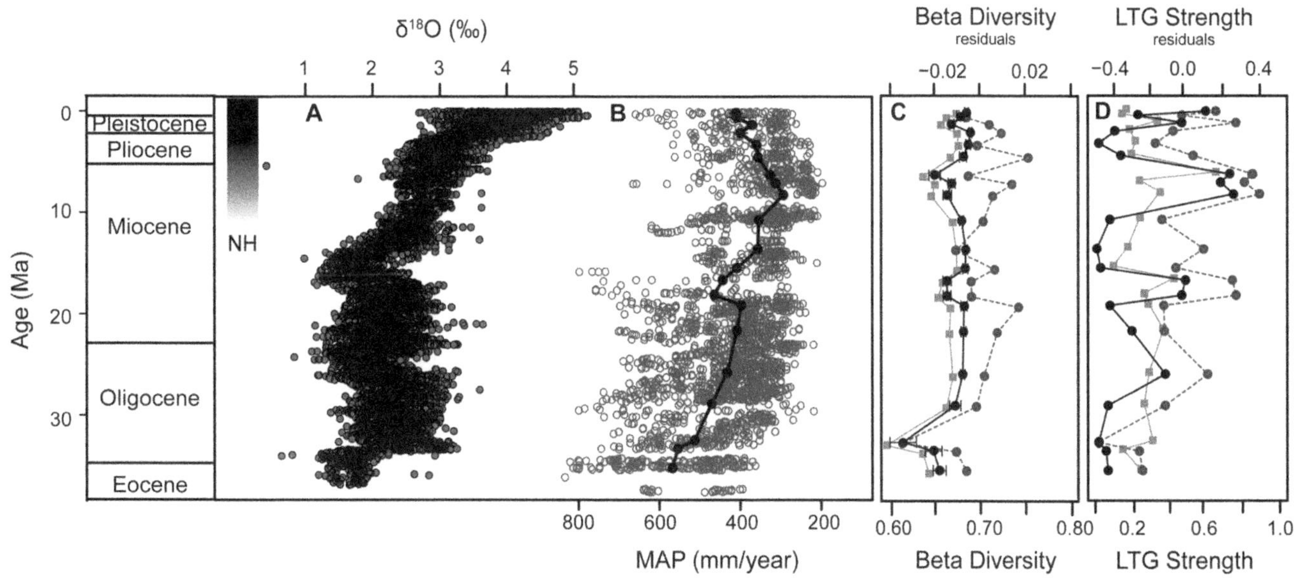

Figure 1. Mid to late Cenozoic trends of (A) $\delta^{18}O$ (‰) from benthic foraminifera (Zachos et al. 2008), (B) mean annual precipitation estimated from paleosols (Retallack, 2007), (C) β diversity of North American mammal species measured using multivariate dispersion (average distance from the centroid), and (D) strength of latitudinal turnover gradients (LTGs) measured as gradient strength for North American fossil mammals. Black lines are raw values, gray lines are residuals from significant sampling bias predictors, and gray dashed lines are re-sampled. Standard errors for re-sampled data are too small to display.

Contemporary ecological theory predicts that mammal diversity either declines monotonically with productivity or shows a unimodal pattern, declining with both low and high productivity [1,2,70,72]. Further, stronger latitudinal diversity gradients are associated with cooler, less productive environments [71] and steeper latitudinal climate gradients [1,70]. Both sets of predictions assume that changes in climate, productivity, and seasonality influence rates of origination and extinction [72,73], niche breadths [74], as well as the carrying capacity of the ecosystem [75], all factors that change the spatial turnover of terrestrial faunas [70]. Specifically, terrestrial organisms in low latitude, high productivity environments show low rates of speciation and extinction [73], high β diversity [2,76], and weak or absent latitudinal diversity gradients [71]. In contrast, high latitude organisms show high rates of speciation and extinction [73], low β diversity [2,76], and strong latitudinal diversity gradients [71]. Evolutionary history also plays a role in determining rates of spatial community turnover. Modern tropical organisms show

Table 2. Results of best fit generalized least squares models relating β diversity and latitudinal turnover gradient (LTG) strength to mean annual precipitation from paleosols (Retallack, 2007), $\delta^{18}O$ (‰) from benthic forams (mm/year; Zachos et al. 2001; 2008), length of North American Land Mammal Age subdivision, number of taxa sampled, sampling area (km^2), and number of fossil localities.

Dependent Variable	Parameters of Best Fit Model	Variance explained by model (%)	t value	p
Beta Diversity	Mean annual precipitation (quadratic)	66.51	−3.25	0.005
	Length of NALMA subdivision		2.43	0.027
	Number of taxa		5.30	<0.001
Beta Diversity Residuals	Mean annual precipitation (quadratic)	26.48	−3.50	0.002
Beta Diversity Re-sampled	Mean annual precipitation (quadratic)	66.04	−2.39	0.029
	Length of NALMA subdivision		2.51	0.023
	Number of taxa		5.47	<0.001
Latitudinal Turnover Gradient Strength (LTGs)	Mean annual precipitation (quadratic)	46.76	−5.65	<0.001
	Area		−4.62	<0.001
	Number of taxa		−4.36	<0.001
	Area : Number of taxa		4.85	<0.001
LTG Residuals	Mean annual precipitation (linear)	37.48	−3.79	0.001
LTG Re-sampled	Number of taxa	28.59	−2.55	0.020

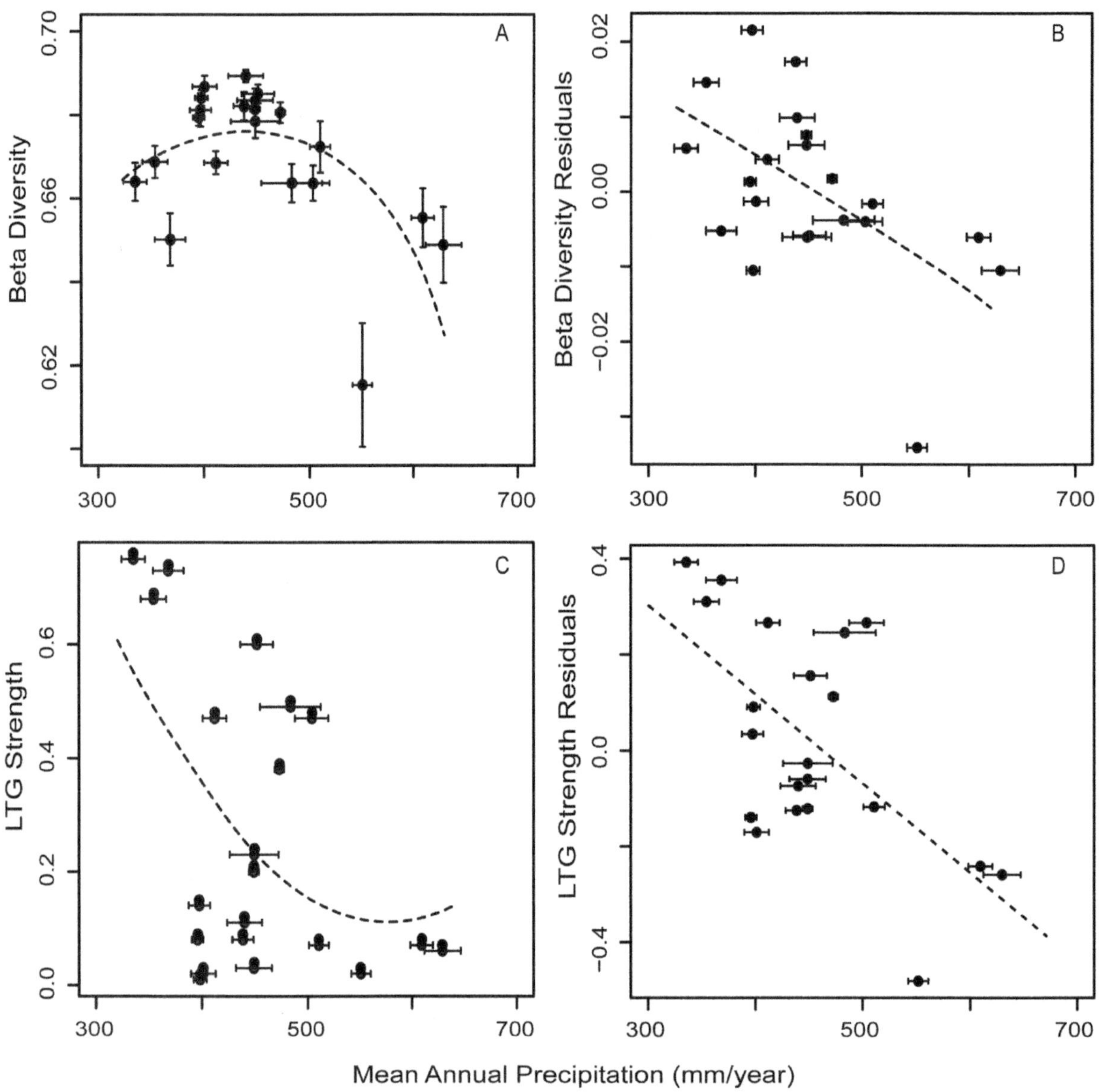

Figure 2. Relationship of mean annual precipitation estimated from paleosols (Retallack, 2007) with North American fossil mammal (A) raw β diversity ($R^2 = 0.43$), (B) residual beta diversity ($R^2 = 0.26$) and (C) raw latitudinal turnover gradient (LTG) strength ($R^2 = 0.25$), and (D) residual LTG strength ($R^2 = 0.37$).

faster turnover than their temperate counterparts regardless of the rate of environmental change [70]. Spatial and, by extension, temporal patterns of β diversity are the result of a mosaic of ecological and evolutionary processes.

Cenozoic fossil mammal β diversity peaked at intermediate values of mean annual precipitation and declined under both drier and wetter conditions (MAP; ~450 mm per year; Fig. 2B), showing a similar shape to latitudinal diversity curves for modern mammals [71]. Mammal β diversity was similarly lowest during periods of relative cooling, including the early Oligocene and late Miocene, coincident with declining atmospheric CO_2 [77–80] and, in the latter case, the expansion of ice sheets in the Northern Hemisphere [27,36], strengthening of thermohaline circulation [27,37,81–84], and transition from C_3 to C_4 dominated ecosystems at middle latitudes [66,85,86]. Declining β diversity during the late Miocene is also coincident with increased maximum body mass [87], an ecologically relevant characteristic linked to lower ecosystem energy [88,89]. Water is a key component in photosynthesis and therefore net primary productivity (NPP) and MAP are correlated at a global scale, showing an asymptotic relationship [90]. Our results therefore suggest that putatively lower energy ecosystems (e.g. early Oligocene, late Miocene) supported more spatially homogenous mammal faunas than putatively higher energy ecosystems (e.g. late Eocene, mid Miocene, mid Pliocene). Temporal changes in fossil mammal β diversity (this study) are therefore conceptually similar to spatial patterns observed in extant mammals.

Early Oligocene mammals had lower β diversity than expected based on MAP (Fig. 1C; Fig. 2A). The early Oligocene is associated with rapid global cooling [64] and expansion of open grassy ecosystems [91], which may have resulted in lower ecosystem energy. However, our taxonomic sample is the poorest

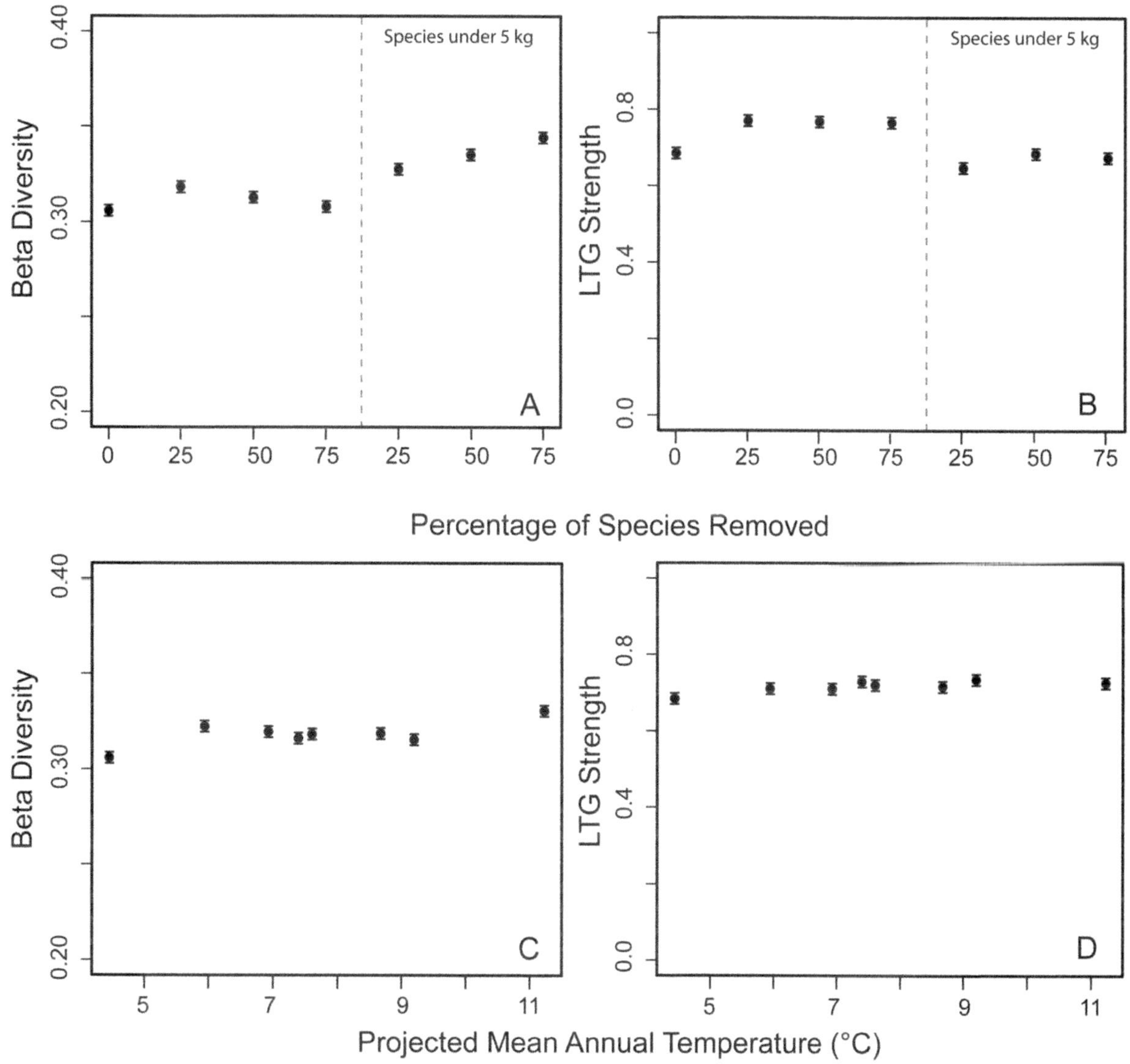

Figure 3. (A) β diversity (distance from centroid) and (B) latitudinal turnover gradients (LTG) strength of extant North American mammals under incomplete taxonomic sampling (removal of 25, 50, and 75% of species in sample) and body mass bias (removal of 25, 50, 75% of species smaller than 5 kg) and (C) β diversity (distance from centroid) and (D) latitudinal turnover gradients (LTG) strength of extant North American mammals under several International Panel on Climate Change scenarios (Special Reports on Emissions Scenarios).

for the early Oligocene; number of taxa is a significant predictor of fossil mammal β diversity (Table 2), suggesting some variation in preservation of species among NALMA subdivisions. Rarefied diversity also shows little change from the late Eocene to the early Oligocene [10]. However, our incomplete sampling trials show that removing even 75% of species reduces β diversity by a negligible amount (Fig. 3A), suggesting that at least some (but not all) of the observed decline in early Oligocene β diversity may have been climatically driven.

The magnitude of the latitudinal turnover gradient (LTG) for fossil mammals shows a temporally cyclic pattern that increases in amplitude during the late Cenozoic as well as a general trend toward stronger LTGs (Fig. 1D), coincident with the formation of ice on Svalbard at ~15 Ma and perennial Arctic sea ice at ~14 Ma, declining atmospheric CO_2 [37], and declining terrestrial MAP (Fig. 2B). Specifically, LTGs are strongest when precipitation is lowest (putatively lower productivity environments) and weakest at when precipitation is highest (putatively high productivity environments; Fig. 2B), similar to modern mammals that show weak or absent latitudinal diversity gradients in the tropics and strong diversity gradients at mid to high latitudes [71]. Climate gradients are steeper at mid to high latitudes in North America due to the albedo of high latitude glaciation. Northern glaciation is an important means by which solar radiation is reflected from high latitudes, resulting in cool, low productivity Arctic environments [92,93]. Mammal communities are sorted along a latitudinal axis according to their climatic tolerances and the process of abiotic filtering, whereby taxa meet the limits of their environmental tolerances and are excluded from communities farther north [94]. Although late Miocene sea and land ice thickness and extent were reduced compared to the modern, increasing northern albedo and strengthening of thermohaline circulation are coincident with that strengthening of mammal

LTGs during the late Miocene (25–60% stronger than for any preceding NALMA; Fig. 1D) [27,81–84].

At first glance, the Pliocene appears to be anomalous because the magnitude of the mammalian LTG declines dramatically (60–70% reduction in the magnitude of the LTG; Fig. 1D). However, evidence from fossil deposits on Ellesmere Island show that approximately 3.5 Ma the Pliocene Arctic was ~14–22°C warmer than present [83,95,96] with an associated reduced volume of Arctic sea ice [27,82]. Pliocene Arctic warming is similarly coincident with reduced richness gradients of marine zooplankton [81]. The Pliocene might therefore be the "exception" that proves the rule.

Under modern global warming, Arctic winter temperatures have increased at a greater rate than at southern latitudes [97]. Long-term projections suggest boosts in high latitude net primary productivity due to increasing nitrogen fertilization and increases in mean annual precipitation of 100–150 mm per year or 5–20% at middle to high latitudes [98]. From our analyses of fossil North American mammals and published studies of beta diversity [18], we therefore expect weakened climate gradients and thus weakened LTGs due to northward range shifting, and, in the long-term, declining β diversity under the influence of modern anthropogenic climate change. β diversity decline may be facilitated by the homogenization of communities due to any of the following (note the lack of mutual exclusivity): i) extinction of species with small geographic ranges and replacement with wide-ranging species, ii) evolution toward larger range sizes within species, and, iii) invasion by wide-ranging species even without the extinction of residents [18]. However, our climate space models that are based on SRES scenarios corresponding to absolute mean annual temperatures of 4.4–11.2°C (averaged across North and South America) did not show changes in mammal LTGs or β diversity (Fig. 3C–D). We suggest that climate space models (CSMs) are unlikely to accurately forecast the outcomes of anthropogenic climate change for modern mammals because current CSM algorithms do not incorporate microevolutionary, macroevolutionary, or ecological processes, such as niche shifts, niche creation, and differences in dispersal abilities that are inherent in the response of animals to climate change. However, even on modern ecological timescales, rapid evolutionary changes and niche shifts have been observed in native and invasive populations [41], and this local adaptation complicates the prediction of range shifts. On longer timescales, taxa adapt to new climates and the processes of speciation and extinction help form new terrestrial communities. Without the explicit inclusion of evolutionary parameters and historical data for the taxa of interest, we are unlikely to accurately predict long-term changes in terrestrial biodiversity patterns.

We have shown here that macroecological patterns of North American mammal community composition varied considerably over the past 35 million years in response to changes in global climate change and Arctic glaciation (Fig. 1C–D). Furthermore, our comparison of fossil evidence with climate-space forecast models (CSMs) suggests that CSMs (in which species are modeled to simply track climate variables) may distort the degree of community composition change we should expect in the future. A unifying ecological theory relating diversity to climate must address both the spatial and temporal dimensions of diversity, as well as both richness and community composition. However, studies of organismal richness are far more common than studies of community composition (β diversity), despite the importance of the latter in conservation and their vast potential for contributing to our understanding of the processes underlying modern biodiversity. Studying the community composition of fossil animals represents a new frontier in paleontological research with potential to truly inform modern conservation.

Supporting Information

Figure S1 Maps of North America showing the distribution of fossil localities for all sampled North American Land Mammal Age subdivisions.

Figure S2 Model fit statistics for climate space models of extant North American mammals. Model performance was tested using area under the operating curve (A; AUC), the true skill statistics (B; TSS), and the proportion of correct classification (C).

Table S1 Summary of Special Emissions Report Scenarios (SERs) to which we fit climate models for extant mammalian species.

Table S2 List of mammalian taxa included and excluded from the species distribution models.

Appendix S1 Sources for the majority of mammal occurrence data downloaded from the Fossilworks database.

Acknowledgments

We thank John P. Hunter for a thorough review of this paper. Further, we thank John Alroy for his substantial contributions to the fossil data used in this analysis, accessed via his Fossilworks website, and his detailed review of the paper. We would also like to thank two anonymous reviewers, D. Currie, M. Clementz, M. Churchill, R. Haupt, J. Hoffmann, and E. Lightner for reviewing earlier versions of this manuscript, as well as L. Fahrig and S. Kim for constructive comments on this project.

Author Contributions

Conceived and designed the experiments: DF CH NR. Performed the experiments: DF. Analyzed the data: DF. Contributed reagents/materials/analysis tools: CH RG. Contributed to the writing of the manuscript: DF. Manuscript copyediting: CH RG NR.

References

1. Hawkins BA, Field R, Cornell HV, Currie DJ, Guegan JF, et al. (2003) Energy, water, and broad-scale geographic patterns of species richness. Ecology 84: 3105–3117.
2. Qian H, Badgley C, Fox DL (2009) The latitudinal gradient of beta diversity in relation to climate and topography for mammals in North America. Global Ecology and Biogeography 18: 111–122.
3. Condit R, Pitman N, Leigh EG Jr, Chave J, Terborgh J, et al. (2002) Beta-diversity in tropical forest trees. Science 295: 666–669.
4. Baselga A, Lobo JM, Svenning JC, Aragón P, Araújo MB (2012) Dispersal ability modulates the strength of the latitudinal richness gradient in European beetles. Global Ecology and Biogeography 21: 1106–1113.
5. Engle VD, Summers JK (1999) Latitudinal gradients in benthic community composition in Western Atlantic estuaries. Journal of Biogeography 26: 1007–1023.
6. Condamine FL, Sperling FAH, Wahlberg N, Rasplus JY, Kergoat GJ (2012) What causes latitudinal gradients in species diversity? Evolutionary processes and ecological constraints on swallowtail biodiversity. Ecology Letters 15: 267–277.

7. Currie DJ, Fritz JT (1993) Global patterns of animal abundance and species energy use. Oikos 67: 56–68.
8. Currie DJ, Francis AP, Kerr JT (1999) Some general propositions about the study of spatial patterns of species richness. Ecoscience 6: 392–399.
9. Mittelbach GG, Schemske DW, Cornell HV, Allen AP, Brown JM, et al. (2007) Evolution and the latitudinal diversity gradient: speciation, extinction and biogeography. Ecology Letters 10: 315–331.
10. Figueirido B, Janis CM, Pérez-Claros JA, Renzi MD, Palmqvist P (2012) Cenozoic climate change influences mammalian evolutionary dynamics. Proceedings of the National Academy of Sciences USA 109: 722–727.
11. Sepkoski JJ (1998) Rates of speciation in the fossil record. Philosophical Transactions of the Royal Society of London B 353: 315–326.
12. Mayhew PJ, Bell MA, Benton TG, McGowan AJ (2012) Biodiversity tracks temperature over time. Proceedings of the National Academy of Sciences USA 109: 15141–15145.
13. Kent R, Bar-Massada A, Carmel Y (2011) Multiscale analyses of mammal species composition-environment relationship in the contiguous USA. PLoS One 6: e25440.
14. Legendre P, Borcard D, Peres-Neto PR (2005) Analyzing beta diversity: partitioning the spatial variation of community composition data. Ecological Monographs 75: 435–450.
15. Whittaker RJ, Willis KJ, Field R (2001) Scale and species richness: towards a general, hierarchical theory of species diversity. Journal of Biogeography 28: 453–470.
16. Soininen J (2010) Species turnover along abiotic and biotic gradients: patterns in space equal patterns in time? BioScience 60: 433–439.
17. Hassall C, Hollinshead J, Hull A (2012) Temporal dynamics of aquatic communities and implications for pond conservation. Biodiversity and Conservation 21: 829–852.
18. Dornelas M, Gotelli NJ, McGill B, Shimadzu H, Moyes F, et al. (2014) Assemblage time series reveal biodiversity change but not systematic loss. Science 344: 296–299.
19. Janis CM, Damuth J, Theodor JM (2000) Miocene ungulates and terrestrial primary productivity: where have all the browsers gone? Proceedings of the National Academy of Sciences USA 97: 7899–7904.
20. Atwater AL, Davis EB (2011) Topographic and climate change differentially drive Pliocene and Pleistocene mammalian beta diversity of the Great Basin and Great Plains provinces of North America. Evolutionary Ecology Research 13: 833–850.
21. Finarelli JA, Badgley C (2010) Diversity dynamics of Miocene mammals in relation to the history of tectonism and climate. Proceedings of the Royal Society of London, Series B 277: 2721–2726.
22. Davis EB (2005) Mammalian beta diversity in the Great Basin, western USA: palaeontological data suggest deep origin of modern macroecological structure. Global Ecology and Biogeography 14: 479–490.
23. Barnosky AD, Hadly EA, Bell CJ (2003) Mammalian response to global warming on varied temporal scales. Journal of Mammalogy 84: 354–368.
24. Barnosky AD (2005) Effects of Quanternary climatic change on speciation in mammals. Journal of Mammalian Evolution 12: 247–264.
25. Eberle J, Fricke H, Humphrey J (2009) Lower-latitude mammals as year-round residents in Eocene Arctic forests. Geology 37: 499–502.
26. Eberle JJ, Fricke HC, Humphrey JD, Hackett L, Newbrey MG, et al. (2010) Seasonal variability in Arctic temperatures during early Eocene time. Earth and Planetary Science Letters 296: 481–486.
27. Polyak L, Alley RB, Andrews JT, Brigham-Grette J, Cronin TM, et al. (2010) History of sea ice in the Arctic. Quaternary Science Reviews 29: 1757–1778.
28. Clementz MT, Sewall JO (2011) Latitudinal gradients in greenhouse seawater δ18O: evidence from Eocene sirenian tooth enamel. Science 332: 455–458.
29. Micheels A, Bruch A, Mosbrugger V (2009) Miocene climate modelling sensitivity experiments for different CO_2 concentrations. Palaeontologia Electronica 12: 5A.
30. Post E, Forchhammer MC, Bret-Harte MS, Callaghan TV, Christensen TR, et al. (2009) Ecological dynamics across the Arctic associated with recent climate change. Science 325: 1355–1358.
31. Primack RB, Ibáñez I, Higuchi H, Lee SD, Miller-Rushing AJ, et al. (2009) Spatial and interspecific variability in phenological responses to warming temperatures. Biological Conservation 142: 2569–2577.
32. Bradshaw WE, Holzapfel CM (2006) Evolutionary response to rapid climate change. Science 312: 1477–1478.
33. Parmesan C (2006) Ecological and evolutionary responses to recent climate change. Annual Review of Ecology and Systematics 37: 637–639.
34. Chen IC, Hill JK, Ohlemüller R, Roy DB, Thomas CD (2011) Rapid range shifts of species associated with high levels of climate warming. Science 333: 1024–1026.
35. Parmesan C, Yohe G (2003) A globally coherent fingerprint of climate change impacts across natural systems. Nature 421: 37–42.
36. Foster GL, Lunt DJ, Parrish RR (2009) Mountain uplift and the threshold for sustained Northern Hemisphere glaciation. Climate of the past discussions 5: 2439–2464.
37. Foster GL, Lear CH, Rae JWB (2012) The evolution of pCO_2, ice volume and climate during the middle Miocene. Earth and Planetary Science Letters 341–344: 243–254.
38. Lawler JJ, White D, Neilson RP, Blaustein AR (2006) Predicting climate-induced range shifts: model differences and model reliability. Global Change Biology 12: 1568–1584.
39. Hoffmann AA, Sgró CM (2011) Climate change and evolutionary adaptation. Nature 470: 479–485.
40. Thuiller W, Münkemüller T, Lavergne S, Mouillot D, Mouquet N, et al. (2013) A road map for integrating eco-evolutionary processes into biodiversity models. Ecology Letters 16: 94–105.
41. Lavergne S, Mouquet N, Thuiller W, Ronce O (2010) Biodiversity and climate change: integrating evolutionary and ecological responses of species and communities. Annual Review of Ecology, Evolution, and Systematics 41: 321–350.
42. Sepkoski JJ (1997) Biodiversity: past, present, and future. Journal of Paleontology 71: 533–539.
43. Rose PJ, Fox DL, Marcot J, Badgley C (2011) Flat latitudinal gradient in Paleocene mammal richness suggests decoupling of climate and biodiversity. Geology 39: 163–166.
44. Secord R, Bloch JI, Chester SGB, Boyer DM, Wood AR, et al. (2012) Evolution of the earliest horses driven by climate change in the Paleocene-Eocene thermal maximum. Science 335: 959–962.
45. Patterson BD, Ceballos G, Sechrest W, Tognelli MF, Brooks T, et al. (2007) Digital distribution maps of the mammals of the Western Hemisphere, version 3.0. NatureServe, Arlington, Virginia, USA.
46. Carrasco MA, Kraatz BP, Davis EB, Barnosky AD (2005) Miocene mammal mapping project (MIOMAP). University of California Museum of Paleontology.
47. McCoy ED, Connor EF (1980) Latitudinal gradients in the species diversity of North American mammals. Evolution 34: 193–203.
48. Lewin-Koh NJ, Bivand R (2008) maptools package version 0.8–16.
49. Pebesma EJ, Bivand RS (2005) Classes and methods for spatial data in R. R News 5.
50. Bivand RS, Pebesma EJ, Gomez-Rubio V (2008) Applied spatial data analysis with R. New York: Springer.
51. Peng RD (2007) The gpclib package version 1.5–5.
52. Wickham H (2009) ggplot2: elegant graphics for data analysis. New York: Springer.
53. Bivand R, Rundel C (2012) rgeos: interface to geometry engine version 0.2–16.
54. Venables WN, Ripley BD (2002) Modern and applied statistics with S. New York: Springer.
55. Nakicenovic N, Swart R (2000) Emissions scenarios: a special report of Working Group III of the Intergovernmental Panel on Climate Change. Cambridge: Cambridge University Press.
56. Girvetz EH, Zganjar C, Raber GT, Maurer EP, Kareiva P, et al. (2009) Applied climate-change analysis: the climate wizard tool. PLoS One 4: e8320.
57. Thuiller W, Georges D, Engler R (2012) BIOMOD: Ensemble platform for species distribution modeling. Ecography 32: 369–373.
58. Anderson MJ, Ellingsen KE, McArdle BH (2006) Multivariate dispersion as a measure of beta diversity. Ecology Letters 9: 683–693.
59. Oksanen J, Blanchet FG, Roeland Kindt PL, Minchin PR, O'Hara RB, et al. (2012) Package vegan version 2.0–7.
60. Tuomisto H, Ruokolainen K (2006) Analyzing and explaining beta diversity? understanding the targets of different methods of analysis. Ecology 87: 2697–2708.
61. Benson RBJ, Mannion PD (2012) Multi-variate models are essential for understanding vertebrate diversification in deep time. Biology Letters 8: 127–130.
62. Mannion PD, Upchurch P, Carrano MT, Barrett PM (2011) Testing the effect of the rock record on diversity: a multidisciplinary approach to elucidating the generic richness of sauropodomorph dinosaurs through time. Biological Reviews 86: 157–181.
63. Benton MJ, Dunhill AM, Lolyd GT, Marx FG (2011) Assessing the quality of the fossil record: insights from vertebrates. In: A. J McGowan and A. B Smith, editors. Comparing the geological and fossil records: implications for biodiversity studies. London: Geological Society of London. 63–94.
64. Zachos JC, Dickens GR, Zeebe RE (2008) An early Cenozoic perspective on greenhouse warming and carbon-cycle dynamics. Nature 451: 279–283.
65. Zachos J, Pagani M, Sloan L, Thomas E, Billups K (2001) Trends, rhythms, and aberrations in global climate 65 Ma to present. Science 292: 686–693.
66. Retallack GJ (2007) Cenozoic paleoclimate on land in North America. Journal of Geology 115: 271–294.
67. Development core team R (2012) R: A language and environment for statistical computing. Vienna, Austria: Foundation for Statistical Computing.
68. Dornelas M, Magurran AE, Buckland ST, Chao A, Chazdon RL, et al. (2013) Quantifying temporal change in biodiversity: challenges and opportunities. Proceedings of the Royal Society B 280: 1–10.
69. Bartoń K (2013) Multi-model inference package 'MuMIn' version 1.10.0 (http://cran.r-project.org/web/packages/MuMIn/MuMIn.pdf).
70. Buckley LB, Jetz W (2008) Linking global turnover of species and environments. Proceedings of the National Academy of Sciences USA 105: 17836–17841.
71. Currie DJ (1991) Energy and large-scale patterns of animal- and plant-species richness. American Naturalist 137: 27–49.
72. VanderMeulen MA, Hudson AJ, Scheiner SM (2001) Three evolutionary hypotheses for the hump-shaped productivity–diversity curve. Evolutionary Ecology Research 3: 379–392.

73. Weir JT, Schluter D (2007) The latitudinal gradient in recent speciation and extinction rates of birds and mammals. Science 315: 1574–1576.
74. Vázquez DP, Stevens RD (2004) The latitudinal gradient in niche breadth: concepts and evidence. American Naturalist 164: E1–E19.
75. Buckley LB, Davies J, Ackerly DD, Kraft NJB, Harrison SP, et al. (2010) Phylogeny, niche conservatism and the latitudinal diversity gradient in mammals. Proceedings of the Royal Society, Series B 277: 2121–2138.
76. Qian H, Xiao M (2012) Global patterns of the beta diversity energy relationship in terrestrial vertebrates. Acta Oecologica 39: 67–71.
77. Franks PJ, Beerling DJ (2009) Maximum leaf conductance driven by CO_2 effects on stomatal size and density over geologic time. Proceedings of the National Academy of Sciences USA 106: 10343–10347.
78. DeConto RM, Pollard D, Wilson PA, Pälike H, Lear CH, et al. (2008) Thresholds for Cenozoic bipolar glaciation. Nature 455: 652–657.
79. Tripati AK, Roberts CD, Eagle RA (2009) Coupling of CO_2 and ice sheet stability over major climate transitions of the last 20 million years. Science 326: 1394–1397.
80. Zhang YG, Pagani M, Liu Z, Bohaty SM, DeConto R (2013) A 40-million-year history of atmospheric CO_2. Philosophical Transactions of the Royal Society, Series A 371: 1–20.
81. Yasuhara M, Hunt G, Dowsett HJ, Robinson MM, Stoll DK (2012) Latitudinal species diversity gradient of marine zooplankton for the last three million years. Ecology Letters 15: 1174–1179.
82. Haywood AM, Valdes PJ, Sellwood BW, Kaplan JO, Dowsett HJ (2001) Modelling middle Pliocene warm climates of the USA. Palaeontologia Electronica 4: 1–21.
83. Ballantyne AP, Greenwood DR, Damsté JSS, Csank AZ, Eberle JJ, et al. (2010) Significantly warmer Arctic surface temperatures during the Pliocene indicated by multiple independent proxies. Geology 38: 603–606.
84. Ballantyne AP, Rybczynski N, Baker PA, Harington CR, White D (2006) Pliocene Arctic temperature constraints from the growth rings and isotopic composition of fossil larch. Palaeogeography, Palaeoclimatology, Palaeoecology 242: 188–200.
85. Fox DL, Honey JG, Martin RA, Peláez-Campomanes P (2012) Pedogenic carbonate stable isotope record of environmental change during the Neogene in the southern Great Plains, southwest Kansas, USA: Oxygen isotopes and paleoclimate during the evolution of C_4-dominated grasslands. Geological Society of America Bulletin 124: 431–443.
86. Strömberg CAE, McInerney FA (2011) The Neogene transition from C_3 to C_4 grasslands in North America: assemblage analysis of fossil phytoliths. Paleobiology 37: 50–71.
87. Smith FA, Boyer AG, Brown JH, Costa DP, Dayan T, et al. (2010) The evolution of maximum body size of terrestrial mammals. Science 330: 1216–1219.
88. Freckleton RP, Harvey PH, Pagel M (2003) Bergmann's rule and body size in mammals. American Naturalist 161: 821–825.
89. Blackburn TM, Gaston KJ, Loder N (1999) Geographic gradients in body size: a clarification of Bergmann's rule. Diversity & Distributions 5: 165–174.
90. Del Grosso S, Parton W, Stohlgren T, Zheng D, Bachelet D, et al. (2008) Global potential net primary production predicted from vegetation class, precipitation, and temperature. Ecology 89: 2117–2126.
91. Jacobs BF, Kingston JD, Jacobs LL (1999) The origin of grass-dominated ecosystems. Annals of Missouri Botanical Garden 86: 590–643.
92. Alexeev VA, Langen PL, Bates JR (2005) Polar amplification of surface warming on an aquaplanet in "ghost forcing" experiments without sea ice feedbacks. Climate Dynamics 24: 655–665.
93. Holland MM, Bitz CM (2003) Polar amplification of climate change in coupled models. Climate Dynamics 21: 221–232.
94. Soininen J, McDonald R, Hillebrand H (2007) The distance decay of similarity in ecological communities. Ecography 30: 3–12.
95. Csank AZ, Tripati AK, Patterson WP, Eagle RA, Rybczynski N, et al. (2011) Estimates of Arctic land surface temperatures during the early Pliocene from two novel proxies. Earth and Planetary Science Letters 304: 291–299.
96. Rybczynski N, Gosse JC, Harington CR, Wogelius RA, Hidy AJ, et al. (2013) Mid-Pliocene warm-period deposits in the High Arctic yield insight into camel evolution. Nature Communications 4: 1–9.
97. Kaplan JO, Bigelow NH, Prentice IC, Harrison SP, Bartlein PJ, et al. (2003) Climate change and Arctic ecosystems: 2. Modeling, paleodata-model comparisons, and future projections. Journal of Geophysical Research 108: 1–17.
98. Oechel WC, Vourlitis GL (1994) The effects of climate change on land-atmosphere feedbacks in arctic tundra regions. Trends in Ecology and Evolution 9: 324–329.

7

Impact of Climate Change on Potential Distribution of Chinese Caterpillar Fungus (*Ophiocordyceps sinensis*) in Nepal Himalaya

Uttam Babu Shrestha[1]*, Kamaljit S. Bawa[2,3]

1 Institute for Agriculture and the Environment, University of Southern Queensland, Toowoomba, Queensland, Australia, **2** Department of Biology, University of Massachusetts, Boston, Massachusetts, United States of America, **3** Ashoka Trust for Research in Ecology and Environment (ATREE), Bangalore, India

Abstract

Climate change has already impacted ecosystems and species and substantial impacts of climate change in the future are expected. Species distribution modeling is widely used to map the current potential distribution of species as well as to model the impact of future climate change on distribution of species. Mapping current distribution is useful for conservation planning and understanding the change in distribution impacted by climate change is important for mitigation of future biodiversity losses. However, the current distribution of Chinese caterpillar fungus, a flagship species of the Himalaya with very high economic value, is unknown. Nor do we know the potential changes in suitable habitat of Chinese caterpillar fungus caused by future climate change. We used MaxEnt modeling to predict current distribution and changes in the future distributions of Chinese caterpillar fungus in three future climate change trajectories based on representative concentration pathways (RCPs: RCP 2.6, RCP 4.5, and RCP 6.0) in three different time periods (2030, 2050, and 2070) using species occurrence points, bioclimatic variables, and altitude. About 6.02% (8,989 km^2) area of the Nepal Himalaya is suitable for Chinese caterpillar fungus habitat. Our model showed that across all future climate change trajectories over three different time periods, the area of predicted suitable habitat of Chinese caterpillar fungus would expand, with 0.11–4.87% expansion over current suitable habitat. Depending upon the representative concentration pathways, we observed both increase and decrease in average elevation of the suitable habitat range of the species.

Editor: Helge Thorsten Lumbsch, Field Museum of Natural History, United States of America

Funding: Rufford Small Grants for Nature Conservation (RSGs), UK and Conservation Trust Grant of National Geographic Society, USA supported for the field data collection of this work. The funders had no role in study design, data collection and analysis, decision to publish, or preparation of the manuscript.

* Email: ubshrestha@yahoo.com

Introduction

The climate of our planet is changing at an unprecedented rate. Global average temperature has increased by 0.85°C from 1880 to 2012 and it is likely to increase further by a minimum of 0.3°C–1.7°C (RCP 2.6) to a maximum of 2.6°C–4.8°C (RCP 8.5) by the end of this century relative to 1986–2005 temperature [1]. Ecosystems and species have already responded to global climate change [2]. Changes in community structure, composition, and dynamics have been observed at ecosystem levels [3–4] whereas alteration in phenology, modification of physiology, and shifts in distribution have been documented at the level of species [5]. Shifts in plant species distribution may increase vulnerability to extinction [6]. Therefore, understanding the potential impacts of climate change on the distribution of species is important for mitigation of future biodiversity losses [7].

The Himalaya, a region of unique biodiversity, rich cultural and ethnic diversity, and varied topography has warmed up by about 1.5°C, three times more than the global average, during the last 25 years from 1982 to 2006 [8]. Climate change in the Himalaya has already impacted hydrology, agriculture, ecosystems, and species [9]. Similarly, changes in phenology of vegetation [8] and altitudinal shifts in vegetation communities [10] and species such as *Rheum nobile*, *Saussurea stella*, *Rhodiola bupleuroides*, *Ponerorchis chusua*, *Microgynaecium tibeticum*, *Meconopsis simplicifolia*, and *Pedicularis trichoglossa* [11] have been noted. Comparatively higher resolution (50×50 km) regional climate models show that temperature and precipitation in the Himalayan region will continue to increase in future [12], and these changes are further likely to impact the distribution of biodiversity, as for example, predicted for Rhododendrons [13]. However, potential shift of many other plant species caused by past and future climate changes in this biologically diverse region of the world are largely unknown [11,13].

Historical records have been largely used in documenting potential shifts in the distribution of species due to the past climate. However, in many cases, historical data on species distribution are lacking in the Himalaya. Moreover, such approaches might not be sufficient to predict the changes in current predicted distribution due to future changes in climate. Species distribution modeling (SDM), based on the quantitative relationship between environmental variables and species occurrence points, has made it possible to recognize species' niche requirements [14] and to understand the impacts of climate change on species distribution [15]. Various SDM methods, statistical, machine learning, and classification and distance are in current use to model species distribution [16]. Among them, a general purpose machine learning method—MaxEnt, developed by Phillips et al. (2006) [17]—is one of the most popular methods [18] and used to model a wide range of plant and animal species such as Malayan Sun Bear [19], Snow leopard [20] Lantana [21], Yew [22], Rhododendrons [13] and habitats globally [18–23]. It performs very well to estimate current as well as future distributions of species due to climate change [24].

Here we use the MaxEnt model to map the current potential distribution of Chinese caterpillar fungus (*Ophiocordyceps sinensis*) and to predict changes in the potential distribution of the species under future climate change trajectories in Nepal Himalaya. We also use the model to predict changes in the current potential elevation range of Chinese caterpillar fungus due to future climate change. We assume that both the extent of suitable habitat and range of Chinese caterpillar fungus will change with the change in climate in the area.

Materials and Methods

Study Area

Nepal, a mountainous country is situated on the southern slope of the central Himalaya. The country covers an area of 147,181 km^2 and has the widest elevational gradient of any country in the world, ranging from 60 m.a.s.l (meters above sea level) to Mount Everest, the highest point of the world at 8,848 m.a.s.l.. This vast altitudinal gradient creates a wide variation in physiographic, climatic, topographic, and edaphic conditions resulting in rich biodiversity. The Chinese caterpillar fungus, which is reported from 27 mountain districts of Nepal, is one of the most important biological and socio-economic components of Nepal's rich biodiversity.

Nepal Himalaya has been experiencing climate change and its impact; mean annual maximum temperature increased by 0.6°C per decade during 1977–2000 [25]. Similarly, climate model projections for Nepal show temperature increases of 0.5–2.0°C (mean 1.4°C) by the 2030s; 1.7–4.1°C (mean 2.8°C) by the 2060s; and 3.0–6.3°C (mean 4.7°C) by the 2090s [26]. Although there is no clear trend in precipitation, different models show an increase in monsoon precipitation and a decrease in winter precipitation in the future [26].

Species

Chinese caterpillar fungus is an endemic species of the Himalayan countries: Bhutan, China, India and Nepal. It a parasitic complex formed by a parasitic relationship between the fungus, *Ophiocordyceps sinensis* and the caterpillar of 'ghost' moth species belonging to the genus *Thitarodes* [27]. In Nepal, this caterpillar-fungus complex occurs in the open grasslands at elevations between 3,500–5,000 m.a.s.l. [28–29]. Although it has various therapeutic usages, the major trade of Chinese caterpillar fungus, with the popular name "Himalayan Viagra", is due to its presumed effects as an aphrodisiac and powerful tonic [27]. It is one of the most expensive natural medical resources of the world [30–31]. In September 2012, the highest price of Chinese caterpillar fungus in China—the major trade destination—was $140,000/kg (for the best quality product), two times more expensive than gold [32]. Although the fungus has an extraordinarily high price, it is collected extensively by the poorest mountain communities to support their livelihood. In recent years, the availability of the Chinese caterpillar fungus in its natural habitat has been dwindling and may threaten the livelihoods of the poorest mountain people [29]. The decline has been attributed, in part, to climate change [27,29,33].

Species location data

The species occurrence data used for this study were collected from our own field surveys conducted in six districts (Dolpa, Manang, Dolakha, Gorkha, Darchula, Bajhang) of Nepal between 2011 and 2013. Altogether, 37 species occurrence points distributed in central, western and eastern Nepal were used to model the potential distribution (Figure 1). We secured the necessary permits from the National Trust for Nature Conservation (NTNC) to conduct field studies in protected areas of Manang and Gorkha.

Environmental and bioclimatic data

To model potential distribution of Chinese caterpillar fungus across our study area, we first collected 19 bioclimatic layers, and one topographic (altitude) layer. The 19 grid-based bioclimatic variables (Table 1) were downloaded from the worldclim datasets (www.worldclim.com). Bioclimatic variables calculated from monthly temperature and precipitation values were generated through interpolation of average monthly climate data from weather stations at 2.5 arc minute spatial resolution [34]. They are ecologically meaningful variables that describe annual trends, seasonality, and extremes of temperature and precipitation. Altitude was derived from digital elevation data from the Shuttle Radar Topographic Mission (SRTM); this was then resampled into 2.5 arc minute (~5 km) spatial resolution by using the nearest neighbor resampling technique in ArcGIS.

We calculated pairwise correlations and removed highly correlated variables ($r^2 \leq 0.80$) to minimize the impact of multicollinearity and over-fitting of the model. The remaining seven (Bio 3, Bio 7, Bio 11, Bio 14, Bio 15, Bio 18, and Bio 19) bioclimatic variables (Table 2) were used to model the distribution of Chinese caterpillar fungus in current and future climate conditions. All datasets were changed into ASCII files in ArcGIS as required by the MaxEnt software. We repeated the same process to generate predicted maps in three different future climate scenarios for 2030, 2050 and 2070. We did not take account of land use or land cover change, human disturbances, species dispersal or changes in biotic interactions in the model.

To determine the future distribution of the species under different climate trajectories, we used datasets of future climate from the International Center for Tropical Agriculture (www.ccafs-climate.org), a global agricultural research institution. We selected three future GHG (greenhouse gas) concentration trajectories, also known as representative carbon pathways (RCP 2.6, RCP 4.5 and RCP 6.0), for three different time periods (2030, 2050 and 2070) as adopted by the IPCC in its fifth Assessment Report (AR5) [1]. RCP 2.6 is the lowest GHG concentration pathway in which radioactive forcing (global energy imbalances) levels reach 3.1 W/m^2 by mid-century and drops 2.6 W/m^2 by 2100 [35]. RCP 4.5 is a stabilization scenario in which the total radiative forcing reaches to 4.5 W/m^2 by 2100 and stabilizes due to the employment of a range of technologies and strategies for reducing GHG emissions [36]. Likewise, RCP 6.0 also represents stabilization by 2100, this time at 6.0 W/m^2 by 2100 [37]. We selected a global circulation model, HadGEM2-CC (Hadley Global Environment Model 2 Carbon Cycle) developed by the Hadley Center, United Kingdom [38]. HadGEM2 models have been used to perform all the CMIP5 (Coupled Model Intercomparison Project Phase 5) centennial experiments including ensembles of simulations of the RCPs. HadGEM2-CC is one of the models used by the international governmental panel on climate change (IPCC) in its fifth Assessment Report (AR5). For consistency, we used the same seven bioclimatic variables (Bio 3, Bio 7, Bio 11, Bio 14, Bio 15, Bio 18, and Bio 19) we used for

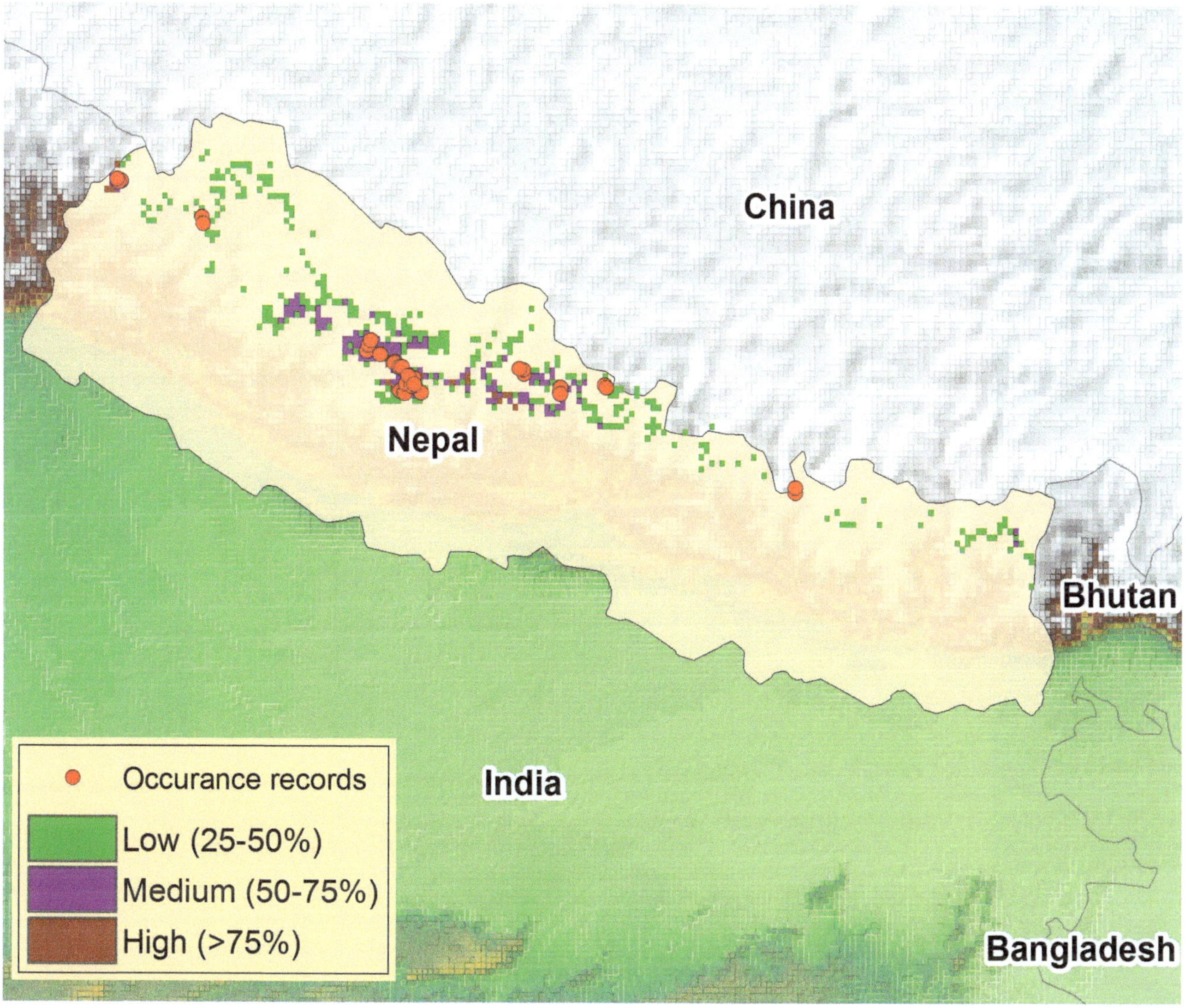

Figure 1. Predicted potential distribution of Chinese caterpillar fungus under current bioclimatic conditions and location of occurrence used for modeling.

modeling current potential distribution of the species to predict future distribution. Those data are statistically downscaled from a Global Circulation Model (GCM) based on the sum of interpolated anomalies to high resolution monthly climate surfaces from Worldclim [34].

Modeling

We used freely available MaxEnt software to model the current and future distributions of Chinese caterpillar fungus. MaxEnt is a general purpose machine learning method that estimates the probability distribution of a species occurrence based on environmental conditions of a location in which the species is found by calculating the distribution of maximum entropy i.e. the most spread out distribution in space for a given set of constraints [17]. It is the most popular species distribution modeling method with more than 1000 published usages since 2005 [18,23]. MaxEnt has also outperformed other methods and has shown higher predictive accuracy than other methods [6,39]. MaxEnt performs well to estimate potential range shifts of species due to climate change [24]. The method is easy to use and has the functionality to use presence only data.

We used the following parameter values in the MaxEnt model: random test percentage = 25%, regularization multiplier = 1, maximum number of backgrounds points = 10,000, maximum iterations = 1,000 and convergence threshold = 0.00001. This means that the model uses 25% of the original presence data against 10,000 random background points (pseudo-absences) for modeling the distribution. Model robustness is commonly evaluated by AUC values that range from 0 to 1; AUC values between 0.5–0.7 are considered low, 0.7–0.9 moderate and >0.9 high [16,35]. We also used AUC values to determine the model accuracy. The output of MaxEnt is continuous data with values ranging from 0 (lowest) to 1 (highest) probability of distribution. We imported the MaxEnt output data and reclassified into three classes of habitat suitability: low suitability (25–50% probability of occurrence), medium suitability (50–75% probability of occurrence) and high suitability (>75% probability of occurrence) by omitting the values below 25% as non-suitable habitat based on the logistic threshold, with suitable conditions predicted above the

Table 1. Bioclimatic variables used for modeling habitat of Chinese caterpillar fungus.

Variables Name	Code	Data Source	Resolution
Altitude	**Alt**	**STRM**	**90 m**
Annual Mean Temperature	BIO 1	Worldclim	~5 km (2.5 arc min)
Mean Diurnal Range [Mean of monthly (Max Temperature - Min Temperature)]	BIO 2	Worldclim	~5 km (2.5 arc min)
Isothermality (BIO 2/BIO 7) (*100)	BIO 3	Worldclim	~5 km (2.5 arc min)
Temperature Seasonality (Standard Deviation*100)	BIO 4	Worldclim	~5 km (2.5 arc min)
Max Temperature of Warmest Month	BIO 5	Worldclim	~5 km (2.5 arc min)
Min Temperature of Coldest Month	BIO 6	Worldclim	~5 km (2.5 arc min)
Temperature Annual Range (BIO 5-BIO 6)	BIO 7	Worldclim	~5 km (2.5 arc min)
Mean Temperature of Wettest Quarter	BIO 8	Worldclim	~5 km (2.5 arc min)
Mean Temperature of Driest Quarter	BIO 9	Worldclim	~5 km (2.5 arc min)
Mean Temperature of Warmest Quarter	BIO 10	Worldclim	~5 km (2.5 arc min)
Mean Temperature of Coldest Quarter	BIO 11	Worldclim	~5 km (2.5 arc min)
Annual Precipitation	BIO 12	Worldclim	~5 km (2.5 arc min)
Precipitation of Wettest Month	BIO 13	Worldclim	~5 km (2.5 arc min)
Precipitation of Driest Month	BIO 14	Worldclim	~5 km (2.5 arc min)
Precipitation Seasonality (Coefficient of Variation)	BIO 15	Worldclim	~5 km (2.5 arc min)
Precipitation of Wettest Quarter	BIO 16	Worldclim	~5 km (2.5 arc min)
Precipitation of Driest Quarter	BIO 17	Worldclim	~5 km (2.5 arc min)
Precipitation of Warmest Quarter	BIO 18	Worldclim	~5 km (2.5 arc min)
Precipitation of Coldest Quarter	BIO 19	Worldclim	~5 km (2.5 arc min)

threshold and unsuitable below. This reclassification of suitable conditions allowed us to compare the change in classes over time and space.

We also estimated the total area of predicted habitat of Chinese caterpillar fungus under current and future climatic conditions by calculating the number of 'presence' grid cells multiplied by their spatial resolution. To observe potential changes in the altitudinal distribution of Chinese caterpillar fungus, we extracted elevation values of the pixels of predicted presence from the predicted and current distribution maps generated by MaxEnt models. We compared the mean elevation values of the areas of predicted presence for current and future climate scenarios using independent sample T-test.

Results

Our study demonstrates for the first time the potential distribution of Chinese caterpillar fungus habitat in Nepal Himalaya. Out of seven predictor bioclimatic variables used for this study, the relative contribution of two variables (Bio 11 and Bio 15) to the model was more than 89%. Mean temperature of the coldest quarter (Bio 11) had the highest contribution (74.5%) to the model followed by precipitation seasonality (Bio 15), with 15.3% contribution. The bioclimatic variable, isothermality (Bio 3) had the lowest contribution (0.3%). The MaxEnt model's jack-knife test of variable importance showed that the mean temperature of the coldest quarter (Bio 11) was the variable with the highest gain when used in isolation and the gain decreased when it was omitted (Figure 2). The validity of the model for current distribution of Chinese caterpillar fungus was high with AUC = 0.98 indicating that the selected variables described the distribution of Chinese caterpillar fungus very well.

Responses of each environmental variable that influenced the predicted suitable distribution of Chinese caterpillar fungus are shown in response curves (Figure 3). These response curves show changes in the logistic prediction when each environmental variable changes by keeping all other environmental variables at their average sample value. Models showed that the distribution of Chinese caterpillar fungus is highly controlled by temperature. It prefers areas where the mean annual temperature remains below 15°C and the mean temperature of the coldest quarters stays slightly lower than zero but not less than minus 10°C. Precipitation in the driest month and precipitation seasonality from 75–150 mm might also be useful predictors but excessive precipitation above 200 mm was not found to be very apposite.

The potential distribution of Chinese caterpillar fungus under current bioclimatic conditions and occurrence locations is shown in Figure 1. Based on our model, about 6.02% (8,989 km^2) of the country is suitable for Chinese caterpillar fungus habitat, and the fungus occurs in 26 mountainous districts of Nepal. The greatest concentration of pixels (potential habitat or predicted presence) is observed in Dolpa, Rukum, Manang, Myagdi, and Jumla districts when the suitable habitat map is overlaid with the district map of Nepal.

To estimate the potential impact of climate change on the distribution of Chinese caterpillar fungus, the predicted distribution of Chinese caterpillar fungus in future climate scenarios is shown in Figure 4. Across all scenarios in three different periods, the predicted area of Chinese caterpillar fungus distribution was predicted to increase. The estimated area of the predicted distribution is given in Figure 5. The maximum expansion (a 4.87% addition to the current potential suitable area) would occur under RCP 2.6 by 2070 whereas the minimum expansion (0.11% addition in current potential suitable area) was predicted for the year 2030 for the same pathway. Much of the expansion

Table 2. Correlation matrix of altitude and bioclimatic variables.

Code	Alt	BIO1	BIO2	BIO3	BIO4	BIO5	BIO6	BIO7	BIO8	BIO9	BIO10	BIO11	BIO12	BIO13	BIO14	BIO15	BIO16	BIO17	BIO18	BIO19	LULC
Alt		−1.00	−0.06	0.13	0.10	−0.99	−0.99	−0.14	−1.00	−0.99	−1.00	−1.00	−0.76	−0.81	0.00	−0.74	−0.79	0.31	−0.58	0.46	0.12
BIO1			0.05	−0.14	−0.10	0.99	0.99	0.13	1.00	0.99	1.00	1.00	0.76	0.80	0.01	0.73	0.78	−0.28	0.57	−0.44	−0.13
BIO2				0.08	0.73	0.15	−0.08	0.38	0.08	0.00	0.09	0.01	−0.20	−0.07	−0.51	0.40	−0.10	−0.44	−0.27	−0.35	0.18
BIO3					−0.54	−0.23	−0.13	−0.40	−0.15	−0.15	−0.17	−0.11	0.05	0.01	−0.07	−0.01	0.04	−0.49	0.30	−0.49	0.01
BIO 4						0.03	−0.21	0.93	−0.07	−0.14	−0.05	−0.16	−0.35	−0.24	−0.32	0.15	−0.27	0.04	−0.52	0.14	0.17
BIO5							0.97	0.26	0.99	0.98	0.99	0.98	0.70	0.76	−0.02	0.73	0.74	−0.24	0.48	−0.38	−0.11
BIO6								0.01	0.98	0.99	0.98	0.99	0.78	0.80	0.09	0.65	0.79	−0.20	0.60	−0.37	−0.16
BIO7									0.16	0.09	0.18	0.08	−0.19	−0.05	−0.43	0.38	−0.09	−0.16	−0.37	−0.09	0.15
BIO8										0.99	1.00	1.00	0.75	0.80	0.00	0.74	0.78	−0.29	0.57	−0.45	−0.12
BIO9											0.99	0.99	0.76	0.80	0.02	0.69	0.78	−0.26	0.58	−0.41	−0.13
BIO10												0.99	0.74	0.79	−0.01	0.74	0.77	−0.28	0.55	−0.43	−0.12
BIO11													0.77	0.80	0.03	0.71	0.79	−0.28	0.60	−0.44	−0.13
BIO12														0.98	0.18	0.63	0.99	−0.22	0.92	−0.38	−0.16
BIO13															0.06	0.74	0.99	−0.32	0.88	−0.47	−0.13
BIO14																−0.33	0.07	0.62	0.14	0.46	−0.14
BIO15																	0.72	−0.64	0.53	−0.73	−0.06
BIO16																		−0.31	0.90	−0.46	−0.14
BIO17																			−0.32	0.96	−0.13
BIO18																				−0.46	−0.14
BIO19																					−0.08

Unrelated variables (Correlation ≤0.80) used for the study BIO3, BIO7, BIO 11, BIO 15, BIO 18, BIO 19, LULC.

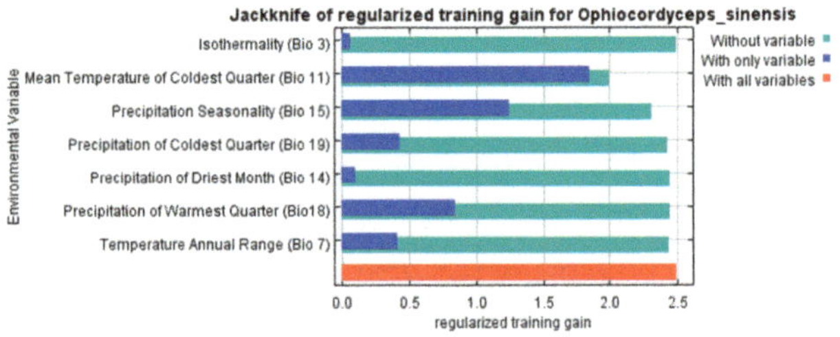

Figure 2. Results of jackknife test of relative importance of predictor variables for Chinese caterpillar fungus.

occurs in western Nepal whereas there are nominal changes only in the eastern parts of the country. To be precise, maximum increases in the area of suitable habitat occurs in Dolpa, Mustang, and Mugu districts under RCP 2.5 in 2050 whereas the greatest reduction in habitat area occurs in the Taplejung district of Eastern Nepal. Taplejung might lose its entire area of suitable habitat under RCP 4.5 by the year 2050. Similar loss of the entire area of Chinese caterpillar fungus habitat was predicted in Humla district under RCP 6.0 by the year 2070. However, expansion of potential habitat might occur in two new districts, Dolpa and Pachthar under RCP 6.0 by 2070. Under both the RCP 4.5 and RCP 6.0 trajectories, the area of predicted distribution increases continuously up to 2050 then decreases by 2070 whereas in the RCP 2.6 trajectory, the distribution continuously increases from 2030 to 2070 and reaching a maximum area in 2070 (Figure 5).

Expansion in both lower and upper altitudinal limits was observed under all future climate change pathways (Figure 6). The average altitude of current distribution and that of future distributions increased under four pathways: RCP 2.6 (2070), RCP 4.5 (2050), RCP 6.0 (2050) and RCP 6.0 (2070) whereas a decrease was found in all other pathways (Table 3). However, none of the changes were statistically significant ($p = 0.01$) except RCP 4.5 (2070) based on the independent sample T-test between the altitude means of current potential distribution range and that of future potential distribution range.

Discussion

Our work represents the first attempt to model the impact of future climate change on the distribution of Chinese caterpillar fungus—a flagship species of the Himalaya. Climate change projections for the Nepal Himalaya predict that average annual temperatures will increase by 3.0–6.3°C (mean 4.7°C) by the 2090s and that the warming would be greater in higher elevation regions—habitats of the Chinese caterpillar fungus [26]. Since climate change has already impacted species' habitats including those of several species of fungi worldwide [40–42], it will likely affect the distribution of this species in Nepal as well. We here provide a model that could be used to predict how the distribution of species might be affected by changes in future climate depending upon the different representative concentration pathways.

It is widely believed that temperature and humidity play important roles in yields [27–28], abundance [43], and the probability of infection and sporulation [44] of Chinese caterpillar fungus. Furthermore, Chinese caterpillar fungus distribution is also assumed to be affected by winter and summer temperatures and the seasonality of precipitation [45–46]. Our observations that the mean temperature of the coldest quarter and the seasonality of precipitation affect the distribution of Chinese caterpillar fungus is in line with observations that temperature and humidity play important roles in the ecology and physiology of the fungus. Furthermore, bioclimatic response variables that affect the predicted distribution of the species concur with the knowledge of local communities about the habitat needs of the species [46].

Although the potential distribution predicted by the model is not the observed distribution, there is close congruence between the two distributions. The districts of Dolpa, Rukum, Manang, Myagdi, and Jumla showed a greater area of potential distribution (maximum concentration of the pixels). These districts are the major producers of Chinese caterpillar fungus in Nepal, contributing approximately 60% of the total volume of Chinese caterpillar fungus traded in Nepal [47]. Spatial overlap of the potential occurrence with districts where Chinese caterpillar fungus is reported is 96.29% (26 districts in the model vs. 27 districts in actual). However, the model slightly overestimates the potential distribution for the Kaski district and underestimates it for Dolakha District. Nonetheless, both the maximum spatial overlapping and high AUC value suggests that the model has a high level of accuracy.

Our models show that the potential distribution of Chinese caterpillar fungus will extend under all future climate scenarios, suggesting that additional new suitable habitats will be created for Chinese caterpillar fungus with the future climate change. However, this expansion may not guarantee an increase in the production of Chinese caterpillar fungus in future. Studies in other parts of the world have shown both expansion and shrinkage of potential habitat in response to future climate change [48–50]. Both the reductions and increase in range size were predicted for the endemic flora of California under future climate change scenarios [48] as well as higher plants in Europe [49]. Range expansion of lemur parasites is predicted for Madagascar [50] whereas shrinkage of Rhododendrons has been predicted in the Himalaya [13]. Likewise a 60% reduction in the range of Ngwayir (*Pseudocheirus occidentalis*) is predicted in Australia [51].

It has been unofficially reported that the altitudinal limit of Chinese caterpillar fungus seems to have shifted upwards by 200–500 m in China due to past climate change [27,33], which may have also caused local extinction of the fungus from Mount Emai in the Chinese province of Sichuan, located on the edge of the Tibetan plateau [52–53]. However, we found no noteworthy changes in the altitudinal ranges of the modeled potential distributions of Chinese caterpillar fungus under predicted future climate change trajectories in this study. Furthermore, it is unclear what methodology Yang (2008) had used to observe the shifts in the altitudinal range of Chinese caterpillar fungus [33]. Due to the unavailability of historical data on the distribution of Chinese caterpillar fungus in Nepal Himalaya, monitoring the impact of past climate change on the distribution of Chinese caterpillar fungus is impossible.

One of the major current drivers of decline of Chinese caterpillar fungus is extensive harvesting [27,29]. Every year about 100 thousand people are involved in harvesting Chinese caterpillar fungus in Nepal [54]. The abundance of Chinese caterpillar fungus in the future will largely depend on human use. Although the area of suitable habitat for Chinese caterpillar fungus is predicted to expand with future climate change, the availability of fungus is likely to be largely governed by human impacts on populations. Future distribution models will have to take into account actual and potential harvesting intensities at different locations. Our model does not account for harvesting pressure or future transformation of land. Furthermore, since the Chinese caterpillar fungus complex involves an array of linked biotic interactions of diverse organisms with different life cycles and

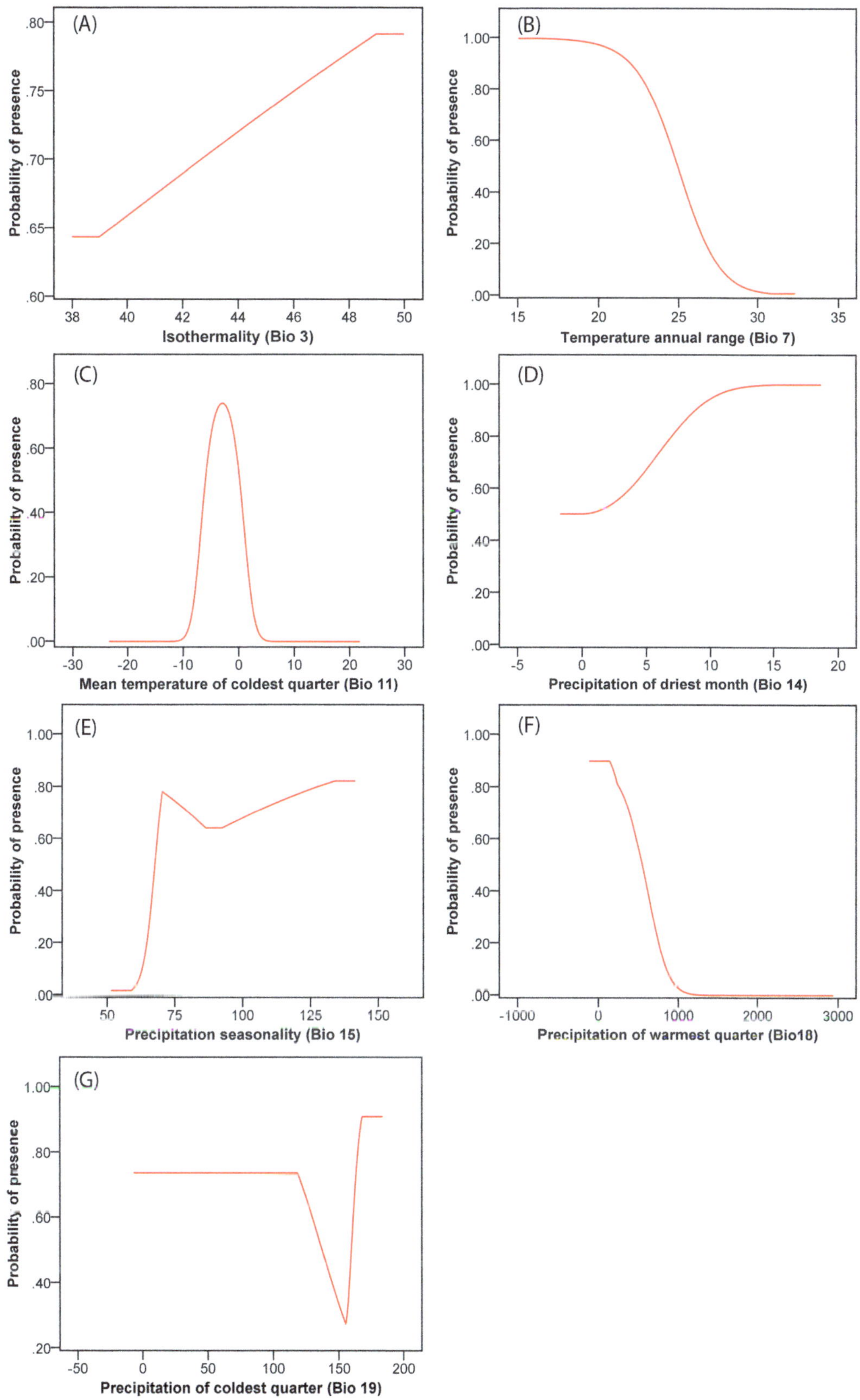

Figure 3. Response curves for the predictors of the MaxEnt Model.

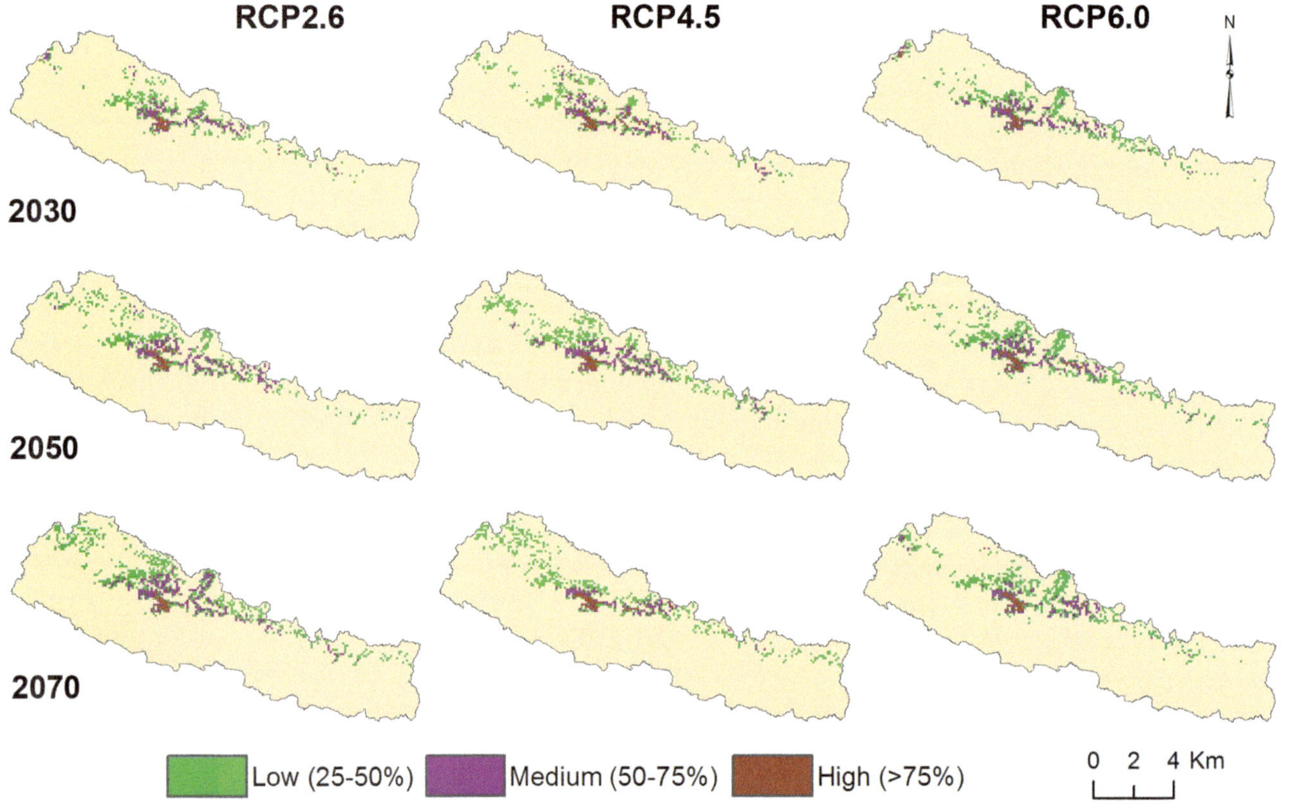

Figure 4. Predicted future distribution of Chinese caterpillar fungus in future climate scenarios.

generation spans, future climate change may affect the interacting species differentially. Future research would therefore need to incorporate harvesting, future land use and land cover change, and biotic interactions in distribution models of not only this species but other species as well.

Although this study lays down the foundation for studying the dynamics of Chinese caterpillar fungus distribution through time and space, it has some limitations. Bioclimatic variables used in this study have a 5×5 km spatial resolution. This resolution may be acceptable for species with broad range sizes including Chinese

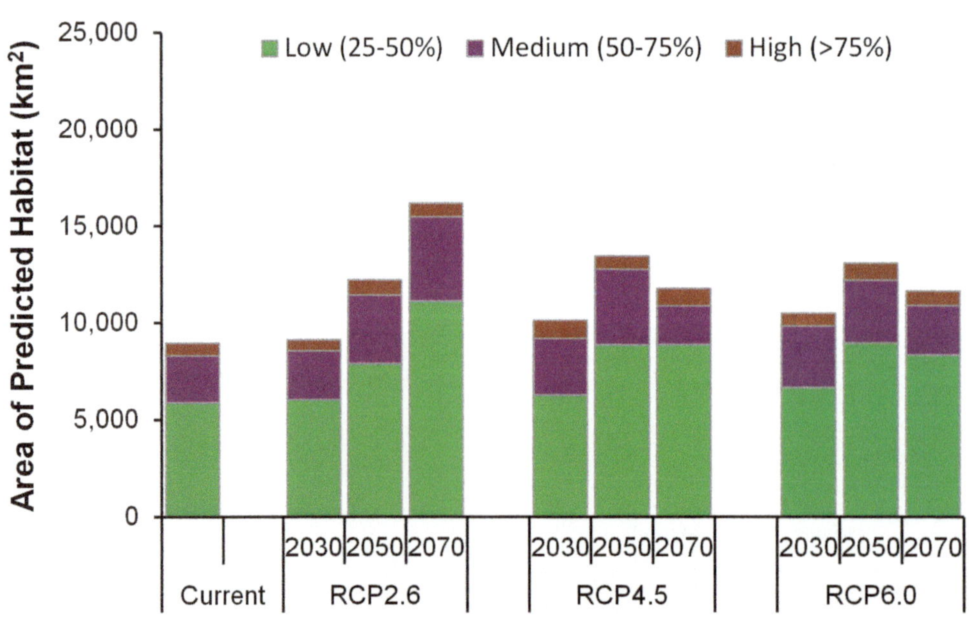

Figure 5. The estimated areas of the predicted distribution of Chinese caterpillar fungus.

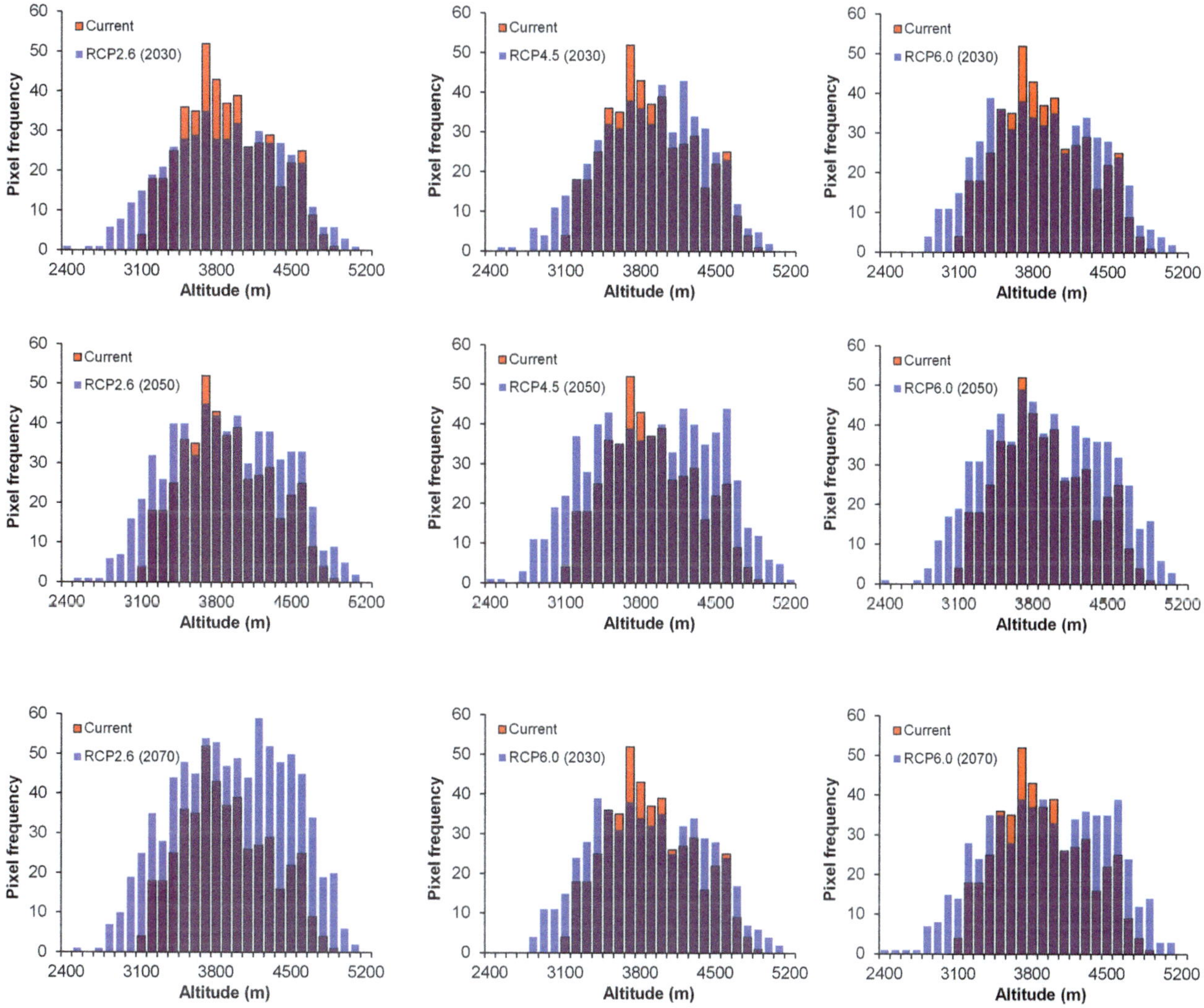

Figure 6. Change in average elevation in distribution of Chinese caterpillar fungus in three future climate scenarios.

caterpillar fungus [55], however, given the diverse topography of the Himalayan region, finer scale resolution of the environmental data might be needed to produce results at scales relevant to species with more restricted habitat requirements. As we used a limited number of samples from only six districts while this species is reported from 27 mountain districts of Nepal, this resolution was used as a compromise between fine scale resolution and the precision of the coordinates for the species' occurrences. More

Table 3. Independent sample T-test between average altitude of current predicted distribution and future predicted distributions.

Climate trajectories	Year	Mean	SD	P value of independent sample T-test
RCP2.6	2030	3921.68	523.03	0.302
RCP2.6	2050	3933.59	523.11	0.500
RCP2.6	2070	3998.86	529.83	0.104
RCP4.5	2030	3948.13	486.09	0.861
RCP4.5	2050	3955.49	566.19	0.938
RCP4.5	2070	3878.31	525.48	0.011
RCP6.0	2030	3929.22	514.55	0.416
RCP6.0	2050	3959.01	538.16	0.841
RCP6.0	2070	3982.11	548.55	0.337

occurrence records spread across the entire country of Nepal and finer resolution of environmental data that also capture microclimates, edaphic conditions, vegetation dynamics, and landscape heterogeneity might enable more sophisticated models to be developed in future.

Despite these limitations, the study provides information on the suitable climate space or bioclimatic envelope responsible for the potential distribution of Chinese caterpillar fungus. Furthermore, understanding the potential distribution of one of the most exploited species of this region is a necessary to design an efficient conservation plan. This study provides evidence of the distribution of suitable habitat for the Chinese caterpillar fungus in Nepal Himalaya using a robust bioclimatic model. It can be further useful in assessing threats and evaluating gaps in protected areas in order to conserve this economically important species. Furthermore, our evaluation of the potential impacts of future climate change on the potential suitable habitat of the species is important in devising adaptive responses and precautionary measures for the sustainable management of the habitat of the species in future.

Acknowledgments

We thank Kamal Nepali, Shivish Bhandari, Bikram Pandey, Chhabi Thapa and Rabindra Parajuli for their support in the field. We thank Kathryn Reardon-Smith and anonymous reviewer for valuable comments for improving the manuscript.

Author Contributions

Conceived and designed the experiments: UBS KB. Performed the experiments: UBS. Analyzed the data: UBS. Wrote the paper: KB UBS.

References

1. IPCC (2013) Climate Change 2013: The physical science basis. Contribution of working group I to the fifth assessment report of the Intergovernmental Panel on Climate Change. In: Stocker TF, Qin D, Plattner GK, Tignor M, Allen SK, Boschung J, Nauels A, Xia Y, Bex V, Midgley PM, editors. Cambridge University Press, Cambridge, United Kingdom and New York, NY, USA. pp. 1535.
2. Hansen AJ, Neilson RP, Dale VH, Flather CH, Iverson LR, et al. (2001) Global change in forests: responses of species, communities, and biomes. Bioscience 51: 765–779.
3. Walther GR, Post E, Convey P, Menzel A, Parmesan C, et al. (2002) Ecological responses to recent climate change. Nature 416: 389–395.
4. Parmesan C (2006) Ecological and evolutionary responses to recent climate change. Ann Rev Ecol Evol Syst 37: 637–669.
5. Bellard C, Bertelsmeier C, Leadley P, Thuiller W, Courchamp F (2012) Impacts of climate change on the future of biodiversity. Ecol Lett 15: 365–377.
6. Summers DM, Bryan BA, Crossman ND, Meyer WS (2012) Species vulnerability to climate change: impacts on spatial conservation priorities and species representation. Glob Change Biol 18(7): 2335–2348.
7. Pressey RL, Cabeza M, Watts ME, Cowling RM, Wilson KA (2007) Conservation planning in a changing world. Trends Ecol Evol 22: 583–592.
8. Shrestha UB, Gautam S, Bawa KS (2012) Widespread climate change in the Himalayas and associated changes in local ecosystems. PLoS ONE 7(5): e36741.
9. Xu J, Grumbine RE, Shrestha A, Eriksson M, Yang X, et al. (2009) The melting Himalayas: Cascading effects of climate change on water, biodiversity, and livelihoods. Conserv Biol 23(3): 520–530.
10. Joshi PK, Rawat A, Narula S, Sinha V (2012) Assessing impact of climate change on forest cover type shifts in Western Himalayan Eco-region. J For Res 23(1): 75–80.
11. TelwalaY, Brook BW, Manish K, Pandit MK (2013) Climate-induced elevational range shifts and increase in plant species richness in a Himalayan biodiversity epicentre. PLoS ONE 8(2): e57103.
12. Kulkarni A, Patwardhan SKKK, Ashok K, Krishnan R (2013) Projected climate change in the Hindu Kush–Himalayan region by using the high-resolution regional climate model PRECIS. Mt Res Dev 33(2): 142–151.
13. Kumar P (2012) Assessment of impact of climate change on Rhododendrons in Sikkim Himalayas using Maxent modelling: limitations and challenges. Biod Conserv 21: 1251–1266.
14. Hirzel AH, Lay GL, Helfer V, Randin C, Guisan A (2006) Evaluating the ability of habitat suitability models to predict species presences. Ecol Model 199(2): 142–152.
15. Beaumont LJ, Pitman AJ, Poulsen M, Hughes L (2007). Where will species go? Incorporating new advances in climate modelling into projections of species distributions. Glob Change Biol 13(7): 1368–1385.
16. Franklin J (2010) Mapping Species Distributions – Spatial inference and prediction, ecology. biodiversity and conservation, Cambridge University Press, Cambridge. 336 p.
17. Phillips SJ, Anderson RP, Schapire RE (2006) Maximum entropy modeling of species geographic distributions. Ecol Model 190(3–4): 231–259.
18. Fourcade Y, Engler JO, Rödder D, Secondi J (2014) Mapping species distributions with MAXENT using a geographically biased sample of presence data: A performance assessment of methods for correcting sampling bias. PLoS ONE 9(5): e97122.
19. Nazeri M, Jusoff K, Madani N, Mahmud AR, Bahman AR, et al. (2012) Predictive modeling and mapping of Malayan Sun Bear (*Helarctos malayanus*) distribution using maximum entropy. PLoS ONE 7(10): e48104.
20. Li J, Wang D, Yin H, Zhaxi D, Jiagong Z, et al. (2014) Role of Tibetan Buddhist monasteries in Snow Leopard conservation. Conserv Biol 28(1): 87–94.
21. Taylor S, Kumar L, Reid N, Kriticos DJ (2012) Climate Change and the Potential Distribution of an Invasive Shrub, *Lantana camara* L.. PLoS ONE 7(4): e35565.
22. Poudel RC, Möller M, Gao L-M, Ahrends A, Baral SR, et al. (2012) Using morphological, molecular and climatic data to delimitate yews along the Hindu Kush-Himalaya and adjacent regions. PLoS ONE 7(10): e46873.
23. Merow C, Smith MJ, Silander Jr. JA (2013) A practical guide to MaxEnt for modeling species' distributions: what it does, and why inputs and settings matter. Ecography 36(10): 1058–1069.
24. Hijmans RJ, Graham CH (2006) The ability of climate envelope models to predict the effect of climate change on species distributions. Glob Chang Biol 12(12): 2272–2281.
25. Shrestha AB, Wake CP, Mayewski PA, Dibb JE (1999) Maximum temperature trends in the Himalaya and its vicinity: an analysis based on temperature records from Nepal for the period 1971–94. J Climate 12: 2775–2786.
26. NCVST (2009) Vulnerability through the eyes of vulnerable. Climate change induced Uncertainties and Nepal's Development Predicaments. Institute for Social and Environmental Transition-Nepal Normal (ISET-N). 95 p.
27. Winkler D (2010) Caterpillar fungus (*Ophiocordyceps sinensis*) production and sustainability on the Tibetan plateau and in the Himalayas. Asian Medicine 5: 291–316.
28. Devkota S (2010) *Ophicordyceps sinensis* (Yarsagumba) from Nepal Himalaya: status, threats and management strategies. In: Zhang PH, editor.Cordyceps Resources and Environment. Grassland Supervision Center by the Ministry of Agriculture, People's Republic of China. pp. 91–108.
29. Shrestha UB, Bawa KS (2013) Trade, harvest, and conservation of caterpillar fungus (*Ophiocordyceps sinensis*) in the Himalayas. Biol Conserv 159: 514–520.
30. Stone R (2008) Last stand for the body snatcher of the Himalayas? Science 322: 1182.
31. Shrestha UB (2012) Asian medicine: A fungus in decline. Nature 482: 35.
32. Xuan C (2012) How long can the Caterpillar fungus craze last? China Dialogue. Available: http://www.chinadialogue.net/article/show/single/en/5143-How-long-can-the-caterpillar-fungus-craze-last. Accessed 24 December 2012.
33. Yang DR (2008) Recommendations on strengthening the protection of the resources of the rare Tibet Plateau fungus—*Cordyceps sinensis* and its ecological environment' (Jia qiang qing zang gao yuan zhen xi zhen jun—Dong cong xia cao zi yuan he sheng tai huan jing bao hu de jian yi) Unpublished report to Chinese Academy of Science (in Chinese).
34. Hijmans RJ, Cameron SE, Parra JL, Jones PG, Jarvis A (2005) Very high resolution interpolated climate surfaces for global land areas. Int J Climatol 25(15): 1965–1978.
35. van Vuuren D, den Elzen M, Lucas P, Eickhout B, Strengers B, et al. (2007) Stabilizing greenhouse gas concentrations at low levels: an assessment of reduction strategies and costs. Climat Chang 81: 119–159.
36. Clarke L, Edmonds J, Jacoby H, Pitcher H, Reilly J, Richels R (2007) Scenarios of Greenhouse Gas Emissions and Atmospheric Concentrations. Sub-report 2.1A of Synthesis and Assessment Product 2.1 by the U.S. Climate Change Science Program and the Subcommittee on Global Change Research. Department of Energy, Office of Biological & Environmental Research, Washington DC, USA. pp. 154.
37. Fujino J, Nair R, Kainuma M, Masui T, Matsuoka Y (2006) Multi-gas mitigation analysis on stabilization scenarios using AIM global model. The Energy Journal Multi-Greenhouse Gas Mitigation and Climate Policy, Special Issue # 3: 343–354.
38. Collins WJ, Bellouin N, Doutriaux-Boucher M, Gedney N, Halloran P, et al. (2011) Development and evaluation of an Earth-System model—HadGEM2. Geosci Model Dev Discuss 4: 997–1062.
39. Elith J, Graham CH, Anderson P, Dudik M, Ferrier S, et al. (2006) Novel methods improve prediction of species distributions from occurrence data. Ecography 29: 129–151.
40. Gange AC, Gange EG, Sparks TH, Boddy L (2007) Rapid and recent changes in fungal fruiting patterns. Science 316: 71.
41. Kauserud H, Heegaard E, Semenov MA, Boddy L, Halvorsen R, et al. (2010) Climate change and spring-fruiting fungi. P Roy Soc B-Biol Sci 277: 1169–1177.

42. Gange AC, Gange EG, Mohammad AB, Boddy L (2011) Host shifts in fungi caused by climate change? Fungal Ecol 4(2): 184–190.
43. Weckerle CS, Yang Y, Huber FK, Li Q (2010) People, money, and protected areas: the collection of the caterpillar mushroom *Ophiocordyceps sinensis* in the Baima Xueshan Nature Reserve, Southwest China. Biod Conserv 19: 2685–2698.
44. Zhang YJ, Sun BD, Zhang S, Wang M, Liu XZ, et al. (2010) Mycobiotal investigation of natural *Ophiocordyceps sinensis* based on culture-dependent investigation. Mycosystema 29: 518–527.
45. Cannon PF, Hywel-Jones NL, Maczey N, Norbu L, Samdup T, et al. (2009) Steps towards sustainable harvest of *Ophiocordyceps sinensis* in Bhutan. Biodivers Conserv 18: 2263–2281.
46. Shrestha UB, Bawa KS (2014) Harvesters' perceptions of population status and conservation of Chinese caterpillar fungus in the Dolpa region of Nepal (under review).
47. Thapa YB, Chettri R (2011) Handbook for sustainable collection of Yarsagumba, Department of Forest Resource, Government of Nepal, Kathmandu, Nepal (Yarsagumba Digo Sankalan Prabidi Sahayogi Pustika) (in Nepali).
48. Loarie SR, Carter BE, Hayhoe K, McMahon S, Moe R, et al. (2008) Climate change and the future of California's endemic flora. PLoS ONE 3(6): e2502.
49. Bakkenes M, Alkemade JRM, Ihle F, Leemans R, Latour JB (2002) Assessing effects of forecasted climate change on the diversity and distribution of European higher plants for 2050. Global Change Biol 8: 390–407.
50. Barrett MA, Brown JL, Junge RE, Yoder AD (2013) Climate change, predictive modeling and lemur health: Assessing impacts of changing climate on health and conservation in Madagascar. Biol Cons 157: 409–422.
51. Molloy SW, Davis RA, Van Etten JB (2013) Species distribution modelling using bioclimatic variables to determine the impacts of a changing climate on the western ringtail possum (*Pseudocheirus occidentals*; Pseudocheiridae). Environm Conserv 1–11.
52. Li QS, Li L, Yin DH, Fu SQ, Huang TF, et al. (1991) Biological characteristics of *Cordyceps sinensis*. Special Wild Econ. Anim Plant Res 1: 42–45 (in Chinese).
53. Li Y, Wang XL, Jiao L, Jiang Y, Li H, et al. (2011). A survey of the geographic distribution of *Ophiocordyceps sinensis*. J Microbiol 49: 913–919.
54. Shrestha UB, Shrestha BB, Shrestha S, Ghimire S, Nepali K (2014). Chasing Chinese caterpillar fungus (*Ophiocordyceps sinensis*) harvesters in the Himalayas: Harvesting practice and its conservation implications in western Nepal. Soc Nat Res In press.
55. Seo C, Thorne JH, Hannah L, Thuiller W (2009). Scale effects in species distribution models: implications for conservation planning under climate change. Biol lett 5: 39–43.

8

Heritability and Evolutionary Potential in Thermal Tolerance Traits in the Invasive Mediterranean Cryptic Species of *Bemisia tabaci* (Hemiptera: Aleyrodidae)

Fang-Zhou Ma[1]◔, Zhi-Chuang Lü[1]◔, Ren Wang[1], Fang-Hao Wan[1,2]*

1 State Key Laboratory for Biology of Plant Diseases and Insect Pests, Institute of Plant Protection, Chinese Academy of Agricultural Sciences, Beijing, P. R. China, **2** Center for Management of Invasive Alien Species, Ministry of Agriculture, Beijing, P. R. China

Abstract

With advancing global climate change, the analysis of thermal tolerance and evolutionary potential is important in explaining the ecological adaptation and changes in the distribution of invasive species. To reveal the variation of heat resistance and evolutionary potential in the invasive Mediterranean cryptic species of *Bemisia tabaci*, we selected two Chinese populations—one from Harbin, N China, and one from Turpan, S China—that experience substantial heat and cold stress and conducted knockdown tests under static high- and low-temperature conditions. ANOVAs indicated significant effects of populations and sex on heat knockdown time and chill coma recovery time. The narrow-sense heritability (h^2) estimates of heat tolerance based on a parental half-sibling breeding design ranged from 0.47±0.03 to 0.51±0.06, and the estimates of cold tolerance varied from 0.33±0.07 to 0.36±0.06. Additive genetic variances were significantly different from zero for both heat and cold tolerance. These results suggest that invasive *B. tabaci* Mediterranean cryptic species possesses a strong ability to respond to thermal selection and develops rapid resistance to climate change.

Editor: Xiao-Wei Wang, Zhejiang University, China

Funding: This work was funded by the National Natural Science Foundation of China (31100269), and the National Basic Research and Development Program (Grant No. 2009CB119200), the Ministry of Science and Technology, China, and Common Wealth Special Fund for the Agricultural Industry (No. 201303019). The funders had no role in study design, data collection and analysis, decision to publish, or preparation of the manuscript.

Competing Interests: The authors have declared that no competing interests exist.

* Email: wanfanghao@caas.cn

◔ These authors contributed equally to this work.

Introduction

The invasion of exotic species has accelerated along with the intensification of global climate change, the invasion of exotic species has accelerated. With respect to ecology, as one of the effects of global climate change, bioinvasion has deeply changed the structure and function of global ecosystems [1,2]. With respect to evolution, biological invasion and global climate change have influenced selection pressures to various degrees and have even affected the evolutionary trajectories of organisms [3,4].

The ability of an organism to remain active under extreme conditions is a critical component of fitness [5]. Thermal tolerance is an important factor in the distribution and range dynamics of insects [6–9]. Thermal tolerance has received much attention because it provides insight into how climate shapes variations in the ecology, distribution, hereditary and evolution of species [10–15]. The thermal tolerance of an alien species is crucial to its successful invasion [16], especially under conditions of climate warming [17]. Invasive species with high thermal tolerance could adapt faster under climate warming [18,19,20]. Determining thermal tolerance is an important first step in understanding the ways in which environmental variation affects fitness and, through changes in fitness, the dynamics of a given population. The differences in thermal tolerance among populations is an inheritable characteristic, representing a means by which insect evolution reacts to environmental stress [21,22]. After an exotic species invades a new environment, new selection pressures may result in genetic variation within the species and may even result in the rapid adaptive evolution of tolerance traits [23,24,25]. Short-term evolutionary potential depends on the additive genetic variance in the population. The additive variance is often measured as heritability, the fraction of the total phenotypic variance that is additive. Thus, heritability is a common measure of evolutionary potential [26]. Therefore, understanding the variations in heritability and evolvability in thermal tolerance can help predict a species' future responses to climate change [5,27,28].

The whitefly *Bemisia tabaci* (Gennadius) is a complex species, containing at least 30 morphologically indistinguishable cryptic species, and undergoes rapid evolution [29]. The concomitant eruption of a group of plant viruses has caused considerable agricultural losses in many countries. The *B. tabaci* Mediterranean cryptic species (MED) can cause rapid and widespread invasions and is considered particularly dangerous [30,31,32]. First discovered in Yunnan Province, China, in 2003, *B. tabaci* MED has become the dominant pest species, expanding its range from tropical areas to frigid areas of China within ten years [36] and threatening to replace *Bemisia tabaci* (Gennadius) Middle East-Asia Minor1 (MEAM1) and indigenous species to become the new "super bug" [32]. The high heat-resistance ability of *B. tabaci* MED is one of the potential mechanisms underlying its invasive

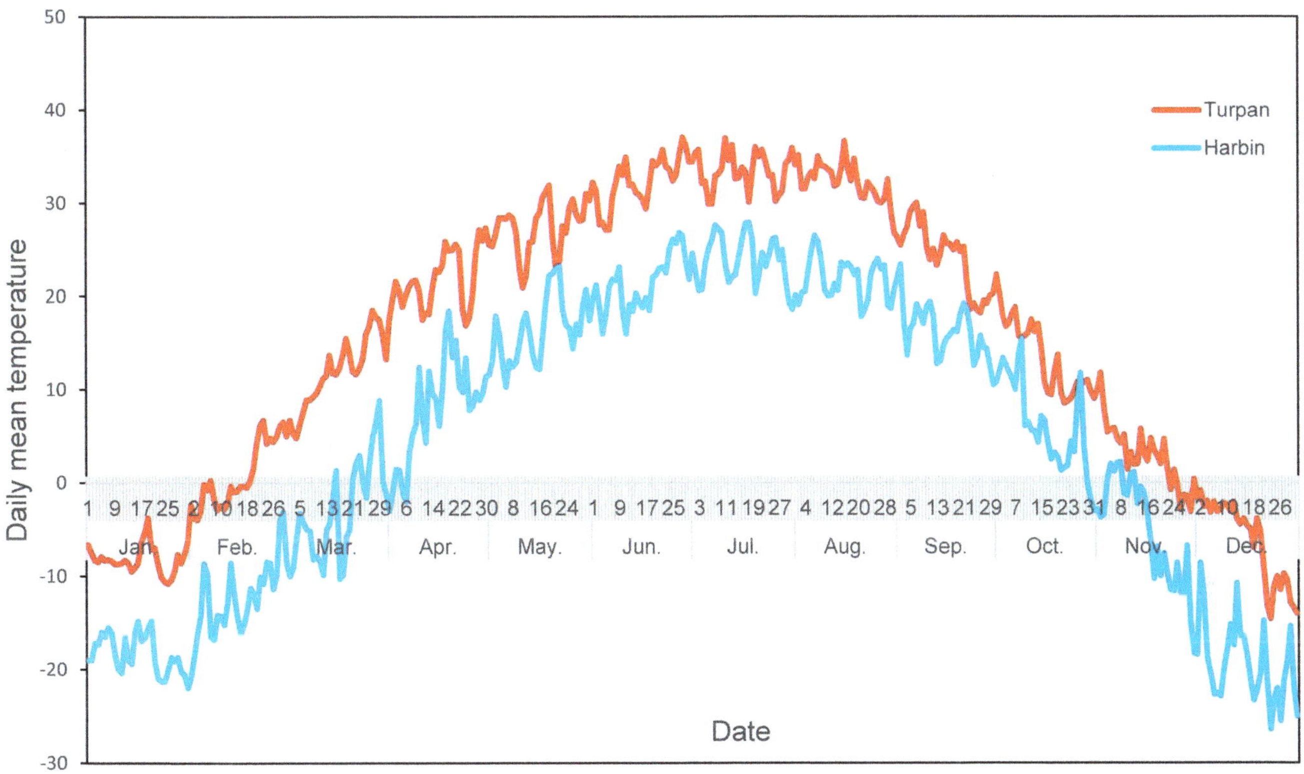

Figure 1. The daily mean air temperatures in Harbin and Turpan, China, in 2012. Data were obtained from the official website of the China Meteorological Data Sharing Service System (www.cdc.cma.gov.cn). The weather stations providing these data are located in Harbin, Heilongjiang (45.56°N, 126.70°E), and Turpan, Xinjiang (42.93°N, 89.13°E).

success [33,34]. Thus, we considered MED to be an excellent research subject for rapid adaptive evolution.

The present study reports the trait means and variations in heat and cold resistance of *B. tabaci* MED to evaluate the evolutionary potential of these fitness components. By estimating the heritability of thermal tolerance, we show that *B. tabaci* MED exhibits high variance in its response to heat stress and has displayed modest evolvability after its successful invasion in China. As a first survey of the quantitative genetics of thermal resistance traits in the whitefly, we studied two introduced populations of the invasive *B. tabaci* MED whitefly distributed across China. We hypothesized that the broad distribution and ecological success of this invasive species could be explained by the high additive genetic variation in the physiological traits related to functional capacities.

Materials and Methods

Ethics Statement

B. tabaci is not a protected species in China; thus, no specific permissions were required for these locations and activities (e.g., the authority responsible for a national park or other protected area of land, the relevant regulatory body concerned with protection of wildlife, *etc*.).

Whitefly Collection and Rearing

Whitefly eggs, nymphs and pseudopupae were collected from cotton, tomato and eggplant crops in fields near Harbin (45.56°N, 126.70°E), Heilongjiang Province, northern China, and Turpan (42.93°N, 89.13°E), Xinjiang Uighur Autonomous Region, western China. Harbin has the coldest weather and longest winters among the major Chinese cities. Due to the Siberian high-pressure system and its location above 45°N, the 24 h average temperature in January is only −18.4°C, and the annual mean temperature is 4.25°C, with the lowest temperatures dropping down to −42.6°C. *B. tabaci* can be found from July to November in the open field and overwinters in greenhouses in the winter months in Harbin [35]. As shown in Figure 1, *B. tabaci* MED experiences a daily mean air temperature (DMAT) range from 18°C to 26°C in July, and the DMAT decreases slowly in August and September in Harbin. Field populations of *B. tabaci* MED suddenly collapse when the temperature drops rapidly in mid- or late October, and field populations then move into greenhouses in Harbin in December; however, adult *Bemisia* can also be found in the open field during this time [39].

Turpan, one of China's four "oven cities", is located at the other climatic extreme where summertime temperatures can soar to 49.6°C. Turpan has a unique temperate continental arid desert climate, with bright sunshine, high temperatures, and large day-night differences in temperature. The average maximum temperature is above 38°C from June to August, with more than 100 days experiencing a daily maximum above 35°C and 35–40 days with a daily maximum temperature above 40°C. *B. tabaci* MED can be found in cotton fields from April to December in Turpan. As shown in Figure 1, *B. tabaci* MED also experiences DMATs ranging from 18°C to 26°C in April, and the DMAT increases gradually in May and June. The DMAT can exceed 30°C in July and August, and *B. tabaci* MED expands its population distribution in the field during this period in Turpan. Field populations in Turpan decrease gradually in September, October and December with the decreasing DMAT.

Figure 1 shows that the temperature has a similar fluctuation trend during the year, but the daily mean temperatures are approximately 10°C lower in Harbin than in Turpan.

Therefore, we expected that whiteflies collected at these locations would be suitable to test our hypotheses regarding adaptation through thermal tolerance in an invasive species.

The Harbin and Turpan field populations (G0) were reared separately. Fifty field-inseminated females were collected from each population and used to establish iso-female lines in 100 specially designed rearing cages in the laboratory; the whiteflies were maintained at 26°C under a 14:10 hour light:dark cycle at 55–60% humidity. The cylindrical rearing cage was 13.5 cm long, with an internal fly-proof plastic screen (11.0 cm in diameter) made of transparent polypropylene to aid in ventilation. Each cage contained a single cotton seedling planted in a pot, which was filled with water. Species/cryptic species identification was confirmed in the G1 generation using the RAPD-PCR method [36] to ensure that all lines were in fact *B. tabaci* MED. This technique has proven to be a highly reliable method for distinguishing MED from other complex species [36]. Only the MED iso-female lines were kept and allowed to expand for an additional generation in the laboratory to ensure large population sizes for each line prior to setting up a mass breeding population. In the second generation (G2) after collection, mass-bred populations were founded with 10 females and 10 males from each of the 50 iso-female lines. The Harbin MED mass-bred population and the Turpan MED mass-bred population were kept separately at 26°C under a 14:10 hour light:dark cycle at 55–60% humidity in two 50×50×50 cm screened cages containing 16 cotton plants, and these populations were reared for three generations to ensure a large, effective population size.

Experimental Design and Data Collection

To compare the heat and cold resistance between the populations and sexes, after three generations of mass rearing, four hundred individuals were randomly selected from each mass-bred population (G5) and scored for knockdown resistance (100 females and 100 males per heat stress and cold stress experiment) on the third day after emergence.

At the same time, we used a standard paternal half-sibling breeding design to estimate the genetic parameters for heat and cold tolerance of both of the Harbin and Turpan MED populations. For both populations (G5), one hundred virgins of parental males (sires) were randomly collected from each mass-bred population immediately after emergence using an air suction insect collector (described below). Each sire was placed in a rearing cage containing a single cotton seedling (with two cotyledons) and five virgin females (dams); the flies were allowed to mate and oviposit for 3 days before the adults were tested individually for knockdown resistance. When their offspring (G6) emerged, 4 females of each dam were collected, reared on cotton plants, mated with males and scored for knockdown resistance (two females for heat stress and 2 females for cold stress) when they were 3 days old. The breeding experiment was performed in the laboratory at 26°C under a 14:10 hour light:dark cycle at 55–60% humidity.

Heat and Cold Stress

To test heat knockdown resistance, female individuals were carefully collected using an air suction insect collector (Mouth Aspirator with HEPA Filter, Model 612, John W. Hock Company, Gainesville, Florida, U.S.A., http://johnwhock.com/products/aspirators/mouth-aspirators).

To expose the insects to heat stress, one adult was confined in a 5 mL centrifuge tube containing a cotton pad through which it could breathe freely. The tubes were moved to a water bath controlled by a high-precision thermoregulator (CC-106A, Germany, Huber Kältemaschinenbau GmbH), which was set at a constant temperature of 45°C. We measured the interval between when the tube was placed in the water bath and when the whitefly lost control of its body and fell to the bottom of the tube.

To test cold resistance, we measured the time to recovery following a chill coma induced by cold shock. One adult, confined as described above, was exposed to −5°C through submersion into a 50% glycol solution cooled by a refrigeration bath circulator (K6-cc-NR, Germany, Huber Kältemaschinenbau GmbH) for a period of 10 minutes, and recovery time was measured at 26°C. Whiteflies assessed for heat and cold stress were tested on the same day.

Statistical Analysis

Our data were generated from a standard paternal half-sibling breeding design (Lynch and Walsh, 1998). The animal model used to analyze the data was

$$y = \alpha + \mathbf{XB} + \mathbf{Z}_s d_s + \mathbf{Z}_d d_d + e \quad (1)$$

where $\mathbf{X}$ is the design matrix for the fixed effect of run $\mathbf{B}$ and $\mathbf{Z}_s$ and $\mathbf{Z}_d$ are the design matrices for the random effects of sire and dam, respectively. The total phenotypic variance (σ^2_P) for the breeding design for the purpose of estimating genetic parameters was represented by

$$\sigma^2_P = \sigma^2_S + \sigma^2_D + \sigma^2_W \quad (2)$$

where σ^2_S, σ^2_D, and σ^2_W, are the sire, dam and within-group level variance components, respectively. Variance and covariance

Table 1. Mean, standard error (1 S.E.), and coefficients of variation for the thermal resistance of the Harbin and Turpan populations of *B. tabaci* MED cryptic species.

		Heat knockdown time, min		Chill coma recovery time, min	
		Mean ± S.E.	CV (%)	Mean ± S.E.	CV (%)
Harbin	Female	9.50±0.33	37.52	7.85±0.28	87.93
	Male	8.07±0.44	39.14	7.58±0.31	88.66
Turpan	Female	12.12±0.53	40.36	9.45±0.42	86.50
	Male	9.83±0.57	41.05	7.90±0.40	88.60

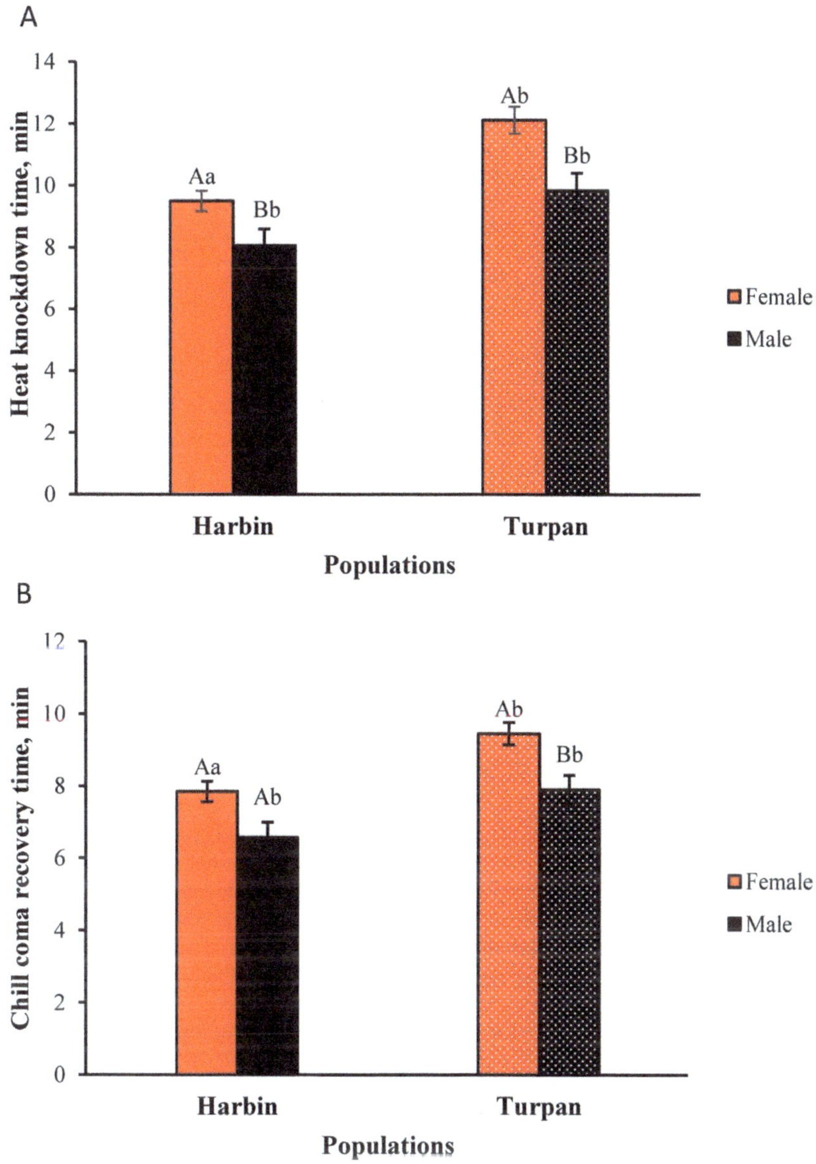

Figure 2. Heat knockdown time (A) and chill coma recovery time (B) (min) for the Harbin and Turpan populations of the *Bemisia tabaci* MED cryptic species. The numbers represent the means ±1 S.E., N = 100 in all cases. Bars topped with a different capital letter indicate a significant difference between the females and males within the same population; the different lowercase letters indicate a significant difference in the same sex between the two populations of *B. tabaci* MED cryptic species (two-way ANOVA followed by *post hoc* Tukey-Kramer multiple comparison test, $P<0.05$).

components were estimated using the restricted maximum likelihood technique via the MIXED procedure in SAS (SAS Institute, Cary, NC). As we used a half-sib full-sib breeding design, the sire variance, σ^2_S, is one-fourth of the additive genetic variance (V_A) [37,38]. Thus, to estimate V_A, we multiplied the sire variance by four.

The additive genetic variance for each trait was first estimated using a univariate model. Log likelihood ratio tests were performed, where the final model for each trait was compared to a model in which σ^2_S was set to zero to determine whether the levels of additive genetic variance for each trait were significantly different from zero [38,39,40]. The phenotypic variance (V_P) in knockdown resistance was computed using all known relationships among individuals. Then, we estimated the narrow-sense heritability for both traits. Narrow-sense heritability for each trait was estimated as the additive genetic variance (V_A) divided by the total phenotypic variance (V_P) [37,41]. We conducted Student's t-tests to determine whether the variance components and heritability estimates differed significantly. Estimates of evolvability, I_A, and the additive genetic coefficient of variance,

$$I_A = V_A/\bar{X}^2 \quad (3)$$

(where $\bar{X}$ is the trait mean) [26,42], were also computed for both traits.

For the population and sexual comparisons, ANOVAs were performed to evaluate the differences in the knockdown and recovery times, followed by the Tukey-Kramer multiple comparison test. The population (Harbin and Turpan) and sex (female

Table 2. ANOVA table of heat knockdown time and chill coma recovery time of *B. tabaci* MED cryptic species.

Source	*df*	Mean Square	*F*	*P*
Heat knockdown time				
Populations	1	417.56	20.50	<0.001
sex	1	308.46	18.97	<0.001
Populations × sex	3	59.91	14.56	0.198
Total	466			
Chill coma recovery time				
Populations	1	36.21	21.77	<0.001
sex	1	27.34	17.92	<0.001
Populations × sex	3	5.90	8.81	0.418
Total	476			

and male) were used as the two factors. $P<0.05$ was considered to be statistically significant.

Results

Differences in Heat and Cold Resistance between the Populations and Sexes

Means, standard errors, and coefficients of variation (%) for heat knockdown time and chill coma recovery time of the Harbin and Turpan populations are listed in Table 1. The mean heat knockdown time was higher in the Turpan population than the higher latitude Harbin population; however, the mean chill coma recovery time was shorter in the Harbin population than the Turpan population (in both females and males). The mean heat knockdown times for the females of both the Harbin and Turpan populations were longer than those of the males, whereas the mean chill coma recovery times were shorter in males than in females. The variation of each trait was similar for both females and males, whereas the coefficients of variation varied considerably between the two traits. The rank order of variation was high for chill coma recovery time (ranging from 86.50% to 88.66%) and moderate for heat knockdown time (ranging from 39.14% to 41.05%).

Figure 2 shows the mean (±1 S.E.) heat knockdown and chill coma recovery times for the two populations of MED. At least 100 individuals of each sex from both populations were examined. An ANOVA indicated the significant effects of population (Harbin and Turpan) ($F_{1,\ 466}=20.50$, $P<0.001$) and sex ($F_{1,\ 466}=18.97$, $P<0.001$) on heat knockdown time (Figure 2a, Table 2), along with the significant effects of population (Harbin and Turpan) ($F_{1,\ 476}=21.77$, $P<0.001$) and sex ($F_{1,\ 476}=17.92$, $P<0.001$) on chill coma recovery time (Figure 2b, Table 2). There was no significant interaction between population and sex ($F_{3,\ 476}=14.56$, $P=0.198$) on heat knockdown time or chill coma recovery time ($F_{3,\ 476}=8.81$, $P=0.418$). Females showed a significantly higher heat resistance than males ($F_{1,\ 233}=8.126$, $P<0.001$, for Harbin; $F_{1,\ 243}=8.198$, $P<0.001$, for Turpan) and a significantly lower cold resistance ($F_{1,\ 233}=8.785$, $P<0.001$, for Harbin; $F_{1,\ 243}=8.779$, $P<0.001$, for Turpan).

Table 3. Narrow-sense heritability (h^2), variance, and coefficient of variation components for heat knockdown time and chill coma recovery time (Harbin and Turpan populations) of *B. tabaci* MED cryptic species.

Mean/Variance*	Harbin population		Turpan population	
	Heat knockdown time, min	Chill coma recovery time, min	Heat knockdown time, min	Chill coma recovery time, min
Mean ± S.E.	9.73±0.38	7.79±0.44	11.99±0.38	9.68±0.44
V_A ± S.E.	9.80±0.002[1]	5.24±0.002[1]	11.69±0.002[1]	6.50±0.002[1]
V_E ± S.E.	13.58±3.13	11.23±2.67	18.69±4.22	14.07±2.85
V_P ± S.E.	19.21	14.56	24.87	19.69
h^2 ± S.E.	0.51±0.06*	0.36±0.06*	0.47±0.03*	0.33±0.07*
I_A × 100	10.35	8.63	8.13	6.94
CV_A	0.322	0.294	0.285	0.263
CV_E	0.143	0.185	0.130	0.150
N	946	957	929	936

Additive genetic variance (V_A), environmental variance (V_E), phenotypic variance (V_P), narrow-sense heritability (h^2), evolvability ($I_A\times100$), coefficient of additive genetic variance (CV_A) and coefficient of environmental variance (CV_E) for heat knockdown time and chill coma recovery time. $N=$ sample size.
[1]$P<0.05$ for log likelihood ratio test of significant differences from zero;
*$P<0.05$.

Heritability and Genetic Variance Components

The estimates of genetic variance were similar between the two protocols, as well as between the two populations (Table 3). Narrow-sense heritability estimates were significant for the heat knockdown times of both the Harbin and Turpan populations. The lower heritability for the heat knockdown time of the Turpan population seems to be driven by higher environmental variance, as significant levels of additive genetic variance were detected (Table 3). For chill coma recovery time, the narrow-sense heritability estimates were significant for both the Harbin and Turpan populations, and the lower heritability for the Turpan population seems to be driven by higher environmental variance (Table 3). This interpretation is supported by the mean standardized estimates of variation (Table 3); the coefficients of additive genetic variation and environmental variation were slightly larger and smaller for the Harbin population, respectively.

Discussion

In our study, we compared the thermal tolerances of the Harbin and Turpan *B. tabaci* populations. The results indicated that genetic differences in thermal tolerance existed between the populations, with an increasing heat tolerance and a decreasing cold tolerance in the Turpan population compared with the Harbin population. Because Turpan experiences a very specific climate, characterized by a long-term, extremely high temperature, the Turpan population may be acclimated to the heat due to its exposure to a generally higher temperature compared with the Harbin population. By contrast, Harbin has the coldest temperatures and longest winter months compared with other major Chinese cities; thus, the Harbin population can survive a colder climate compared with the Turpan population. Our study demonstrated that field populations of *B. tabaci* MED cryptic species possessed a high heat tolerance in response to thermal stresses. This finding implies that there is a significant difference in the adaptive strategies to temperature between the Harbin and Turpan populations as a result of their different experiences with air temperature. Thus, we inferred that high, narrow-sense heritability for both heat and cold resistance is the direct driving force for the adaptive microevolution of *B. tabaci*.

Founder effects, genetic drift, and natural selection can all lead to genetic differentiation in populations of an invading species relative to the invader's source population [43]. In most cases, the alien populations have a high degree of spatial isolation, precluding gene flow among populations. In a laboratory selection experiment using velvetleaf (*Abutilon theophrasti*) and green amaranth (*Amaranthus retroflexus*), significant genetic shifts were observed in just four generations of selection under treatments involving different soil moisture levels, competitor diversity regimes, and intraspecific densities [44]. In the present study, thermal tolerance and phenotypic variance of the invasive whitefly *B. tabaci* were revealed using the static knockdown time method, which has been widely used in *Drosophila* experiments [45,46]. Here, whitefly females of both study populations were more heat tolerant but less cold tolerant than males. These results partially agree with previous studies on MEAM1; however, a different experimental protocol was utilized in this study [47,48,49].

Differences in heat shock proteins (HSPs), especially *Hsp*70, which plays an important role in heat tolerance, might cause the observed differences between females and males. The optimum mRNA expression of HSP genes in females promotes a higher survival rate under heat shock conditions, and there are large between-species differences in HSP gene expression levels [50,51,52].

Given adequate genetic variability, invading species are often able to make rapid evolutionary adjustments. With regard to heat resistance, estimates of heritability in *D. melanogaster* based on knockdown time experiments at 39°C suggest values ranging from 0.03 to 0.11 [52] or up to 0.28 [42], whereas experiments at 38°C suggest values ranging from 0.14 to 0.22 [46]. These estimates might not be directly comparable to those obtained here, which ranged from 0.47 to 0.51 at 45°C, due to the difference in the temperature used to assess the trait as well as the different species examined. Nevertheless, the values obtained here showed that a substantial proportion of the total phenotypic variation in heat resistance was caused by additive genetic variation. Few studies have examined the genetic variance of cold resistance [53–56]. Estimates of heritability on recovery from a chill coma at 0°C suggest values ranging from 0.01 to 0.38 [55]. Our estimates of heritability for *B. tabaci* MED based on chill coma recovery time at −5°C (0.33 to 0.36) also indicated that a substantial proportion of the total phenotypic variation was caused by additive genetic variation.

The values of the evolvability measure, CV_A, which reflects the potential shift in a trait relative to the trait mean, were 0.285–0.322 and 0.263–0.294 for heat knockdown time and chill coma recovery time, respectively. Consequently, the evolvability of cold resistance indicated a stronger evolutionary response than that of heat resistance. For heat resistance, due to the influence of the experimental protocol on estimates, previous estimates of evolvability for *D. melanogaster* based on knockdown time at 38°C suggest broad values, ranging from 0.02 to 14.53. Estimates of additive genetic variance and measures of evolvability were higher when *D. melanogaster* flies were exposed to a static stress rather than a ramping (when temperature was increased to an upper limit) stress when genetic variances were standardized according to the trait means [46].

Several studies have investigated the heritability and evolutionary potential of thermal tolerance traits in invasive species. In the current paper, we provide data for the invasive Mediterranean cryptic whitefly species (*B. tabaci*), which is now widely distributed across China. By measuring the narrow-sense heritability (h^2) of thermal tolerance traits and because of the high CV_A value of knockdown time, we show that natural *B. tabaci* MED cryptic species exhibit a high adaptive potential against both heat and cold stress. The h^2 and CV_A values of heat tolerance traits were higher than the cold tolerance traits in natural MED, which indicated the whiteflies were more able to respond to heat selection in field MED populations than cold selection. We suggest that MED expansion is likely to be favored by climate warming and that the high evolutionary potential of heat tolerance has enabled its invasion into mid- and high-latitude areas.

Acknowledgments

We are very grateful to Prof. Gabor L. Lövei (Department of Agroecology, Aarhus University) and Ms. Lauren Stirling for their English language revisions of this manuscript. We thank Prof. Zhong-Shi Zhou (Institute of Plant Protection, Chinese Academy of Agricultural Sciences) and Ms. Chu-Wei Li (College of Agriculture, Yangtze University) for their assistance during the experiments.

Author Contributions

Conceived and designed the experiments: FZM ZCL FHW. Performed the experiments: FZM. Analyzed the data: FZM ZCL. Contributed reagents/materials/analysis tools: FZM ZCL. Wrote the paper: FZM ZCL FHW RW.

References

1. Huenneke LF (2000) Outlook for plant invasions: interactions with other agents of global change. In: Luken JO, Thieret JW. Assessment and Management of Plant Invasions. New York/Heidelberg/Berlin: Springer-Verlag. pp. 95–103.
2. Simberloff D (2000) Nonindigenous species: a global threat to biodiversity and stability. In: Raven PH. Nature and Human Society: the Quest for A Sustainable World. Washington D C: National Academy Press. pp. 325–334.
3. Lövei GL (1997) Biodiversity – Global change through invasion. Nature 388: pp. 627–628.
4. Mooney HA, Cleland EE (2001) The evolutionary impact of invasive species. Proc Natl Acad Sci USA. 98:5446–5451. doi:10.1073/pnas.091093398.
5. Loeschcke V, Hoffmann AA (2007) Consequences of heat hardening on a field fitness component in *Drosophila* depend on environmental temperature. Am Nat 169: 175–183.
6. Hoffmann AA, Blows MW (1994) Species borders: ecological and evolutionary perspectives. Trends Ecol Evol 9: 223–227.
7. Addo-Bediako A, Chown SL, Gaston KJ (2000) Thermal tolerance, climatic variability and latitude. Proc R Soc Lond B Biol Sci 267: 739–745.
8. Chown SL (2001) Physiological variation in insects: hierarchical levels and implications. J Insect Physiol 47: 649–660.
9. Chown SL, Nicolson S (2004) Insect physiological ecology: mechanisms and patterns. Oxford: Oxford University Press. 243 p.
10. Janzen DH (1967) Why mountain passes are higher in the tropics. Am Nat 101: 233–249.
11. Pörtner H (2001) Climate change and temperature-dependent biogeography: oxygen limitation of thermal tolerance in animals. Naturwissenschaften 88: 137–146.
12. Ghalambor CK, Huey RB, Martin PR, Tewksbury JJ, Wang G (2006) Are mountain passes higher in the tropics? Janzen's hypothesis revisited. Integr Comp Biol 46: 5–17.
13. Chown SL, Terblanche JS (2006) Physiological diversity in insects: ecological and evolutionary contexts. Adv In Insect Phys 33: 50–152.
14. Terblanche JS, Sinclair BJ, Jaco Klok C, McFarlane ML, Chown SL (2005) The effects of acclimation on thermal tolerance, desiccation resistance and metabolic rate in *Chirodica chalcoptera* (Coleoptera: Chrysomelidae). J Insect Physiol 51: 1013–1023.
15. Zhou ZS, Rasmann S, Li M, Guo JY, Chen HS, et al. (2013) Cold temperatures increase cold hardiness in the next generation *Ophraella communa* beetles. PloS one 8: e74760, doi: 10.1371/journal.pone.0074760.
16. Kang L, Chen B, Wei J-N, Liu T-X (2009) Roles of thermal adaptation and chemical ecology in *Liriomyza* distribution and control. Annu Rev Entomol 54: 127–145.
17. Dukes JS, Mooney HA (1999) Does global change increase the success of biological invaders? Trends Ecol Evol 14: 135–139.
18. Mooney HA, Hobbs RJ (2000) Invasive species in a changing world. Washington, DC: Island Press. 421 p.
19. Parmesan C (2006) Ecological and evolutionary responses to recent climate change. Annu Rev Ecol Evol Syst 37: 637–669.
20. Zerebecki RA, Sorte CJ (2011) Temperature tolerance and stress proteins as mechanisms of invasive species success. PLoS One 6: e14806, doi:10.1371/journal.pone.0014806.
21. Masaki S (1996) Geographical variation of life cycle in crickets (Ensifera: Grylloidea). Eur J Entomol. 93: 281–302.
22. Sorensen JG, Dahlgaard J, Loeschcke V (2001) Genetic variation in thermal tolerance among natural populations of *Drosophila buzzatii*: down regulation of *Hsp*70 expression and variation in heat stress resistance traits. Funct Ecol 15: 289–296.
23. Blossey B, N tzold R (1995) Evolution of increased competitive ability in invasive non-indigenous plants: a hypothesis. J Ecol 83:887–889.
24. Keane RM, Crawley MJ (2002) Exotic plant invasions and the enemy release hypothesis. Trends Ecol Evol 17:164–170.
25. Strauss SY, Rudgers JA, Lau JA, Irwin RE (2002) Direct and ecological costs of resistance to herbivory. Trends Ecol Evol 17:278–285.
26. Hansen TF, Pélabon C, Houle D (2011) Heritability is not Evolvability, Evol Biol 38: 258–277.
27. Williams SE, Bolitho EE, Fox S (2003) Climate change in Australian tropical rainforests: an impending environmental catastrophe. Proc R Soc Lond B Biol Sci 270: 1887–1892.
28. Ju RT, Gao L, Zhou XH, Li B (2013) Tolerance to high temperature extremes in an invasive lace bug, *Corythucha ciliata* (Hemiptera: Tingidae), in subtropical China. PLoS One 8: e54372, doi:10.1371/journal.pone.0054372.
29. De Barro PJ, Liu SS, Boykin LM, Dinsdale AB (2011) *Bemisia tabaci*: a statement of species status. Annu Rev Entomol 56: 1–19.
30. Muñiz M, Nombela G, Barrios L (2002) Within- plant distribution and infestation pattern of the B- and Q- biotypes of the whitefly, *Bemisia tabaci*, on tomato and pepper. Entomol Exp Appl 104: 369–373.
31. Moriones E, Navas-Castillo J (2000) Tomato yellow leaf curl virus, an emerging virus complex causing epidemics worldwide. Virus Res 71: 123–134.
32. Hu J, De Barro P, Zhao H, Wang J, Nardi F, et al. (2011) An extensive field survey combined with a phylogenetic analysis reveals rapid and widespread invasion of two alien whiteflies in China. PLoS One 6: e16061, doi:10.1371/journal.pone.0016061.
33. Lü ZC, Gao QL, Yu H, Guo JY, Wan FH (2014) Increased survival and prolonged longevity mainly contribute to the temperature-adaptive evolutionary strategy in invasive *Bemisia tabaci* Middle East Asia Minor 1. J Insect Sci In press.
34. Bonato O, Lurette A, Vidal C, Fargues J (2007) Modelling temperature-dependent bionomics of *Bemisia tabaci* (Q- biotype). Physiol Entomol 32: 50–55.
35. Fu X, Ye L-F, Wang G-Q, Ge F (2011) Occurrence and development for *Bemisia tabaci* and Trialeurodes vaporariorum in Heilongjiang Province. Chinese J. Appl. Entomol. 48: 32–37.
36. De Barro PJ, Driver F (1997) Use of RAPD PCR to distinguish the B biotype from other biotypes of *Bemisia tabaci* (Gennadius) (Hemiptera: Aleyrodidae). Aust J Entomol 36: 149–152.
37. Falconer DS, Mackay TFC (1996) Introduction to Quantitative Genetics. Harlow: Pearson Education Limited. 438 p.
38. McGuigan K, Blows MW (2010) Evolvability of individual traits in a multivariate context: partitioning the additive genetic variance into common and specifric components. Evolution 64: 1899–1911.
39. Littell R, Milliken A, Stroup W, Wolfinger R (1996) SAS System for Mixed Models. Cary, NC.: SAS Institute, Incoportated. 828 p.
40. Simonsen AK, Stinchcombe JR (2010) Quantifying evolutionary genetic constraints in the ivyleaf morning glory, *Ipomoea hederacea*. Int. J. Plant Sci. 171: 972–986.
41. Lynch M, Walsh B (1998) Genetics and the Analysis of Quantitative Traits. Sunderland: Sinauer Associates, Inc. 980 p.
42. Houle D (1992) Comparing evolvability and variability of quantitative traits. Genetics 130: 195–204.
43. Daehler C, Strong D (1997) Hybridization between introduced smooth cordgrass (*Spartina alterniflora*; Poaceae) and native California cordgrass (*S. foliosa*) in San Francisco Bay, California, USA. Am J Bot 84: 607–607.
44. Zangerl A, Bazzaz F (1984) Effects of short-term selection along environmental gradients on variation in populations of *Amaranthus retroflexus* and *Abutilon theophrasti*. Ecology 65: 207–217.
45. Loeschcke V, Sorensen JG, Hoffmann A (2003) Adaptation of *Drosophila* to temperature extremes: bringing together quantitative and molecular approaches. J Therm Biol 28: 175–216.
46. Mitchell KA, Hoffmann AA (2010) Thermal ramping rate influences evolutionary potential and species differences for upper thermal limits in *Drosophila*. Funct Ecol 24: 694–700.
47. Terblanche JS, Deere JA, Clusella-Trullas S, Janion C, Chown SL (2007) Critical thermal limits depend on methodological context. Proc R Soc Lond B Biol Sci 274: 2935–2942.
48. Lue ZC, Wan FH (2008) Differential gene expression in whitefly (*Bemisia tabaci*) B- biotype females and males under heat- shock condition. Comp Biochem Physiol Part D Genomics Proteomics 3: 257–262.
49. Dahlgaard J, Hasson E, Loeschcke V (2001) Behavioral differentiation in oviposition activity in *Drosophila buzzatii* from highland and lowland populations in Argentina: plasticity or thermal adaptation? Evolution 55: 738–747.
50. Yu H,Wan FH (2009) Cloning and expression of heat shock protein genes in two whitefly species in response to thermal stress. J Appl Entomol 133: 602–614.
51. Lü ZC, Wan FH (2011) Using double-stranded RNA to explore the role of heat shock protein genes in heat tolerance in *Bemisia tabaci* (Gennadius). J Exp Biol 214: 764–769.
52. McColl G, Hoffmann AA, McKechnie SW (1996) Response of two heat shock genes to selection for knockdown heat resistance in *Drosophila melanogaster*. Genetics 143: 1615–1627.
53. Huey R, Crill W, Kingsolver J, Weber K (1992) A method for rapid measurement of heat or cold resistance of small insects. Funct Ecol 6: 489–494.
54. Charo-Karisa H, Rezk MA, Bovenhuis H, Komen H (2005) Heritability of cold tolerance in Nile tilapia, *Oreochromis niloticus*, juveniles. Aquaculture 249: 115–123.
55. Kellermann V, van Heerwaarden B, Sgro CM, Hoffmann AA (2009) Fundamental evolutionary limits in ecological traits drive *Drosophila* species distributions. Science 325: 1244–1246.
56. Kimura MT (2004) Cold and heat tolerance of drosophilid flies with reference to their latitudinal distributions. Oecologia 140: 442–449.

9

Will Climate Change Affect Outbreak Patterns of Planthoppers in Bangladesh?

M. P. Ali[1]*, Dingcheng Huang[2]*, G. Nachman[3], Nur Ahmed[1], Mahfuz Ara Begum[1], M. F. Rabbi[1]

1 Entomology Division, Bangladesh Rice Research Institute, Gazipur, Bangladesh, **2** International Affairs Office, University of Chinese Academy of Sciences, Beijing, China, **3** Department of Biology, University of Copenhagen, Copenhagen, Denmark

Abstract

Recently, planthoppers outbreaks have intensified across Asia resulting in heavy rice yield losses. The problem has been widely reported as being induced by insecticides while other factors such as global warming that could be potential drivers have been neglected. Here, we speculate that global warming may increase outbreak risk of brown planthopper (*Nilaparvata lugens* Stål.). We present data that demonstrate the relationship between climate variables (air temperature and precipitation) and the abundance of brown planthopper (BPH) during 1998–2007. Data show that BPH has become significantly more abundant in April over the 10-year period, but our data do not indicate that this is due to a change in climate, as no significant time trends in temperature and precipitation could be demonstrated. The abundance of BPH varied considerably between months within a year which is attributed to seasonal factors, including the availability of suitable host plants. On the other hand, the variation within months is attributed to fluctuations in monthly temperature and precipitation among years. The effects of these weather variables on BPH abundance were analyzed statistically by a general linear model. The statistical model shows that the expected effect of increasing temperatures is ambiguous and interacts with the amount of rainfall. According to the model, months or areas characterized by a climate that is either cold and dry or hot and wet are likely to experience higher levels of BPH due to climate change, whereas other combinations of temperature and rainfall may reduce the abundance of BPH. The analysis indicates that global warming may have contributed to the recent outbreaks of BPH in some rice growing areas of Asia, and that the severity of such outbreaks is likely to increase if climate change exaggerates. Our study highlights the need to consider climate change when designing strategies to manage planthoppers outbreaks.

Editor: Gil Bohrer, The Ohio State University, United States of America

Funding: This research was supported by Bangladesh Rice Research Institute (BRRI). The funders had no role in study design, data collection and analysis, decision to publish, or preparation of the manuscript.

Competing Interests: The authors have declared that no competing interests exist.

* E-mail: panna_ali@yahoo.com (MPA); huangdc@ucas.ac.cn (DH)

Introduction

Outbreak frequency of *Nilaparvata lugens* (brown planthopper; BPH) has been increasing in Asian rice growing countries in recent years (2005–2012) [1]. These trends are widely linked to adverse effects on BPH natural enemies of the increased use of broad spectrum insecticides for control of a range of pests [2–11]. They are rarely linked to environmental changes such as global warming that can directly or indirectly influence outbreak trends in rice growing regions [12]. Insecticide resistance is another reason for the increase.

Rice is extremely important for human food supply, yet a comprehensive understanding of how climate change affects rice pests is still premature. As the global climate continues to change, there is a need to understand how planthoppers respond to low-frequency (interannual or longer period) climatic variability. Such low frequency climatic variability is of interest because of its significance for the understanding and prediction of protracted climatic anomalies [13]. Since ongoing climate change has a profound effect on insect pests and changes their pest status [14] and population dynamics [15], climate change probably contributes to recent destructive outbreaks of brown planthopper (BPH) *Nilaparvata lugens* (Stål) (Homoptera: Delphacidae). BPH is one of the most serious pests of rice in both temperate and tropical regions of East and South Asia and has become especially problematic over the past few years [16]. They are active throughout the year in the tropics [17]. In Bangladesh, where rice is grown year-round, they are prevalent throughout the year and can produce 8–11 generations per a year [personal communication, Nur Ahmed].

BPH's life cycle comprises three distinct stages: egg, nymph and adult. Both nymphs and adults directly damage rice plants through sucking the cell sap from the base (stem) of the plants [18] and by transmitting viruses such as rice ragged stunt (RRSV) and rice grassy stunt (RGSV) [19–20] which cause severe losses. In 2005–2006, more than 485,000 hectares of rice in southern Vietnam were severely affected by viral diseases seemingly spread by BPH, resulting in the loss of 828,000 tons of rice valued at US$120 million [21].

Recently, outbreaks of BPH have become ever more common in Asia where farmers experience devastating losses [1], and have threatened rice production in parts of Thailand, Philippines, Indonesia, India, Bangladesh, Malaysia and China in the years of 2009–2010. Global warming may have been a contributing factor because steadily warmer autumns have occurred since the 1990s. Changed climatic conditions of particular interest are milder winters and warmer summers as well as changing patterns of precipitation. The latter includes increased risk of both extreme precipitation and severe dry spells. These conditions are predicted by the IPCC [22] to become more frequent due to changes in the global climate. Overall, temperature is predicted to increase 1.5–4.5°C during the present century [23] and precipitation to rise 10 to 15% because a warmer atmosphere holds more water [24–25]. These changes could profoundly affect the population dynamics and the status of insect pests of crops [24], [26]. At the same time, migrant species may also change status to become major rice pests. Thus, in Japan *N. lugens* was dominant until 1984, *Sogotula furcifera* by 1993 and *Cnaphalocrosis medinalis* up to now [27].

Reported BPH outbreaks are mostly attributed to the application of insecticides in 2008–2012 [2–11]. Other factors that may induce the problem and threaten rice production in major rice producing countries are greatly neglected although temperature has already started to rise with concurrent changes in precipitation patterns. For an array of insects these phenomena may affect e.g. phenology and pest status. Analysis of trapping data from several decades indicates that climate change also causes a change in the phenology of insects [28]. It is suggested that the use of the lower developmental threshold (T_0) and the thermal constant (K) that have been reported for insects would be useful in predicting the phenology of insect communities under global warming [29]. Therefore, a more detailed analysis of correlations between climatic parameters and phenology, abundance and pest status are needed. The hypothesis is that a higher frequency of destructive outbreaks of plant hoppers is related to climatic changes.

In this paper, we use concurrent information on air temperatures, precipitation and abundance of BPH, recorded in Bangladesh over a 10-years period, to analyze the complex relationship between the climate variables and the dynamics of BPH. The aim of the present study is to provide a better understanding of the role of global warming in eliciting destructive outbreaks of BPH during the recent years as well as predicting its future role.

Materials and Methods

Study Site

This study was conducted at a field site belonging to Bangladesh Rice Research Institute (24°0′0″N, 90°25′48″E), which is about 36 km north of Dhaka, the capital of Bangladesh. The site, which is located in a rice-growing area, has been used mainly for insect pest monitoring and forecasting since 1980.

Data Collection

The BPH population was monitored daily by a light trap (Fig. 1) which was placed close to rice fields. The trap used a 100-watt tungsten light bulb positioned at 2 m height. Lighting hours were set for 12 h (from evening to dawn) and a strip with dichlorvos (or Vapona) was put in each collecting cage to kill the insects when caught. A steel drum was placed below the cages to collect all insects including minute ones and to allow for their easy removal from the cages. The bulb and Vapona were changed when needed. The insects in the cages were removed every morning and preserved temporarily in paper bags dated daily before processing. Trapped insects were identified, counted and deposited in the Entomology Division of Bangladesh Rice Research Institute. There were no other lights near the trap. Monthly BPH catches from January to December from 1998 through 2007 were then calculated and used in this study (Appendix to manuscriptS).

Figure 1. The light trap used to collect brown planthopper (BPH) in this study. It was placed adjacent to a rice field belonging to Bangladesh Rice Research Institute (BRRI).

Climate Data Source

Climate variables (daily maximum and minimum temperatures, and precipitation) covering the same interval and study area were provided by the Plant Physiology Division of Bangladesh Rice Research Institute (www.brri.gov.bd).

Statistical Analysis

Trends in climate and abundance of BPH from 1998 through 2007. We analyzed whether the climate data showed any long-term trends with respect to monthly temperatures and precipitation. For this purpose, we applied three climatic variables for every month through the 10 years period: *MinTemp* and *MaxTemp* which represent the average of the daily minimum and maximum temperatures during a month, and *Rain* which is the average amount of rain falling per day during the month. The three dependent variables were analyzed by a general linear model (PROC GLM in SAS [30]) with *Month* as a classification variable and *Year* (number of years since 1998) as a quantitative variable. The model also included the interactions between *Month* and *Year* to test whether trends differed among months. In the following, this model is termed the "null-model". Overall, the model included 12 month parameters, 1 year parameter, and 12 Month X Year interaction terms, to a total degree of freedom of 23.

The monthly variation in BPH was analyzed in the same way as the climate variables, using the dependent variable N calculated as the average number of adult BPH caught per day during a month. Prior to the analyses we checked the dependent variables for normality and variance homogeneity. Variables that did not meet these requirements were transformed with the appropriate transformation obtained by means of Taylor's power law [31].

Effect of weather variables on the abundance of BPH. The null-model predicting the expected number of BPH by means of the independent variables *Month* and *Year* and their interactions was expanded to include temperature and rainfall in order to investigate how much of the residual variance that could be accounted for by taking the actual weather in the month into account. The null-model expanded with the quantitative climate variables (called *Temp* and *Rain*) is termed the "full climatic model". To avoid collinearity between the independent variables, minimum and maximum temperatures were not used in the same model because they tend to be positively correlated.

The full climatic model included first and second order terms of *Temp* and *Rain* plus all interaction terms between these independent variables. *Temp* could be either *MinTemp* or *MaxTemp*. Variables were transformed if needed to meet the criteria of normality and variance homogeneity. Analyses were conducted by means of PROC GLMSELECT in SAS [30] using a backwards elimination procedure based on Akaike's corrected information criterion (AICC). The model with the lowest AICC was regarded as the best.

Predicted effects of climate change on BPH population. We used the general linear model identified above to simulate what will happen to the BPH if climate change results in higher temperatures in combination with a changing pattern of rain. We applied the following scenarios: Daily average temperatures were assumed to increase with 0°C, 1°C or 2°C and daily precipitation was either increased or decreased with 10%. The relative change in abundance of BPH for each month was calculated as $R = 100(N - N_0)/N_0\%$, where N is the predicted catch of BPH for a given scenario and N_0 is the corresponding value in the *status quo* scenario (i.e. the scenario with no climate change).

Results and Discussion

The period from December to February is the cold and dry season with low abundance of planthoppers (Fig. 2A–C). Insect pests often respond rapidly and dramatically to changes in climatic conditions affecting development (such as sudden precipitation and extreme temperatures), leading to large temporal variations in insect pest populations [32].

Trends in Climate and Abundance of BPH from 1998 Through 2007

Figure 2 indicates that standard deviations of rainfall and BPH tend to increase with the average. We therefore applied Taylor's power law [31] to identify the transformation of data that best stabilizes the variance. The appropriate transformations were obtained as $y^* = y^p$ where $p = 1 - b/2$, b is the slope of the regression line fitted to the empirical values of log(variance) plotted against log(average) from each of the 12 months (Fig. 3). As p for daily rainfall is close to one third, the cubic root transformation was used to transform daily rainfall values into $R = Rain^{1/3}$. For daily counts of PBH, b is close to 2, which means that this variable should be log transformed. As N in two cases was 0, we used the log(N+1) transformation. There was no need of transforming temperature values.

The model consisting of *Year*, *Month* and the interaction between them explained 97.2% of the variation in minimum temperatures, 89.0% in maximum temperatures and 76.8% in the transformed values of daily precipitation. However, for all three variables, neither *Year* nor the interactions between *Year* and *Month* were significant ($P > 0.05$). There was a considerable monthly variation in both temperatures and rainfall, but the analysis did not indicate that this pattern has changed over the 10-years period.

With respect to a trend in BPH abundance, the null-model explained 84.8% of the variation in $y^* = \log(N+1)$ ($F_{23,119} = 23.21$; $P < 0.0001$). There was no overall trend over years ($P = 0.2468$), whereas both *Month* and the interactions between *Month* and *Year* were significant (*Month*: $P < 0.0001$; *Year***Month*: $P = 0.0020$). Only April showed a significant trend in BPH over the 10 years with a slope of 0.1540 per year ($P < 0.0057$). *Month* alone explained 79.4%

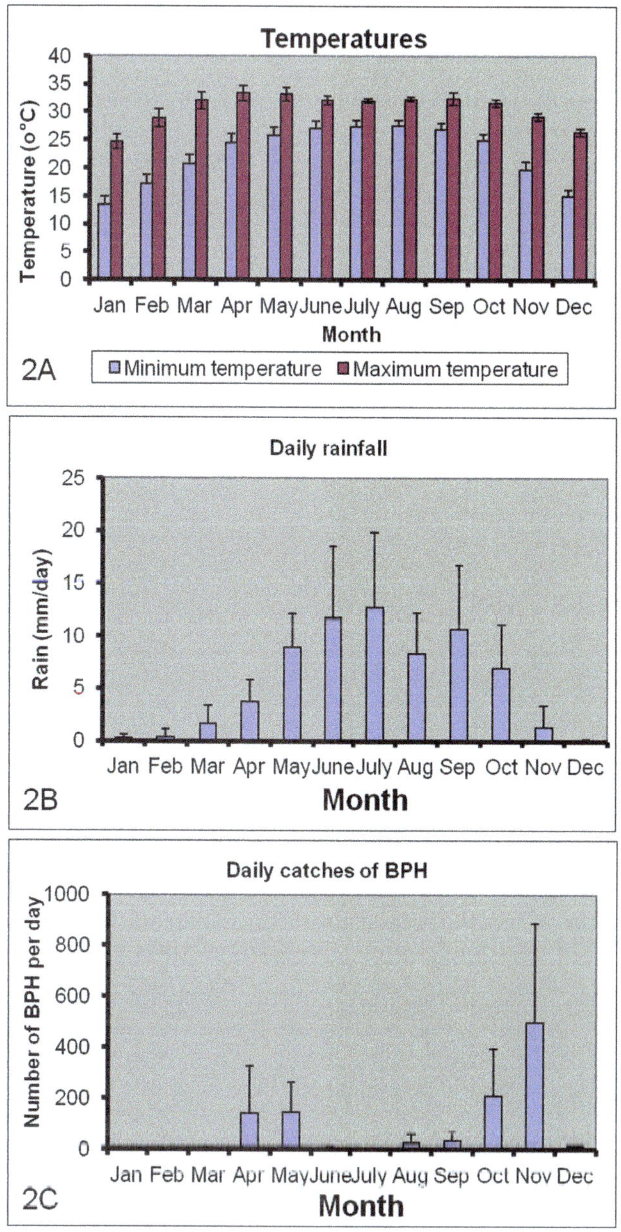

Figure 2. Monthly averages (±SD) of (A) Minimum and maximum temperatures; (B) daily rainfall; and (C) daily catches of Brown Planthoppers (BPH). Data from Bangladesh Rice Research Institute (BRRI).

of the total variation in log (N+1). As we could not demonstrate that the climate had changed, the most likely explanation for the increased abundance of BPH in April is that this month coincides with the growing and harvesting period of Boro rice, and that areas planted with Boro rice have expanded during the recent years [33–34].

Effect of Weather Variables on the Abundance of BPH

When the null-model used above to predict the abundance of BPH was extended by including climatic variables, *MinTemp* had a higher predictive power than *MaxTemp*. We therefore used *MinTemp* to represent temperature in the following analyses. The full climatic model explained 86.3% of the variation in the

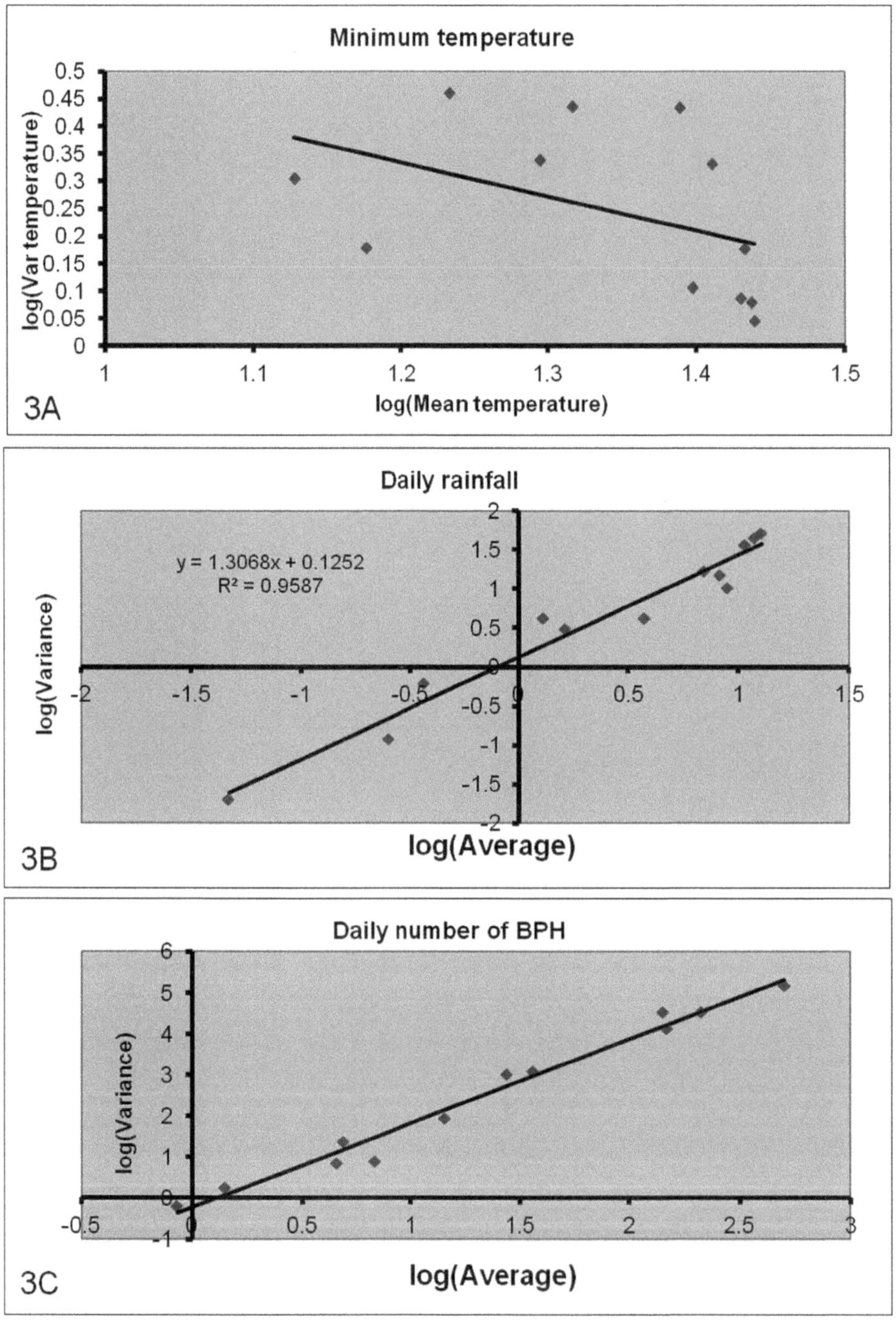

Figure 3. Log (variance) plotted against log(average) for (a) daily rainfall and (b) daily catches of BPH adults. Each dot represents a month. The straight line for daily rainfall is described by $y = 1.3068x+0.1252$ ($R^2 = 0.9587$) and for daily catches of BPH by $y = 2.0512x-0.2258$ ($R^2 = 0.9793$).

observed values of log(N+1) ($F_{31,88} = 17.9$; $P<0.0001$). The AICC of the null-model (i.e. the model without climate variables) was −94.80. The climate-driven model with the lowest value of AICC (−107.12) contained the terms *Month, T*, T^2, R, R^2, and T^2R^2, where T is the monthly minimum temperature (i.e. $T = MinTemp$) and. $R = Rain^{1/3}$. This reduced version of the full model (called the best climatic model) explained 83.6% of the variation in log(N+1) ($F_{16,103} = 32.8$; $P<0.0001$). The parameters associated with the model's variables are shown in Table 1.

Though the best climatic model explained slightly less of the variation than the null-model did (83.6% *vs* 84.8%), the former has 7 parameters fewer than the null-model. However, the main advantage of the climate-driven model is that it provides a tool for analyzing how temperature and rain influence the dynamics of BPH. Thus, the model's parameters indicate that increasing temperatures will have a positive influence on BPH at low temperatures but a negative influence at high temperatures. Low levels of precipitation (e.g. <5 mm) will benefit BPH as long as the minimum temperature is below *ca* 23°C. Above this threshold,

Table 1. Parameter values of the model that best describes the observed values of log(N+1), where N is the average number of BPH caught per day during a month.

Parameter	Estimate	Standard Error	t Value	*P*
January	−2.3351	1.5228	−1.53	0.1282
February	−2.3138	1.7269	−1.34	0.1832
March	−1.6957	1.8130	−0.94	0.3518
April	−0.2333	1.8110	−0.13	0.8978
May	0.1209	1.7931	0.07	0.9464
June	−1.0147	1.7756	−0.57	0.5689
July	−1.2602	1.7753	−0.71	0.4794
August	−0.6212	1.7711	−0.35	0.7265
September	−0.5186	1.7802	−0.29	0.7714
October	0.2358	1.8074	0.13	0.8964
November	0.0983	1.8015	0.05	0.9566
December	−1.4999	1.6344	−0.92	0.3609
T	0.3528	0.1767	2.00	0.0485
T^2	−0.0120	0.0044	−2.69	0.0083
R	0.7169	0.2094	3.42	0.0009
R^2	−0.9471	0.2213	−4.28	<.0001
T^2R^2	0.00119	0.00029	4.09	<.0001

T is the temperature and R the cubic root of the daily rainfall in month i. The predicted value of log(N+1) for month i is obtained as $\log(N+1) = M_i + 0.3528T - 0.0120T^2 + 0.7169R - 0.9471R^2 + 0.00119T^2R^2$ where M_i is the parameter associated with month i. P is the probability that the true parameter value is equal to 0.

more rain will increase BPH. Figure 4 shows the predicted combined effects of temperature and rainfall when added to the monthly value of log(N+1).

Predicted Effects of Climate Change on BPH Population

Figure 5 shows the predicted monthly abundances of BPH under the various scenarios while Figure 6 shows the predicted relative changes in BPH abundance. The model predicts that BPH will become more abundant in January and during June-July, in particular if higher temperatures are associated with more precipitation. As these months are usually characterized by low abundances of BPH, the economic consequences are likely to be negligible. On the other hand, climate change is expected to decrease abundance of BPH during the spring and autumn seasons. These changes in seasonal patterns seem to benefit Boro rice crops (April–May), whereas rice grown in the Aus (July–

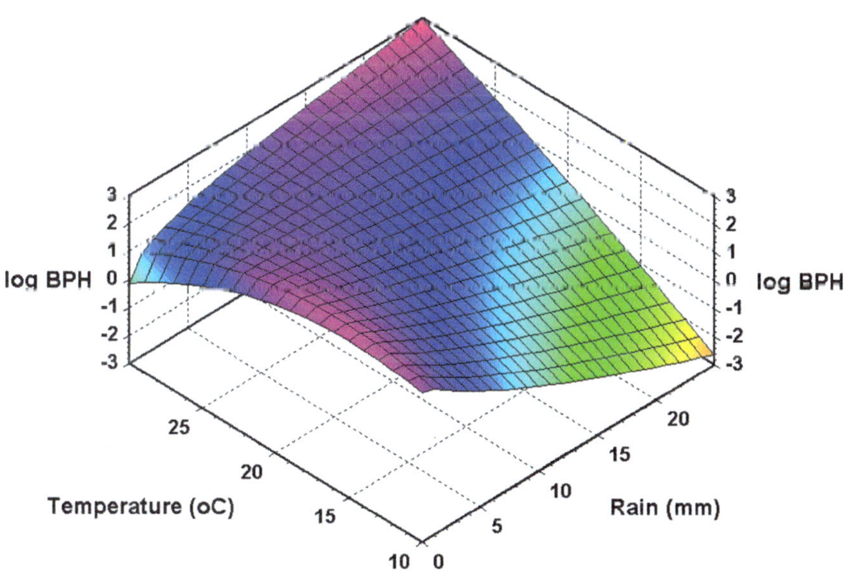

Figure 4. The effects of minimum temperature and rainfall on the predicted abundance of BPH (log(N+1)) (see Table 1 for further explanation).

August) and Aman (November–December) seasons will become more exposed to BPH. Overall, the daily mean number of BPH on a yearly basis is predicted to decline with an increase in temperature (Fig. 7). The only scenario that may increase BPH is the one where precipitation declines without a concurrent increase in temperature.

Even though our model indicates that BPH should become less abundant in Bangladesh if climate change causes an increase in the mean temperatures, it does not imply that this will be the case everywhere. Areas characterized by climatic conditions similar to those prevalent in Bangladesh during January (cool and dry) or July (hot and wet) seems most vulnerable to a changing climate. Using the model in combination with GIS-based climate information could assist in pointing out regions where future rice production is jeopardized.

The results confirm that the combined effect of temperature and rainfall is complex [35–36]. Global warming may increase evaporation and precipitation with 10 to 15% because a warmer atmosphere holds more moisture [24–25]. In warm areas this is expected to increase populations of BPH. In some areas, however, global warming is predicted to result in a drier climate. Thus, Dai [37] suggests severe and widespread droughts in the next 30–90

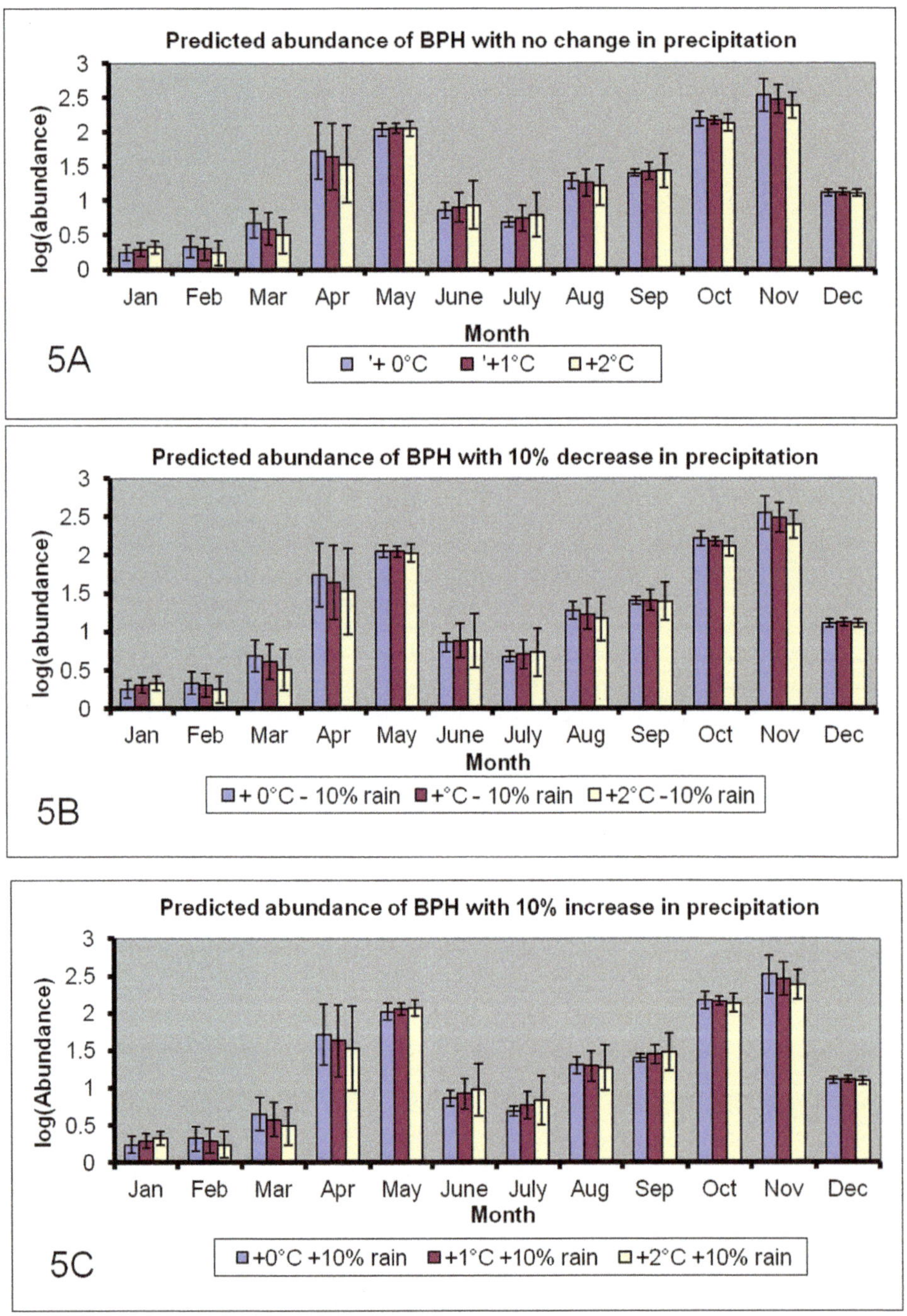

Figure 5. The predicted daily catches of adult BPH (±SD) if minimum temperature increases with either 0°C, 1°C or 2°C, and daily precipitation either decreases or increases with 10%.

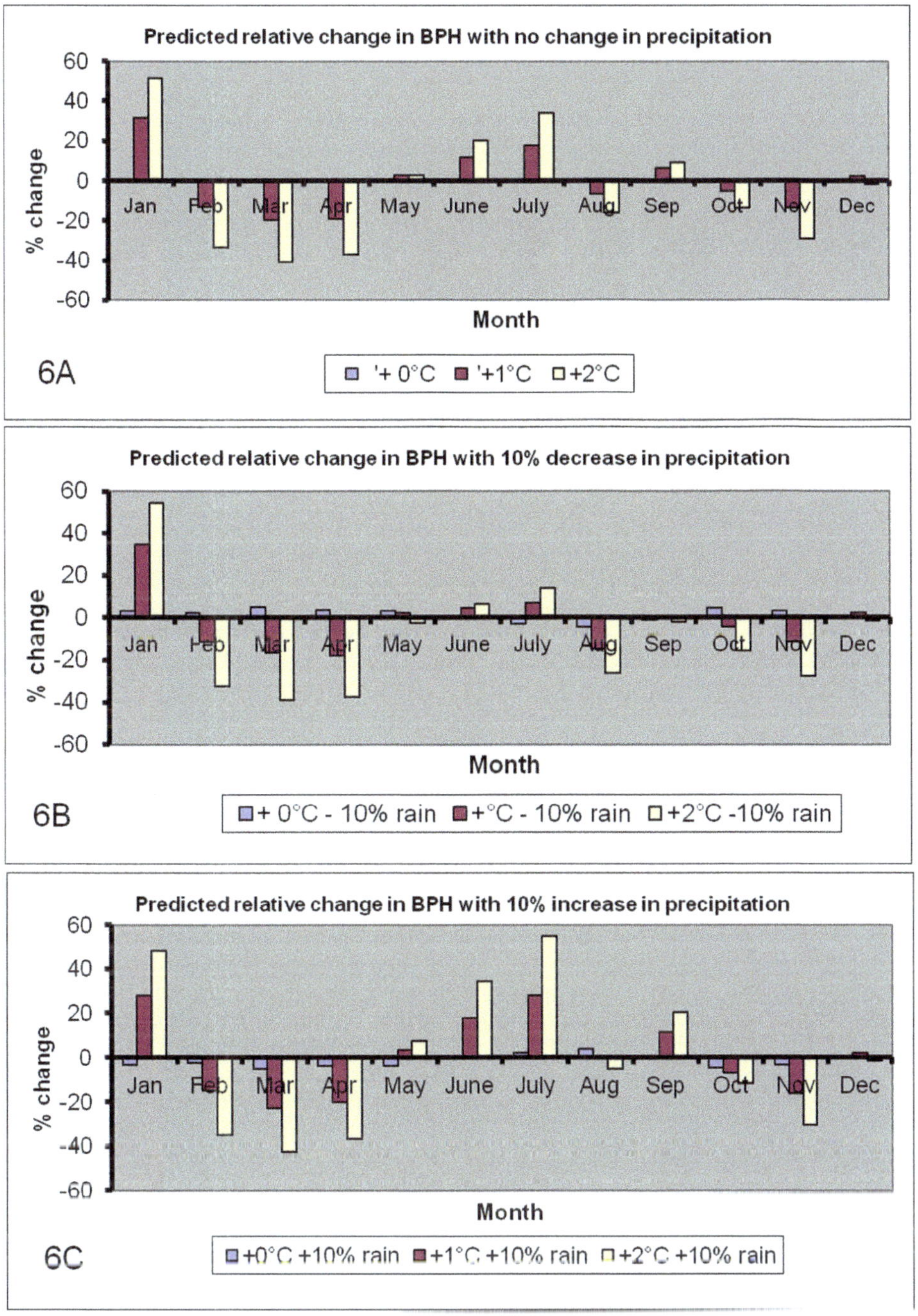

Figure 6. The predicted relative changes in daily catches of BPH under the various scenarios.

years over many land areas. According to the model, this may also exaggerate problems with BPH, but only in areas where minimum temperatures stay below *ca* 23°C.

Climate change, and particularly global warming, may have a dramatic impact on pest insect species [38]. Many authors predict that climate change will have a number of effects on insects: sweeping shifts in herbivory rates; altered distribution and outbreak frequency of key insect pests; unpredictably altered relationships with natural enemies; and a general decrease in biodiversity [39–41]. The findings of the present study indicate that the reported BPH outbreak in rice fields occurring during the warm seasons might be attributed to global warming [42]. However, the model also shows that temperatures exceeding 23°C in a dry climate may have an inhibiting effect on population density, implying that the consequences of global warming on planthoppers are ambiguous. Thus, their long-term abundance may increase during the cold season and decrease during the hot season, depending on how the ambient temperatures change relative to the threshold value of *ca* 23°C. Planthopper outbreaks were observed in the boro rice growing season i.e. the dry season [43] and significant BPH outbreaks were recorded during the dry season in the early 1990s in the Vientiane Plain [44] along with higher temperatures. The population subsequently declined in January due to low temperatures and peaked again in March and

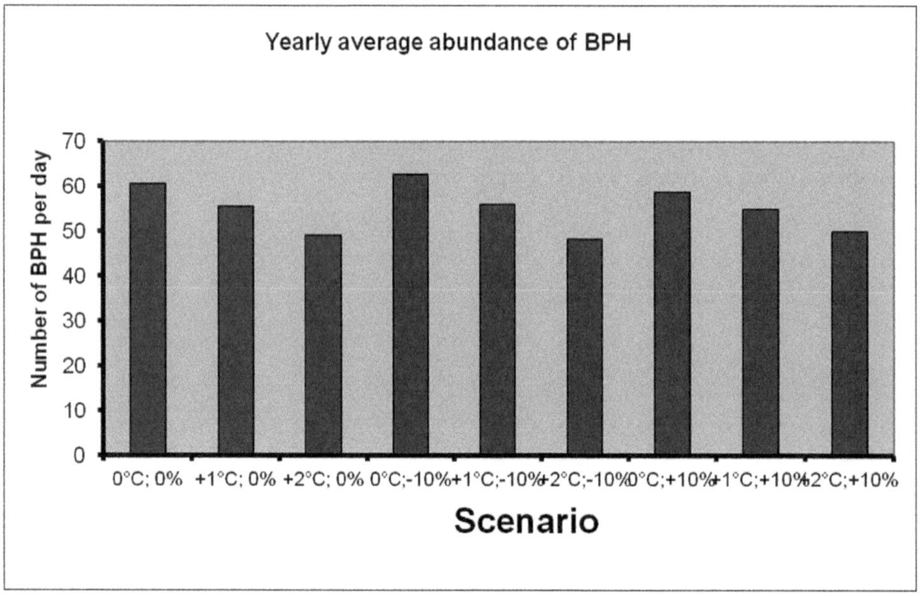

Figure 7. The predicted mean daily catches of BPH averaged over a year if daily minimum temperature changes with either 0°C, +1°C or +2°C, and if daily rainfall changes with either 0%, −10% or +10%.

April 2008 during the dry season [45] when temperatures were high. These weather conditions seem to promote BPH outbreaks, which is in accordance with the predictions based on our model, because minimum temperatures in the area during the dry season rarely exceed 23°C [12].

Wide-scale outbreaks of BPH in tropical Asian rice in the 1970s and 1980s [46–47] were attributed to the insecticidal destruction of natural enemies [48]. One of the main causes of the recent series of outbreaks in Thailand and Indonesia was also insecticide misuse [49]. Although insecticide applications to rice do not always trigger BPH outbreaks [46], they often disrupt the actions of BPH's natural enemies either by their direct killing action or by disrupting food chains [50]. Spraying of insecticide in rice field for controlling other pests destroys essential ecosystem services that regulate invading planthopper adults, thus increasing the risk of hopperburn on the farm [11]. Sublethal insecticide applications could theoretically increase BPH's capacity for migration because the planthoppers would acquire more fat and sugar, which provide fuel for flight, when feeding on insecticide-treated rice plants than when feeding on untreated plants [49]. The applications of sublethal doses of certain insecticides could enhance both BPH's reproductive and migratory capacity and theoretically increase the threat of BPH outbreaks even if the insecticides do not harm natural enemies [51].

It is tempting to conclude that human-controlled inputs such as insecticides or insecticides in combination with nitrogen fertilizer, for example, have been totally responsible for the synchronous BPH outbreaks in tropical rice observed across many areas of Asia shortly after the beginning of the green revolution and again more recently [49], but there is a widespread concern that outbreaks of herbivorous insects will increase in frequency and severity as a consequence of global warming. These predictions, however, are untested and still very uncertain.

In an environment conducive to more rapid BPH increases, natural enemies and BPH-resistant cultivars would be expected to have less effect in regulating the density of BPH, and pesticides harming natural enemies would therefore be expected to have an above-average negative impact [49]. Kiritani [27] also reported that every 1°C rise from around $X = 4$°C would result in a decrease in winter mortality of about 16.5% based on a regression model linking winter mortality (Y) to the mean temperature in January (X).

N. lugens is a long-range migrant species that immigrates into Japanese paddy fields from mainland China in early summer [52]. Our study does not consider immigration of BPH into the study area because this phenomenon has not yet been observed in Bangladesh. Moreover, we accentuate that our model is a simple measure of the importance of climate compared to all other factors (such as pesticides, natural enemies etc). It does not focus on the overall causes for pest outbreaks, nor does it independently attribute pest trends to the many technological, climatic and biotic factors that may affect such trends. However, it discloses a pragmatic approach to assessing the combined importance of climate variables with special reference to evaluating the possible impact of climate change on pest outbreaks.

Long time series like ours provide a good basis for linking pest outbreaks to climate change and to separate the influence of other factors, such as pesticides, from that of climate. Our study indicates that climate change may have contributed to the recent destructive outbreaks of planthoppers in rice crops [53]. However, more research is needed to predict more accurately the long-term consequences of a changing climate, and especially how such changes will affect pest dynamics at both regional and local scales.

Conclusion

Data collected from 1998 through 2007 did not indicate that temperature and precipitation patterns have changed significantly during this 10-years period. Therefore, we attribute the significant increase in BPH during April to changes in agricultural practice rather than climate change. However, the statistical models demonstrate that temperature and rainfall are driving variables for the dynamics of BPH, so any changes in these factors due to climate change are likely to affect the outbreak patterns of BPH. We tested this by simulating different climatic scenarios and found a quite complex picture depending on the severity and direction of the expected climatic changes. Thus, our study emphasizes the importance of taking future climate into consideration when deciding which, where and when the different rice cultivars should be planted and harvested so as to minimize crop losses due to pests.

Given the challenge of linking pest impacts and directional climate change for well-studied agricultural, maricultural (cultivation of marine organisms in their natural habitats, usually for commercial purposes), and human diseases, it is not yet possible to predict the consequences for biodiversity. Very few empirical studies directly explore the relationship between climate and rice pest problems. Even fewer explore interactions between temperature and various components of an insect's life cycle [54–56]. We therefore identify three priorities for research to improve our ability to predict impacts of climate change on rice pests:

1. *Collect baseline data on insects of both major and minor importance*: Baseline data are critical to predict changes in a warmer climate, but such data are rarely collected for rice ecosystems. Monitoring programs for the prevalence and severity of pests and their population- and community-level impacts must be implemented for a wider range of natural systems.
2. *Separate the effects of multiple climate variables on insects*: To predict accurately future responses to climate change, we must quantify the direct and synergistic effects of multiple climate variables, such as temperature and precipitation, on insects. Laboratory and/or field experiments are crucial for studying the specific effects of these variables [57].
3. *Forecast epidemics*: Forecasting models using climate variables can effectively predict outbreaks for some crops and pests. Crop disease programs have long been in effect as, e.g., rice blast (*Pyricularia oryzae*) models based on temperature and moisture forecasts when an epidemic will start and when to apply fungicide for optimal control [58]. Such forecasting programs are also in development for human diseases with climate sensitivity, such as Rift Valley fever, which is associated with warm El Niño events of high rainfall [58–59], and cholera, which is predictable from sea surface temperature associations with El Niño [59]. Similar forecasting models will be very useful for predicting epidemic outbreaks of insect pests in rice and we hope that the present paper will contribute to achieving this goal.

Acknowledgments

We greatly acknowledge the Bangladesh Rice Research Institute (www.brri.gov.bd) for supporting the staff to operate the data monitoring device including logistic support to collect field data, Entomology Division Staff for technical support, and Plant Physiology Division for supplying climate data.

Author Contributions

Conceived and designed the experiments: MFR NA MPA. Performed the experiments: MPA NA MAB. Analyzed the data: MPA GN DH. Contributed reagents/materials/analysis tools: MFR GN MPA. Wrote the paper: MPA GN DH.

References

1. Heong KL (2010) Assessing BPH outbreak risks of commonly used insecticides. Available: http://ricehoppers.net/2010/09/farmers-in-central-thailand-remain-trapped-by-the-bph-problem/. Accessed 30 Sept 2010.
2. Chien HV, Cuong L (2009) Farms that apply insecticides for leaf folder control are 10 times more at risk to hopperburn. Available: http://ricehoppers.net/2009/09/farms-that -apply-insecticides-for-leaf-folder-control-are-10-times-more-at-risk-to-hopperburn/. Accessed 13 Sept 2009.
3. Heong KL (2009) Planthopper outbreaks in 2009. Available: http://ricehoppers.net/2009/09/. Accessed 25 Sept 2009.
4. Islam Z, Haque SS (2009) Rice planthopper outbreaks in Bangladesh. Available: http://ricehoppers.net/2009/08/rice-planthopper-outbreaks-in-bangladesh/planthopper-outbreaks-in-2009. Accessed 20 Aug 2009.
5. Moni (2011) BPH continues to threaten Thai rice farmers – heavy losses expected. Available: http://ricehoppers.net/2011/04/bph-continues-to threaten-thai-rice-farmers heavy-losses-expected/. Accessed 20 April 2011.
6. Moni (2012) Planthopper problems intensify in Thailand's rice bowl. Available: http://ricehoppers.net/2012/03/planthopper-problems-intensify-in-thailands-rice-bowl/. Accessed 21 March 2012.
7. Luecha M (2010) Farmers' insecticide selections might have made their farms vulnerable to hopperburn in Chainat, Thailand. Available: http://ricehoppers.net/2010/01/farmers%e2%80%99. Accessed 17 Jan 2010.
8. Matsumura M, Sanada-Morimura S (2010) Recent status of insecticide resistance in Asian rice planthoppers. JARQ 44: 225–230.
9. Soitong K, Sriratanasak W (2012) Will planthoppers continue to threaten Thailand's rice production in 2012? Rice Department, Bangkok, Thailand. Available: http://ricehoppers.net/2012/02/will-planthopper-continue-to-threaten-thailands-rice-production-in-2012/. Accessed 13 Feb 2012.
10. Heong KL, Schoenly KG (1998) Impact of insecticides on herbivore-natural enemy communities in tropical rice ecosystems. In: Haskell PT, McEwen P, editors. Ecotoxicology: Pesticides and Beneficial Organisms. Chapman and Hall, London, UK. pp. 381–403.
11. Teo C (2011) Insecticide abuse in rice production causes planthopper outbreaks. Available: http://www.asianscientist.com/topnews/irri-ban-insecticides-rice-production-due-to-planthopper-outbreaks-2011/. Accessed 21 Dec 2011.
12. BRRI (2012) Bangladesh Rice Research Institute annual report, 2012.
13. Upmanu L, Michael M (1994) The Great Salt Lake: A barometer of low frequency climate variability. Water Resources Res 31: 2383–2635.
14. Estay SA, Lima M, Labra FA (2009) Predicting insect pest status under climate change scenarios: combining experimental data and population dynamics modelling. J Appl Entomol 133: 491–499.
15. Karuppaiah V, Sujayanad GK (2012) Impact of climate change on population dynamics of insect pests. World J Agri Sci 8: 240–246.
16. Hu G, Cheng XN, Qi GJ, Wang FY, Lu F, et al. (2011) Rice planting systems, global warming and outbreaks of *Nilaparvata lugens* (Stål). Bull Entom Res 101: 187–99.
17. Kulshreshtha JP, Anjaneyulu A, Padmanabhan SY (1974) The disastrous brown plant-hopper attack in Kerala. Indian Farming 24: 5–7.
18. Ali MP, Alghamdi SS, Begum MA, Alam MZ, Huang D, et al. (2012) Screening of rice genotypes for resistance to the brown planthopper, *Nilaparvata lugens* Stal. Cereal Res Commun 40: 502–508.
19. Hibino H (1989) Insect-borne viruses in rice. In: Harris KF editor. Advances in disease vector research. Springer-Verlag, New York, USA, pp. 209–241.
20. Hibino H (1996) Biology and epidemiology of rice viruses. Annu Rev Phytopathol 34: 249–274.
21. Du PV, Cabunagan RC, Cabauatan PQ, Choi HS, Choi IR, et al. (2007) Yellowing syndrome of rice: Etiology, current status, and future challenges. Omonrice 15: 94–101.
22. IPCC (2007) Climate Change 2007. In: Solomon S, et al., editors. The Physical Science Basis. Contribution of Working Group I to the Fourth Assessment Report of the Intergovernmental Panel on Climate Change. Cambridge Univ. Press, Cambridge, UK and New York, NY, USA.
23. Meehl GA, Stocker TF, Collins W, Friedlingstein P, Gaye A, et al. (2007) Global climate projections. In: Solomon S, et al., editors. Climate Change 2007: The Physical Science Basis. Contribution of Working Group I to the Fourth Assessment Report of the Intergovernmental Panel on Climate Change. Cambridge University Press, Cambridge, pp. 747–845.
24. Crosson P (1997) Impacts of climate change on agriculture, Climate Issues Brief No. 4.
25. Cannon R (1998) The implications of predicted climate change for insect pests in the UK, with emphasis on non-indigenous species. Glob Change Biol 4: 785–796.
26. Woiwod I (1997) Detecting the effects of climate changes on Lepidoptera. J Insect Conserv 1: 149–158.
27. Kiritani K (2006) Predicting impacts of global warming on population dynamics and distribution of arthropods in Japan. Popul Ecol 48: 5–12.
28. Esbjerg P (2011) Phenology and pest status change of *Agrotis segetum* as related to the climate change in Northwestern Europe. IOBC/WPRS Bull 65: 43–49.
29. Kiritani K (1988) Effects of climate change on the insect fauna (in Japanese). Meteorol Res Rept 162: 137–141.

30. SAS (2008) SAS Enterprise Guide 4.3. SAS Institute Inc. SAS Campus Drive, Cary, NC 27513.
31. Taylor LR (1961) Aggregation, variance, and the mean. Nature 189: 732–735.
32. Bale JS, Masters GJ, Hodkinson ID, Awmack C, Bezemer TM, et al. (2002) Herbivory in global climate change research: direct effects of rising temperatures on insect herbivores. Glob Change Biol 8: 1–16.
33. Singh UP (2002) Boro rice in Eastern India. Rice-wheat consortium regional technical coordination committee meeting. 10–14 February 2002. Rice-wheat consortium for the Indo-Gangetic Plains, New Delhi, India. Available http://www.Scirp.org/journal/jep/. Accessed Nov 2009.
34. Rahman MK, Parvin L (2009) Impact of irrigation on food security in Bangladesh for the past three decades. J Environ Protect 1: 40–49.
35. Escalada MM (2010) Thailand: Insect influx threatens rice production. Available: http://www.irinnews.org/report/90269. Accessed 25 Aug 2010.
36. Petzoldt C, Seaman A (2008) Climate change effects on insects and pathogens. In: University of Vermont Extension, New York State College of Agriculture and Life Sciences, editors. Climate change and agriculture: promoting practical and profitable responses. New York State IPM Program, New York State Agricultural Extension Station, Geneva, New York.
37. Dai A (2013) Increasing drought under global warming in observations and models. Nature Climate Change 3: 52–58.
38. Logan JA, Regniere J, Powell JA (2003) Assessing the impacts of global warming on forest pest dynamics. Front Ecol Environ 1: 130–137.
39. Williams DW, Liebhold M (1995) Herbivorous insects and global change: potential changes in the spatial distribution of forest defoliators. J Biogeogr 22: 665–71.
40. Fleming RA, Volney WJA (1995) Effects of climate change on insect defoliator population processes in Canada's boreal forest: some plausible scenarios. Water Air Soil Poll 82: 445–54.
41. Coley PD (1998) Possible effects of climate change on plant/herbivore interactions in moist tropical forests. Climatic Change 39: 455–72.
42. Schiller JM, Chaphengxay MB, Linquist B, Rao SA (2006) Rice in Laos. Loss Banos (Philippine), IRRI. 435p.
43. Win SS, Muhamad R, Ahma ZAM, Adam NA (2011) Population fluctuations of brown planthopper, *Nilaparvata lugens* Stal. and white backed planthopper, *Sogatella furcifera* Horvath on rice. J Entomol 8: 183–190.
44. IRIN (2012) Humanitarian news and analysis. Thailand: Rice pests multiply post-floods. Available: http://www.irinnews.org/report/95058/Thailand-rice-pests-multiply-post-floods. Accessed 12 March 2012.
45. Kontodimas DC, Eliopoulos PA, Stathas GJ, Economou LP (2004) Comparative temperature-dependent development of *Nephus includens* (Kirsch) and *Nephus bisignatus* (Boheman) (Coleoptera: Coccinellidae) preying on *Planococcus citri* (Risso) (Homoptera: Pseudococcidae): Evaluation of a linear and various nonlinear models using specific criteria. Environ Entomol 33: 1–11.
46. Heinrichs EA, Aquino GB, Chelliah S, Valencia SL, Reissig WH (1982) Resurgence of *Nilaparvata lugens* (Stål) populations as influenced by method and timing of insecticide applications in lowland rice. Environ Entomol 11: 78–84.
47. Shepard BM, Khan ZR, Pathak MD, Heinrichs EA (1991) Management of insect pests of rice in Asia. In: Pimentel D, editor. CRC Handbook of Pest Management in Agriculture. CRC Press, Boca Raton, FL, pp. 255–278.
48. Matteson PC (2000) Insect pest management in tropical Asian irrigated rice. Annu Rev Entomol 45: 549–574.
49. Bottrell DG, Schoenly KG (2012) Resurrecting the ghost of green revolutions past: The brown planthopper as a recurring threat to high-yielding rice production in tropical Asia. J Asia-Pacific Entomol 15: 122–140.
50. Heong KL, Schoenly KG (1998) Impact of insecticides on herbivore-natural enemy communities in tropical rice ecosystems. In: Haskell PT, McEwen P, editors. Ecotoxicology: Pesticides and Benefical Organisms. Chapman & Hall, London, UK, pp. 381–403.
51. Yin JL, Xu HW, Wu JC, Hu JH, Yang GQ (2008) Cultivar and insecticide applications affect the physiological development of the brown planthopper, *Nilaparvata lugens* (Stal) (Hemiptera: Delphacidae). Environ Entomol 37: 206–212.
52. Kiritani K (2001) Insects and cimate (in Japanese). Seizando, Tokyo.
53. Hart AJ, Bale JS, Fenlon JS (1997) Developmental threshold, day-degree requirements and voltinism of the aphid predator *Episyrphus balteatus* (Diptera: Syrphidae). Ann appl Biol 130: 427–437.
54. Jaramillo J, Chabi-Olaye A, Kamonjo C, Jaramillo A, Vega FE, et al. (2009) Thermal tolerance of the coffee berry borer *Hypothenemus hampei*: Predictions of climate change impact on a tropical insect pest. Plos One 4(8): e6487. doi:10.1371/journal.pone.0006487.
55. Harvell CD, Mitchell CE, Ward JR, Altizer S, Dobson AP, et al. (2002) Climate warming and disease risks for terrestrial and marine biota. Science 296: 2158–2162.
56. Ishigua K, Hashimoto A (1991) Computer-based forecasting of rice blast epidemics in Japan. In: Teng PS, Polland LR, Argosino G, editors. Rice blast modeling and forecasting. International Rice Research Institute, Seoul, South Korea. pp. 39–53.
57. Linthicum KJ, Anyamba A, Tucker CJ, Kelley PW, Myers MF, et al. (1999) Climate and satellite indicators to forecast rift Valley fever epidemics in Kenya. Science 285: 397.
58. Anyamba A, Linthicum KJ, Tucker CJ (2001) Climate-disease connections: Rift valley fever in Kenya. Cad Saude Publica 17: 133–140.
59. Pascual M, RodoX, Ellner SP, Colwell R, Bouna MJ (2000) Cholera dynamics and El Niño-Southern oscillation. Science 289: 1766.

Strong Discrepancies between Local Temperature Mapping and Interpolated Climatic Grids in Tropical Mountainous Agricultural Landscapes

Emile Faye[1,2,3]*, Mario Herrera[3], Lucio Bellomo[4], Jean-François Silvain[1], Olivier Dangles[1,3,5]*

1 Institut de Recherche pour le Développement (IRD), UR 072, Laboratoire Evolution, Génomes et Spéciation, UPR 9034, Centre National de la Recherche Scientifique (CNRS), Gif sur Yvette, France et Université Paris-Sud 11, Orsay, France, **2** UPMC Univ Paris06, Sorbonne Universités, Paris, France, **3** Facultad de Ciencias Exactas y Naturales, Pontificia Universidad Católica del Ecuador, Quito, Ecuador, **4** Mediterranean Institute of Oceanography (MIO) CNRS/INSU, IRD, UM 110, Université de Toulon, La Garde, France, **5** Instituto de Ecología, Universidad Mayor San Andrés, Cotacota, La Paz, Bolivia

Abstract

Bridging the gap between the predictions of coarse-scale climate models and the fine-scale climatic reality of species is a key issue of climate change biology research. While it is now well known that most organisms do not experience the climatic conditions recorded at weather stations, there is little information on the discrepancies between microclimates and global interpolated temperatures used in species distribution models, and their consequences for organisms' performance. To address this issue, we examined the fine-scale spatiotemporal heterogeneity in air, crop canopy and soil temperatures of agricultural landscapes in the Ecuadorian Andes and compared them to predictions of global interpolated climatic grids. Temperature time-series were measured in air, canopy and soil for 108 localities at three altitudes and analysed using Fourier transform. Discrepancies between local temperatures vs. global interpolated grids and their implications for pest performance were then mapped and analysed using GIS statistical toolbox. Our results showed that global interpolated predictions over-estimate by 77.5±10% and under-estimate by 82.1±12% local minimum and maximum air temperatures recorded in the studied grid. Additional modifications of local air temperatures were due to the thermal buffering of plant canopies (from −2.7°K during daytime to 1.3°K during night-time) and soils (from −4.9°K during daytime to 6.7°K during night-time) with a significant effect of crop phenology on the buffer effect. This discrepancies between interpolated and local temperatures strongly affected predictions of the performance of an ectothermic crop pest as interpolated temperatures predicted pest growth rates 2.3–4.3 times lower than those predicted by local temperatures. This study provides quantitative information on the limitation of coarse-scale climate data to capture the reality of the climatic environment experienced by living organisms. In highly heterogeneous region such as tropical mountains, caution should therefore be taken when using global models to infer local-scale biological processes.

Editor: Michael Sears, Clemson University, United States of America

Funding: This work was partly conducted within the project "Adaptive management in insect pest control in thermally heterogeneous agricultural landscapes" (ANR-12-ISV7-0013-01) funded by the Agence Nationale pour la Recherche (ANR, http://www.agence nationale-recherche.fr/). A financial support of the McKnight Foundation (http://www.mcknight.org/) to EF during the fieldwork of this study is greatly acknowledged. The funders had no role in study design, data collection and analysis, decision to publish, or preparation of the manuscript.

Competing Interests: The authors have declared that no competing interests exist.

* Email: ehfaye@gmail.com (EF); olivier.dangles@ird.fr (OD)

Introduction

Bridging the gap between the predictions of coarse-scale climate models and the fine-scale climatic reality of species is increasingly recognized as a key issue of climate change biology research [1,2,3,4]. Despite decades of study on microclimates [5,6,7,8] and evidence for habitat-related and topographical variations in local temperatures and their relevance for species ecology [2,9,10,11,12,13], most attempts to understand and model species distributions still do not integrate spatially-explicit fine-scale climatic data (e.g. [14,15,16]). Many work use global model of temperature interpolation to examine species vulnerability to climate change and, doing so, ignore the critical issue of habitat complexity in climate buffering [4,5,17]. Indeed, climate surfaces used in species distribution models (SDMs) are rarely generated or interpolated to a resolution finer than 1 km^2 (e.g. WorldClim database), a resolution that is still very coarse relative to the home ranges or body size of most species [13,18]. For instance, [8] showed that climate grid lengths used in SDMs are, on average, ~10,000-fold larger than studied animals, and ~1,000-fold larger than studied plants. Their meta-analysis showed that the WorldClim was the most widely used climatic dataset in global SDMs. As this commonly used coarse scale climatic data in SDMs overlook the spatiotemporal thermal heterogeneity experienced by organisms, there is an urgent need for a more sophisticated use of these datasets for making inferences about biological processes that are driven by hour to hour operative temperatures of organisms.

An important yet poorly studied issue in climate change biology is to quantify to what extent climatic conditions differ between widely used 1 km^2 interpolated grid cells of global climatic database and real-world landscapes of similar areas. While it is now well-known that most organisms, especially tiny ectotherms such as insects and other arthropods, do not experience the climatic conditions recorded at weather stations [9,12,18], there is little quantitative information on the spatial and temporal heterogeneity at the landscape scale of local climatic conditions (i.e. conditions at biologically relevant scales, e.g., from cm to km for insects) and their consequences for organisms' performance. A better quantification of the climatic conditions of ecologically-relevant habitats over relatively large landscape scales (e.g., 1 km^2) is therefore a necessary first step to better incorporate dynamical microclimate into global distribution models.

Here, we investigate the sources of variance between global interpolated and local temperatures by examining 1) how well WorldClim predicts local air temperatures in our study region (the tropical Andes), 2) to what extent temperatures in crop canopies and soils differ from local air temperatures, and 3) how relevant is to use WorldClim to infer the potential performance of an insect crop pest. Addressing these questions is not an easy task as the mosaic of climatic habitats relevant for small ectothermic species at a 1-km^2 scale in real-world landscapes may be outstandingly complex. In this study, we focused on highland agricultural landscapes of the tropical Andes as most prior similar data came from low elevation and temperate agroecosystems. In such systems, most crop pests experience, over their entire life cycle, climatic conditions in three well-defined environmental layers (air, air inside-canopy and soil) and these conditions are remarkably stable over the year [19]. In this context, we firstly decided to map over replicated 1-km^2 climatic grid cells the ecologically relevant local temperatures for ectothermic crop pests in agricultural landscapes, and to compare these maps to interpolated temperature grid cells of the widely used WorldClim database. We used Fourier analysis applied to local temperature time-series as a tool to fit daily variations of temperature and to feature microclimate discrepancies in space and in time (both in terms of amplitude and phase). We then explored the implication of our thermal landscape mapping for pest performance by comparing temperature frequencies in our grid cells with the temperature-dependent growth curve of the potato tuber moth (*Phthorimaea operculella*) a major crop pest species in the region and worldwide.

Materials and Methods

1. Study area

The Ecuadorian Andes are characterized by a low seasonality, with mean temperatures varying more within days (up to 30°K variation) than within months and years (less than 0.6°K and 0.2°K variations, respectively, see [19]). This region exhibits a marked altitudinal gradient in temperatures (between 2000 and 4000 m) with mean monthly air temperature roughly decreasing by 0.6°K every 100 m of elevation [20]. Agricultural landscapes dominate the altitudinal belt between 2600 and 3800 m, and are typically composed by small field crops (mainly potato *Solanum tuberosum* L., broad bean *Vicia faba* L., corn *Zea mays* L., alfalfa *Medicago sativa* L., and pasture), natural grasslands (páramos) and a few forest patches [21]. Under the climatic conditions of the region, crops can be planted and harvested all year round, thereby creating a landscape mosaic of a wide variety of crops at different phenological stages.

Our study area was located 115 km south from the equatorial line (01°01′36″S, 78°32′16″W) in the Cotopaxi province of Ecuador. It spread out on a 20-km^2 elevation transect (2.35×8.5 km), ranging from 2,600 to 3,800 m a.s.l. The gradient had a Southwest exposure and an average slope of 9.5° (±5.2) (based on a 30 m resolution digital elevation model). To investigate the elevation effect on local vs. global interpolated temperature variations, we divided our study area into three 400 m altitudinal belts which correspond to natural floors in the hillside (2,600–3,000 m, 3,000–3,400 m, and 3,400–3,800 m) with a mean monthly temperature of 13.2±0.4°C, 10.8±0.6°C, and 9.3±0.4°C, respectively. Beyond temperature, these belts also differed in terms of landscape composition (Appendix S1 in Supporting Information), with lower elevations dominated by small fields (0.3±0.1 Ha) of potato, corn, broad bean, and pasture while the higher band had larger fields (0.7±0.3 Ha) of mainly potato and pasture. Working in these agricultural landscapes no requires specific permissions expect the kind agreement of the field owner. The presented study did not involve endangered or protected species.

2. Temperature data collection

In each of the three-altitudinal belts, we measured temperature regimes in six habitats (five crops and natural grasslands) where insect pests can be found. In each habitat, we defined three layers: air, air inside-canopy (referred as "air canopy" in the text) and soil. These layers are all used by most insect pests over their life cycle: air layer by adults, air canopy layer by adults and leaf-eating larvae and pupae, soil layer by tuber feeding larvae and pupae. In each layer of each habitat, temperature was recorded with a 1 min time step using data loggers (Hobo U23-001-Pro-V2 internal temperature loggers, Onset Computer Corporation, Bourne, USA) with an accuracy of ±0.21°K over the 0–50°C range and a resolution of 0.02°K at 25°C. According to [4], 1) air loggers were fixed on a wooden stake at 1 m high to overstep most crop canopies and sheltered by a 20 cm^2 white plastic roof to minimize solar radiation heating; the roof was itself placed 5 cm above the logger to avoid warming by greenhouse effect, 2) air canopy loggers were placed 0.3 m high inside vegetation 5 cm bellow large leaves to minimize the effect of direct solar radiation and 3) soil loggers were buried 0.1 m into the ground where roots and tubers grow (see Appendix S2 for photographs). In each field, only one logger per layer measured the temperatures. Those triplets of loggers were located at the centre of the field to avoid edge effect (see Appendix S3 for an analysis of the spatial variability of temperatures within a field and [22]). As vegetation land cover influences microclimate beneath and around plants, see [5,6], we repeated these 54 measurements (3 elevations ×6 habitats ×3 layers) for three classes of leaf area index (LAI) [23] defined as follows: 0 (bare soil), 0.01–0.5 for and >0.5 of LAI. Minimum LAI was fixed to 0.01 to avoid confusion with bare soil and allowed enough leaf area to place the loggers underneath. At each measurement site, LAI values were visually estimated (twice) measuring the ratio of leaf area within a 1-m^2 quadrant sub-divided into 0.1 m^2 cells delimited by strings. This indirect method did not account for leaves that lie on each other however it relates to shaded areas that influence inside-canopy and soil microclimates [23].

Each of the 162 measurement combinations (3 altitudinal belts ×6 habitats ×3 layers × 3 LAI classes) was replicated 1–3 times depending on availability of habitats at a given elevation and phenology stage. In total 324 independent temperature time series were acquired over 15 days between September and December 2011 (data available in Appendices S9, S10 and S11). Importantly, under the climatic conditions of the study area, 15-days time series characteristics did not differ from those obtained over one year (see Appendix S4 for details). At each measurement site, we

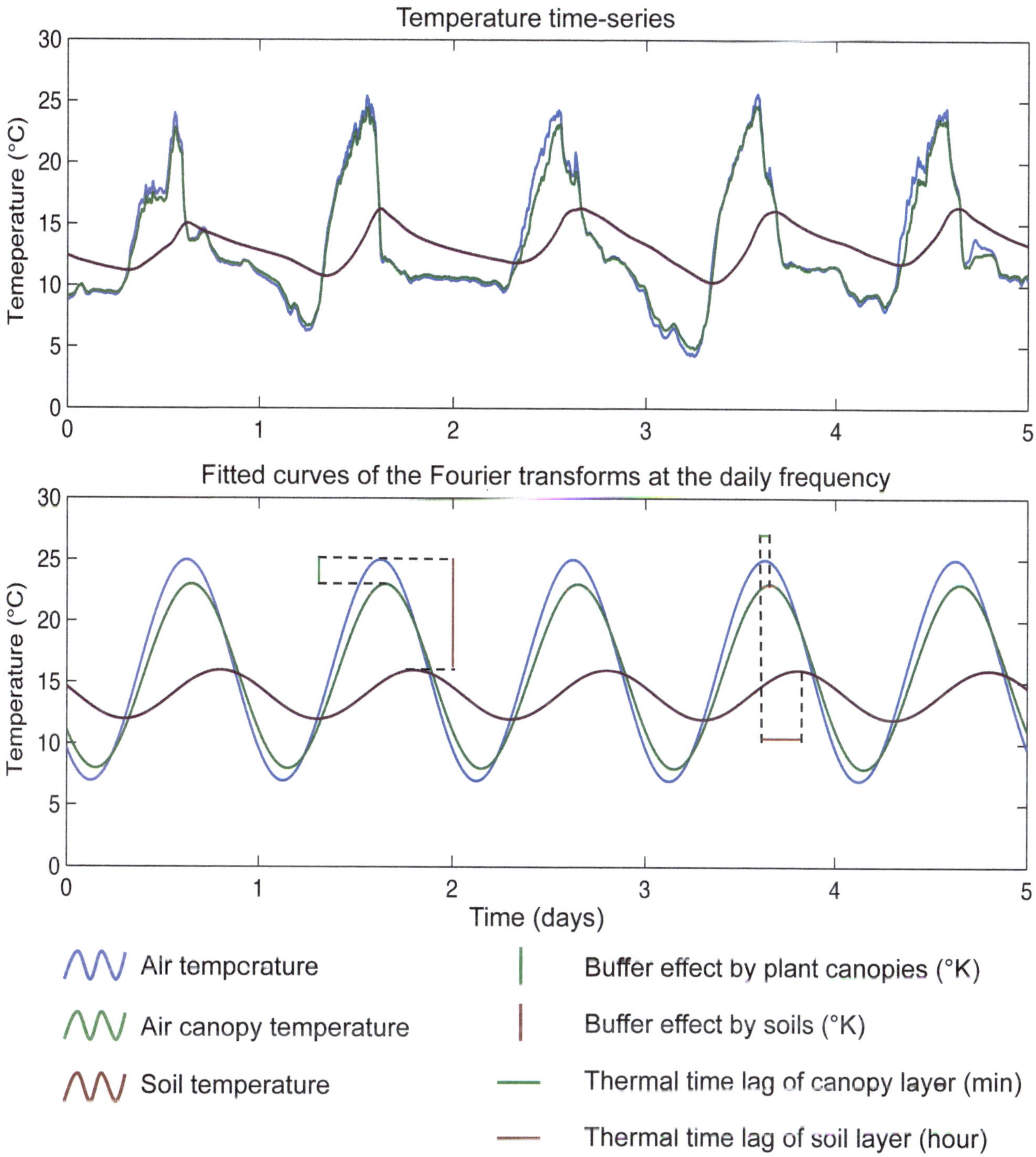

Figure 1. Fit of temperature time series with discrete Fourier transforms at the daily frequency K_d. Air temperatures are in blue, crop canopy temperatures are in green and soil temperatures are in brown.

recorded the UTM-WGS84 geographic coordinates with a handheld GPS Garmin Oregon 550 (Garmin, Olathe, USA).

3. Global solar radiations

Infrared and visible radiations (expressed in Watt/m^2) were monitored in each altitudinal belts using a LI-1400 LI-COR datalogger equipped with a LI-200 pyranometer sensor (LI-COR, Lincoln, USA) placed perpendicular to gravity. Between 9:00 AM and 4:00 PM, mean global solar radiations ranged from 500 to 1000 watts/m^2, with temporal variability mainly induced by short-term changes in cloud cover.

4. Data analyses

4.1. Times series analyses using Fourier transforms. Air and air canopy temperature time series showed extreme events during a few minutes that were certainly due to strong radiations experienced at the study sites – these affected loggers recording despite their plastic roofs. Therefore, we found

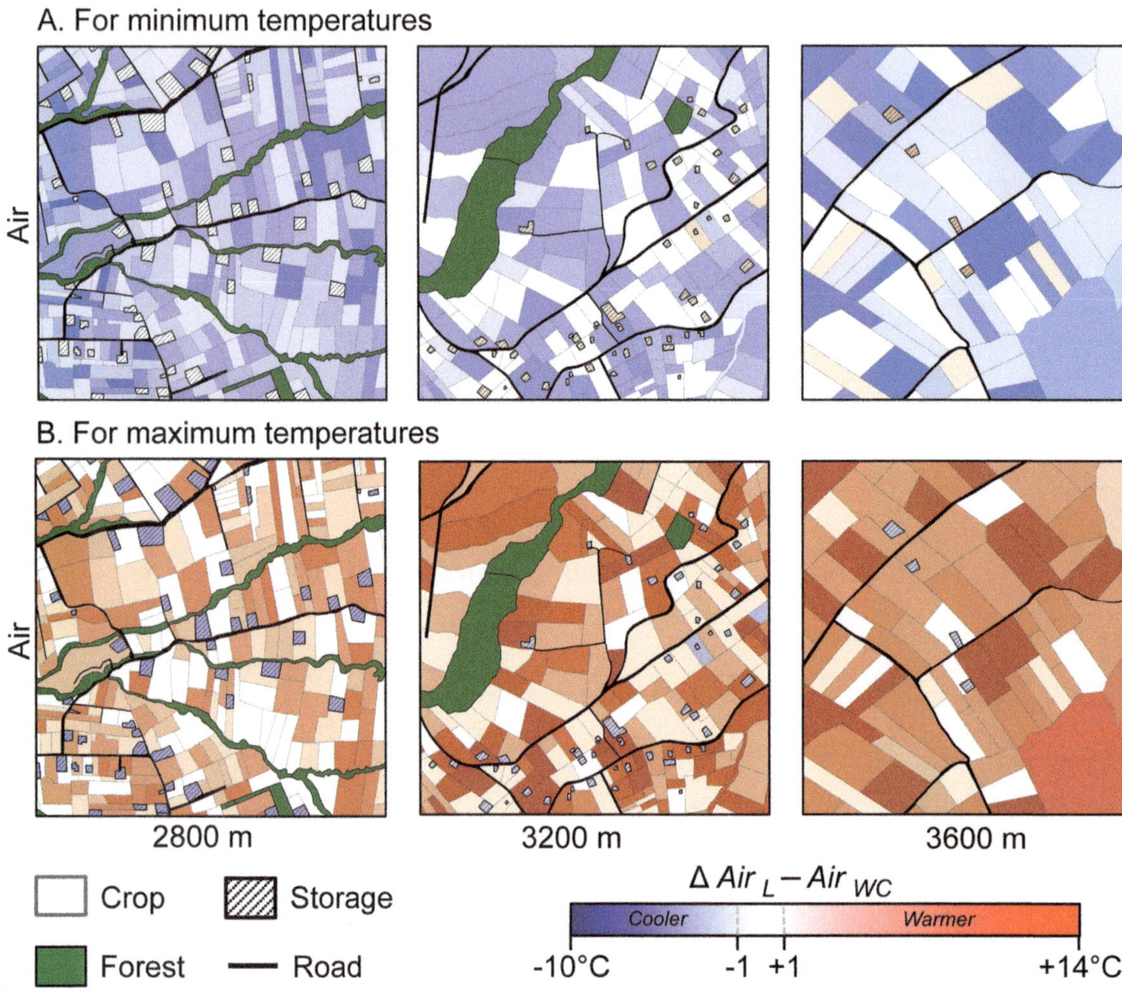

Figure 2. Maps showing the differences between local air temperatures and the WorldClim interpolated minimum (A) and maximum (B) (Δ $Air_L - Air_{WC}$). Blue colours indicate Δ $Air_L - Air_{WC} <0$, i.e. area where local air temperatures are cooler than those gave by WorldClim. Red colours indicate Δ $Air_L - Air_{WC} >0$, i.e. area where air local temperatures are warmer than the ones gave by the WorldClim. White colours Δ $Air_L - Air_{WC}=0$ indicate areas where air WorldClim temperatures equate air local temperatures ($\pm 1°C$). The extent and position of each square is equal to the spatial resolution of the WorldClim database: 30-arc sec that is the equivalent of 0.86 km^2 for the study area. Temperatures in storages were obtained from [26].

relevant to fit our time series data with a discrete Fourier transform (DFT) at the daily frequency k_d (Fig. 1) as this allowed averaging daily minimum and maximum temperatures while limiting the effect of short extremes (mainly for maximum). Moreover fitting temperature time series with the DFT allowed us to circumvent (or partially resolve) the issue of comparing time series with different temporal resolution: a sinusoid built from a daily time step time series will be accurate enough to compare with another sinusoid built from a one minute time step time series (our operative temperatures vs. global climatic models).

DFT analyses allowed us estimating two important descriptors of the time series at the daily frequency k_d: the amplitude A_d and the phase ϕ_d of the DFT (see Appendix S5 for details). The thermal amplitude allowed us to measure the thermal buffer effect in Kelvin between air and canopy layers and air and soil layers (Fig. 1 and Appendix S5). The phase allowed us to measure the thermal time lag expressed in minute in inside-canopy and soil layers with respect to the air layer (Fig. 1 and Appendix S5). Thermal time lag therefore quantifies the time delay in time series to reach their maximum between air vs. canopy and air vs. soil layers. This is an important climatic parameter to test whether microclimate conditions below canopy (canopy and soil layers) would track air conditions with some time lag depending on habitat characteristics.

We also ran DFT analyses on a four-year monitoring (2008–2012) of air temperatures (recorded at one meter high with half an hour time step with the same shelter process described above) to measure the seasonality. Analyses were performed for the three-altitudinal belts of the study area (2800, 3200, 3600 m) by reading the amplitude at the seasonal frequencies (91, 182 and 364 days, see Appendix S6). On average the Fourier transform amplitudes at 91, 182 and 364 days were 0.14 (+/−0.01), 0.44 (+/−0.04), 0.97 (+/−0.03)°K indicating that the seasonality was negligible in the study area [24].

All Fourier analyses were performed in MATLAB R2011a (Mathworks, Natick, USA). The effects of habitat, elevation, LAI classes and the interaction "elevation × LAI classes" on daytime and nigh-time DFT amplitudes and on DFT thermal time lag were assessed using a two-way ANOVA with Bonferroni corrections. When habitat was found significant, we ran post-hoc

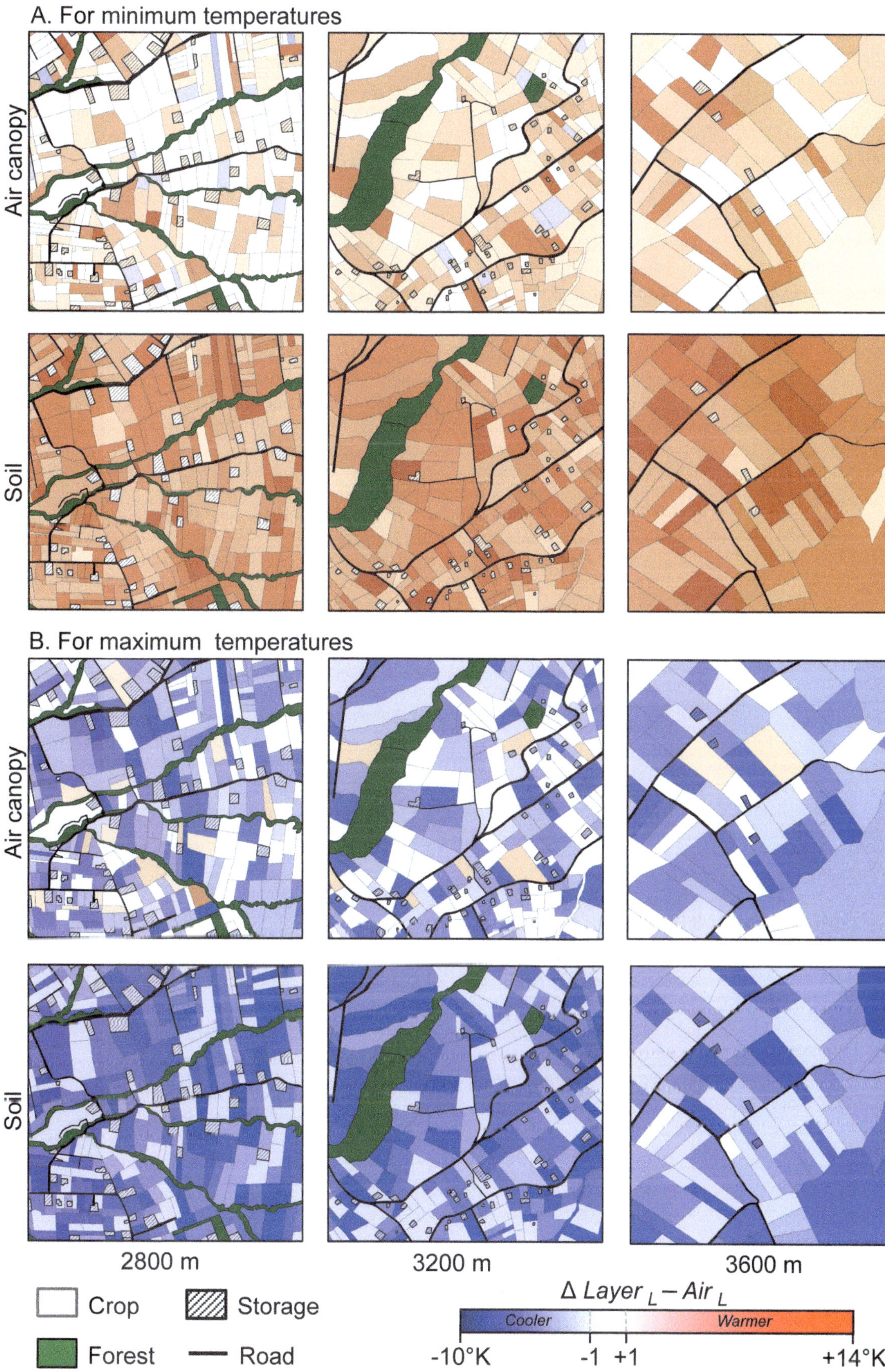

Figure 3. Maps showing the differences between local air canopy and soil temperatures with the air local for minimum (A) and maximum (B) *(Δ Layer $_L$ – Air $_L$)*. Colour code is given in Figure 2.

multiple comparisons using a Tukey HSD test to identify differences among habitats. All statistical analyses were performed in R version 3.0.0 (R Development Core Team 2012).

4.2. Thermal landscape analyses. To compare local temperatures with global interpolated climate data employed in species distribution models, we considered one of the most widely used and readily available climate database, WorldClim [25]. The WorldClim database is a set of global climate layers (interpolated averages of monthly minimum, maximum and mean 1.5 m high air temperatures from weather stations spread out worldwide) with a spatial resolution of 30 arc seconds. Close to equator, this resolution is equivalent to squares of 0.86 km. In each altitudinal belt, we selected one WorldClim grid cell with homogenous slope (between 5.4° and 7.9°), micro-topography and exposition (south-west). Based on a digitized municipal cadastre (from the town council of Salcedo, Cotopaxi province) and a 5-m resolution digital orthophoto (Ecuadorian Military Geographical Institute, www.igm.gob.ec/site/index.php), we built the digital landscape of each grid cell in ArcGIS 10.01 (ESRI, Redlands, USA). In addition to the six studied habitats, crop storage infrastructures were also included into the digital maps as they significantly modify air temperature patterns, offering optimal conditions for crop pest development [26]. Outside air vs. inside air storage-temperature relationships for different elevations were derived from measurements made by [26] within the same area with

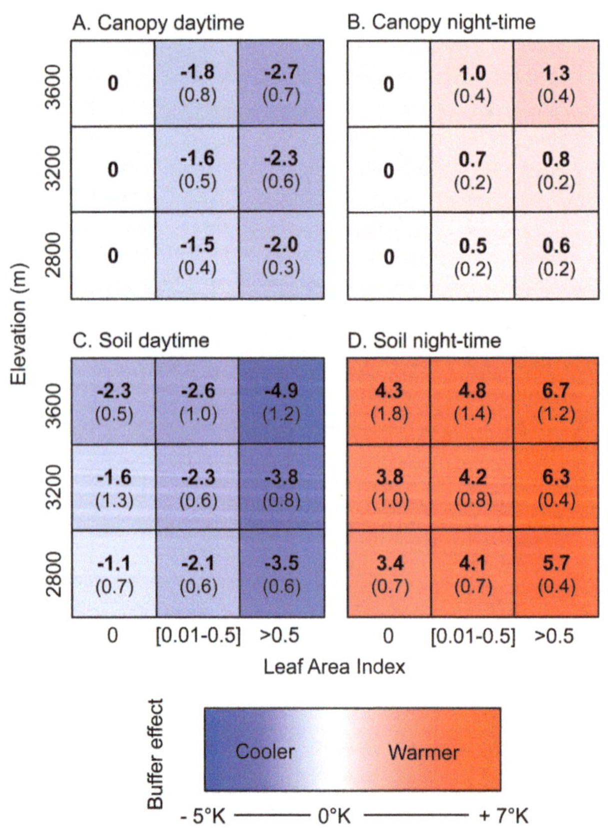

Figure 4. Mean thermal buffering from Fourier transforms at the daily frequency for canopy (A, B) and soil temperatures (C, D) as a function of elevation and leaf area index. (A, C) show the daytime temperature excursion with respect to air, whereas (B, D) are the equivalent results for night-time temperatures. The 95% interval of confidence is given between brackets. Blue colours show colder temperatures than air. Red colours show warmer temperatures than air.

similar temperature data design (see Fig. 1 in Appendix A2 of their paper). Roads and woodlots were also indicated on the maps even if they were not included in the temperature comparison analysis, as they do not constitute relevant habitats for crop pests.

In order to simulate landscape thermal heterogeneity, crop habitats were attributed with one crop type (potato, broad bean, corn, alfalfa or pasture) and one LAI classes (0, 0.01–0.5, >0.5) based on a survey of 85 sites in the region, in which we quantified landscape composition (% of each crop and LAI classes) in 100-m radius sampling circles (see Appendix S7). For each habitat, we assigned the corresponding air, air canopy and soil temperature values at each elevation. Finally, since we were particularly interested in minimal and maximal values, as they are the most biologically relevant for ectothermic crop pests [4], we focused on minimum and maximum temperatures obtained from the DFT analyses and the WorldClim database.

Afterwards, we decomposed the variance of temperatures between global interpolated grids and local temperatures measured in agricultural landscapes by mapping the differences in minimum and maximum temperatures between the air local temperatures (Air_L) and the WorldClim interpolated temperatures (Air_{WC}) for the three studied grid cells. Then, to illustrate the part of the variance due to microclimate effects, we mapped the differences in minimum and maximum temperatures between measured local air canopies, soil temperatures ($Layer_L$) and the air local temperatures (Air_L) for the three studied grid cells.

4.3. Pest performance in thermal landscape. As a final step of our analysis, we explored the implication of our thermal landscape mapping for pest performance by comparing temperature frequencies in our grid cells with the temperature-dependent growth curve of a major crop pest species in the region: *Phthorimaea operculella* (Lepidoptera: Gelechiidae). This pest is considered one of the most important potato pests worldwide, but also attacks a wide variety of other crops such as tomato (*Solanum lycopersicum* L.), eggplant (*Solanum melongena* L.) or tobacco (*Nicotiana tabacum* L.) (see [27] for a review). *P. operculella* feeds on different part of the plant (leaves, stems, and tubers) and also tubers in storage structures [26,28]. In agricultural landscapes, *P. operculella* is abundant in virtually all types of habitats (even far from its host plant) because 1) this pest is able to fly over large distances (100–250 m) to infest suitable host plants [29] and 2) a significant quantity of tubers are left in the field after harvest, and are rapidly colonized by the moth before the following crop is planted. It is therefore common to observe infested potato plants in corn or broad bean fields. These left-over potatoes are well know by farmers and agronomists as significant obstacle to the control of these pests [28].

The temperature-dependent growth rate curve of *P. operculella* larvae (in day-1) over a 0–40°C range was obtained using published temperature-response data of laboratory experiments performed in the Andean region (see [30] for details). PTM development rate data were then modeled with the [31] equation as modified by [32]:

$$D(T) = \frac{\frac{dT}{298.16}\exp\left[\frac{e}{R}\left(\frac{1}{298.16} - \frac{1}{T}\right)\right]}{1 + \exp\left[\frac{f}{R}\left(\frac{1}{g} - \frac{1}{T}\right)\right] + \exp\left[\frac{h}{R}\left(\frac{1}{i} - \frac{1}{T}\right)\right]} \quad (1)$$

where T is temperature in Kelvin (°C+273.15), $R = 1.987$, and d, e, f, g, h, and i estimated parameters. This model has been widely used to describe the kinetics of insect development based on several assumptions about the underlying developmental control

Table 1. Results of the two-way ANOVA with a Bonferroni correction on the effects of habitat, elevation, LAI and elevation × LAI terms on daytime and nigh-time DFT amplitudes and thermal time lag on inside-canopy and soil temperature time series.

Effect	Canopy				Soil			
	Df	Mean sq	F value	P value	Df	Mean sq	F value	P value
Daytime amplitude								
Habitat	5	6.282	3.370	**0.007***	5	5.745	2.466	0.036
Elevation	2	12.491	6.701	**0.002***	2	5.722	2.456	0.089
LAI	1	40.171	21.551	**<0.001***	1	136.78	58.705	**<0.001***
Elevation × LAI	2	2.513	1.348	0.263	2	0.292	0.125	0.882
Residuals	132	1.864			127	2.330		
Night-time amplitude								
Habitat	5	0.936	3.895	**0.002***	5	1.390	0.839	0.525
Elevation	2	4.539	18.896	**<0.001***	2	2.143	1.293	0.278
LAI	1	1.083	4.509	0.035	1	157.52	95.041	**<0.001***
Elevation × LAI	2	1.754	7.302	0.010	2	0.097	0.059	0.943
Residuals	132	0.240			127	1.657		
Thermal Time Lag								
Habitat	5	0.001	1.297	0.269	5	0.009	3.881	**0.003***
Elevation	2	0.001	5.777	**0.004***	2	0.024	10.139	**<0.001***
LAI	1	0.001	29.322	**<0.001***	1	0.022	9.165	**0.003***
Elevation × LAI	2	0.001	2.374	0.097	2	0.005	2.334	0.101
Residuals	132	0.001			127	0.002		

Bold* indicates significant results ($P<0.05$).

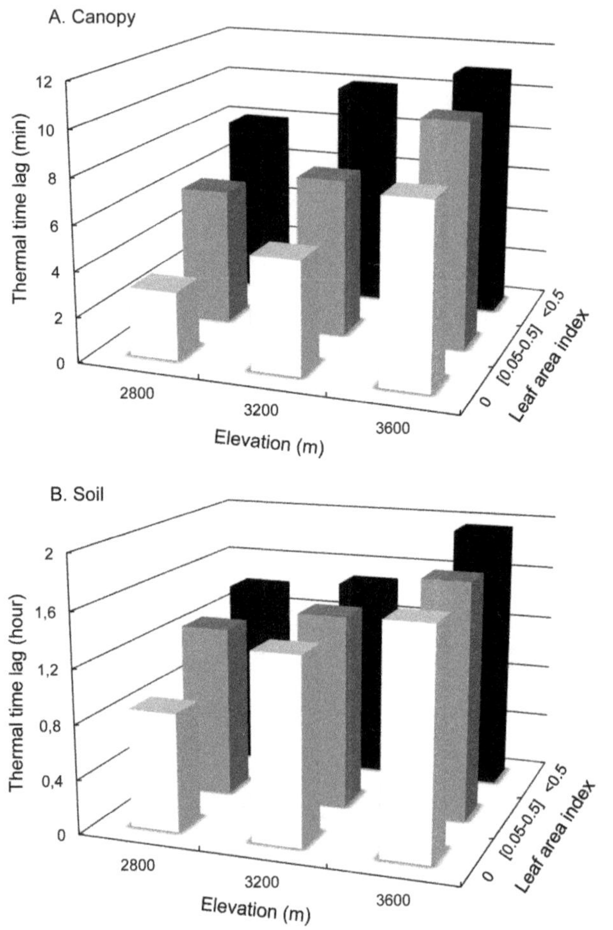

Figure 5. Thermal time lag from Fourier transforms at the daily frequency for canopy (A) and soil temperatures (B) as a function of elevation and leaf area index. The z-axis (log+1 transformed) is expressed in minutes (A) and in hours (B).

enzymes. For instance, it has been used to describe poikilotherms' temperature-dependent development [33].

We then compared the growth rate performance curve of *P. operculella* for local temperature distribution (canopy and soil layer temperatures) and for global interpolated ones (e.g., Fig. 3 in [3]). Distributions of canopy and soil minimum, maximum and mean temperatures were extracted from the three digitized landscapes using the geostatistical analyst extension of ArcGIS. Canopy and soil temperature frequencies were expressed as the percent of total grid cell area. The growth performance model of *P. operculella* given by Eqn. 1 was implemented with WorldClim minimum and maximum temperatures and the local minimum, maximum and mean temperature distribution. This allowed estimating insect growth rate within the range of WorldClim and measured field data.

Results

1. Local vs. global air temperature discrepancies in thermal landscapes

Differences in average minimum and maximum temperatures between local air temperatures and the global coarse grain interpolated air temperatures from the WorldClim (Δ Air $_L$ − Air $_{WC}$) were mapped for the three studied grid cells (Fig. 2). While minimum local air temperatures were cooler than those predicted by WorldClim in 77.5±10% of the studied areas (blue areas, average min Δ Air $_L$ − Air $_{WC}$ = −2.9°K) maximum local air temperatures were warmer than extrapolated temperatures in 82.1±12% of the studied areas (red areas, average max Δ Air $_L$ − Air $_{WC}$ = +5.6°K). This pattern was not influenced by elevation. Notably, for all elevations, local mean air temperatures were quite well predicted by the WorldClim (+/−1°K) as in average 55.3±3.4% of the studied areas felt in the range of Air $_L$ − Air $_{WC}$ ≤1°K (Appendix S8).

2. Temperature discrepancies due to microclimate in agricultural landscapes

Differences in average minimum and maximum temperatures between local canopy and soil temperatures and local air temperatures (Δ Layer $_L$− Air $_L$) were mapped for the three studied grid cells (Fig. 3). Overall, canopy and soil areas were always cooler than maximum air temperature and were always warmer than air minimum temperatures resulting in a general buffer effect of minimum and maximum air temperatures by canopy and soil layers. The buffer effect on air temperatures was significantly stronger for soil than for canopy layer (see Fig. 4, Student's t-test, $t = -27.10$ and $t = 4.52$, $P<0.001$ for night-time and daytime, respectively). Interestingly, the buffer effect on air temperatures by soil was higher during night-time than daytime (Fig. 4D) while the opposite pattern was found in crop canopy (Fig. 4A).

Elevation had a significant effect on air temperature buffering in the canopy layer but not in the soil layer (Table 1). Contrastingly, LAI had a highly significant thermal buffering effect in both soils (night and daytime) and canopies (daytime, see Table 1). Buffer effect on air temperatures by bare soil (e.g. without plant cover, LAI = 0) ranged from −1.1°K to −2.3°K for daytime and from 3.4°K to 4.3°K for night-time. Crop type had no significant effect on buffering patterns except for potato in which higher buffer effects were recorded (Post-Hoc HSD test, $P<0.05$).

Overall, thermal time lag was much shorter in canopies (7.5±2.6 min) than in soils (1.5±0.3 hours, Fig. 5). LAI classes had a significant positive effect on thermal time lag for both canopy and soil layers (Table 1). On average, thermal time lag increased by 2 min. in canopies and 30 min. in soils between two LAI classes. Similarly, elevation had a significant positive effect on thermal time lag for both canopy and soil layers (Table 1) with an average increase of 2±0.3 min. in canopies and of 60±31 min. in soil between two altitudinal belts (Fig. 5).

3. Thermal performance curve using local vs. interpolated temperatures

To assess the implication of local vs. global interpolated temperature discrepancies for crop pest performances, we plotted the frequency distribution of the minimum (blue bars), maximum (red bars) and mean local (stripped bars) temperatures and those given by WorldClim (from minimum to maximum temperature, shaded region in the background) with the temperature-dependent growth rate curve of the potato moth *P. operculella* (Fig. 6). As a general pattern, global interpolated temperature ranges predicted lower growth rates of *P. operculella* than those predicted by local temperatures at all elevations, in both inside-canopy and soil layers (where the pest lives most of their time). While mean temperature distribution generally fell within the WorldClim min-max range, extreme temperatures (and especially maximum ones) largely exceeded this range.

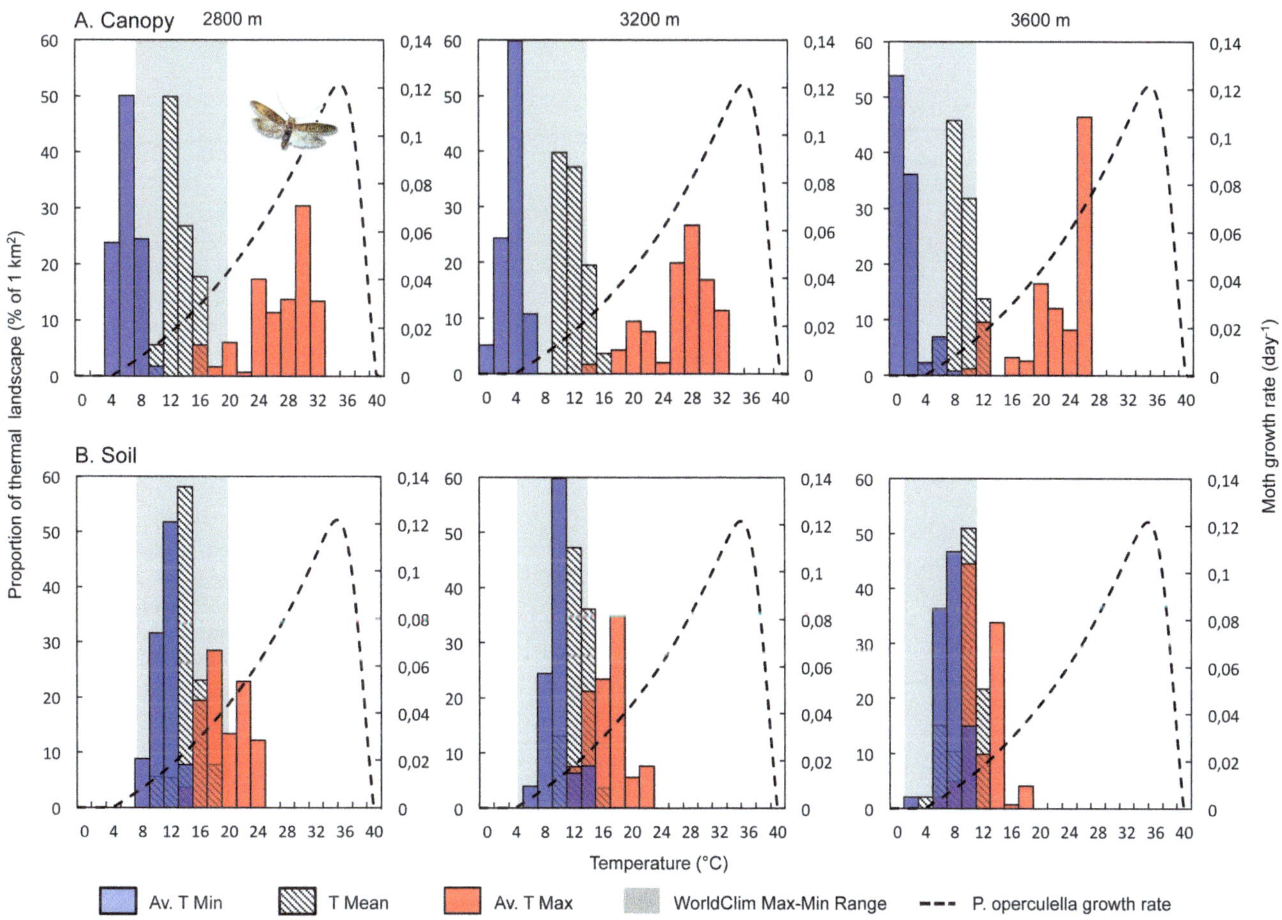

Figure 6. Superimposed plot of the temperature-dependent growth rate curve of the potato moth Phthorimaea operculella (dashed line) and the frequency distribution (% of area) of average minimum (blue), maximum (red) and mean (striped) temperatures for canopy and soil layers at the three studied elevations. Grey (shaded) bands in the background represent the WorldClim minimum and maximum temperature range.

The WorldClim estimations predicted *P. operculella* growth rates ranging between 0.007 and 0.045 day^{-1} at 2800 m, and between 0 and 0.018 day^{-1} at 3600 m, the maximum rates being slightly lower than those predicted by soil temperatures (0.068 day^{-1} at 2800 m and 0.037 day^{-1} at 3600 m). These differences were exacerbated in canopy layers where estimated maximum growth rates were 2.6–4.3 times higher than those predicted by WorldClim (0.118 day^{-1} at 2800 m and 0.079 day^{-1} at 3600 m). Discrepancies between WorldClim and local temperature-based growth rate estimations were not significantly affected by elevation (One-way ANOVA, $F = 7.79$, $P = 0.219$ and $F = 1.67$, $P = 0.419$ for canopies and soils, respectively).

Discussion

Accurate predictions of the responses of organisms to climate change using SDMs require knowledge of microclimates at spatial and temporal scales relevant for studied organisms [13,34,35]. To our knowledge, our study is the first to quantify the thermal heterogeneity among a set of agricultural habitats at fine spatial and temporal scales and to compare those thermal microhabitats to the most widely used global climatic dataset in SDMs. By documenting the mosaic of thermal habitats found in tropical agricultural landscapes, our study confirms previous evidence that microclimates strongly differ from nearby macroclimates due to the variability of air motion and solar radiation patterns created by complex topographies with heterogeneous elevation, slope angle, exposure or roughness [1,7,18,36]. Our study therefore supports the view that results from the long tradition of agrometeorological studies on microclimates (e.g. [6,17,22]) have to be revived in the new context of microhabitat modelling for predicting the response of organisms to climate change.

1. LAI-based and elevation-based climate heterogeneity

In contrast to many previous studies (see [7] for a review), our objective was not to examine the well-documented effect of topography on local temperatures but rather to examine the less-known effects of habitat types and vegetation land cover on thermal landscape features. We found significant thermal time lag and buffer effects on air temperatures by plant and soil layers below crop canopies during night-time and daytime. The top of canopies reflects and absorbs part of the solar radiation during the day, allowing less energy to reach the layers (plants and soils) below canopies. During the night, infrared heat released from both the ground and plants is partly held back by the canopy above [5]. As a consequence plants and soils limit night-time cooling and daytime warming [6], leading to a significant buffer effect of minimum and maximum temperatures [1,4,17]. That is also why we found a buffer effect on air temperatures by soil higher during night-time than daytime and the opposite pattern for crop canopies.

Our results indicate a strong effect of elevation on thermal buffering and thermal time lag by canopy and soil layers. This could result from the combination of a negative relationship between elevation and air temperature and a positive relationship between elevation and solar radiation exposure, part of which is absorbed by plants and soils [6]. As a result, the difference between air temperature and canopy and soil temperature increased with elevation. Interestingly, the modifications of local temperatures by habitats and LAI were of the same magnitude (from −2.70 to 4.82°C in average) than that generated by topography-related factors [7,36], supporting the need to better consider habitat effects on microclimates.

2. Fine scale variations in temperature vs. climatic units

Our findings show that the complex agricultural mosaic resulting from habitat types and LAI classes at the landscape scale was a major modifier of the thermal patterns in the studied tropical highlands. More importantly, our findings revealed that, at best, 55% of landscape habitats had real mean air temperatures that were well estimated by WorldClim predictions while in average less than 20% of these areas had minimum and maximum air temperatures well estimated. Additional thermal discrepancies between large and fine-scale temperatures resulted from heterogeneity in crop types and phenologies. This strongly supports the view that the common use of the WorldClim database arrayed into 1-km^2 grids may not adequately capture the reality of the climatic environment experienced by living organisms, in particular tiny ectothermic species [2,3,13,18]. It is important to note that to obtain the highest level of thermal heterogeneity we chose a complex mountainous agricultural study area that provided boundary conditions for climate modelling. Indeed, these mountainous areas provide strong climatic gradients and extreme habitat fragmentation which combined with un-seasonal agrosystem make up a mosaic of thermal patches that expanded the difficulties to faithfully assess climatic parameters for modelling [25]. In view of the urgent need of fine scale climate data with large extent [2,8,35] more research is necessary to develop accurate up- or down-scaling methods, in mountainous locations where thermal heterogeneity is large, and may be needed to properly describe the ecologically significant microclimates [7,37].

3. Microclimates and species distribution models

From tiny insects to mega-herbivores, it is well recognized that species ecology is strongly influenced by micro-climatic features of the landscape [2,10,11,12,13] yet quantitative information on how thermal landscape heterogeneity may affect species performance is scarce. Short-scale differences in temperatures may provide opportunities for individual organisms, even with limited dispersal capabilities, to escape unfavourable microclimates or to maximize physiological performances by selecting preferred microclimates [38,39]. Our analysis showed that predictions on *P. operculella* growth rates strongly differed between Wordclim-based and locally-measured temperatures, suggesting that global species distribution models using global coarse-scale climatic datasets without further microclimate modelling could be strongly limited to accurately predict species occurrence and performance, in particular that of ectotherms living in habitats such as mountain slopes. Such a spatial heterogeneity in thermal patches, where climatic conditions are strongly modified, provides a mosaic of favourable, sub-optimal or lethal thermal habitats that directly influences the performance of natural populations of ectotherms.

Coarse-extent modeling of microclimate is currently one of the major obstacles to predicting how organism will react to their experienced environments and forecast their distribution under climate change [8]. To date, two main types of models have been shown to provide relatively accurate, continent-wide calculations of microclimate: statistical model and mechanistic model [13]. The first one is statistical as the variables are not deterministically but stochastically related. These models perform statistical correlation of species occurrences with climatic data and have proven to be powerful interpolative tools for defining and projecting climatic envelopes [40,41]. A disadvantage of these statistical models is that they can only be applied to the conditions under which they are fitted. On the other hand, mechanistic models of the climatic responses of organisms [13,34] use fundamental knowledge of the interactions between process variables to define the model structure. Therefore they do not require much data for model development and validation. One of them is the Microclim model recently developed by [35,42] for all terrestrial landmasses − except Antarctica− which quantify key microclimatic parameters at macro-scales, with a relatively fine spatial (15 km^2) and temporal resolution (hours). The microclimatic parameters such as wind velocity, humidity, and solar radiation allow building energy and mass budgets of organisms, and therefore serve as key inputs for biophysical models of species distributions.

It is important to highlight that a better spatiotemporal resolution in temperature patterns should go in pair with the development of more accurate temperature-based population dynamics models to integrate it [2,13,34,43]. Existing predictions of models based on insect response measured in constant temperatures may yield different and less realistic results than those from predictions of models that include the effect of real temperature fluctuation on insect biology [33]. For example, to date, we still do not know the impact of a few hours of warm temperature for the performance of ectotherm species at longer time scales [33]. In this context, fine-scale spatiotemporal temperature mapping has revealed a key step for any studies aiming at understanding, predicting and managing the responses of species distributions to climate change.

Supporting Information

Appendix S1 Habitat and field size distribution in the three studied altitudinal belts.

Appendix S2 Photos of the temperature recording experiment.

Appendix S3 Spatial variability of temperatures within a field.

Appendix S4 Comparison of time series analysis outputs using 15 days vs. 1-year temperature data.

Appendix S5 Fourier analysis description.

Appendix S6 Seanonality measured on four year air temperature time series with Discrete Fourier Transform.

Appendix S7 Crop habitat composition survey used in the study area.

Appendix S8 Local and global air mean temperature discrepancies mapping.

Appendix S9 Microclimate temperature time-series used in this work #1.

Appendix S10 Microclimate temperature time-series used in this work #2.

Appendix S11 Microclimate temperature time-series used in this work #3.

Acknowledgments

We are grateful to the city council of Salcedo (Cotopaxi, Ecuador) for providing the digital shape files of the study area cadastre. We also thank all farmers who collaborated with us during fieldwork. And finally, we gratefully acknowledge Dr. Pincebourde S. and Dr. Duyck F. for their constructive scientific comments and suggestions.

Author Contributions

Conceived and designed the experiments: EF OD. Performed the experiments: EF MH. Analyzed the data: EF LB OD. Contributed reagents/materials/analysis tools: EF LB OD. Contributed to the writing of the manuscript: EF LB JFS OD.

References

1. Scherrer D, Korner C (2011) Topographically controlled thermal-habitat differentiation buffers alpine plant diversity against climate warming. J Biogeogr 38: 406–416.
2. Bennie J, Hodgson JA, Lawson CR, Holloway CTR, Roy DB, et al. (2013) Range expansion through fragmented landscapes under a variable climate. Ecol Lett 117: 285–229.
3. Logan ML, Huynh RK, Precious RA, Calsbeek RG (2013) The impact of climate change measured at relevant spatial scales: new hope for tropical lizards. Global Change Biol 19: 3093–3102.
4. Scheffers BR, Edwards DP, Diesmos A, Williams SE, Evans TA (2013) Microhabitats reduce animal's exposure to climate extremes. Global Change Biol 20: 495–503.
5. Geiger R (1965) The climate near the ground. Cambridge, USA.
6. Jones HG (1992) Plants and microclimate: a quantitative approach to environmental plant physiology. Cambridge University Press, Cambridge.
7. Dobrowski SZ (2011) A climatic basis for microrefugia: the influence of terrain on climate. Global Change Biol 17: 1022–1035.
8. Potter KA, Woods HA, Pincebourde S (2013) Microclimatic challenges in global change biology. Global Change Biol 19: 2932–2939.
9. Cloudsley-Thompson JL (1962) Microclimates and the distribution of terrestrial arthropods. Annu Rev Entomol 7: 199–222.
10. Tracy CR (1977) Minimum size of mammalian homeotherms: role of the thermal environment. Science 198: 1034–1035.
11. Willmer PG (1982) Microclimate and the environmental physiology of insects. Adv Insect Physiol 16: 1–57.
12. Unwin DM, Corbet SA (1991) Insects, plants and microclimate. Richmond Publishing Company Ltd.
13. Kearney M, Porter W (2009) Mechanistic niche modelling: combining physiological and spatial data to predict species' ranges. Ecol Lett 12: 334–350.
14. Beaumont LJ, Pitman AJ, Poulsen M, Hughes L (2007) Where will species go? Incorporating new advances in climate modelling into projections of species distributions. Global Change Biol 13: 1368–1385.
15. Deutsch CA, Tewksbury JJ, Huey RB, Sheldon KS, Ghalambor CK, et al. (2008) Impacts of climate warming on terrestrial ectotherms across latitude. P Natl Acad Sci USA 105: 6668–6672.
16. Warren RJ, Chick L (2013) Upward ant distribution shift corresponds with minimum, not maximum, temperature tolerance. Global Change Biol 19: 2082–2088.
17. Suggitt AJ, Gillingham PK, Hill JK, Huntley B, Kunin WE, et al. (2011) Habitat microclimates drive fine-scale variation in extreme temperatures. Oikos 120: 1–8.
18. Sears MW, Raskin E, Angilletta MJ (2011) The world is not flat: defining relevant thermal landscapes in the context of climate change. Integr Comp Biol 51: 666–675.
19. Dangles O, Carpio C, Barragan AR, Zeddam JL, Silvain JF (2008) Temperature as a key driver of ecological sorting among invasive pest species in the tropical Andes. Ecol Appl 18: 1795–1809.
20. McCain CM (2007) Could temperature and water availability drive elevational species richness patterns? A global case study for bats. Global Ecol Biogeogr 16: 1–13.
21. Dangles O, Carpio FC, Villares M, Yumisaca F, Liger B, et al. (2010) Community-based participatory research helps farmers and scientists to manage invasive pests in the Ecuadorian andes. Ambio 39: 325–335.
22. Baldocchi DD, Verma SB, Rosenberg NJ (1983) Microclimate in the soybean canopy. Agr Meteorol 28: 321–337.
23. Wilhelm WW, Ruwe K, Schlemmer MR (2000) Comparisons of three-leaf area index meters in a corn canopy. Crop Sci 40: 1179–1183.
24. Fitzpatrick EA (1964) Seasonal distribution of rainfall in Australia analysed by Fourier methods. Archiv Meteorologie, Geophysik und Bioklimatologie 13: 270–286.
25. Hijmans RJ, Cameron SE, Parra JL, Jones PG, Jarvis A (2005) Very high resolution interpolated climate surfaces for global land areas. Int J Climatol 25: 1965–1978.
26. Crespo-Perez V, Rebaudo F, Silvain JF, Dangles O (2011) Modeling invasive species spread in complex landscapes: the case of potato moth in Ecuador. Landscape Ecol 26: 1447–1461.
27. Rondon S (2010) The potato tuberworm: a literature review of its biology, ecology, and control. Am J Potato Res 87: 149–166.
28. Hanafi A (1999) Integrated pest management of potato tuber moth in field and storage. Potato Res 42: 373–380.
29. Cameron PJ, Walker GP, Penny GM, Wigley PJ (2002) Movement of potato tuberworm within and between crops, and some comparisons with diamondback moth. Environ Entomo 31: 65–75.
30. Crespo-Pérez V, Dangles O, Régnière J, Chuine I (2013) Modeling temperature-dependent survival with small datasets: insights from tropical mountain agricultural pests. Bul Entomol Res 103(03): 336–343.
31. Sharpe PJH, DeMichele DW (1977) Reaction-kinetics of poikilotherm development. J Theor Biol 64: 649–670.
32. Schoolfield RM, Sharpe PJH, Magnuson CE (1981) Non-linear regression of biological temperature-dependent rate models based on absolute reaction-rate theory. J Theor Biol 88: 719–731.
33. Gilbert E, Powell JA, Logan JA, Bentz BJ (2004) Comparison of three models predicting developmental milestones given environmental and individual variation. B Math Biol 66(6): 1821–1850.
34. Buckley LB, Urban MC, Angilletta MJ, Crozier LG, Rissler LJ, et al. (2010) Can mechanism inform species' distribution models? Ecol Lett 13: 1041–1054.
35. Kearney M, Shamakhy A, Tingley R, Karoly DJ, Hoffmann AA, et al. (2013) Microclimate modelling at macro scales: a test of a general microclimate model integrated with gridded continental-scale soil and weather data. Global Change Biol Doi : 10.1111/2041-210X.12148.
36. Scherrer D, Korner C (2010) Infra-red thermometry of alpine landscapes challenges climatic warming projections. Global Change Biol 16: 2602–2613.
37. Fridley JD (2009) Downscaling climate over complex terrain: high finescale (< 1000 m) spatial variation of near-ground temperatures in a montane forested landscape. J Appl Meteorol Clim 48: 1033–1049.
38. Kinahan AA, Pimm SL, van Aarde RJ (2007) Ambient temperature as a determinant of landscape use in the savanna elephant, Loxodonta africana. J Therm Biol 32: 47–58.
39. Dillon ME, Liu R, Wang G, Huey RB (2012) Disentangling thermal preference and the thermal dependence of movement in ectotherms. J Therm Biol 37: 631–639.
40. Guisan A, Thuiller W (2005) Predicting species distribution: offering more than simple habitat models. Ecol Lett 8: 993–1009.
41. Elith J, Leathwick JR (2009) Species distribution models: ecological explanation and prediction across space and time. Annu Rev Ecol Evol S 40: 677–697.
42. Kearney MR, Isaac AP, Porter WP (2014) microclim: Global estimates of hourly microclimate based on long-term monthly climate averages. Scientific Data 1.
43. Bakken GS, Angilletta MJ (2014) How to avoid errors when quantifying thermal environments. Func Ecol 8: 96–107.
44. Bloomfield P (2004) Fourier analysis of time series: an introduction. Wiley-Interscience, New York, USA.

Mapping the Abundance and Distribution of Adélie Penguins Using Landsat-7: First Steps towards an Integrated Multi-Sensor Pipeline for Tracking Populations at the Continental Scale

Heather J. Lynch[1]*, Mathew R. Schwaller[2]

1 Department of Ecology and Evolution, Stony Brook University, Stony Brook, New York, United States of America, **2** Mesoscale Atmospheric Processes Laboratory, NASA/ Goddard Space Science Center, Greenbelt, Maryland, United States of America

Abstract

The last several years have seen an increased interest in the use of remote sensing to identify the location of penguin colonies in Antarctica, and the estimation of the abundance of breeding pairs contained therein. High-resolution (sub-meter) commercial satellite imagery (e.g., Worldview-1, Quickbird) is capable of colony detection and abundance estimation for both large and small colonies, and has already been used in a continental-scale survey of Adélie penguins. Medium-resolution Landsat imagery has been used successfully to detect the presence of breeding penguins, but has not been used previously for abundance estimation nor evaluated in terms of its minimum colony size detection threshold. We report on the first comprehensive analysis of the performance of these two methods for both detection and abundance estimation, identify the sensor-specific failure modes that can lead to both false positives and false negatives, and compare the abundance estimates of each method over multiple spatial scales. We find that errors of omission using Landsat imagery are low for colonies larger than ~10,000 breeding pairs. Both high-resolution and Landsat imagery can be used to obtain unbiased estimates of abundance, and while Landsat-derived abundance estimates have high variance for individual breeding colonies relative to estimates derived from high-resolution imagery, this difference declines as the spatial domain of interest is increased. At the continental scale, abundance estimates using the two methods are roughly equivalent. Our comparison of these two methods represents a bridge between the more developed high-resolution imagery, which can be expensive to obtain, and the medium-resolution Landsat-7 record, which is freely available; this comparison of methodologies represents an essential step towards integration of these disparate sources of data for regional assessments of Adélie population abundance and distribution.

Editor: Guy J-P. Schumann, University California Los Angeles, United States of America

Funding: HJL was supported by National Science Foundation/Office of Polar Programs - Award No. 1255058; http://www.nsf.gov. HJL and MRS were supported by the National Air and Space Administration - Award No. NNX14AC32G; http://www.nasa.gov. The funders had no role in study design, data collection and analysis, decision to publish, or preparation of the manuscript.

Competing Interests: The authors have declared that no competing interests exist.

* Email: heather.lynch@stonybrook.edu

Introduction

Though the idea of using satellite imagery to map penguin colonies is over thirty-years old [1], the use of these approaches for ecological research and long-term monitoring has been in active development only in the last few years (e.g., [2–4]). Recent data policy changes, such as the free availability of archival Landsat imagery, have opened access to vast amounts of medium resolution satellite imagery over Antarctica, and while access to very high-resolution (VHR) imagery is hampered by licensing and cost, increased tasking of satellites for imagery in the Antarctic have rapidly increased the quantity of imagery available. Recent studies demonstrate that penguin colony presence and abundance can be detected on a continental scale by manual interpretation of very high resolution (VHR) satellite imagery [5] and that automated algorithms that exploit the spectral signature of penguin guano can be used to detect the presence of penguin colonies in medium-resolution Landsat data [2,6]. Taken together, these new analytic methods—and the Terabytes of data to which they can be applied—afford the possibility of routine, continent-wide surveys of Adélie penguin presence and abundance, although to date no attempt has been made to use Landsat for regional abundance estimation or to compare the detection and abundance information derived from these two types of satellite images. Our analysis represents a bridge between abundance estimates based on very high-resolution imagery and the medium-resolution data record available from Landsats-4, -5 and -7. Forging this connection will facilitate integrated abundance estimates in the era when both of these satellite data types overlap, and will advance the development of a continuous time series of abundance that leverages a Landsat data record dating back to the 1980s.

This paper presents the results of an automated retrieval of Adélie penguin colonies along the entire continent of Antarctica from medium-resolution Landsat-7 imagery (30 m pixel size), and its comparison with data on presence and abundance based on manual interpretation of VHR satellite imagery (~0.60 m pixel size). Landsat-7 launched on April 15, 1999 and has collected imagery over Antarctica continuously since that time. Unfortunately, the ETM+ Scan Line Corrector failure in 2003 occurred just as some of the first very-high resolution images became available for the Antarctic, precluding direct contemporaneous comparison of abundance estimates from these two types of sensors. We leverage the recently published global Adélie survey [5] to assess whether Landsat-7 can be used to estimate abundance as well as occupancy, and to estimate the lower detection threshold for Landsat-7. Since the uncorrupted Landsat-7 imagery dates from 1999–2003 and the global Adélie survey [5] estimates predominately stem from 2010–2013, the correlation between the abundance estimates will be highly conservative, since the correlation is degraded by changes in true abundance in the intervening decade. Nevertheless, the relatively slow dynamics of penguin colonies means that abundances between these two periods should be fairly consistent within the margin of error of the abundance estimation itself, allowing us to investigate whether the number of pixels retrieved in the earlier period scales with the abundance reported by [5] for the latter period. This cross-validation approach is the best currently available to quantify the accuracy and identify sources of error in using medium-resolution Landsat imagery for occupancy (presence vs. absence) and abundance. Controlling or correcting these errors is a key consideration for future automated surveys of Adélie penguins in Antarctica.

In our analysis and discussion, we explicitly distinguish between the probability of detection and the estimation of abundance conditional on the detection of a colony. The detection of a colony based on the spectral characteristics of its guano stain is a binary classification subject to errors of omission (false negative) and commission (false positive). Abundance estimation is a separate challenge in which success is judged by the accuracy (bias and variance) of estimates of population size. We use the detection probability model along with a probability model for colony size to estimate the total abundance of penguins missing given the results of a Landsat survey. This methodology allows us to correct future Landsat surveys for detection failures and paves the way for achieving regular unbiased estimates of the global Adélie population. We considered continental Antarctica (in which the Adélie penguin is the only breeding *Pygoscelis* spp. penguin) separate from the Antarctic Peninsula in our analysis. Guano stains on the Antarctic Peninsula are ambiguous with respect to species in Landsat imagery and the three *Pygoscelis* spp often nest in colonies mixed at Landsat's larger spatial resolution. Additionally, as described in the Discussion, these two regions have different 'failure modes' with respect to detection and imagery interpretation.

Our goals for this analysis were three-fold: (1) To estimate the lower detection threshold for Landsat-7 imagery, (2) to demonstrate that the number of pixels classified as penguin guano in Landsat-7 scales with abundance, and (3) to estimate variance in abundance estimates at varying scales of spatial aggregation. Our survey includes both continental Antarctica, where Landsat imagery has been used for detection of Adélie penguin colonies previously [2], as well as the Antarctic Peninsula, where its use is novel. The comparison and cross-validation of multiple data streams is a necessary step in the full integration of these data into a single remote sensing retrieval and analysis pipeline for monitoring Adélie penguin populations. This approach represents a significant improvement over direct field counts alone, which lack the scalability of remote sensing surveys, and are subject to underestimation bias in regional or global estimates due to the remoteness and inaccessibility of sections of the Antarctic coastline [7]. Population estimates that scale in space and time are expected to more seamlessly interoperate with trophic and climate models, so that changes in Adélie penguin populations can be clearly tied to harvesting in the Southern Ocean fisheries and climate change.

Methods

Automated Landsat retrievals

Retrieval of Adélie penguin colony location and spatial extent over the Antarctic Peninsula was conducted using Landsat-7 Enhanced Thematic Mapper-Plus ETM+ data. The Landsat data used in this study were obtained as a Climate Data Record (CDR), which was generated using the methods developed by [8], and distributed by the U.S. Geological Survey Earth Resources Observation and Science Center (https://espa.cr.usgs.gov). This CDR applies a uniform processing to the entire dataset, and includes a host of products including top of atmosphere (TOA) reflectance, surface reflectance, brightness temperature, as well as masks for clouds, cloud shadows, adjacent clouds, land and water. The Landsat CDR TOA product was used for the automated retrieval of Adélie penguin colonies following the methods described in [2]. One difference between our Landsat survey and that described in [2] is that additional exposed soil and rock surfaces were sampled and included in the training dataset for the AP retrieval algorithm because the AP has a proportionally larger expanse of exposed soil and a different suite of soil types than the rest of the continent. The training dataset consisted of 473 Landsat-7 pixels selected from known Adélie colonies plus an additional 10,688 pixels selected from areas of exposed rock, soil and vegetation. The retrieval algorithm used this training dataset to define an ellipsoid in the spectral data space that surrounds pixels from Adélie colonies, with the assumption that the surface of the ellipsoid separates the pixels within its bounds from other surface types. The equation of the ellipsoid is captured in a "transition matrix" which is used to build a decision rule for determining whether or not new pixels belong to the Adélie colony class. All of the Landsat-7 ETM+ reflective bands (bands 1–5 plus band 7) were used in the AP retrieval. The retrieval on the AP and surrounding islands was conducted on 62 scenes covering this region but did not include the South Shetlands or the South Orkney Islands because cloud-free imagery was unavailable for the era investigated. The methods described, however, can be directly applied to imagery from these areas in future surveys. Additional details on the AP retrieval algorithm, including the transition matrix and decision rules used in this study, can be found in the additional information within the file Table S2. An Antarctic continent-wide retrieval of Adélie penguin colonies was obtained by combining the AP retrievals with those previously generated by [2]. In all cases, imagery was selected from the Landsat-7 era in the years from 1999–2003, prior to the failure of the ETM+ Scan Line Corrector.

Abundance and distribution from high-resolution commercial satellite imagery

We used the Adélie penguin survey by [5], which was based on VHR imagery supplemented with recent field surveys, as the basis for studying the performance characteristics of Landsat retrievals.

Details of the detection and abundance estimation of Adélie penguins in VHR imagery may be found in [4] and [5]. From their global survey of Adélie penguins, [5] produced a detailed list of all Adélie penguin colonies (location and estimated abundance) known to exist at the time of the survey, which we used as a baseline for assessing successful retrieval and abundance estimation using the lower resolution Landsat survey method. The comparison of our automated detection algorithm to the Adélie survey results from [5] yields four possible outcomes for each potential Adélie penguin breeding site.

1. A penguin colony is detected using both methods.
2. No penguin colony is detected by either approach. This situation encompasses the vast majority of the Antarctic coastline.
3. A penguin colony is identified in [5] but not identified by our automated detection algorithm applied to Landsat imagery. These incidents are considered errors of omission for the Landsat detection algorithm because the colonies identified by [5] were, with a few exceptions, validated by direct field surveys. In these cases, we consulted the available imagery to determine what characteristics may have led to this error of omission.
4. An area is flagged as containing a colony of breeding penguins by the automated detection algorithm but was not reported in [5]. These cases may represent an error of omission for [5] or an error of commission by the automated detection algorithm. In these cases, we searched for additional VHR imagery that might resolve these two possible scenarios. If no additional imagery was available, we were not able to determine the nature of the discrepancy. All such cases are reported in the hopes that future field surveys can validate the true occupancy status of these locations.

The conversion between sensor pixels flagged as guano and the abundance of penguins breeding within that area is determined by an 'apparent density' that is not necessarily equal to the actual density of breeding pairs as measured on the ground [4]. Nevertheless, it has been demonstrated that the closely packed nature of Adélie penguin colonies allows for the estimation of abundance based on the area of guano staining in very high-resolution imagery [4]. As a first pass, we used Kendall's τ rank correlation coefficient to investigate whether the number of pixels retrieved by Landsat-7 scales with penguin abundance as reported by [5]. In other words, we ask whether the colony with the largest number of pixels retrieved by Landsat-7 is also the largest colony as reported in [5], and so forth. We then use the abundance estimates of [5] to develop a predictive model for estimating the total abundance of Adélie penguins from medium-resolution Landsat imagery, recognizing that the ~10 year time lag between the abundance estimates reported in [5] and the older Landsat imagery significantly increases the uncertainty of these estimates. Nevertheless, such 'ballpark' estimates can be invaluable in the near term, particularly for estimating approximate abundance for previously undiscovered colonies, and provide a road map for re-analysis when matched-pairs of Landsat-8 imagery and field counts become available in the future. This methodology, applied here to Landsat-7 imagery and in the future to Landsat-8 or other medium resolution sensors, can be used to automate future continental scale surveys yielding global estimates of Adélie distribution and abundance with little to no manual interpretation. This abundance model involves two components: (1) the estimation of breeding abundance represented by the pixels identified as guano in the Landsat survey, and (2) the estimation of the abundance of breeding Adélie penguins not detected.

Abundance estimation for detected colonies

For the first component of total abundance, involving the estimation of abundance conditional on detection, we considered all breeding populations found by the Landsat-7 automated retrieval process whose population had been estimated previously by [5]. The data used for our abundance estimates are detailed in Table S1. Seven colonies were not included in the development or validation of the abundance model because they either included a significant number of flying birds (which inflated the estimated size of the Adélie colony) or because a discrepancy between the Landsat retrieval and the location reported in [5] made it impossible to know whether the same breeding population was being considered in each method (Table S2). We used a Poisson regression model for abundance as a function of guano area

$$Abundance_i \sim Pois(\lambda_i) \tag{1}$$

$$\log(\lambda_i) = \beta_0 + \log(Area_i)$$

where λ_i is the expected abundance of the ith colony and e^{β_0} can be interpreted as an "apparent density" (nests per 'detected guano area'). Since many applications, particularly within the context of the Convention on the Conservation of Antarctic Marine Living Resources, require abundance estimates at scales larger than an individual breeding colony (e.g., Convention Subareas), we also quantified predictive accuracy of our final abundance model for spatial units of increasing scale, from individual colonies to the entire continent.

Estimation of abundance in colonies not detected by Landsat

The total abundance of Adélie penguins not detected in a Landsat survey is a convolution of the probability distribution of colony abundance and the probability of detecting a colony given (i.e. conditional on) its size. The expected size of a colony that is not detected by Landsat imagery can be calculated as the following conditional expectation

$$\begin{aligned} E[N|M=0] &= \sum_{n=0}^{\infty} n * P(N=n|M=0) \\ &= \sum_{n=0}^{\infty} n * \frac{P(M=0|N=n)P(N=n)}{P(M=0)} \end{aligned} \tag{2}$$

where N is the number of breeding pairs, and M represents the binary probability that a colony is missing (M = 0) from a Landsat survey. The probability that a colony is missing conditional on its size (i.e., $P(M=0|N=n)$ = *1-detection probability*) can be estimated with a logistic regression model fit using the abundances of each colony as reported in [5]. In the absence of future matched VHR-Landsat surveys, the detection probability can be assumed a function of the sensor that would not change for future surveys using Landsat-7 imagery. The probability distribution of colony sizes $P(N)$ is a long-tailed distribution best fit by a log-normal distribution. In theory, we could estimate the best-fit parameters for the colony size probability distribution $P(N)$ using the partially-censored Landsat survey detections by inverting Eq. 2 for $P(N)$;

however, the low rates of detection for very small colonies make it difficult to establish the shape of the colony size distribution from the Landsat detections alone. We therefore fit a log-normal distribution to the abundance data in [5] and used a discretized version of this best-fitting model for $P(N)$ in Eq. 2 to estimate the total abundance of Adélie penguins missing from the Landsat survey due to non-detection.

The sum in Eq. 2 can be calculated using Markov Chain Monte Carlo integration, and the resulting expectation can be multiplied by expected number of missing colonies to estimate the total missing abundance. We modeled the total number of undetected Adélie colonies in the surveyed area as a sample from a Negative Binomial distribution

$$\text{Number of undetected colonies} \sim \text{NegBin}(\text{Number detected}, p_{detected})$$

$$\text{Number of undetected colonies} \sim \text{NegBin}(\text{Number detected}, p_{detection}) \quad (3)$$

where the overall probability of detection $p_{detection}$ was estimated using the survey data in [5] as a basis for comparison.

Estimation of abundance at aggregated spatial scales

To estimate the improvement in predictive accuracy at regional scales containing multiple penguin colonies, we compared total abundance estimated by VHR and Landsat for subsets of the original data. Our goal was to determine how many colonies $k=1,2,3,\ldots$ would have to be enclosed within an area for the estimate of total abundance by Landsat-7 to be statistically indistinguishable from the corresponding total abundance estimated by VHR. For example, to compare total abundance estimates for a (hypothetical) area containing $k=5$ colonies, we would draw 5 colonies at random from the total set of colonies in our study (continental and Peninsula areas combined) and use Eq. 1 (with best-fit parameters; see Results) to estimate the total abundance that would be estimated for those 5 colonies. The total abundance would then be compared to the expected total abundance from the VHR survey of [5], and the scaled difference recorded as

$$\textit{Scaled difference} = \frac{VHR - Landsat}{VHR + Landsat}. \quad (4)$$

Repeating this procedure 1000 times, we generated a distribution of scaled differences between the two methods. The null distribution consisted of calculating the same scaled difference but with the Landsat abundance replaced by a second independent draw from the VHR abundance distribution. (The abundance of each colony is associated with a sampling distribution, so independent draws from that distribution yield different estimates of abundance. This variability represents our uncertainty in the true abundance of the colony.) In other words, we wanted to differentiate between differences due to Landsat's relatively poorer performance characteristics, and differences due to inherent uncertainty for the VHR counts used as "true" abundance. The empirical cumulative distribution functions (VHR-Landsat vs. VHR-VHR) were compared using a one-sided Komolgorov-Smirnov test.

Results

Using Landsat data over the Antarctic Peninsula, the automated retrieval flagged 143 areas as belonging to the "Adélie penguin colony" class. Based on the site-by-site comparisons described above, these retrievals included 16 confirmed Adélie penguin colonies. Among these retrieved areas were 17 additional areas identified as likely Adélie penguin colonies that could not be confirmed because VHR data were not available and the sites were not previously mentioned in the literature (see Table S3). Using Landsat data over continental Antarctica, the automated retrieval flagged 187 areas as belonging to the "Adélie penguin colony" class, within which were 91 confirmed Adélie penguin colonies, and 8 areas identified as likely, but unconfirmed, Adélie penguin colonies (see Table S3). The Landsat retrieval of Adélie penguin colonies over the entire continent of Antarctica can be quickly visualized via the gzip-compressed Keyhole Markup Language (kmz) formatted file provided (see additional information within the file Table S2).

Probability of detection

The detection performance of the automated Landsat retrieval improved with increasing Adélie penguin colony population size (Figure 1). The probability of detecting an Adélie penguin colony using Landsat ($p_{detection}$) was modelled using the logistic transformation

$$\log\left(\frac{p_{detection}}{1-p_{detection}}\right)=\alpha+\beta * N \quad (5)$$

where N is the abundance of breeding pairs, $\hat{\alpha}=-1.00$ (SE = 0.32) and $\widehat{\beta}=0.0004$ (SE = 0.0001) for the continental colonies and $\widehat{\alpha}=-1.72$ (SE = 0.45) and $\widehat{\beta}=0.0003$ (SE = 0.0001) for the Peninsula colonies. The smallest colony detected in the Landsat retrieval had 322 pairs (Clark Island in Marie Byrd Land), but the threshold for 50% detection probability (as estimated by the logistic model) was 2240 breeding pairs for continental colonies and 6405 breeding pairs for Peninsula colonies (Table 1).

While nearly all of the smallest colonies go undetected using the automated interpretation of Landsat imagery, small Adélie penguin colonies (1 to 3000 pairs) contribute only ~2% to the total Adélie population [5]. Along the coastline of continental Antarctica, the Landsat method successfully identified Adélie penguin colonies making up 98.3% of the population total for this region. The AP has a significantly larger proportion of small colonies when compared to the rest of the continent, and consequently the Landsat method missed a larger number of these colonies in both absolute numbers and in proportion to the total (Figure 1). On the AP, the colonies found by the Landsat method account for 90.2% of the sample population. When all data are pooled together, the Landsat method retrieved those colonies that account for 97.3% of the population total. Furthermore, the Landsat retrieval algorithm detected a number of potential Adélie colonies not captured by the manual imagery search of [5], see Table S3.

There were three Adélie colonies with populations greater than 4000 pairs that were missed in the Landsat retrievals: on Haswell Island (4011 pairs), Coulman South (14786 pairs), and Penguin Point (19377). As described in [2] it is likely that Haswell Island was abandoned during the period when the Landsat imagery were collected (2001 and 2002). The Coulman South colony was covered by deep shadow in the Landsat imagery and is therefore a true error of omission. The only useable Landsat imagery over

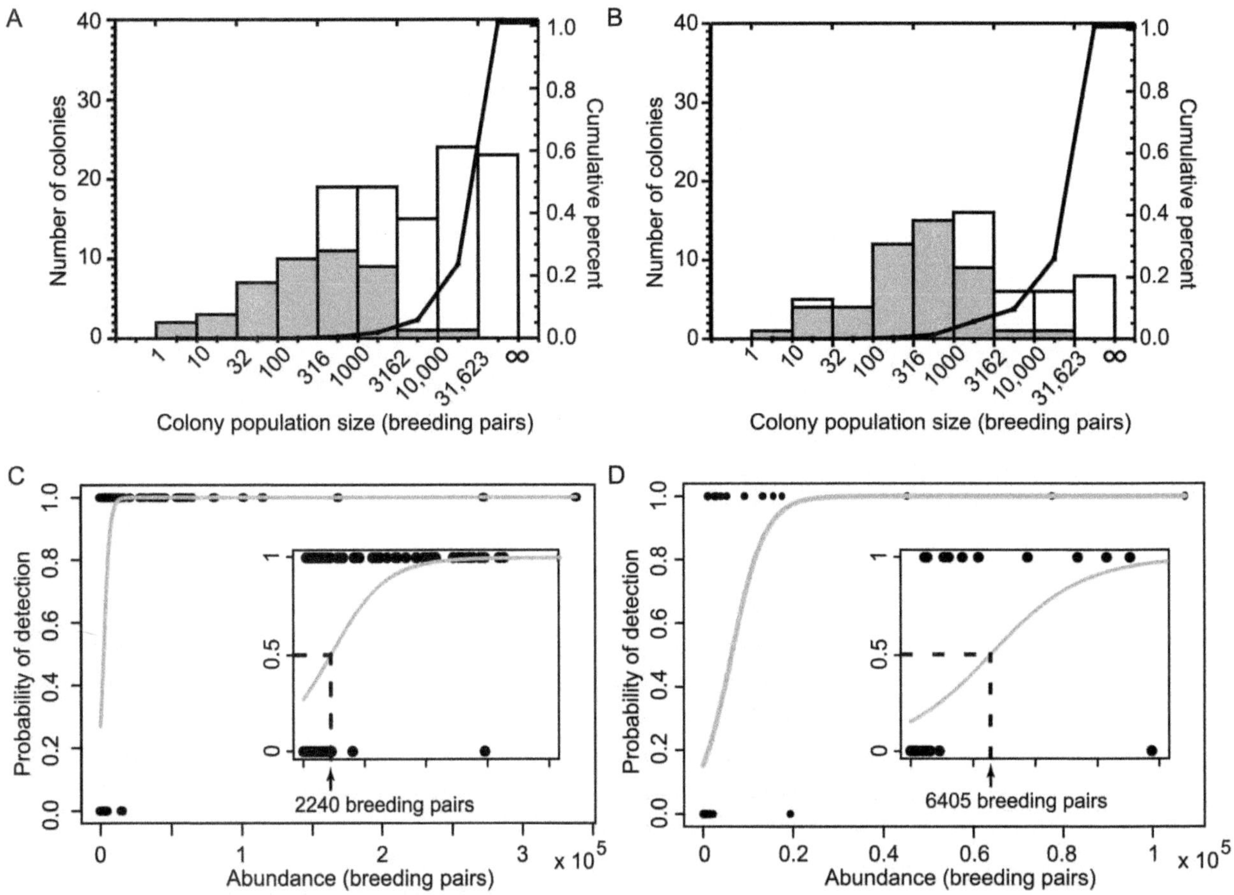

Figure 1. Probability of detection using Landsat-7 imagery. (A, B) The number of Adélie penguin colonies located (white) and missed (shaded) by the Landsat retrieval method for continental Antarctica (A) and the Antarctic Peninsula (B). The percent contribution of each bin to the total population is also illustrated. The horizontal axis is logarithmic, with boundaries equal to 10^0, 10^1, $10^{1.5}$, 10^2, $10^{2.5}$, etc. (C, D) Probability of detection as a function of colony size (as reported by [5]) along the coast of continental Antarctica (C) and on the Antarctic Peninsula (D). Gray line represents best-fitting logistic model. Inset represents portion of the plot shaded in light gray.

Penguin Point available during the survey era was collected on November 24, 2001, which is very early in the Adélie penguin breeding season. A second cloud-free image from January 8, 2001 was available but it was unusable due to band saturation. It is likely that cloud-free Landsat imagery from later in the breeding season would resolve the Penguin Point colony (Figure S1).

Abundance estimation for detected colonies

The number of Landsat-7 pixels retrieved as penguin guano is significantly correlated to the abundance of breeding penguins as reported in [5] (Kendall's $\tau = 0.73$ [$n = 88$; $p < 0.001$]). The Poisson regression model in Eq. 1 yielded estimates of $\widehat{\beta}_0 = -1.0844$ (SE = 0.0007) and $\widehat{\beta}_0 = -0.661$ (SE = 0.002), for the continental and Peninsula data, respectively, corresponding to estimates of apparent density of 0.34 nests/m^2 and 0.52 nests/m^2,

Table 1. Probability of detection using Landsat-7 retrieval algorithm.

$p_{detection}$	Colony size	
	Continent	**AP**
0.50	2240	6405
0.75	4705	10485
0.90	7190	14575
0.95	8870	17355
0.99	12590	23495

Colony sizes (breeding pairs) at various probabilities of detection using Landsat-7 retrieval algorithm.

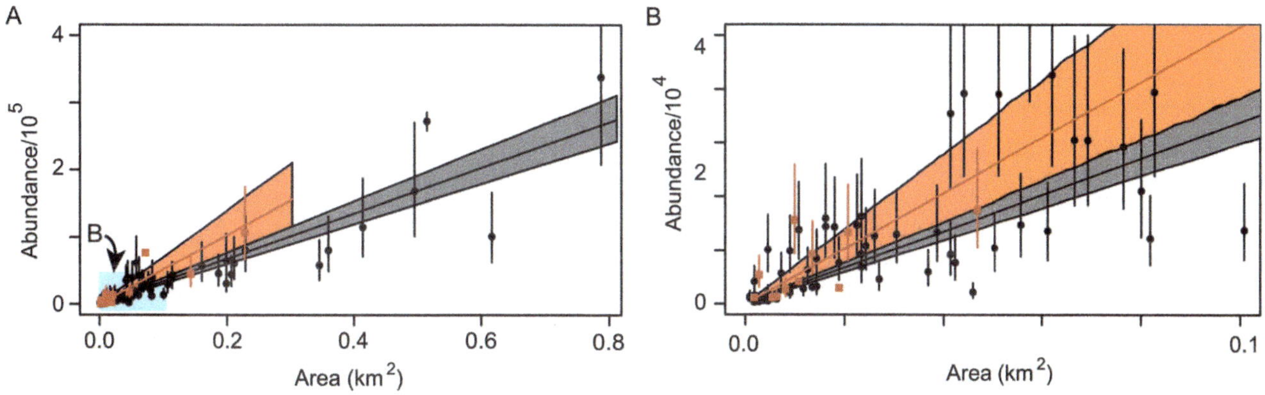

Figure 2. Colony abundance vs. guano stain area. (A) Colony abundance (and 95th percentile confidence intervals; as reported by [7] as a function of the area identified as guano in the Landsat-7 survey (black circles = continental Antarctica, orange squares = Antarctic Peninsula), with best-fitting Poisson regression model (and associated 95th percentile prediction envelop; gray-shaded envelop = continental Antarctica, orange-shaded envelop = Antarctic Peninsula). (B) Zoomed in portion shown as blue box in Panel A.

respectively. As expected from a Poisson regression, the predictive envelope grows with colony size (Figure 2).

Estimation of abundance of colonies not detected by Landsat

We used the distribution of colony abundances reported in [5], to find the best estimates of the log-normal distribution parameters for $P(N)$ (Figure 3)

$$P(N)\sim Lognormal(\mu,\sigma) \quad (6)$$

and found $\widehat{\mu}=8.06$ (SE = 0.21) and $\widehat{\sigma}=2.34$ (SE = 0.15) for the continental colonies and $\widehat{\mu}=6.64$ (SE = 0.32) and $\widehat{\sigma}=2.19$ (SE = 0.23) for the Peninsula colonies. Using Eq. 2, we estimate the total (Peninsula + continent) number of Adélie breeding pairs missing from the Landsat survey to be 119738 (95th percentile CI: 70364–188070), which is higher but generally consistent with the empirical value of 75582 breeding pairs found in the VHR survey [5] but not found by Landsat-7.

Estimation of abundance in colonies with no prior estimates

The Landsat survey found eight colonies that were not detected in the VHR survey reported by [5] but which are outside the breeding range of either gentoo penguins (*Pygoscelis papua*) or chinstrap penguins (*P. antarcticus*), which can appear similar in satellite imagery [9]. While other errors of commission are possible, we can be confident that these are not conflated with mis-identification of penguin species. Using Eq. 1, we estimate the total abundance of Adélie penguins in these eight colonies as ~38000 breeding pairs (Table S3), although these estimates, and in fact the existence of Adélie penguins at these eight locations, needs to be confirmed by higher-resolution satellite imagery or direct observation.

Table S3 lists 17 additional locations flagged as penguin colonies by the Landsat method that are within the breeding range of all three *Pygoscelis* species and thus cannot be identified to species, three of which are large penguin colonies in the Danger Islands. The Danger Islands, a collection of islands off the tip of Joinville Island at the end of the Antarctic Peninsula, are poorly surveyed due to frequent heavy ice cover and steep terrain that preclude easy access, but are known to contain extremely large populations of Adélie and gentoo penguins [10,11]. Within the

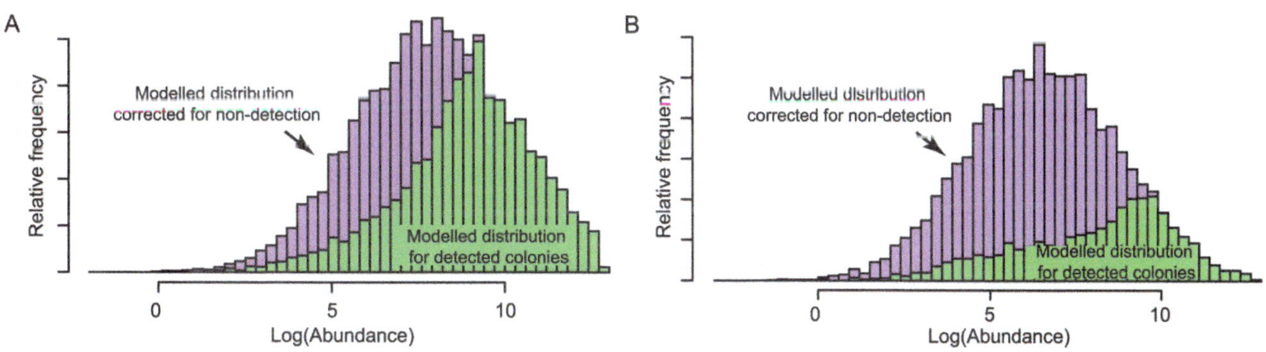

Figure 3. Colony size distribution censored by Landsat-7 detection limit. We used a log-normal distribution model for colony size to calculate the total abundance of Adélie penguins not captured by our Landsat survey (see Eq. 2). The influence of size-dependent non-detection on the distribution of colony sizes detected by Landsat in continental Antarctica (A) and on the Antarctic Peninsula (B) can be seen in the difference between the modelled distribution of colony sizes (log-scale) for all Adélie penguin colonies (purple) and the distribution of colony sizes as detected by Landsat (green).

Danger Islands, the Landsat survey identified three major penguin populations at Brash Island, Earle Island, and Darwin Island, containing an estimated 166078 (95th percentile CI: 123666–228268), 23649 (95th percentile CI: 17361–32163), and 7419 (95th percentile CI: 5384–9931) breeding pairs of penguins, respectively. Based on prior surveys of the area, these are likely to be overwhelmingly composed of Adélie penguins, making Brash Island potentially one of the largest Adélie penguin colonies in the world.

Jorge Island was identified by [10] as one location for which no census information existed, and was another location identified by the Landsat survey as containing breeding penguins. We estimate 3733 (95th percentile CI: 2710–5064) breeding pairs of penguins (of unknown species) breeding at this location.

Errors of commission

An earlier study [2] of Adélie penguin detection in continental Antarctica reported commission errors for the Landsat method on the order of 1% or less in the retrieval of Adélie colonies along the southern coastline of Antarctica. In a similar analysis for the retrievals on the Antarctic Peninsula, we used required prior knowledge of these locations and, where possible, confirmation by high-resolution satellite imagery. We find that commission errors for the Antarctic Peninsula fall into several categories including: pixels that identify potentially new colonies as determined by visual examination of high-resolution satellite imagery but which have not yet been confirmed by other methods, pixels that identify other penguin colonies or those with mixed species including Adélie penguins, pixels that identify flying bird breeding areas (known to exist, or previously unknown to exist), and pixels that are "pure errors" where various soil types are incorrectly categorized as Adélie penguin colonies. The commission error for soils incorrectly classified as penguin colony was calculated to be 3% for the Antarctic Peninsula, and 2% for the entire continent.

Aggregated estimates of abundance

While abundance estimates of individual colonies are variable using Landsat retrievals, our model for estimating abundance from the number of pixels identified as guano appears unbiased and thus neither consistently overpredicts or underpredicts abundance. Consequently, abundance estimates using Landsat retrievals improve at larger scales of aggregation. We find that abundance estimates using Landsat retrievals are statistically indistinguishable from those obtained using very-high resolution methods for aggregations containing >c. 40 colonies.

Discussion

As described below, the retrievals of Adélie colony extent from medium-resolution imagery correlate well with comparable datasets, and exhibit relatively low errors of omission and commission. As has been demonstrated also for high-resolution imagery [4], the retrievals of colony extent (i.e. guano area) as captured by automated classification of Landsat imagery correlate well with abundance. As a result, this area-abundance relationship can be used as the basis for estimating abundance in areas where the population is otherwise unknown. As illustrated in Figure 1, the performance of the Landsat retrieval for colony detection improved with increasing colony population size. While nearly 100% of the smallest Adélie penguin colonies were undetected by the Landsat method, these contribute very little to regional and continental aggregated abundance and the Landsat retrievals are thus able to capture the vast majority of the Adélie penguin population. The Antarctic Peninsula has a significantly larger proportion of small colonies when compared to the rest of the continent and a higher threshold for detection, and consequently the Landsat method missed a larger number of these colonies in both absolute numbers and in proportion to the total.

The Antarctic Peninsula remains a challenge

The higher threshold for colony detection on the Antarctic Peninsula likely stems from several factors. These include a larger ratio of exposed surface material compared to snow cover, a greater variability of rock and soil types compared to the rest of the continent, and increased surface weathering. The AP also experiences significantly more rainfall than the continental portions of Antarctica. Weathering may wash away guano making it more difficult to detect smaller colonies, and it may contribute to estimation errors as the guano stain grows or shrinks within a given breeding season. At the same time, the topography of the Antarctic Peninsula is highly complex, with steep terrain that may obscure smaller colonies. Finally, higher cloud cover results in fewer usable scenes each year.

One of the major challenges of using satellite imagery to map Adélie penguin colonies on the Antarctic Peninsula is the existence of two other, closely related, species of penguins (gentoo and chinstrap) whose guano displays similar spectral signatures to that of the Adélie. Moreover, these three species frequently nest in mixed-species colonies, and while differentiation of species has been shown possible in high-resolution imagery [9], the spatial resolution of Landsat is larger than the scale at which Adélie penguins may cluster within a mixed-species colony. In other words, a single Landsat pixel of a mixed-species colony likely contains multiple species that cannot be distinguished. Other *Pygoscelis* spp. penguins are not the only source of errors of commission. Flying bird species, such as the blue-eyed shag (*Phalacrocorax atriceps*), also produce guano that can be incorrectly identified as penguin guano and can cause errors of commission in both Landsat and VHR imagery.

Errors of commission vary across regions and sensors

Our experience with both VHR and Landsat imagery interpretation has shown that the causes of false positives (errors of commission) are both region and sensor specific. In the Antarctic Peninsula region, errors of commission can stem from breeding blue-eyed shags, as well as from large areas of snow algae, which is often red or pink in color, and iron-tinted scree on steep sloping ridges (Lynch Unpublished data). As well, areas used by penguins for molting can contain significant accumulations of guano and confuse mapping of breeding territories. Landsat imagery retrievals may contain areas used by flying birds either alone or with breeding Adélie penguins. We have also found abiotically-driven errors of commission in Landsat retrievals due to large wet alluvial fans, and others apparently due to sediment runoff or chlorophyll in the water just offshore.

Future research

Increased weathering, complex terrain, and the existence of congeneric species make it difficult to estimate omission and commission error rates for the Antarctic Peninsula. While a Landsat retrieval of a gentoo colony is an error of commission in the context of an Adélie survey, it still has biological significance, particularly if it identifies a colony that was previously undiscovered. The use of prior information regarding the distribution of penguins and the Antarctic's flying bird species may allow us to use adaptive classification methods that minimize both errors of commission and omission by essentially 'raising the bar' for the detection of previously unknown Adélie penguin colonies and simultaneously 'lowering the bar' for the re-detection of known Adélie penguin colonies. In the same vein, prior estimates (e.g., from direct field surveys) of the fraction of penguins that are of each species (Adélie, gentoo, and chinstrap) at mixed colonies may allow us to estimate Adélie abundance to produce less biased estimates at the regional of continental scale.

While Landsat imagery lacks the sensitivity to detect small colonies, and is therefore inappropriate for studying processes such as colonization and range expansion, we have demonstrated that it can be used to obtain estimates of Adélie abundance over regional scales, and that bias in regional estimates can be corrected for non-detection. Corrections for non-detection will be made more accurate by better statistical models for colony size, but the existing precision of regional and global Adélie abundance is minimally affected by non-detection of the smallest colonies.

Acknowledgments

We thank Claire Porter for her expertise and tireless efforts processing the satellite imagery required for this analysis.

Author Contributions

Conceived and designed the experiments: HJL MRS. Performed the experiments: HJL MRS. Analyzed the data: HJL MRS. Wrote the paper: HJL MRS.

References

1. Schwaller MR, Benninghoff WS, Olson CE (1984) Prospects for satellite remote-sensing of Adélie penguin rookeries. International Journal of Remote Sensing 5: 849–853.
2. Schwaller MR, Southwell CJ, Emmerson LM (2013) Continental-scale mapping of Adélie penguin colonies from Landsat imagery. Remote Sensing of Environment 139: 353–364.
3. Fretwell PT, LaRue MA, Morin P, Kooyman GL, Wienecke B, et al. (2012) An emperor penguin population estimate: The first global, synoptic survey of a species from space. PLoS ONE 7: e33751.
4. LaRue MA, Lynch HJ, Lyver P, Barton K, Ainley DG, et al. (2014) Establishing a method to estimate Adélie penguin populations using remotely-sensed imagery. Polar Biology 37: 507–517.
5. Lynch HJ, LaRue M (2014) First global survey of Adélie penguin populations. The Auk 131: 457–466.
6. Mustafa O, Pfeifer C, Peter H-U, Kopp M, Metzig R (2012) Pilot study on monitoring climate-induced changes in penguin colonies in the Antarctic using satellite images. Project FKZ 3711 85 199. Federal Ministry of the Environment.
7. Southwell C, Smith D, Bender A (2009) Incomplete search effort: a potential source of bias in estimates of Adélie penguin breeding populations in the Australian Antarctic Territory. Polar Record 45: 375–380.
8. Masek JG, Vermote EF, Saleous N, Wolfe R, Hall FG, et al. (2006) A Landsat surface reflectance data set for North America, 1990-2000. Geoscience and Remote Sensing Letters 3: 68–72.
9. Lynch HJ, White R, Black AD, Naveen R (2012) Detection, differentiation, and abundance estimation of penguin species by high-resolution satellite imagery. Polar Biology 35: 963–968.
10. Lynch HJ, Ratcliffe N, Passmore J, Foster E, Trathan PN (2013) Sensitivity analysis identifies high influence sites for estimates of penguin krill consumption on the Antarctic Peninsula. Antarctic Science 25: 19–23.
11. Woehler EJ (1993) The distribution and abundance of Antarctic and Subantarctic penguins. Cambridge: Scientific Committee on Antarctic Research. 76 p.

12

Elevated Temperature Alters the Lunar Timing of Planulation in the Brooding Coral *Pocillopora damicornis*

Camerron M. Crowder[1]*, Wei-Lo Liang[2], Virginia M. Weis[1], Tung-Yung Fan[2,3]*

1 Department of Integrative Biology, Oregon State University, Corvallis, Oregon, United States of America, **2** Institute of Marine Biology, National Dong Hwa University, Pingtung, Taiwan, R.O.C., **3** National Museum of Marine Biology and Aquarium, Pingtung, Taiwan, R.O.C.

Abstract

Reproductive timing in corals is associated with environmental variables including temperature, lunar periodicity, and seasonality. Although it is clear that these variables are interrelated, it remains unknown if one variable in particular acts as the proximate signaler for gamete and or larval release. Furthermore, in an era of global warming, the degree to which increases in ocean temperatures will disrupt normal reproductive patterns in corals remains unknown. *Pocillopora damicornis,* a brooding coral widely distributed in the Indo-Pacific, has been the subject of multiple reproductive ecology studies that show correlations between temperature, lunar periodicity, and reproductive timing. However, to date, no study has empirically measured changes in reproductive timing associated with increased seawater temperature. In this study, the effect of increased seawater temperature on the timing of planula release was examined during the lunar cycles of March and June 2012. Twelve brooding corals were removed from Hobihu reef in Nanwan Bay, southern Taiwan and placed in 23 and 28°C controlled temperature treatment tanks. For both seasons, the timing of planulation was found to be plastic, with the high temperature treatment resulting in significantly earlier peaks of planula release compared to the low temperature treatment. This suggests that temperature alone can influence the timing of larval release in *Pocillopora damicornis* in Nanwan Bay. Therefore, it is expected that continued increases in ocean temperature will result in earlier timing of reproductive events in corals, which may lead to either variations in reproductive success or phenotypic acclimatization.

Editor: Mónica Medina, Pennsylvania State University, United States of America

Funding: This research was supported by a Ministry of Science and Technology grant NSC-101-2611-M0291-002 awarded to T-Y.F., a National Science Foundation East Asian and Pacific Summer Institute grant awarded to C.M.C., and a National Science Foundation OISE award (#1042509) awarded to V.M.W. The funders had no role in study design, data collection and analysis, decision to publish, or preparation of the manuscript.

Competing Interests: The authors have declared that no competing interests exist.

* Email: Crowderc@onid.oregonstate.edu (CMC); tyfan@nmmba.gov.tw (T-YF)

Introduction

Reproductive timing is a critical factor in coral reproductive success and has been correlated to multiple environmental variables including those associated with seasonality such as temperature, solar irradiance, tidal cycles, nocturnal illumination associated with lunar periodicity, and light-dark cycles corresponding with diel fluctuations [1–8]. Previous studies have effectively demonstrated correlations between environmental variables and coral reproduction; however, examining the direct causality of individual variables on the timing of reproduction is critical to understanding the mechanisms controlling reproductive timing. Determining which variables are vital for the coordination of reproductive events will reveal information underlying coral reproductive function. Corals are both the bio-engineers and foundational primary producers of coral reef ecosystems and therefore, understanding how environmental variables affect timing of reproduction is essential for predicting future impacts of climate change on coral reef ecosystem stability.

Temperature has been shown to be a critical variable affecting coral reproductive success. Corals of all life stages are negatively affected by increasing sea-surface temperature attributed to global climate change [9–11]. While significant effort has been dedicated to describing the effects of temperature on the physiology and ecology of corals, less attention has focused on the effects of elevated temperature on coral reproductive timing and success [9,12]. Recent studies have revealed that temperatures exceeding tolerance thresholds reduce polyp fecundity, gametic quality [13,14] and the number of reproductive events in corals [15]. While it is evident that elevated temperatures can impair reproductive processes, predictable seasonal fluctuations in temperature might be a key component controlling the timing of reproduction. Multiple field studies have observed correlations between temperature and reproductive timing [16,17]. This correlation could demonstrate that some corals are displaying reproductive plasticity or alterations in the timing of release to adapt to change, which could be an important mechanism for larval survival and fitness with rising ocean temperatures associated with climate change. Understanding how temperature affects the timing of reproduction will provide information as to how continued increases in sea-surface temperature may further alter reproductive processes and patterns in corals.

Approximately 15% of coral species brood internally fertilized larvae (planulae) that are released during a process known as planulation. Planulation typically occurs multiple times annually and in some cases on a monthly basis [18,19]. Planulae are often buoyant and have the capacity for wide dispersal, but are also able to quickly settle upon release leading to fast rates of colonization

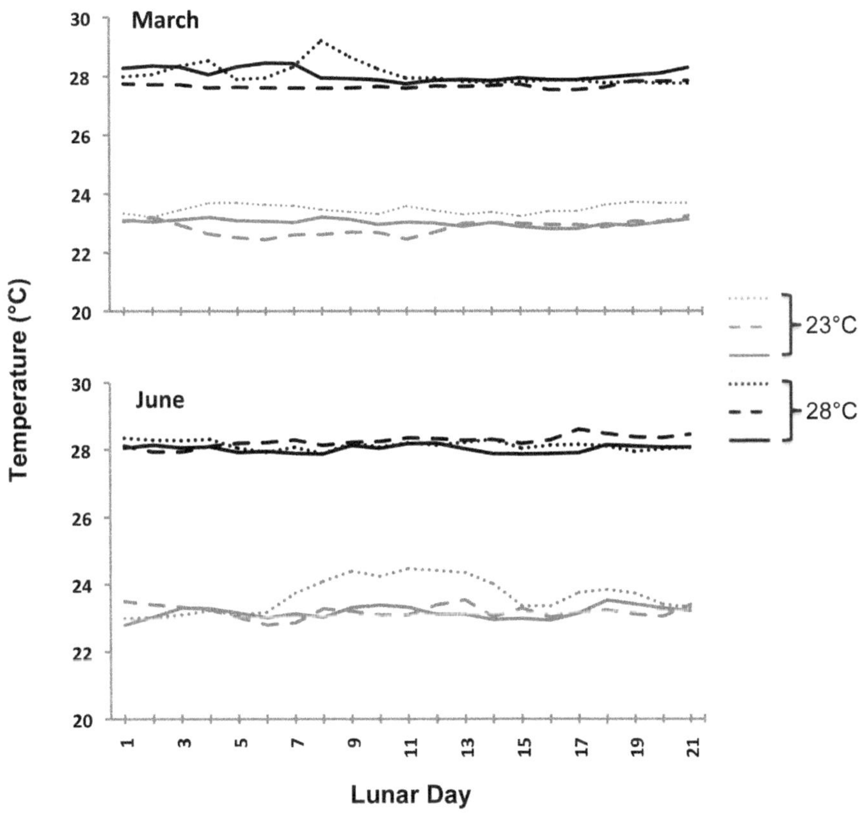

Figure 1. Average daily tank temperatures for both 23 and 28°C temperature treatments in lunar March and June 2012. Three lines represent temperature values in the three separate treatment tanks. Measurements were taken every 10 minutes and averaged for each lunar day.

[2,20]. The timing of planulation has a direct influence on larval survival, dispersal, and recruitment [2,19]. Reproductive events optimally timed to environmental variables such as temperature and light, can affect larval survival particularly in larvae containing symbiotic dinoflagellates [21–24].

The widely studied brooding coral *Pocillopora damicornis*, has been the subject of multiple studies investigating reproduction, particularly reproductive synchrony associated with lunar periodicity [19,20,25–27]. Collectively, these observations show variability in planulation patterns. *P. damicornis* has been shown to release planulae at every lunar phase during monthly lunar reproductive cycles with some consistencies observed within similar geographical locations. This diversity in the timing of planulation between different geographical regions may be the result of phenotypic plasticity that increases fitness [28]. However, to date, little is known about phenotypic plasticity of coral reproductive timing, and specifically how this plasticity may be driven by environmental variables, such as rising ocean temperatures.

In the context of the ongoing debate as to the effect of individual environmental variables on the timing of planulation, it is timely to ask if changes in temperature can directly alter planulation patterns in *P. damicornis*. The purpose of this study was to determine the effect of increased seawater temperature on the timing of planulation within a single lunar cycle and between lunar cycles of different seasons. To investigate this, *P. damicornis* colonies were monitored in temperature-controlled tanks exposed to natural lunar cycles with seawater temperature set to either 23°C (low) or 28°C (high). For both seasons, the timing of planulation was found to be plastic, with elevated temperature treatments resulting in significantly earlier peaks of planula release. This suggests that temperature modifies the lunar timing of larval release in *Pocillopora damicornis* in Nanwan Bay.

Materials and Methods

Coral Collection

Twelve adult colonies of *P. damicornis* were randomly collected with permission from the Kenting National Park, Taiwan from a 10-meter area at Hobihu reef (21°56.799′N, 120°44.968′E) in Nanwan Bay, Taiwan on two separate occasions in lunar March and June 2012. Colonies ranged in size from 7–15 cm in diameter and were removed from reefs 4–5 meters in depth. Within an hour after collection, colonies were transported back to The National Museum of Marine Biology and Aquarium (NMMBA) in Checheng, Taiwan.

Experimental Design

Six 150 liter tank mesocoms situated within a larger (30 ton) aquarium system as described by Mayfield et al. (2013) were used in this study. Three of the tanks were maintained at 23°C (mean winter ocean temperature) and three were maintained at 28°C (mean summer ocean temperature). Two coral colonies were randomly selected and assigned to each tank. Tanks were covered with shade cloth and exposed to natural outdoor sunlight and received a constant supply of sand-filtered seawater. Corals were allowed to acclimate to the experimental temperature conditions for one week prior to monitoring of planulation.

Monitoring of planulation began on lunar day 1 (new moon), March 22, 2012 (lunar March) and July 19, 2012 (lunar June).

Every evening during the lunar cycle, corals from each tank were individually placed into flow-through containers surrounded by 100 μm mesh plankton netting. Flotation devices were placed inside the netting to prevent nets from touching coral colonies. Each morning, nets were removed and the total number of planulae inside the netting was counted. Tank temperatures were recorded every 10 minutes using a HOBO temperature logger throughout lunar March and June 2012 (Figure 1).

Data Analysis

The percent of the total number of planulae released each day was calculated as the total number of planulae released per individual colony per day (mean ± standard deviation, n = 6 colonies per temperature treatment) divided by the total number of planulae released by each colony during the monthly reproductive cycle. Raleigh's test [29] was used to test the null hypothesis that planula release occurred uniformly throughout the lunar cycle. If the null hypothesis was rejected, then the mean lunar day (MLD) and angular deviation of planulation were calculated using circular statistics [19,27,29] to determine lunar timing of planulation. Circular statistic calculations were completed directly in an excel spreadsheet based on the equations and methods described in Zar et al. (1999).

A 1-way ANOVA was performed to compare MLD and angular deviation (AD) for each tank (n = 3), with average MLD and AD calculated for the two corals within each tank, within temperature treatments. A general linear mixed model with repeated measures was completed using a Poisson distribution with colony size treated as a covariate. Statistical analyses were completed using the statistical package R (R Foundation for Statistical Computing Vienna, Austria).

Results

Raleigh's test for uniform distribution indicated that planula release did not occur uniformly throughout the lunar cycle for either temperature treatment over both lunar March and June ($p<0.001$ for all tests) (Table 1). Differences were observed in the timing of planulation, measured as the percentage of planulae released per day with corals at 28°C releasing planulae earlier in the lunar cycle than those at 23°C for both lunar March and June (Figure 2). In lunar March, peak percentage of release occurred on lunar day 8 for the 28°C treatment, compared to lunar day 19 for the 23°C treatment. In lunar June, peak percentage of release was observed as two smaller peaks on lunar day 6 and 10 for the 28°C treatment, compared to a more significant peak at lunar day 12 for the 23°C treatment.

MLD of planulation closely resembled patterns in percentage of planulae released whereas angular deviation, a measure of dispersion, was not significant in lunar March or lunar June, where there was a larger spread in days of release (Table 1). In lunar March, the average MLD and angular deviation of release were 17.5±1.1 and 23.7±5.4 for the 23°C treatment and 8.4±0.7 and 16.1±1.9 for the 28°C treatment, respectively. In lunar June, the average MLD and angular deviation of release were 12.5±1.0 and 23.8±8.6 for the 23°C treatment and 7.7±1.7 and 26.5±10.4 for the 28°C treatment, respectively. One-way nested ANOVAs showed significant differences in MLD ($p<0.001$) between temperature treatments in lunar March and significant differences in MLD ($p = 0.014$), between temperature treatments in lunar June (Table 2).

Additionally, our results indicate that there were substantially more planulae released in lunar March (1,454±890 at 23°C and 3,593±1,359 at 28°C) compared to lunar June (186.3±55 at 23°C and 127±93 at 28°C) (Table 1). Significant differences in the total number of planulae released per lunar day were found to be associated with lunar day in both lunar March ($p<0.001$) and June ($p<0.001$) and colony size was not found to have a significant effect within individual lunar months (Table 3). Additionally, significant differences in the total number of planulae released per day were associated with temperature for lunar March ($p = 0.001$) but not lunar June (Table 3).

Discussion

Influence of Seasonality

In this study we observed a substantial difference in the total number of planulae released between lunar March and June 2012. This observed difference in the total number of planule released is likely due to seasonality. This hypothesis is consistent with previous findings showing that there were differences in planulae abundance between seasons in Pocilloporid corals in the Northwestern Philippines with dry seasons (March–May) having higher numbers of planulae released then wet seasons in (June–October) [20]. Differences in total number of planulae released in our study could also be attributed to colony size, since corals used in lunar March were, on average, 1 cm larger than those used in lunar June (Table 1). However, differences in planulation abundances are not expected to affect our results because timing of planulation was examined individually for each month.

Table 1. Results of Raleigh's test for uniform distribution of planula release by *P. damicornis* colonies during March and June lunar cycles.

	Lunar March		Lunar June	
	23°C	**28°C**	**23°C**	**28°C**
MLD	17.5±1.1	8.4±0.7	12.5±1.0	7.74±1.7
Angular Deviation	23.7±5.4	16.1±1.9	23.8±8.6	26.5±10.4
z	386–7261	2133–4638	93–208	11–291
p	<0.001	<0.001	<0.001	<0.001
Planulae	1,454±890	3593±1359	186.3±55	127±93
Colony Size (cm)	10.4±1.3	11.1±1.8	9.7±1.1	9.8±1.4

Mean lunar day (MLD) ± standard deviation, angular deviation of release ± standard deviation, range of Raleigh's test statistic (z), p-value for Raleigh's test (p), mean number of planulae released per tank ± standard deviation, and average colony diameter (cm) ± standard deviation.

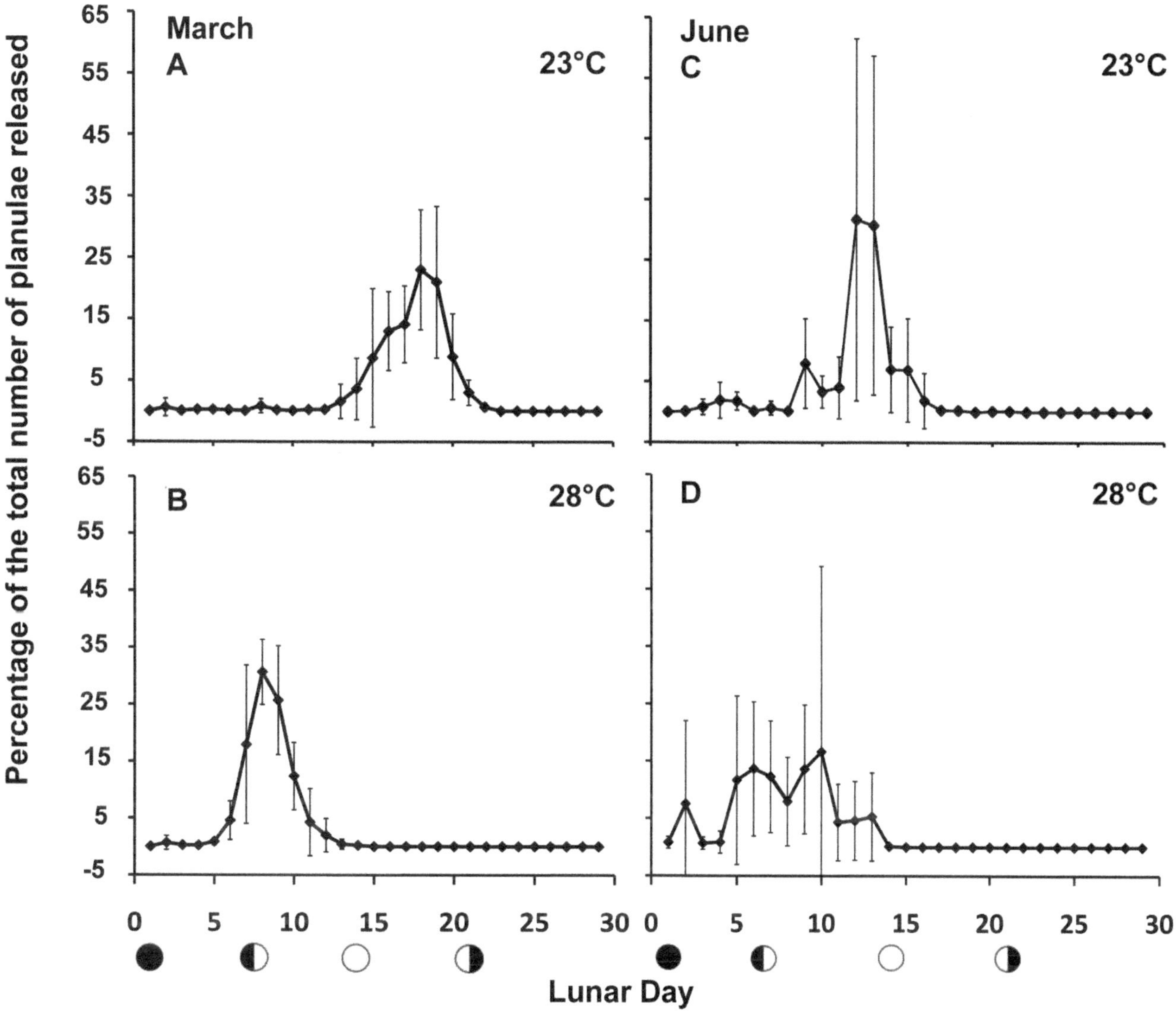

Figure 2. Percent of *P. damicornis* planulae released each day during lunar March (A and B) and Lunar June (C and D) 2012 reproductive cycles for colonies incubated at 23 (A and C) and 28°C (B and D). Points represent means of six colonies ± SE. Moon symbols represent lunar phases (new, 1st quarter, full, and last quarter).

Timing of Planulation

Lunar periodicity has been correlated with the timing of reproduction for multiple coral species [2,18] especially *P. damicornis*, where the timing of planulation is consistently linked to lunar phases [19,20,26,27]. While lunar periodicity likely does play a role in the timing of reproduction, in some cases with good predictability, other environmental factors may be able to disrupt these cycles. In this study we directly tested the influence of low (23°C) and high (28°C) temperatures on reproductive timing, and show that changes in temperature have the capacity to significantly alter the timing of planulation. Clear differences were observed in the timing of planulation, shown as the percentage of planulae released each day (Figure 2), and temperature was found to have a significant impact on the MLD of planulation for both lunar March and June (Table 1 & 2). The observed variations in lunar day of release between temperature treatments suggests that elevated temperature modifies the affect of other cues, such as lunar periodicity, to drive timing of release.

Early release in the high temperature treatment suggests that temperature affects the reproductive physiology of the adult coral and/or the developing planulae, resulting in an acceleration of release [30]. This hypothesis is supported by a previous study on a broadcast spawning coral, *Echniopora lamellosa*, in Taiwan showing that reproductive processes, such as gametogenesis and spawning are plastic and can be accelerated by increasing seawater temperature [31]. Another study on Caribbean corals, within the genus *Madracis*, found that maturation of gametes was positively correlated with increases in seawater temperature, indicating that observed changes in timing could be attributed to internal cues associated with gametogenesis [32].

Alternatively, shifts in the timing of planulation could be a result of negative changes in planula physiology. This hypothesis is supported by multiple studies that have observed decreases in larval survival with elevated temperature. Larvae of the Hawaiian coral *Fungia scutaria* exposed to 27, 29, and 31°C showed gradual decreases in survivorship with animals incubated at 31°C having the highest rates of mortality [24]. Another study conducted on *P.*

Table 2. Results from a 1-way ANOVA reporting the mean lunar day and angular deviation of planula release for lunar March and June.

	Lunar March					Lunar June				
Mean Lunar Day	**DF**	**SS**	**MS**	**F**	**P**	**DF**	**SS**	**MS**	**F**	**P**
Temperature	1	125.68	125.68	175.80	1.90 E-04	1	29.22	29.22	17.66	1.40E-02
Residuals	4	2.86	0.71			4	6.62	1.65		
Angular Deviation										
Temperature	1	85.43	85.43	5.93	7.2 E-02	1	11.00	10.96	0.107	0.76
Residuals	4	57.62	14.41			4	409.20	102.29		

Table 3. Results from a general linear mixed model with repeated measures using a Poisson distribution with colony size treated as a covariate for the total number of planulae released per lunar day.

	Lunar March				Lunar June			
	Estimate	**SE**	**Z**	**P**	**Estimate**	**SE**	**Z**	**P**
Intercept	2.866	2.735	1.050	0.295	5.762	3.576	1.611	0.107
Lunar Day	−0.062	0.001	−61.030	2.00E-16	−0.082	0.004	−19.050	2.00E-16
Temperature	0.256	0.078	3.250	0.001	−0.116	0.093	−1.251	0.211
Colony Size	−0.400	0.292	−1.370	0.171	−0.026	0.262	−0.101	0.919

damicornis larvae in Taiwan found 28°C to be the thermal threshold for maximum respiration in planulae, with higher temperatures leading to reduced respiration and likely metabolic depression [33]. Additionally, a 5-fold decrease in survival was observed with a 1.5°C increase in temperature in *Acropora palmata* embryos and larvae from the Caribbean [11]. Although success of planulae after release was not assessed in this study, it has been shown that early-released planulae have lower settlement rates than those released later within a single reproductive cycle [19].

Reproductive Plasticity

Our results reveal that temperature can act as a driver for plasticity in reproductive timing. Reproductive plasticity can enhance individual success in harsh or fluctuating environments [28]. While our study did not examine reproductive plasticity specifically in an adaptive context, we postulate that the phenotypic plasticity observed in this study may suggest capacity for an adaptive response to elevated temperatures in corals. This ability to shift reproductive timing in high temperature environments may indicate that climate change induced increases in ocean temperature may not be detrimental to reproduction, but rather simply alters its timing. Our findings also indicate that such shifts in timing can occur relatively quickly, as reproductive plasticity was observed over a single reproductive cycle. This is similar to findings observed in the coral *Echinopora lamellosa*, that show early spawning, when transplanted from colder northern to warmer southern Taiwan [19]. Understanding reproductive plasticity in corals is important because plasticity may provide corals the flexibility they need to be successful in a changing climate. However, many questions remain about how reproductive plasticity will influence the fate of corals, and their ecosystems, in the long term.

Conclusions

Our results provide empirical evidence that a 5°C increase in temperature, accelerates the timing of planula release in *P. damicornis* in Nanwan Bay, Taiwan.

It important to note that the corals sampled for this study are of unknown genetic origin, and therefore could be clones of the same genotype. If so, this could decrease the variability in temperature response. Nonetheless, our findings reveal that there is plasticity in the timing of reproduction and these changes can occur rapidly, within a single lunar reproductive cycle. These results highlight the reality that increases in ocean temperature have the capacity to disrupt patterns of planulation in corals. Depending on how shifts in the timing of planulation correlate with other environmental variables and conditions, these alterations could lead to increased planulae mortality and decreased recruitment success, or may be indicative of the potential for adaptation to warming ocean temperatures. Understanding the affect of temperature on reproductive timing in *P. damicornis* provides information that can be used to predict patterns in reproductive success and colonization in a future of rapidly changing ocean climate conditions.

Acknowledgments

We thank those who provided field support and feedback on analysis and writing: Dr. Li-Hsueh Wang, Steve Doo, Crystal McRae, Chih-Jui Tan, Sylvia Zamudio, Lorenzo Bramanti, Joleen Tseng, Robbin Chen, Alex Smith, Zach Haber, Peter Greene, Julia Stevens, Sarah Morton, Luna Sun, Nate Kirk, Sheila Kitchen, Angela Poole, Emily Weis, Hannah Tavalire, Jessica Reimer, Stevan Arnold, Eli Meyer, Dee Denver, and Patrick Chappell.

Author Contributions

Conceived and designed the experiments: T-YF W-LL CMC. Performed the experiments: CMC W-LL. Analyzed the data: CMC W-LL T-YF. Contributed reagents/materials/analysis tools: T-YF VMW. Contributed to the writing of the manuscript: CMC VMW T-YF.

References

1. Babcock R, Bull G, Harrison P, Heyward A, Oliver J, et al. (1986) Synchronous spawnings of 105 scleractinian coral species on the Great Barrier Reef. Marine Biology 90: 379–394.
2. Harrison P, Wallace C (1990) Reproduction, dispersal and recruitment of scleractinian corals. Ecosystems of the world 25: 133–207.
3. McGuire M (1998) Timing of larval release by *Porites astreoides* in the northern Florida Keys. Coral Reefs 17: 369–375.
4. Jokiel PL, Guinther EB (1978) Effects of temperature on reproduction in the hermatypic coral *Pocillopora damicornis*. Bulletin of Marine Science 20: 706–789.
5. Brady AK, Hilton JD, Vize PD (2009) Coral spawn timing is a direct response to solar light cycles and is not an entrained circadian response. Coral Reefs 28: 677–680.
6. Villanueva RD, Baria MVB, dela Cruz DW, Dizon RM (2011) Diel timing of planulation and larval settlement in the coral *Isopora cuneata* (Scleractinia: Acroporidae). Hydrobiologia 673: 273–279.
7. Goodbody-Gringley G, de Putron S (2009) Planulation patterns of the brooding coral *Favia fragum* (Esper) in Bermuda. Coral Reefs 28: 959–963.
8. Goodbody-Gringley G (2010) Diel planulation by the brooding coral *Favia fragum* (Esper, 1797). Journal of Experimental Marine Biology and Ecology 389: 70–74.
9. Hoegh-Guldberg O, Mumby P, Hooten A, Steneck R, Greenfield P, et al. (2007) Coral reefs under rapid climate change and ocean acidification. Science 318: 1737–1742.
10. Donner SD (2009) Coping with commitment: projected thermal stress on coral reefs under different future scenarios. PLoS One 4: e5712.
11. Randall CJ, Szmant AM (2009) Elevated temperature affects development, survivorship, and settlement of the elkhorn coral, *Acropora palmata* (Lamarck 1816). The Biological Bulletin 217: 269–282.
12. Mayfield A, Fan T-Y, Chen C-S (2013) Physiological acclimation to elevated temperature in a reef-building coral from an upwelling environment. Coral Reefs 32: 909–921.
13. Michalek-Wagner K, Willis BL (2001) Impacts of bleaching on the soft coral *Lobophytum compactum*. I. Fecundity, fertilization and offspring viability. Coral Reefs 19: 231–239.
14. McClanahan TR, Weil E, Cortes J, Baird A, Ateweberhan M (2009) Consequences of coral bleaching for sessile reef organisms. Berlin: Springer-Verlag.
15. Howells EJ, Berkelmans R, van Oppen MJ, Willis BL, Bay LK (2013) Historical thermal regimes define limits to coral acclimatization. Ecology 94: 1078–1088.
16. De Putron SJ, Ryland JS (2009) Effect of seawater temperature on reproductive seasonality and fecundity of *Pseudoplexaura porosa* (Cnidaria: Octocorallia): latitudinal variation in Caribbean gorgonian reproduction. Invertebrate Biology 128: 213–222.
17. Nozawa Y, Harrison PL (2007) Effects of elevated temperature on larval settlement and post-settlement survival in scleractinian corals, *Acropora solitaryensis* and *Favites chinensis*. Marine Biology 152: 1181–1185.
18. Harrison PL (2011) Sexual reproduction of scleractinian corals. Coral reefs: an ecosystem in transition. Springer. 59–85.
19. Fan T-Y, Li J-J, Ie S-X, Fang L-S (2002) Lunar periodicity of larval release by pocilloporid corals in southern Taiwan. Zoological Studies-Tapei 41: 288–294.
20. Villanueva RD, Yap HT, Montano MNE (2008) Timing of planulation by pocilloporid corals in the northwestern Phillipines. Marine Ecology Progress Series 370: 111–119.
21. Fan T-Y, Lin K-H, Kuo F-W, Soong K, Liu L-L, et al. (2006) Diel patterns of larval release by five brooding scleractinian corals. Marine Ecology Progress Series 321: 42.
22. Edmunds PJ, Gates RD, Leggat W, Hoegh-Guldberg O, Allen-Requa L (2005) The effect of temperature on the size and population density of dinoflagellates in larvae of the reef coral *Porites astreoides*. Invertebrate Biology 124: 185–193.
23. Yakovleva IM, Baird AH, Yamamoto HH, Bhagooli R, Nonaka M, et al. (2009) Algal symbionts increase oxidative damage and death in coral larvae at high temperatures. Marine Ecology Progress Series 378: 105–112.

24. Schnitzler CE, Hollingsworth LL, Krupp DA, Weis VM (2012) Elevated temperature impairs onset of symbiosis and reduces survivorship in larvae of the Hawaiian coral, *Fungia scutaria*. Marine Biology 159: 633–642.
25. Jokiel PL, Ito RY, Liu PM (1985) Night irradiance and synchronization of lunar release of planula larvae in the reef coral *Pocillopora damicornis*. Marine Biology 88: 167–174.
26. Richmond RH, Jokiel PL (1984) Lunar periodicity in larva release in the reef coral *Pocillopora damicornis* at Enewetak and Hawaii. Bulletin of Marine Science 34: 280–287.
27. Tanner JE (1996) Seasonality and lunar periodicity in the reproduction of pocilloporid corals. Coral Reefs 15: 59–66.
28. Via S, Gomulkiewicz R, De Long G, Scheiner S, Schlichting C, et al. (1995) Adaptive phenotypic plasticity: consensus and controversy. Trends in Ecology and Evolution 10: 212–217.
29. Zar JH (1999) Biostatistical Analysis. Prentice-Hall, New Jersey. 422–469.
30. Schmidt-Nielsen K (1997) Animal physiology: adaptation and environment. Cambridge University Press.
31. Fan T-Y, Dai C-F (1999) Reproductive plasticity in the reef coral *Echinopora lamellosa*. Marine Ecology Progress Series 190: 297–301.
32. Vermeij MJA (2004) The reproductive biology of closely related coral species: gametogenesis in *Madracis* from the southern Caribbean. Coral Reefs 24: 206–214.
33. Edmunds PJ, Cumbo V, Fan T-Y (2011) Effects of temperature on the respiration of brooded larvae from tropical reef corals. The Journal of Experimental Biology 214: 2783–2790.

13

Coping with Daily Thermal Variability: Behavioural Performance of an Ectotherm Model in a Warming World

José M. Rojas[1,2], Simón B. Castillo[1], Guillermo Folguera[3], Sebastián Abades[1,4], Francisco Bozinovic[1]*

1 Departamento de Ecología and Center of Applied Ecology & Sustainability (CAPES), Facultad de Ciencias Biológicas, P. Universidad Católica de Chile, Santiago, Chile, **2** Centro de Investigación e Innovación para el Cambio Climático, Universidad Santo Tomás, Santiago, Chile, **3** Filosofía de la Biología Group, Facultad de Ciencias Exactas y Naturales & Facultad de Filosofía y Letras, Universidad de Buenos Aires, Buenos Aires, Argentina, **4** Instituto de Ecología y Biodiversidad, Santiago, Chile

Abstract

Global climate change poses one of the greatest threats to species persistence. Most analyses of the potential biological impacts have focused on changes in mean temperature, but changes in thermal variance will also impact organisms and populations. We assessed the effects of acclimation to daily variance of temperature on dispersal and exploratory behavior in the terrestrial isopod *Porcellio laevis* in an open field. Acclimation treatments were 24±0, 24±4 and 24±8°C. Because the performance of ectotherms relates nonlinearly to temperature, we predicted that animals acclimated to a higher daily thermal variation should minimize the time exposed in the centre of open field, – i.e. increase the linearity of displacements. Consistent with our prediction, isopods acclimated to a thermally variable environment reduce their exploratory behaviour, hypothetically to minimize their exposure to adverse environmental conditions. This scenario as well as the long latency of animals after releases acclimated to variable environments is consistent with this idea. We suggested that to develop more realistic predictions about the biological impacts of climate change, one must consider the interactions between the mean and variance of environmental temperature on animals' performance.

Editor: Bayden D. Russell, The University of Adelaide, Australia

Funding: This study was funded by Fondo Nacional de Ciencia y Tecnología grant #1130015 and Centro de Ecología Aplicada y Sustentabilidad, Line 3 to FB. The authors thank L. A. Ebensperger and F. M. Jaksic for helpful comments. Experimental protocols followed the rules of The Ethics and Biosafety Committee of the Faculty of Biological Sciences at the Catholic University of Chile, permit No. CBB-100/2012. The funders had no role in study design, data collection and analysis, decision to publish, or preparation of the manuscript.

Competing Interests: The authors have declared that no competing interests exist.

* Email: fbozinovic@bio.puc.cl

Introduction

Current environmental climate variability associated to global change poses one of the greatest threats to organismal functional diversity [1,2]. Anthropogenic impacts on the earth's climate and habitats will likely increase not only in mean temperature but in the frequency of extremely high temperatures and seasonal/daily variability in certain regions [3]. Although ecologists widely recognize the potential impacts of warming, less attention has been paid to changes in thermal variation on a scale that pertains directly to organisms [4,5,6]. Observations and experiments that quantify not only the effect of increases in mean environmental variables but also the effect of environmental variation on organismal traits are important for inferring ecological and evolutionary responses to climate change and the mechanisms by which organisms cope with this variation [7,8]. In this vein, how behavioural traits are affected by acclimation to temperature variability and their consequences for population persistence is poorly known [9]. Understanding the behavioural responses of animals, and their fitness consequences in variable environments, is important to predict the consequences of global warming on biodiversity. Indeed, as recently suggested by Sih [10] and Wingfield [11], the capability to explain and predict how organism respond behaviourally to human-induced rapid environmental changes has important implication for wildlife management as well as is becoming a focal point of basic and applied research within scenarios of global warming.

Terrestrial isopods are good models to test hypotheses in global change biology, since they exhibit a broad distribution under diverse abiotic conditions [12,13,14,15]. As suggested by Warburg et al. [16] and Hassall et al. [17] the main challenge for terrestrial isopod' fitness is to cope with desiccation stress and temperature fluctuation. In this way, some authors have evidenced the plasticity of diverse traits – behaviour included – that would help the persistence of woodlouse along different geographic as well as - climatic conditions [18,19,20,21,22,23,24]. It is common to observe seasonal changes in its daily activity pattern, together with the use of shelters and the formation of aggregations that hypothetically allows woodlice to avoid unfavourable environmental conditions [16,25,26]. Formally, the locomotory activity of terrestrial isopod is described as an uncorrelated discrete Random Walk [27,28,29,30]. This kind of displacement indicate that these animals moves straight ahead for a certain distance, then turn over a random angle (turning angle), and again moves straight ahead to turn without directional persistence [27,28]. The setting of this type of movement is mainly determined by the degree of autocorrelation of the angles of rotation. An increase in this variable determines a more linear movement of individuals. In laboratory studies, an increment in the autocorrelation level is associated mainly to an escape response, characterized by the temporal persistence of autocorrelation [28,30,31,32,33]. In the field, the search of suitable patches (e.g. soil moisture) or a dispersion behaviour is also characterized by a significant autocorrelation of turning angle but with a more unpredictable temporal memory, establishing more complex activity patterns such as correlated random walks or spiral displacement[29,30,34,35,36].

Here we experimentally tested the effect of acclimation to environmental thermal variability on dispersal behaviour in the terrestrial isopod *Porcellio laevis* in an open field. Traditionally, the inclusion of the behaviour in studies of thermal acclimation involves two kind of approaches, namely temperature selection [37,38,39] and locomotory performance [40]. In all cases, behaviour is used as descriptor of the flexibility or sensitivity of a whole-individual to a new thermal regime, contributing to understand process such as adaptation [37,41]. Nevertheless, behavioural responses behind ecological relationship of individuals or its implication on fitness still puzzling. Organismal studies in *P. laevis* evidence plasticity in physiological and life history traits in response to different geographic-climatic conditions [19,20,21,23,24]. Particularly, when the climatic conditions imply thermal variation, a decrease in locomotor and development performance is described in animals acclimated to an increment in the amplitude of ambient temperature variation [7,8,40]. Based on these data, and assuming that dispersal behaviour of *P. leavis* fits to a discrete random walk model, going from linear displacements under stressful condition to a more complex at normal condition [27,28,32,34], we predict that animals acclimated to a higher daily thermal variation should minimize the time exposed in the center of open field or in other words to increase the linearity of displacement. Consequently, Fig 1 shows that a dispersion trajectory is made by steps and the minimal length is determined by the interval sampling of each behavioural record, i.e. - the black line between open circles. Circles represent a sampling event. The absolute angle (α) is defined between the horizontal plane (dx) and the step. The relative angle (ρ) or turning, is the angle between successive steps. Distance (d), is the total distance during each record. The mean distance (R_n) is the net distance between the current location and the first relocation of the trajectory (Fig. 1). In our view this kind of approaches will help to understand if species physiological and behavioral flexibility may buffer (or not and how) the effect of global warming.

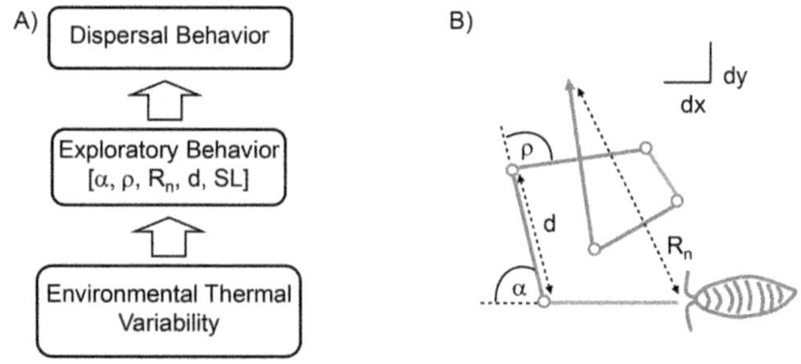

Figure 1. Dependence of behaviour of space use on the environmental thermal variability. (A). Dispersion trajectory is made by steps or SL (B).The minimal length is determined by the interval sampling of each behavioural record, i.e. gray line between open circles. Circles represent a sampling event. The absolute angle (α) is defined between the horizontal plane (dx) and the step. The relative angle (ρ) or turning, is the angle between successive steps. Distance (d), is the total distance during each record. The mean distance (R_n) is the net distance between the current location and the first relocation of the trajectory. Modified from [63].

Material and Methods

Ethics Statement

The study was performed with *P. laevis* specimens collected in the Mediterranean habitats of central Chile at San Carlos de Apoquindo, a Field Station of the Department of Ecology, Catholic University of Chile (33°23′ S, 70°31′ W at 1,230 m above sea level).

Experimental protocols followed the rules of The Ethics and Biosafety Committee of the Faculty of Biological Sciences at the Catholic University of Chile, permit No. CBB-100/2012. *Porcellio leavis* is not an endangered or protected species. We used animals of similar body size and both sexes.

Animals and experimental design

The mean minimum air temperature at San Carlos de Apoquindo is nearly 6°C below the mean and the maximum goes 7°C higher, on average [42]. Respect to yearly seasonality, the broadest range of temperatures is observed in Summer (16–17°C between mean maximum and mean minimum temperature), and the narrowest in Autumn and Winter (9–10°C).

After collection, animals were sorted by sex and maintained in plastic Petri dish (50 mm diameter; with a base layer of plaster-of-paris to maintain humidity). Food, in the form of dry spinach, and water was provided ad libitum [7]. Pregnant females were identified in the laboratory and placed under standard conditions of light (L:D = 12:12) and temperature (24°C) in culture boxes (2.2×2.2×2.4 cm) with a layer of damp sand one cm thick. Previous reports indicated that *P. laevis* exhibit a life cycle of about 12–18 months [43]. Thus, at first 10 days of development, F1 were maintained at same conditions as parents at 24°C to avoid higher rates of mortality [7]. Following Folguera et al. [7] methodology during 74 days, 60 individuals were randomly assigned to one of three thermal treatments in climatic chambers (20 replicates); namely constant temperature regime at 24°C ($\delta = 0$) and two treatments with variance in temperatures, reaching a maximum during daytime and a minimum at night with alternating temperature regimes of 28–20°C ($\delta = 4$) and 32–16°C ($\delta = 8$). These experimental temperatures were chosen because they are within the thermal tolerance range and daily variability of *P. laevis* [20,21,22,44].

Dispersion behaviour of woodlice was evaluated in an open-field setup. The open field consisted in a circular surface of 45 cm radius made of dry plaster and surrounded by a wall of opaque acrylic. Above the open-field we placed a webcam connected to a notebook which allow us to record animals behaviour. Studies were conducted in an acclimated room at 24°C. Displacement of animals was recorded during 15 minutes. Its location (X-Y coordinate) into the open field was determined using a video tracking system (Smart 2.0 – Panlab, Barcelona Spain) at 0.2 intervals seconds (event sampling). Previously to each record, each individual was placed in the centre of field within a dark chamber during 5 minutes to reduce handling stress. Upon the end of these observations, all experimental woodlice subjects were released back at their original sites of capture.

Movement analysis

To compare dispersal behaviour among treatments, we considered that displacement of woodlice follow basically a discrete random walk, without directional correlation. Therefore, a core point of analysis corresponded to checking randomness in the lineal and angular parameters of each paths (Fig. 1) Thus, each woodlouse was characterized by a trajectory. From the trajectory we estimated the parameters indicated in Fig. 1. Net displacement was quantified by the net square displacement (R^2_n) see [29,45]. Each movement variable included the elapsed time since each animal started the displacement until they reached the wall of the open field. As an additional descriptor of spatial behaviour we used the latency to first displacement (seconds). The exploration level or tortuosity of each path we estimated using the mean fractal dimension (D). This value varied from 1 (straight line) to 2 (a travel path which is so tortuous that it completely covers a two-dimensional plane, see [46,47]). The length step, R^2n, and turn angle (relative and absolute) for each path on each interval were estimated using the adehabitat package for R-Cran software [48]. The fractal dimension was estimated with the Fractal Mean estimator using a window range (or frame) of 0.25 to correct the border effect (see [49]) implemented in the Fractal software (Fractal ver. 5, V. Nams, Nova Scotia Agricultural College).

Statistical analysis

A generalized lineal model was used to test differences in the fractal dimension of displayed by isopods. Thermal treatment was used as categorical variable and body size as weight variable [50,51]. A similar statistical approach was used to evaluate the effect of acclimation temperature regime on the turning angle, absolute angle, step size and net distance. For each trajectory the temporal sequence of each parameter was evaluated as a time series. The median value of each time series was used as a statistical descriptor the each parameter, assuming that the median follows a gamma distribution with a log function as link [52]. The fractal dimension was also evaluated using a generalized lineal model, but in this case, using the mean as the central tendency descriptor. The significance of models was tested using loglikelihood ratios between the adjusted model and a null model (intercept). The significance of coefficients of models was estimated using an unpaired two tails t-test. Latency and velocity variables were rescaled to log(x). These variables were assessed using a one way ANOVA. Previously the assumptions of each test were evaluated. The temporal autocorrelation level of behavioural descriptors was evaluated using an autocorrelation function analysis (ACF). These analyses were conducted using the autocorrelation descriptors based in the square differences between successive steps to lineal parameters and chord distance between successive angles to test relative angle, estimated by Adehabitat package for the R-Cran software [47]. The significance level of these descriptors was assessed using the 95% confidence intervals estimated from permutation procedure implemented in Adehabitat package. Previously to each test, random and observed values were centred using the average of random values, and then transformed into absolute values [47,53]. We used correlograms to show mean values by thermal treatment for differences between ACF observed and the lower confidence interval estimated by individual. Finally, the central tendency of the dispersal behaviour was evaluated on frequency distribution of step length and turning angle using the Hartigan's dip test to unimodality implemented in the diptest package for R-Cran [48]. This analysis was pooled all individuals by treatment All parametric tests used were implemented using the R-Cran software (version 2.12.2).

Results

Experimental treatments did not affect body size of individuals (Table 1). Differences in dispersion behaviour as a result of acclimation to different thermal regimes were observed. Animals that experienced the largest thermal variation (δ8) showed a positive autocorrelation in the steps length to a delay of 3 lag. Similarly, the turning angle of these animals showed a lag autocorrelation to 4 (Fig. 2). Animals acclimated to δ0 and δ4 showed positive autocorrelation for the length of the steps to lag 3 only (Fig.2). No autocorrelation on turning behaviour was recorded (Fig. 2). These results were consistent with those observed in the frequency distribution of the relative angles. Animals exposed to a δ0 and δ4 thermal regime showed a symmetrical unimodal distribution centred at 0°. A different pattern is observed in animals acclimated to the larger thermal variation (δ8), that while it is centred at 0°, its distribution is nearly bimodal (Fig. 3).

In the case of the frequency distribution of the length of steps, a tendency to a bimodal distribution was observed, but mainly in animals acclimated to a daily variable temperature (Fig. 4). Particularly, isopods acclimated to highly variable environments (δ8) showed a distribution characterized by a predominance of short steps (Fig. 4). Animals acclimated to δ4 exhibited similar results described for δ8, but with a smaller representation of short steps. Finally, animals acclimated to a stable environment showed a more homogeneous distribution of length steps

A significant adjust of the generalized linear models was only detected for the descriptors of fractal dimension and absolute angle (Table 2). Thus, no differences were observed in the relative angle (turning) of paths among experimental groups (Table 2). Likewise, no significant differences in step length and net distance travelled (R^2n) were detected among experimental conditions (Table 2). Regarding to absolute angles, animals exposed to a constant thermal regime showed a significantly lower absolute angle in comparison to animals facing δ4 regimes (Tables 2, 3). Animals acclimated to a δ8 thermal regime did not change the magnitude of absolute angle (Tables 2, 3). A congruent for the absolute angle was observed for fractal dimension. That is, animals that did not

Table 1. Results for one-way analysis of variance for testing the homogeneity in individual body attributes and the general dispersal performance of woodlice among treatments.

	δ0	δ4	δ8	Statistics
Body length (mm)	10.9 (0.3)^a	11.5 (0.3)^a	10.8 (0.3)^a	$F_{(2,57)} = 1.6$; SS = 5.12
Body mass (mg)	71.7 (3.2)^a	78.4 (3.2)^a	69.8 (3.2)^a	$F_{(2,57)} = 1.9$; SS = 820.9
Speed (cm/s)	3.6 (0.4)^a	5.0 (0.6)^a	3.8 (0.4)^a	$F_{(2,57)} = 1.0$; SS = 0.9
Latency (s)	81.1 (16.3)^a	163.2 (25.4)^b	100.8 (20.4)^{ab}	$F_{(2,57)} = 4.5$; SS = 7.4

Different letters indicate significant differences among treatments (Tukey test at $P<0.05$). Values shown the mean estimated in each thermal variability treatments: δ0, δ4 and δ8 (i.e. 24±0, 24±4 and 24±8°C) for each parameter. In parentheses ±1 standard error.

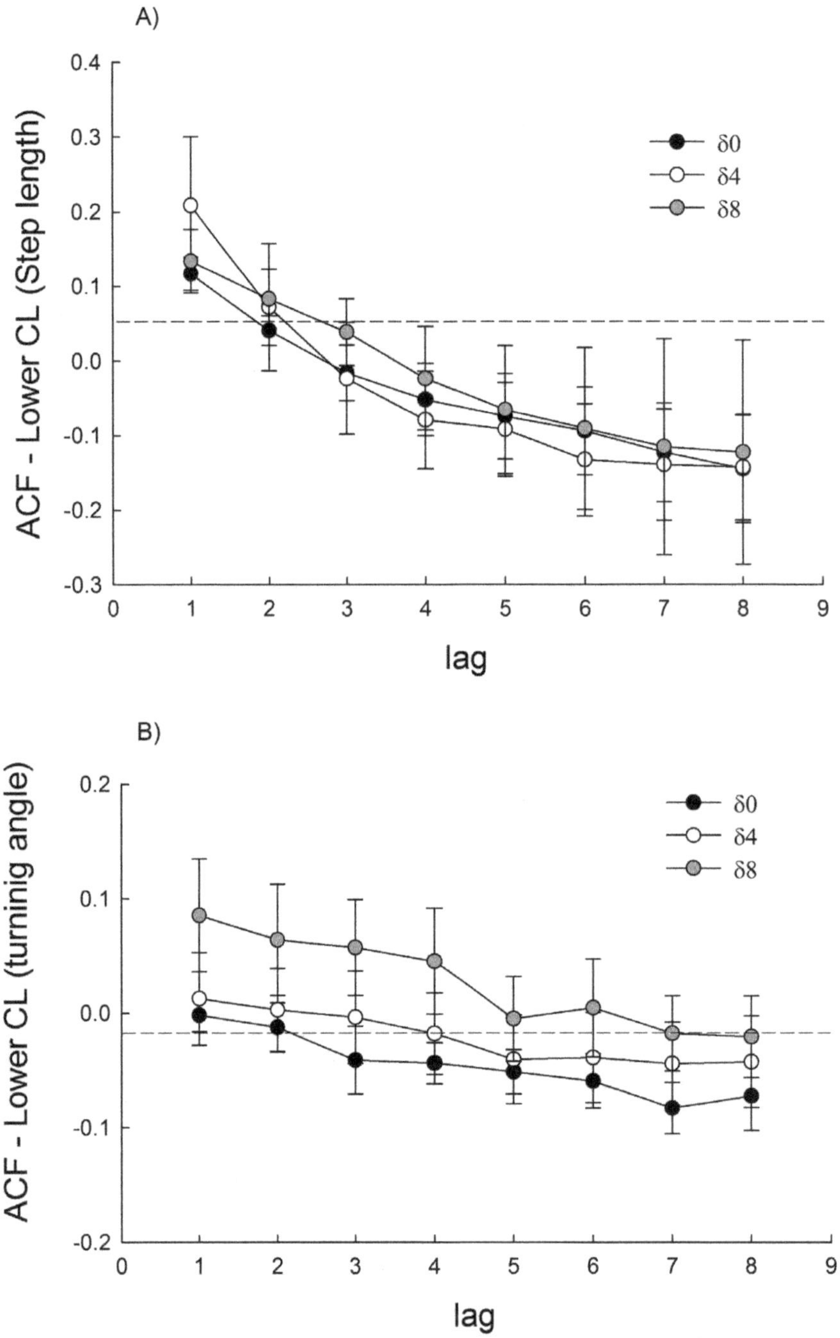

Figure 2. Autocorrelation mean (± standard error) estimated for three thermal treatments of acclimation for step length (A) and relative angle (B). Daily thermal variability treatments are: δ0 = 24±0°C, δ4 = 24±4°C and δ8 = 24±8°C. The ACF were calculated for lag = 0–8. Segmented line indicated the significance limits. Values correspond only to positive correlations. See text for details.

experienced thermal variation showed significant greater displacement (i.e. fractality) in the open field compared with animals that experienced a thermal variability of medium intensity (δ4) (Tables 2, 3, Fig. 5). No differences in the degree of fractality of movement between animals that were exposed to δ4 and δ8 were observed (Tables 2, 3, Fig. 5). When assessing the performance of animals, isopods acclimated to an environment without variability showed a latency to first displacement significantly lower than that observed in animals exposed to the δ4 treatment and similar to those acclimated to δ8 conditions (Table 1). Animals acclimated at δ4 and δ8 exhibited no difference in the latency time (Table 1). Finally, no statistical differences were recorded in speed among animals acclimated to different thermal regimes (Table 1).

Discussion

To predict responses to climate change, physiological and behavioural ecologists must understand the patterns of thermal variation and the mechanisms by which animals cope with this variation [6,10,11,54,55]. As pointed out before, impacts on the earth's climate will likely increase the frequency of extremely high

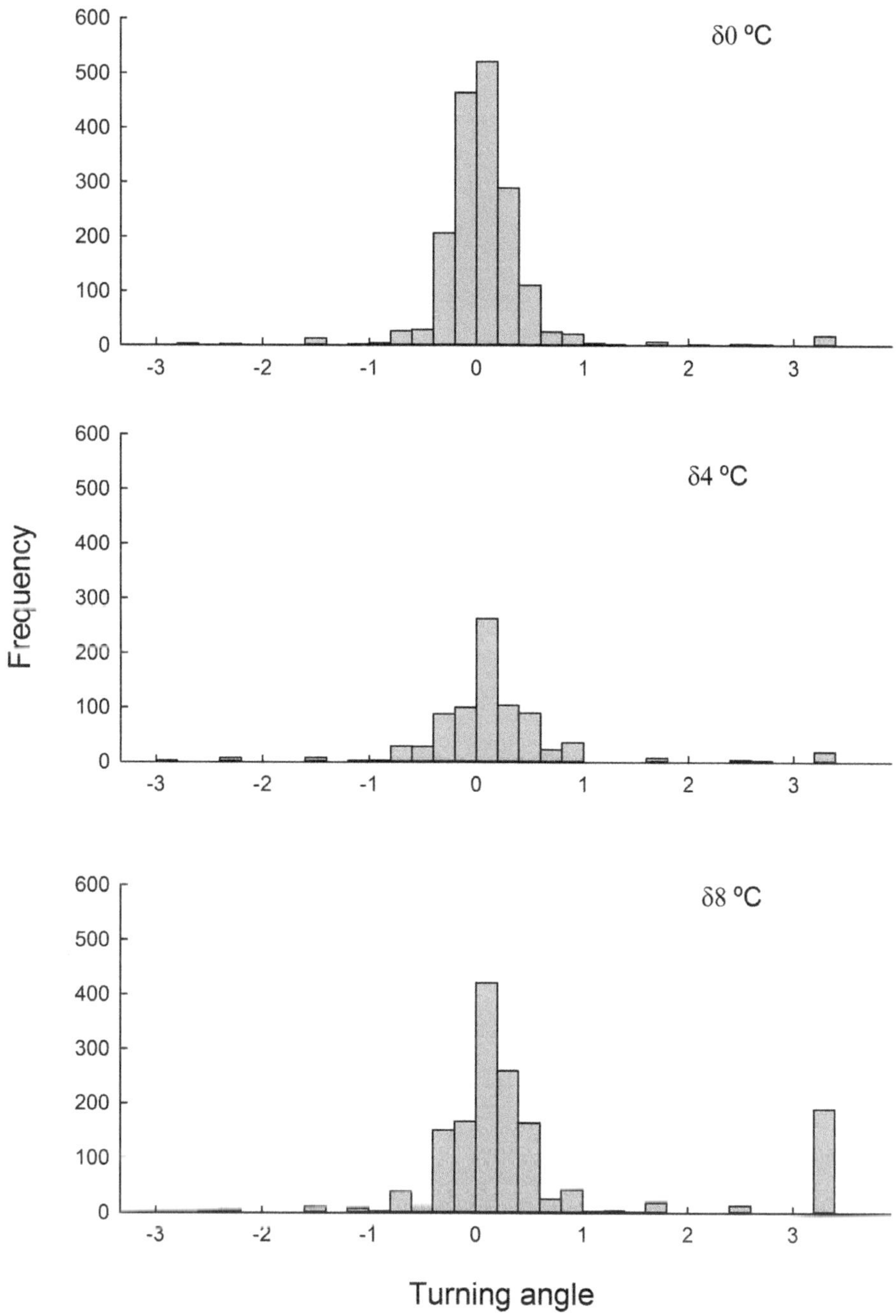

Figure 3. Frequency distribution of turning angle for different thermal treatments. Daily thermal variability treatments are: δ0 – 24±0°C, δ4 – 24±4°C and δ8 = 24±8°C. The Hartigan's dip statistical test revealed a significant bimodality pattern among all treatments. Namely, δ0: D = 0.0214, $P<0.05$; δ4: D = 0.0345, $P<0.05$ and δ8: D = 0.0396, $P<0.05$.

temperatures in certain regions [56]. Many organisms are expected to suffer a decrement in performance and fitness [57,58], but some may preserve their performance through behavioural responses [59]. For instance, we concur with Kearney et al [59] who enlighten the importance of behavioural traits to buffer the impact of global warming, arguing that this is an important and missing element from models of climatic change and predictions of impacts on biodiversity (see also [55]). Accordingly our results with terrestrial isopods represent an example of how changes in behavior allow animals to cope with variable habitats but also how environmental thermal variability may affect behavior.

Helmuth *et al.* [60] analyzed environmental variability at scales relevant to organisms to predict the responses of individuals and populations to thermal variability in the intertidal, where as previously suggested [61] some behavioral traits act as critical phase transition triggered by ambient temperatures. For instance, in hypervariable environments as the intertidal [60,62], the gastropod *Nerita atramentosa* behaviorally avoid the high temperatures through selection of thermal refuges as well as through huddling. Thus, as terrestrial isopods (this study), the

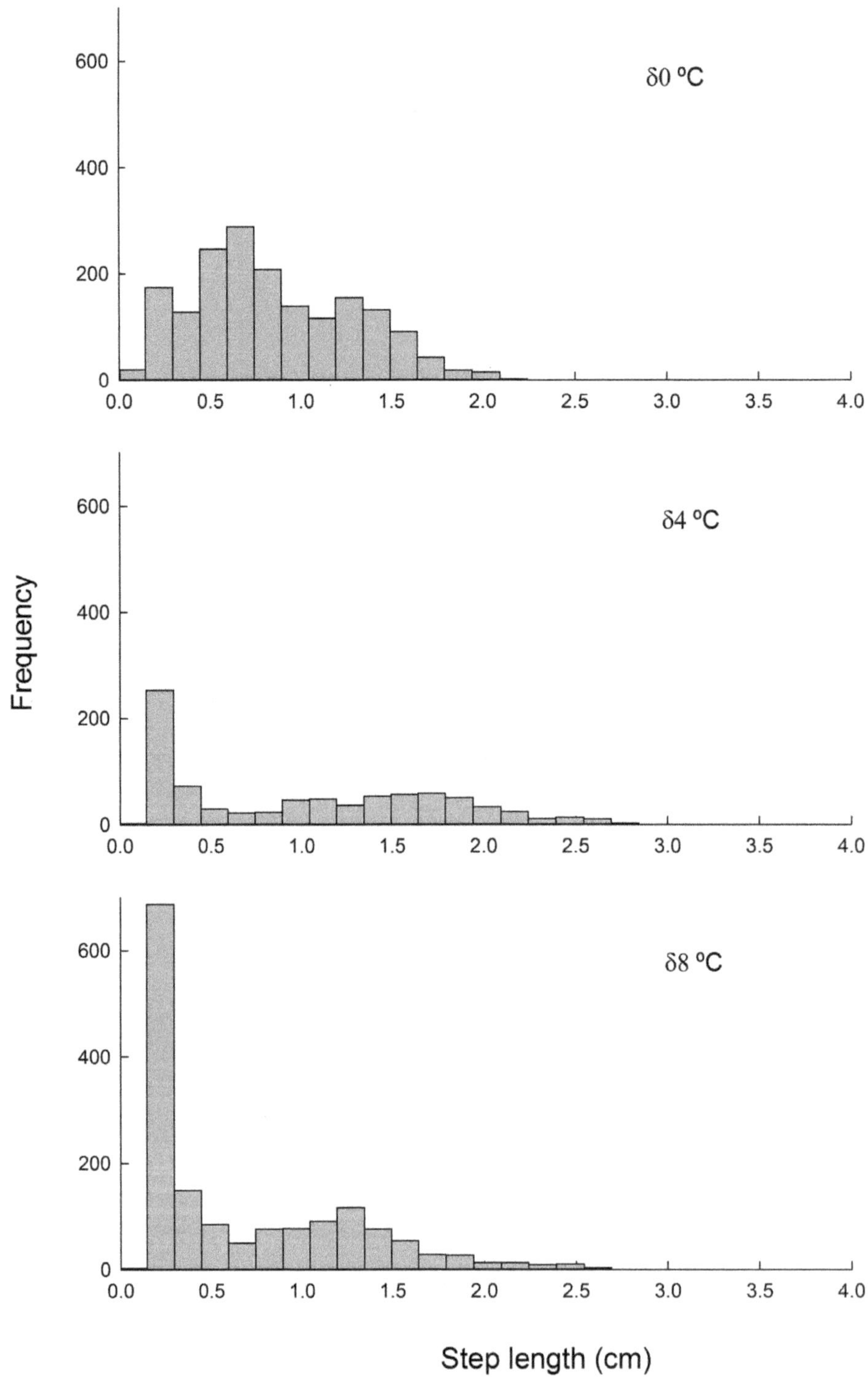

Figure 4. Frequency distribution of step length for different thermal acclimation conditions. Daily thermal variability treatments are: δ0 = 24±0°C, δ4 = 24±4°C and δ8 = 24±8°C. The Hartigan's dip statistical test revealed a significant bimodality pattern among all treatments. Namely, δ0: D = 0.013, $P>0.05$; δ4: D = 0.0123, $P>0.05$ and δ8: D = 0.0614, $P<0.05$.

behavior of marine gastropod during emersion seems to be related by local thermal stability or variability of the environment [63]. Theoretically, in variable habitats it is beneficial for ectotherms to select temperatures below thermal physiological optimum [64]. This hypothesis was tested by [65] in the intertidal snail *Chlorostoma funebralis* throughout experimental studies of behavioral thermoregulation in a thermal gradient. They found a "cold-biased" behavior which seems to guide snails to refuges in crevices. Indeed this snail performed a biased random walk along the gradient and huddle together at temperatures near their lower thermal limits. [66,67] also demonstrated that operative temperature in the intertidal may be one of the leading determinants of

Table 2. Results of generalized linear model for displacement descriptors.

Movement Descriptors	Null Model		Adjusted Model			P-value	AIC
	df	Loglik	Df	Loglik	L ratio		
Fractal dimension	5	49.3	7	7.9	15.91	0.001	−89.9
Absolute angle	5	−83.4	7	−79.5	7.92	0.02	172.9
Turning angle	5	48.1	7	48.7	1.16	0.55	−83.3
Step length	5	−51.9	7	−50.7	2.60	0.27	115.4
Net distance (R^2n)	5	−314.7	7	−.313.5	2.34	0.31	641.0

AIC corresponds to the Akaike information criterion; df: degree of freedom; Loglik: Loglikelihood estimator; L ratio: loglikelihood ratio Chi-square.

time/space variation of behavioral traits in the periwinkle *Echinolittorina peruviana*.

The constraints to behaviour of animals – though changes in ambient temperatures - are probably extremely important in determining when and how a habitat should be used. Climatic effects are of paramount importance since predicted increases in global thermal mean as well as in variance will change many behavioural displays including habitat use and time, among others. Therefore, thermal tolerance should depend on the variance of temperature as well as the mean. Consistent with this view, we found that woodlice exposed to highly variable thermal conditions showed a more conservative displacement. The long correlations of both the turning behaviour as well as the length of steps were greater than those observed in animals exposed to less variable or constant thermal conditions. Particularly for length step, which was not observed in the other test groups were noted. Thus, animals acclimated to a highly variable thermal environment display a more systematic or persistent displacement behaviour when exposed to a new environment, adjusting their behaviour to a more dispersive behaviour, namely less random and more linear [68]. Accordingly, the frequency distributions observed for the linear and angular parameters are consistent with a reduction of the random component in the movement. Among the parameters used to describe the movement of isopods, we consider the use of the fractal dimension, which allows describing the degree of homogeneity in which space was used or explored by the animals during their displacement [49]. Animals acclimated to stable thermal conditions showed a higher degree of scanning in comparison to the other experimental groups. The estimated magnitude for this group is similar to that reported for other terrestrial invertebrates [69], implying that the behaviour of isopods during our study resembles behaviour under natural conditions. This result is consistent with our previous suggestion that animals acclimated to variable conditions exhibit less exploratory behaviour. In this scenario interestingly, animals acclimated to an intermediate thermal variation are situated in an intermediate position. These animals showed a degree of fractalitiy, and bimodality in the frequency distribution of step length, as observed in animals acclimated to highly variable environments. Nevertheless, they also showed a temporal auto-correlation pattern for linear and angular parameters similar to those estimated for animals acclimated to stable environments.

Our results supported that thermal acclimation has significant effects on the behaviour of isopods, a finding not previously reported. Isopods exposed to a thermally variable environment reduce their exploratory behaviour, probably as a way to minimize their exposure to adverse environmental conditions [63,65]. This scenario as well as the long latency of animals acclimated to variable environments is consistent with this proposal. Clearly, further studies are needed to determine some of the ecological/evolutionary consequences of these behavioural and physiological adjustments, yet our results highlight to the importance of the behavioural component when making predictions about population viability under a climate change scenario. Overall, changes in animals phenologies that result from global warming may produce changes in activity patterns, migration, and breeding timing, and ultimately all of these altered synchronization between trophic levels as well as changed species competitive ability [70]. The emerging picture is that some specific behaviours represent an integrated response to the abiotic environment and its variation and the behavioural ability of animals to respond to new climatic scenarios. According to Sih [10], the observed variation in the behavioural responses in our studied animals may be the result of a cue-response system or because unpredictable environmental

Table 3. Coefficients for significant models from Table 2.

Movement Descriptors		Thermal treatments		
		δ0	δ4	δ8
Fractal dimension	Estimate	1.07 (0.01)	−0.03 (0.01)	−0.02 (0.05)
	t-test; *P*-value	127.7; 0.001	−2,84; 0.006	−1.97; 0.05
Absolute angle	Estimate	−0.32 (0.15)	0.86 (0.30)	0.39 (0.26)
	t-test; *P*-value	−2.10; 0.04	2.79; 0.001	1.46; 0.14

In parentheses standard error of coefficient. Student t-test and *P*-values correspond to hypotheses test of the estimated coefficients (df = 57, 4).

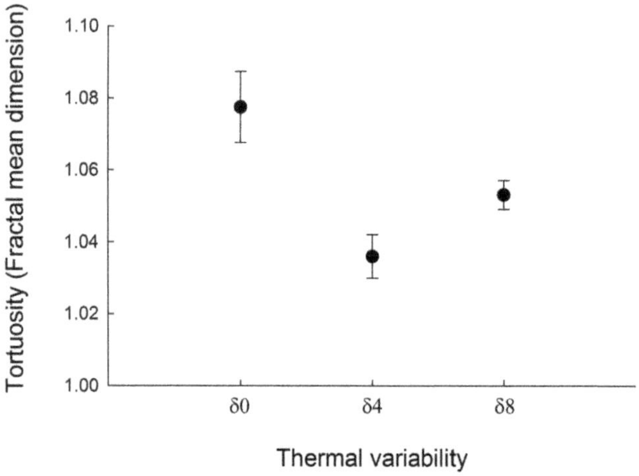

Figure 5. Fractal mean dimension recorded for animals acclimated to different variability thermal regimen. Daily thermal variability treatments are: δ0 = 24±0°C, δ4 = 24±4°C and δ8 = 24±8°C. Bar showed ±1 standard error.

information may affect woodlice responses to novel situations or because just behaviour plasticity may affect an animal's response. Probably a signal detection or a cost-benefit hypotheses on variation in learning may be the causes behind our results [5].

A main principle of physiological and behavioural ecology is that populations exposed to environmental change may crash when most individuals deteriorate, and that individuals decline when they reach a state that prevents them from maintaining homeostasis and behaviour –that is, the proper equilibrium through time with internal processes and foreign stimuli. It follows that the effects of climate change on species cannot be assessed from the type, magnitude or time scale of the perturbation, but from the physiological and behavioural states caused by it. Thus, a same set of environmental conditions may be deleterious for one species or population but harmless to another. The relatively narrow approaches used to the study of climate change on biodiversity are limited because they ignore the mechanisms that organisms use to cope with environmental changes [10,55].

Acknowledgments

We thank L.A. Ebensperger and F. M. Jaksic for helpful comments. Experimental protocols followed the rules of The Ethics and Biosafety Committee of the Faculty of Biological Sciences at the Catholic University of Chile, permit No. CBB-100/2012.

Author Contributions

Conceived and designed the experiments: JMR FB GF. Performed the experiments: SBC GF. Analyzed the data: JMR SA FB. Contributed reagents/materials/analysis tools: FB. Wrote the paper: JMR FB.

References

1. Hansen J, Kharecha P, Sato M, Masson-Delmotte V, Ackerman F, et al. (2013) Assessing "dangerous climate change": required reduction of carbon emissions to protect young people, future generations and nature. PLoS ONE 8: e81648.
2. Sunday JM, Bates AE, Kearney MR, Colwell RK, Dulvy NK, et al. (2014) Thermal-safety margins and the necessity of thermoregulatory behavior across latitude and elevation. Proc Natl Acad Sci USA 111: 5610–5615.
3. Rahmstorf S, Coumou D (2011) Increase of extreme events in a warming world. Proc Natl Acad Sci USA 108: 17905–17909.
4. Bozinovic F, Calosi P, Spicer JI (2011) Physiological correlates of geographic range in animals. Ann Rev Ecol Evol Syst 42: 155–179.
5. Sih A, Ferrari MCO, Harris DJ (2011) Evolution and behavioural responses to human-induced rapid environmental change. Evol Appl 4: 367–387.
6. Estay SA, Lima M, Bozinovic F (2014) The role of temperature variability on insect performance and population dynamics in a warming world. Oikos 123: 131–140.
7. Folguera G, Bastías DA, Caers J, Rojas JM, Piulachs MD, et al. (2011) An experimental test of the role of environmental temperature variability on ectotherm molecular, physiological and life-history traits: implications for global warming. Comp Biochem Physiol A Mol Integr Physiol 159: 242–246.
8. Folguera G, Mensch J, Munoz JL, Ceballos SG, Hasson E, et al. (2011) Ontogenetic stage-dependent effect of temperature on developmental and metabolic rates in a holometabolous insect. J Insect Physiol 56: 1679–1684.
9. Frost AJ, Thomson JS, Smith C, Burton HC, Davis B, et al. (2013) Environmental change alters personality in the rainbow trout, *Onchorhyncus mykiss*. Anim Behav 85: 1199–1207.
10. Sih A (2013) Understanding variation in behavioural responses to human-induced rapid environmental change: a conceptual overview. Anim Behav 85: 1077–1088.
11. Wingfield JC (2013) The comparative biology of environmental stress: behavioural endocrinology and variation in ability to cope with novel, changing environments. Anim Behav 85: 1127–1133.
12. Schultz GA (1972) Ecology and systematics of terrestrial isopod crustaceans from Bermuda (Oniscoidea). Crustaceana 3: 79–99.
13. Schultz GA (1961) Distribution and establishment of a land isopod in North America. Syst Zool 10: 193–196.
14. Miller RH, Cameron GN (1987) Effects of temperature and rainfall on populations of *Armadillidium vulgare* (Crustacea: Isopoda) in Texas. Am Midl Nat 117: 192–198.
15. Helden AJ, Hassall M (1998) Phenotypic plasticity in growth and development rates of *Armadillidium vulgare* (Isopoda: Oniscidea). Isr J Zool 44: 379–394.
16. Warburg MR, Linsenmair KE, Bercovitz K (1984) The effect of climate on the distribution and abundance of isopods. In: Sutton SL, Holdich DM, editors. The Biology of Terrestrial Isopods. Oxford: Clarendon Press. pp. 339–367.
17. Hassall M, Helden A, Goldson A, Grant A (2005) Ecotypic differentiation and phenotypic plasticity in reproductive traits of *Armadillidium vulgare* (Isopoda: Oniscidea). Oecologia 143: 51–60.
18. Greenaway P, Warburg MR (1998) Water fluxes in terrestrial isopods. Isr J Zool 44: 473–486.
19. Warburg MR, Calahorra Y, Amar K-O (2001) Non-seasonal breeding in a Porcellionid isopod. J Crust Biol 21: 375–383.
20. Castañeda LE, Lardies MA, Bozinovic F (2004) Adaptive latitudinal shifts in the thermal physiology of a terrestrial isopod. Evol Ecol Res 6: 1–15.
21. Lardies MA, Bozinovic F (2006) Geographic covariation between metabolic rate and life history traits. Evol Ecol Res 8: 455–470.
22. Lardies MA, Bozinovic F (2008) Genetic variation for plasticity in physiological and life-history traits among populations of an invasive species, the terrestrial isopod *Porcellio laevis*. Evol Ecol Res 10: 747–762.
23. Lardies MA, Catalán TP, Bozinovic F (2004) Metabolism and life-history correlates in a lowland and highland population of a terrestrial isopod. Can J Zool 82: 677–687.
24. Bacigalupe LD, Araya NM, Carter MJ, Catalán TP, Lardies MA, et al. (2007) Maternal effects, maternal body size and offspring energetics: a study in the common woodlouse *Porcellio laevis*. Comp Biochem Physiol A Mol Integr Physiol 147: 349–354.
25. Hassall M, Tuck JM (2007) Sheltering behavior of terrestrial isopods in grasslands. Invertebrate Biology 126: 46–56.
26. Hassall M, Edwards DP, Carmenta R, Derh AM, Moss A (2010) Predicting the effect of climate change on aggregation behaviour in four species of terrestrial isopods. Behaviour 147: 151–164.
27. Hoffman G (1983) The search behavior of the desert isopod *Hemilepistus reaumuri* as compared with a systematic search. Behav Ecol Sociobiol 13: 93–106.
28. Hoffmann G (1983) The random elements in the systematic search behavior of the desert isopod *Hemilepistus reaumuri*. Behav Ecol Sociobiol 13: 81–92.
29. Turchin P (1998) Quantitative analysis of movements. Measuring and modeling population redistribution in animals and plant: Sinauer Associates.
30. Heinz SK, Conradt L, Wissel C, Karin F (2005) Dispersal behaviour in fragmented landscapes: Deriving a practical formula for patch accessibility. Landsc Ecol 20: 83–99.
31. Hughes RN (1967) Turn alternation in woodlice (*Porcellio scaber*). Anim Behav 15: 282–286.
32. Hughes RN (1978) Effects of blinding, antennectomy, food deprivation and simulated natural conditions on alternation in woodlice woodlice (*Porcellio scaber*). Journal Biological Psychology 20: 35–40.
33. Carbines GD, Dennis RM, Jackson RR (1992) Increased turn alternation by woodlice (*Porcellio scaber*) in response to a predatory spider, *Dysdera crocata*. Int J Comp Psychol 5: 138–144.

34. Paris O (1963) The ecology of *Armadillidium vulgare* (Isopoda: Oniscoidea) in California grassland: food, enemies and weather. Ecol Monog 33: 1–22.
35. Warburg MR (1968) Adaptation of terrestrial isopods. Am Zool 8: 545–559.
36. Baker M (2005) Experience influences settling behaviour in desert isopods, *Hemilepistus reaumuri*. Anim Behav 69: 1131–1138.
37. Hutchison VH, Maness JD (1979) Role of behavior in temperature acclimation and tolerance in ectotherms. Am Zool 19: 367–384.
38. Pörtner HO, Bennett AF, Bozinovic F, Clarke A, Lardies MA, et al. (2006) Trade-offs in thermal adaptation: the need for a molecular to ecological integration. Physiol Biochem Zool 79: 295–313.
39. Marais E, Chown SL (2008) Beneficial acclimation and the Bogert effect. Ecol lett 10: 1027–1036.
40. Folguera G, Bastías DA, Bozinovic F (2009) Impact of experimental thermal amplitude on ectotherm performance: Adaptation to climate change variability? Comp Biochem Physiol A Mol Integr Physiol 154: 389–393.
41. Angilletta Jr MJ (2009) Thermal adaptation: a theoretical and empirical synthesis. Oxford: Oxford University Press.
42. Jaksic FM (2001) Spatiotemporal variation patterns of plants and animals in San Carlos de Apoquindo, central Chile. Rev Ch Hist Nat 74: 477–502.
43. Nair GA (1976) Life cycle of *Porcellio laevis* (Latreille) (Isopoda, porcellionidae). Proc Indian Natl Sci Acad B Biol Sci 84: 165–172.
44. Husain MZ, Alikhan MA (1979) Physiological adaptations in crustacea to the environment: Oxygen consumption as a function of body weight and environmental temperature in the terrestrial isopod, *Porcellio laevis* Latreille (Isopoda, Oniscoidea). Crustaceana 36: 277–286.
45. Codling EA, Plank MJ, Benhamou S (2008) Random walk models in biology. J R Soc Interface 5: 813–834.
46. Nams VO (1996) The Vfractal: a new estimator for fractal dimension of animals movement paths. Landsc Ecol 11: 289–297.
47. Dray S, Royer-Carenzi M, Calenge C (2010) The exploratory analysis of autocorrelation in animal-movement studies. Ecological Research 25: 673–681.
48. Team RDC (2008) R: A language and environment for statistical computing. Version 2.12 ed: R Foundation for Statistical Computing.
49. Nams VO (2006) Improving accuracy and precision in estimating fractal dimension of animal movement path. Acta Biotheor 54: 1–11.
50. Bolker BM, Brooks ME, Clark CJ, Geange SW, Poulsen JR, et al. (2009) Generalized linear mixed models: a practical guide for ecology and evolution. Trends Ecol Evol 24: 127–135.
51. Zuur A, Ieno EN, Walker N, Saveliev AA, Smith GM (2009) Mixed effects models and extensions in ecology with R. New York: Springer Science+Business Media.
52. Westerberg L, Lindström T, Nilsson E, Wennerger U (2008) The effect on dispersal from complex correlations in small-scale movement. Ecol Modell 213: 263–272.
53. Calenge C (2006) The package adehabitat for the R software: A tool for the analysis of space and habitat use by animals. Ecol Modell 197: 516–519.
54. Bozinovic F, Bastías DA, Boher F, Clavijo-Baquet S, Estay SA, et al. (2011) The mean and variance of environmental temperature interact to determine physiological tolerance and fitness. Physiol Biochem Zool 84: 543–552.
55. Lankau R, Jorgensen PS, Sih A (2011) Incorporating evolutionary principles into environmental management and policy. Evol Appl 4: 315–325.
56. Solomon S, Qin D, Manning M, Chen Z, Marquis M, et al. (2008) Climate Change 2007: The Physical Science Basis. Cambridge: Cambridge University Press.
57. Deutsch CA, Tewksbury JJ, Huey RB, Sheldon KS, Ghalambor CK, et al. (2008) Impacts of climate warming on terrestrial ectotherms across latitude. Proc Natl Acad Sci USA 105: 6668–6672.
58. Bozinovic F, Catalán TP, Estay SA, Sabat P (2013) Acclimation to daily thermal variability drives the metabolic performance curve. Evol Ecol Res 15: 579–587.
59. Kearney MR, Shine R, Porter WP (2009) The potential for behavioral thermoregulation to buffer cold-blooded animals against climate change. Proc Natl Acad Sci USA 106: 3835–3840.
60. Helmuth B, Broitman BR, Yamane L, Gilman SE, Mach K, et al. (2010) Organismal climatology: analyzing environmental variability at scales relevant to physiological stress. J Exp Biol 213: 995–1003.
61. Canals M, Bozinovic F (2011) Huddling behavior as critical phase transition triggered by low temperatures. Complexity 17: 35–43.
62. Helmuth B (2009) From cells to coastlines: how can we use physiology to forecast the impacts of climate change? J Exp Biol 212: 753–760.
63. Chapperon C, Le Bris C, Seuront L (2013) Thermally mediated body temperature, water content and aggregation behaviour in the intertidal gastropod *Nerita atramentosa*. Ecological Research 28: 407–416.
64. Martin Tara L, Huey Raymond B (2008) Why "suboptimal" is optimal: Jensen's inequality and ectotherm thermal preferences. Am Nat 171: E102–E118.
65. Tepler S, Mach K, Denny MW (2011) Preference versus performance: body temperature of the intertidal snail *Chlorostoma funebralis*. Biol Bull 220: 107–117
66. Muñoz JLP, Camus PA, Labra FA, Finke R, Bozinovic F (2008) Thermal constraints on daily patterns of aggregation and density along an intertidal gradient in the periwinkle *Echinolittorina peruviana*. Journal of Thermal Biology 33: 149–156.
67. Muñoz JLP, Finke R, Camus PA, Bozinovic F (2005) Thermoregulatory behavior, heat gain and thermal tolerance in the periwinkle *Echinolittorina peruviana* in central Chile. Comp Biochem Physiol A Mol Integr Physiol 142: 92–98.
68. Bartumeus F, Da Luz MGE, Viswanathan G.M., Catalan J (2005) Animal strategies: a quantitative random-walk analysis. Ecology 86: 3078–3087.
69. Weins JA, Crist TO, With KA, Milne B (1995) Fractal patterns of insect movement in microlandscape mosaic. Ecology 76: 663–666.
70. Root TL, Hughes L (2005) Present and future phenological changes in wild plants and animals. In: Lovejoy TE, Hanah L, editors. Climate change and biodiversity. London: Yale Press. pp. 61–74.

14

The Importance of Biologically Relevant Microclimates in Habitat Suitability Assessments

Johanna Varner*, M. Denise Dearing

Department of Biology, University of Utah, Salt Lake City, Utah, United States of America

Abstract

Predicting habitat suitability under climate change is vital to conserving biodiversity. However, current species distribution models rely on coarse scale climate data, whereas fine scale microclimate data may be necessary to assess habitat suitability and generate predictive models. Here, we evaluate disparities between temperature data at the coarse scale from weather stations versus fine-scale data measured in microhabitats required for a climate-sensitive mammal, the American pika (*Ochotona princeps*). We collected two years of temperature data in occupied talus habitats predicted to be suitable (high elevation) and unsuitable (low elevation) by the bioclimatic envelope approach. At low elevations, talus surface and interstitial microclimates drastically differed from ambient temperatures measured on-site and at a nearby weather station. Interstitial talus temperatures were frequently decoupled from high ambient temperatures, resulting in instantaneous disparities of over 30°C between these two measurements. Microhabitat temperatures were also highly heterogeneous, such that temperature measurements within the same patch of talus were not more correlated than measurements at distant patches. An experimental manipulation revealed that vegetation cover may cool the talus surface by up to 10°C during the summer, which may contribute to this spatial heterogeneity. Finally, low elevation microclimates were milder and less variable than typical alpine habitat, suggesting that, counter to species distribution model predictions, these seemingly unsuitable habitats may actually be better refugia for this species under climate change. These results highlight the importance of fine-scale microhabitat data in habitat assessments and underscore the notion that some critical refugia may be counterintuitive.

Editor: Corrie S. Moreau, Field Museum of Natural History, United States of America

Funding: J.V. is supported by a National Science Foundation Graduate Research Fellowship (#0750758). Field assistants were supported by a REU supplement to National Science Foundation award #1111533 (PI: T. Schmid). Additional funding for dataloggers and fieldwork was provided by: University of Utah Global Change and Sustainability Center, Society for Integrative and Comparative Biology, The Oregon Zoo Future for Wildlife Fund, The Wilderness Society Gloria Barron Scholarship and American Society of Mammalogists. The funders had no role in study design, data collection and analysis, decision to publish, or preparation of the manuscript.

Competing Interests: The authors have declared that no competing interests exist.

* Email: johanna.varner@utah.edu

Introduction

Anthropogenic climate change has already profoundly affected range and community structure for many taxa [1–5]. Predicting future changes in species distributions and interactions is a challenge with great consequences for conserving biodiversity. However, realistic predictions and viable conservation plans depend on accurate forecasts of future habitat suitability.

Species distribution models (SDMs) are the most common approach to predicting habitat suitability under climate change. In this approach, a species' bioclimatic envelope is statistically determined from its present distribution, and suitable habitats are then predicted where similar climatic conditions will occur in the future [6]. Correlative SDMs have been criticized for failing to incorporate important factors like dispersal, biotic interactions, adaptation and behavioral plasticity [7]. However, the predictive power of these models is improved when used in conjunction with other approaches such as phylogenetics [8], mechanistic heat transfer models [9] or eco-physiological parameters [10].

In spite of recent criticism, few SDMs account for fine scale microhabitat features, which can profoundly influence microclimates and therefore habitat suitability [6,11,12]. In fact, a recent meta-analysis demonstrated that average climate grid lengths in SDMs were 10,000-fold larger than the animals they study [13]. In addition, few studies explicitly incorporate data from the periphery of a species' range, though these populations may be particularly informative of climatic tolerance or critical habitat features [14].

The concept of refugia from climate change has received a great deal of recent attention in the literature [15–18]. Refugia are "safe-haven" habitats where species can persist in times of environmental change and may offer hope for *in situ* persistence, particularly for species with poor dispersal capacity [15]. A key feature of refugia is that their microclimates must be relatively stable and must buffer species against climate variability [17]. However, refugia can be difficult to identify, particularly when they are surrounded by areas of less suitable habitat [17]. Indeed, a key aim of conservation science is identifying and protecting habitat features that can serve as refugia through environmental changes.

In this study, we examined how micro-refugia may allow a climate-sensitive species to persist in habitat that appears unsuitable by the bioclimatic envelope approach. American pikas (*Ochotona princeps*) are small mammalian herbivores that have become widely considered indicators of climate change [19]. We hypothesized that vegetation features may create favorable microclimates at low elevations by insulating against summer heat stress, which is known to affect pikas. Uncovering microhabitat variation in seemingly unsuitable climates provides important information about the thermal tolerance, capacity for plasticity and potential vulnerability of a species to future climate change. This information will therefore be vital for refining SDMs and targeting critical habitat refugia for conservation action.

Materials and Methods

Study organism

American pikas (*Ochotona princeps;* order *Lagomorpha*) are small mammalian herbivores that are typically distributed in high elevation mountains of western North America because they require short, cool summers and winters with extended snowpack [20]. Pikas are obligate talus specialists and are rarely found outside of rockslides and boulder fields that provide these suitable microclimates. In addition, pikas are extremely sensitive to high ambient temperatures [21,22] and possess a limited dispersal ability [23,24]. Unlike most alpine mammals, pikas do not hibernate during the winter but spend the short alpine summer amassing food caches called haypiles [25]. It has been hypothesized that warm summer temperatures may prevent pikas from constructing adequate haypiles, resulting in over-winter mortality [26]. This effect may be compounded by reduced snowpack, which exposes the animals to colder winter temperatures [27–29].

With climate change, parts of the pika's range are becoming unsuitable, resulting in upslope range retractions [27], population declines [28], and localized extinctions [27,30,31]. However, pikas persist in some regions with warm climates, including the Columbia River Gorge (CRG) in Oregon and Washington. Little study has been devoted to CRG pikas, though they exist over a thousand meters below the species' previously recognized bioclimatic envelope [20,32]. Given this species' demonstrated sensitivity to aspects of climate, marginal populations like the CRG have great potential to elucidate crucial features of micro-refugia and tractable conservation targets.

Study sites

We collected microhabitat data at four talus patches (ca. 15,000 m^2 each; hereafter, "sites") occupied by pikas in the CRG during June 2012–June 2014 (Table 1, Fig. S1). All sites were 32–35° in steepness, north to northwest facing and surrounded by a dense forest. One of the unique features of this region is a thick layer of moss that covers 25–80% of the surface of each rockslide. The moss also serves as a substrate for other plant cover, including graminoids, forbs and ferns. Sites ranged in elevation from 94 m to 437 m and varied in moss cover from over 65% ("high moss cover": sites 1 & 2) to less than 30% ("low moss cover": sites 3 & 4; Table 1, Fig. S1). Each of these sites is about a thousand meters lower in elevation than pikas are predicted to occur, based on the climate envelope approach [32].

We estimated potential solar exposure at each site using a previously-described insolation index [33]. The aspect and slope angle of each site were measured using a compass equipped with an inclinometer, and potential solar insolation was calculated as sine(slope) × cosine(aspect). This index ranges from −1 to 1, where values of 1 indicate steeper north-facing slopes with little solar exposure, and values of −1 indicate steep south-facing slopes with high exposure [33].

We also sampled alpine microhabitats typical for pikas at two additional sites on the north face of Mt. Hood, approximately 30 km from sites 1–4 (Table 1). Clast size was similar across CRG and Mt. Hood sites, with most rocks at each talus site having

Table 1. Study sites in the CRG (elev. <500 m) and Mt. Hood (elev. >900 m).

Site	Latitude	Longitude	Elev. (m)	Veg. Cover (%)[a]	Moss Cover (%)	Avg. Clast Size (cm)	Aspect (deg.)	Slope (deg.)	Insolation[b]
Site 1	45°41'49"N	121°39'10"W	94	75.6	61.0	24.6	342	33	0.52
Site 2	45°41'16"N	121°47'45"W	194	71.2	68.6	40.1	2	32	0.53
Site 3	45°40'25"N	121°50'17"W	281	37.0	35.3	53.0	314	35	0.40
Site 4	45°33'50"N	122°09'14"W	437	35.7	25.5	48.7	313	33	0.37
Site 5	45°27'31"N	121°40'24"W	905	36.2	33.3	50.7	349	27	0.45
Site 6	45°24'41"N	121°41'59"W	1682	9.8	1.1	53.8	36	28	0.38

[a]Vegetation cover includes moss, ferns, grass, forbs and shrub cover on the talus.
[b]Insolation = sine(slope) × cosine(aspect), from [33]. See Materials and Methods for interpretation of values.

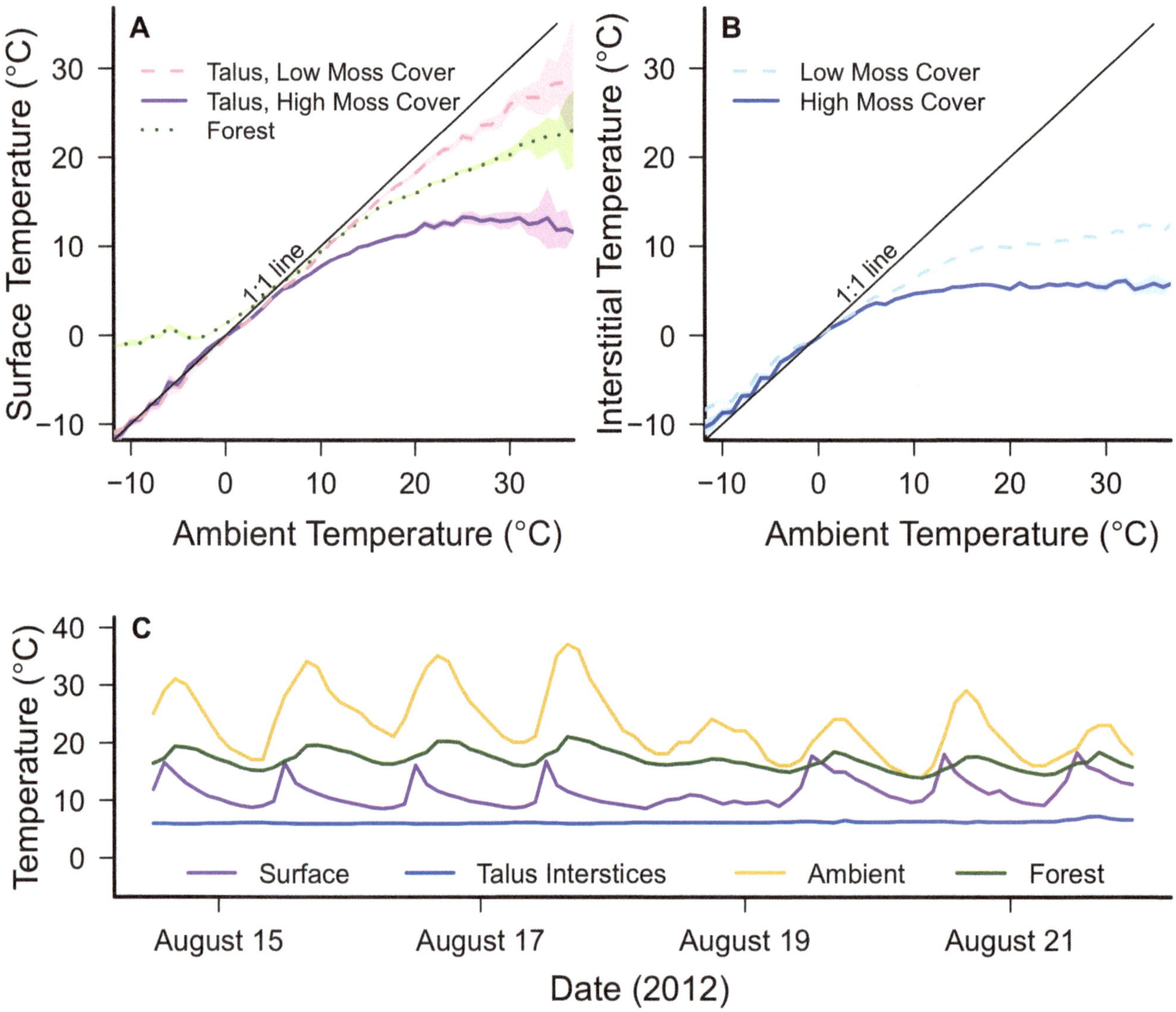

Figure 1. Pika microclimates compared to ambient temperatures. Temperature data were collected every two hours from June 2012 to June 2014 at four sites in the Columbia River Gorge. (A) Lines represent mean temperatures and shaded areas represent 95% confidence intervals. At high ambient temperatures, talus surface temperatures were lowest at sites with high moss cover. Forest surface temperatures were also cooler than ambient temperatures, but talus surface temperatures at sites with low moss cover closely tracked ambient temperatures. (B) Talus interstitial temperatures were functionally decoupled from variation in ambient temperature and remained a cool and constant 4–7°C at sites with high moss cover. (C) Time-trace of ambient, talus surface, forest surface, and talus interstitial temperatures measured at site 2 (high moss cover) during August 2012.

dimensions in the range of 20–100 cm, which is preferred for pikas (Table 1). All six sites were along publicly accessible trails on land owned by the United States Forest Service or Oregon Parks and Recreation Department. Appropriate permits for deploying dataloggers in Oregon State Parks were acquired from Oregon Parks and Recreation Department, permit no. 012-11.

Macroclimate temperature measurements

We downloaded ambient temperature data of the type used for SDMs from the Western Regional Climate Center (available: http://www.raws.dri.edu/wraws/orF.html) at the Cascades Locks station (45°40′10″N, 121°52′54″W, elevation 128 m). This weather station is centrally located in the CRG, 10.6 km from our four low elevation field sites, on average. It is closest (3.4 km) to site 3. To further characterize ambient temperatures at CRG sites, HOBO Pendant temperature dataloggers (model UA001-08; Onset Computer, Bourne, MA) were suspended from a tree branch approximately 2 m above the ground and protected from solar radiation by a white plastic shield. All measurements labeled "ambient" reflect these 2 m shade air temperatures.

Microclimate temperature measurements

We measured temperatures in pika-relevant microhabitats with HOBO Pendant temperature dataloggers, which were housed in waterproof, plastic cases. At each site, we placed loggers at the talus surface in vegetated areas near the bottom of the talus slope where we had observed pikas foraging [34]. All surface loggers were shielded from direct solar radiation by moss. To characterize sub-surface temperatures, we threaded a second, paired sensor into the talus interstices to reach a final depth (i.e., vertical distance directly below the surface logger) of 80–100 cm. In some cases, small sub-surface rocks were removed for sensor insertion and

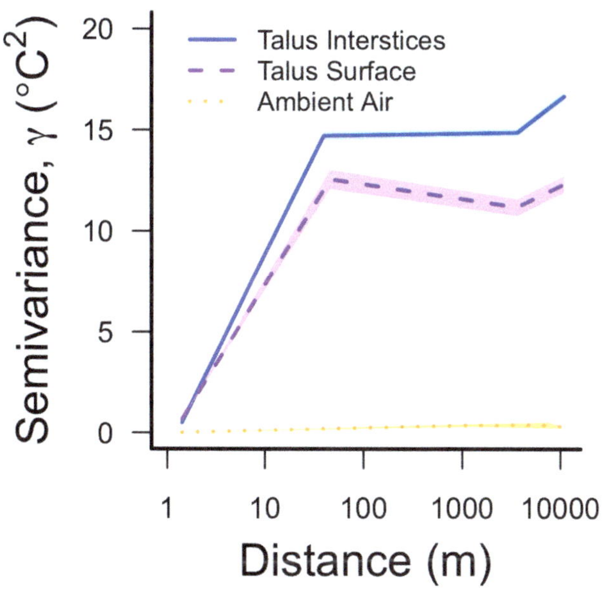

Figure 2. Spatial variation in pika-relevant microclimates and ambient temperatures. Thermal semivariance is shown as a function of distance between dataloggers at the 4 CRG sites during June–August 2013. Lines represent mean semivariance and shaded areas represent 95% confidence intervals. Temperatures in pika-relevant microclimates are far more heterogeneous and less spatially correlated than ambient shade temperatures measured 2 m above the ground.

immediately replaced, but vegetation cover at the surface was not disturbed. Finally, at sites 2 and 3, we deployed a third logger in the surrounding forest to characterize a potential midday thermal refuge. For consistency, these forest surface loggers were also shielded from solar radiation by moss.

Average summer temperature is known to impact pikas [27–29,33]. Furthermore, summer temperatures in the CRG are particularly unusual for this species [32]. To adequately capture average summer temperatures and to characterize spatial variation in summer temperatures, we deployed three additional surface-interstitial temperature logger pairs at each site during June–August 2013. To control for daily patterns of shade cover on the talus, one logger pair was placed near the top of the talus slope, and the other two logger pairs were placed approximately halfway up the talus on the east and west sides. Thus, at each site, we collected temperature measurements every two hours at four locations at the talus surface and four locations in the talus interstices during this time period. All datalogger locations were placed in areas of the talus where pikas were observed foraging, determined from behavioral observations [34].

Finally, to provide an initial characterization of humidity patterns between sites, we deployed HOBO relative humidity logger (model U12-011; one per site) at sites 2 and 3 during June 2012–August 2013. Loggers were placed in plastic housing with two mesh walls that permitted sufficient air exchange for accurate humidity readings but prevented the logger from direct contact with substrates or precipitation. This housing was then threaded approximately 80 cm into the talus interstices.

Spatial variation in temperature

To characterize thermal variation at very small scales, we deployed additional surface, interstitial and ambient loggers at site 2 and site 3 from July 14–August 30, 2013 in close proximity (1 m) to the sensor network described above. We then characterized spatial variation in temperatures within and between sites with variograms, which are widely used in geostatistics to represent autocorrelation between measurements observed at different spatial locations. Variograms were recently recommended as a spatially explicit tool for examining thermal variance in microclimates as a function of distance between measurements [13]. The semivariance (γ) of temperature measurements at a given time point is half the average squared difference between logger-values (x), separated by a distance h, as given by: $\gamma(h)=\frac{1}{2N_h}\sum_{(i,j)|h_{ij}=h}(x_i-x_j)^2$, where i, j are specific logger pairs separated by distance h, and N_h is the number of logger pairs that are separated by this distance [35]. Lower semivariance therefore indicates higher autocorrelation between temperature measurements at a given separation distance. We computed semivariances with the *variogram* function in the R package *gstat* [36,37] for temperature measurements during June–August 2013. This function calculated semivariance at four average separation distances: 1 m (between closely placed loggers at sites 2 and 3), 39 m (between loggers within a site), 3,622 m (between loggers at nearby sites) and 10,908 m (between loggers at distant sites).

Vegetation cover and microclimate

To provide an initial test of the hypothesis that vegetation insulates against extreme surface temperatures, we manipulated vegetation cover at an additional talus patch (ca. 1 km from site 1). A patch of moss approximately 1 m in diameter was experimentally removed and relocated nearby where moss did not naturally grow (Fig. S2). We then measured temperatures every two hours from May 28–June 13, 2012 at four locations: under an unmanipulated patch of moss, under the transplanted moss, under a pile of bare control rocks that did not naturally have any moss, and under the pile of bare rocks where we removed moss (Fig. S2). The bare rocks in this experiment were approximately 15–20 cm in average dimension. All four locations were within 2 m of each other, and therefore should experience similar patterns of shade and sun exposure throughout the day. We compared daily average and maximum temperatures with two-way, repeated measures ANOVA, using treatment (i.e., manipulation vs. control) and cover (i.e., moss vs. rocks) as main effects and date as a repeated measure.

Microclimate and elevation

Temperature should decrease as elevation increases according to region-specific lapse rates, which may also vary by season [38]. In this region, previously reported lapse rates are $-2°C/m$ for minimum temperatures and $-7°C/m$ for maximum temperatures [39]. To investigate the impact of elevation on microclimate and calculate pika-relevant lapse rates, we placed paired surface-interstitial dataloggers at each high elevation site, which recorded temperatures at the same two-hour intervals as CRG loggers during June–August 2013. To test the effect of elevation on temperatures in pika-relevant microclimates, we computed Pearson correlation coefficients between elevation and average, maximum and minimum temperatures when all sites were free of snow.

Data deposition

The temperature and relative humidity measurements collected for this study are freely available in USPACE (http://uspace.utah.edu), the University of Utah's institutional repository. The accession number for these data is ark:/87278/s6b583mp

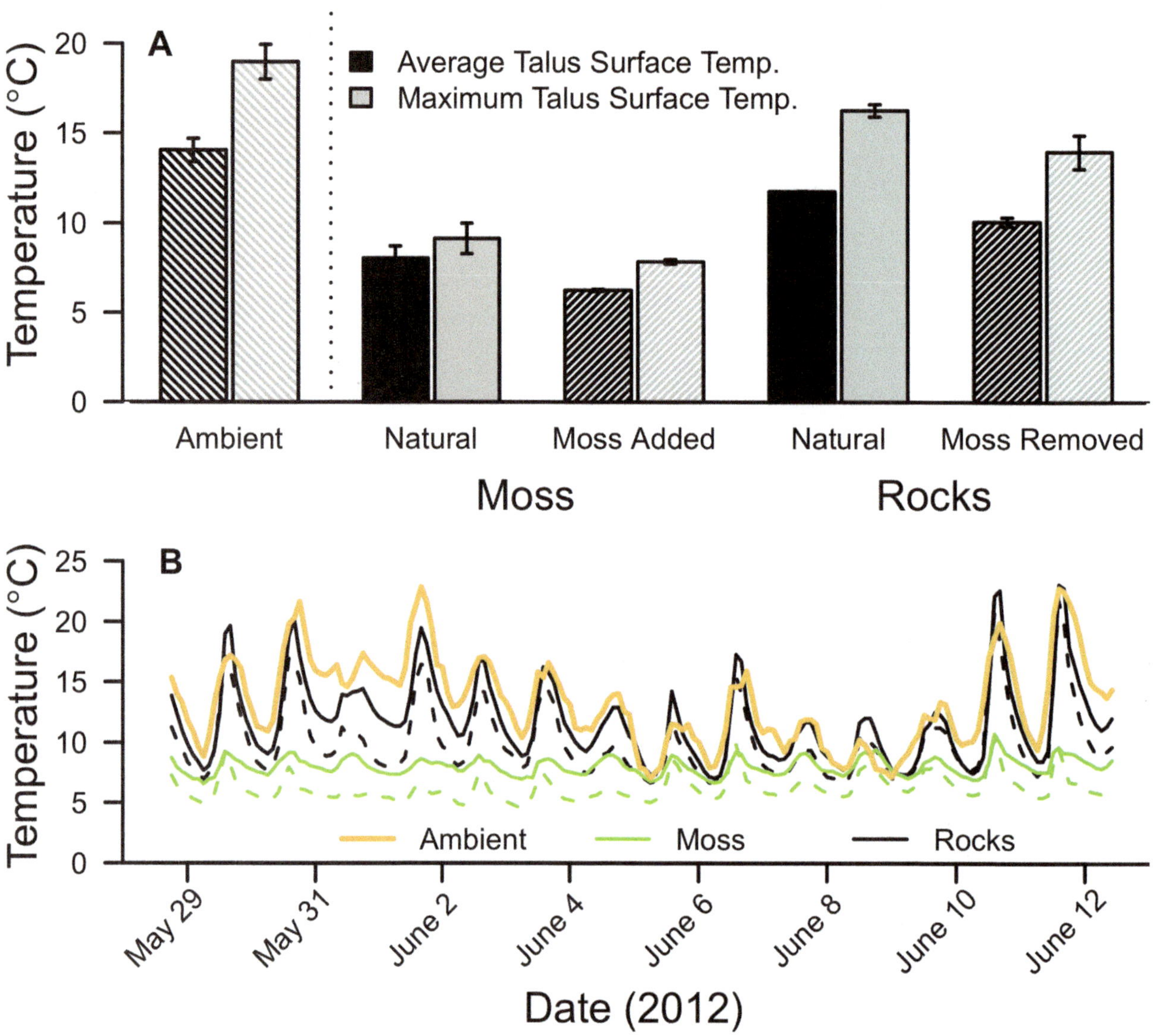

Figure 3. The effect of vegetation cover on summer surface temperatures. (A) Average and maximum summer surface temperatures were significantly cooler under moss, compared to rocks. Both types of surface temperature were cooler than ambient temperatures. (B) Time-trace of temperatures during this experiment.

(available: http://content.lib.utah.edu/cdm/ref/collection/uspace/id/10610).

Results

Macroclimate temperature measurements

We found no difference between 2 m shade ambient temperatures collected at sites 1–3 and temperatures measured at the Locks weather station for daily averages ($F_{3,1288} = 0.654$, $p = \text{NS}$), maxima ($F_{3,1288} = 2.162$, $p = \text{NS}$) or minima ($F_{3,1288} = 0.654$, $p = \text{NS}$) collected between October 2012 and August 2013. We therefore present data from our ambient loggers in future analyses because these data were collected at the same two-hour intervals as microclimate data.

Microclimate temperature measurements

We observed substantial divergence in microclimates relevant to pikas compared to 2 m shade ambient temperatures (Fig. 1). Although surface sensors were below moss cover at all sites, patterns in talus surface temperature were dependent upon overall moss cover at the site. At sites with high moss cover (>65% coverage), talus surface temperatures rarely exceeded 20°C, even at ambient temperatures over 35°C (Fig. 1A, C). In contrast, at sites with low moss cover (≤35% coverage), talus surface temperatures were only a few degrees cooler than ambient temperatures, on average (Fig. 1A). Temperature at the forest floor was also 4–9°C cooler at high ambient temperatures, but rarely dipped below 0°C on very cold days (Fig. 1A).

At all sites, interstitial temperatures were strongly divergent from ambient temperatures (Fig. 1B). Even at ambient temperatures over 35°C, interstitial temperatures at sites with high moss cover remained a cool and constant 6–8°C, resulting in instantaneous disparities of up to 31.5°C between 2 m shade ambient temperatures (measured on-site) and microhabitats relevant to pikas (Fig. 1C). Daily variation in interstitial temperature was <1°C during the summer at sites with high moss cover, essentially decoupling interstitial temperatures from temperatures at the surface (Fig. 1B, C). At sites with low moss cover, interstitial

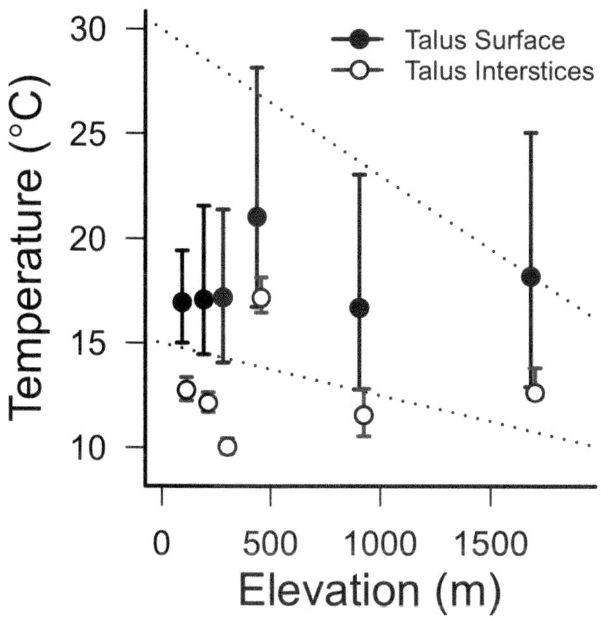

Figure 4. Average summer temperatures by elevation. Points represent mean summer temperatures at the talus surface (filled) and 1 m into talus interstices (open). Bars range from mean daily maxima to mean daily minima in each of these habitats. To show bars that overlap, interstitial points are shifted right by 20 m at some sites. Dotted lines represent previously-calculated lapse rates for the Pacific Northwest, from [39].

temperatures were more variable but remained 5–11°C cooler on average than the talus surface.

Finally, relative humidity varied with ambient temperature (Fig. S3). At ambient temperatures below 10°C, both sites 2 and 3 had uniformly high relative humidity (>90%). At warmer ambient temperatures, relative humidity at both sites became more variable. However, site 2 (high moss cover) had consistently higher relative humidity (75–95%) than site 3 (low moss cover; 55–75% humidity).

Spatial variation in temperature

Semivariance, or the average squared difference in temperature measurements separated by a certain distance, remained very low (<0.4°C^2) among ambient temperatures, indicating high autocorrelation between these measurements at all scales (Fig. 2). In contrast, spatial variation in pika-relevant microclimates far exceeded spatial variation in ambient temperatures at all scales (Fig. 2). At a 1 m separation distance, semivariances for temperatures recorded at the talus surface and in talus interstices were relatively low (<1°C^2), indicating that temperature measurements separated by 1 m were highly correlated. However, at larger spatial scales, semivariance quickly increased to 12°C^2 for temperatures recorded at the talus surface and 16°C^2 for temperatures recorded in the talus interstices, indicating a very low degree of spatial autocorrelation. At separation distances larger than 1 m, microclimate temperatures recorded within the same site were not more similar to each other than temperatures recorded by sensors at the most distant sites.

Vegetation cover and microclimate

During the summer, average talus surface temperatures were 5–9°C cooler under both naturally occurring and transplanted moss, compared to bare rocks (Cover: $F_{(1,31)} = 239.2$, $p<0.0001$; Fig. 3A). Surprisingly, there was also a significant effect of treatment, such that unmanipulated moss and rocks were also 1–2°C warmer than manipulated moss and rocks (Treatment: $F_{(1,15)} = 118.6$, $p<0.001$), but there was no interaction between cover and treatment. Maximum surface temperatures exhibited the same pattern (Cover: $F_{(1,31)} = 131.1$, $p<0.0001$; Treatment: $F_{(1,15)} = 79.1$, $p<0.001$; Fig. 3A). Over the duration of this experiment, both types of surface (moss and rocks) were cooler, on average, than ambient air temperature (Fig. 3A). However, surface measurements under rocks more closely tracked ambient temperatures; whereas surface measurements under moss were often up to 15°C cooler than ambient air temperature (Fig. 3B).

Microclimate and elevation

Summer temperatures in the CRG were substantially cooler than predicted by region-specific lapse rates ([39]; Fig. 4). In fact, elevation was not predictive of talus interstitial temperatures or surface temperatures. All Pearson correlation coefficients between elevation and average, minimum or maximum temperatures were ≤0.6 and non-significant ($p>0.2$). The lowest elevation sites also had less than half the daily temperature range (i.e., daily maximum – daily minimum) of high elevation sites.

Discussion

We investigated the effects of vegetation and elevation on microclimates relevant to a sensitive species living in an atypical habitat. On the basis of climate envelope models alone, the Columbia River Gorge (CRG) appears to be unsuitable habitat for pikas because it does not have short, cool summers or long winters, which are commonly recognized as criteria for pika persistence [e.g., 20,32,40]. However, our results demonstrate that cool microclimates are available in the CRG and that these low elevation rockslides may actually be more suitable for pikas than the nearby alpine talus that is more typical for this species.

Interstitial temperatures in the talus typically depend on heat-transfer processes at the surface (i.e., convection, conduction, solar and long-wave thermal radiation). However, during the summer in the CRG, daily fluctuations in this microhabitat are frequently less than 0.2°C and instantaneous disparities with ambient temperatures may be up to 31.5°C (Fig. 1C), suggesting that, at some locations, the surface heat-transfer processes do not penetrate more than 1 m into the talus. As a result, interstitial microclimates are functionally decoupled from heat-transfer at the surface and are likely governed by other processes. CRG taluses have several features that are atypical for pika habitat, including a thick layer of vegetation covering the talus, high relative humidity, and reduced solar insolation (i.e., due to slope orientation and/or shade from the surrounding forest canopy). Each of these features may contribute to the significant disparities observed between the microclimate relevant to pikas and the macroclimate observed at weather stations.

Vegetation features, and moss cover in particular, appear to buffer against climate variability, differentiating talus microclimates at some sites from the surrounding macroclimate (Fig. 1). Our high moss cover sites likely experience more shade (Table 1) and higher humidity (Fig. S3). These features may facilitate moss growth because mosses are predisposed to grow in shady, humid conditions [41]. However, our experimental results suggest that mosses may also actively buffer talus microclimates from ambient temperature extremes (Fig. 3). In other ecosystems and urban environments, vegetation can similarly lower surface temperatures of dark substrates by 5–20°C [42,43]. This cooling effect may be caused by changes in diffuse reflectivity (albedo), which can strongly affect surface temperatures. Specifically, mosses may raise the albedo of the dark basalt that makes up the talus in this region, increasing the amount of solar radiation that is reflected. The albedo of basalt is 0.1, whereas albedo estimates of moss and lichen-dominated surfaces are two to four times higher [0.2–0.4; 44]. For comparison, a change of albedo in 0.2 is capable of producing a 10°C change in midday surface temperatures [45]. In addition to increasing albedo, vegetation cover may also act as a layer of insulation [44], or it may have an evaporative cooling effect.

Relative humidity and/or precipitation may also play an important role in pika persistence and abundance in this habitat by maintaining high quality vegetation or water for evaporative cooling. During our study, the Locks weather station near our sites averaged ~150 mm of annual rainfall and 70–80% relative humidity. Similarly, relative humidity in the talus rarely dipped below 75% at our high moss cover sites (Fig. S3). Although differences in relative humidity may be responsible for the distribution of vegetation cover on talus in this region, the vegetation itself may also locally raise humidity through evapotranspiration. Although we did not explicitly measure their effects on microclimates, sub-surface water or ice may also influence microclimate temperatures in this region. Water availability is a strong determinant of pika persistence and abundance in other parts of their range [30,46], and we regularly experienced cold air flowing out of talus cavities, which may indicate subsurface rock-ice or water features [47].

These low elevation sites also experience relatively little solar radiation due to northerly slope orientations and daytime shade cover provided by the surrounding forest canopy. This reduced insolation likely has a large effect on both the magnitude and variability of temperatures in pika habitat. For example, although we did not directly measure solar radiation during this study, site 4 likely received more sunlight because its surrounding forest canopy is shorter than other sites due to a wildfire in 1991 (Fig. S1). Indeed, although they are similar in elevation, slope angle, aspect and moss cover, summer surface and interstitial temperatures were approximately 5°C warmer and much more variable at site 4, compared to site 3 (Fig. 4). Similarly, a shorter forest canopy and higher insolation indices at our high elevation sites may have also contributed to greater temperature variability, compared to the CRG (Fig. 4). Unlike high elevations, CRG microclimates rarely experienced temperature extremes beyond the best estimates of thresholds for acute heat stress (25.5°C) [22] or cold stress (−10°C) [27] in this species.

Shade and vegetation cover are also likely responsible for the mild surface temperatures that we observed in the forest at sites 2 and 3 (Fig. 1A). Consistent with our findings, forest surface temperatures in other studies are typically 5–8°C lower than ambient temperatures measured above the ground [48,49]. Forest surface temperatures were typically a few degrees warmer than talus surface temperatures at sites with high moss cover (Fig. 1A, C), but the forest may represent an important thermal refuge for pikas to remain active at midday at sites with low moss cover. Similarly, the forest may also serve as a winter refuge for CRG pikas, as forest temperatures also remained warmer than talus temperatures on the few very cold days in 2012–2014 (<−10°C; Fig. 1A). Finally, thick understory vegetation in the forest may also make pikas less visible to predators or serve as an additional food resource [34].

Regardless of mechanism, the temperature disparities that we observed between weather stations and talus interstices in the CRG far exceed those in the microclimate literature. Previous studies examining ambient temperatures and microclimates at or below the soil surface report disparities up to 10°C between these measurements [12]. Similarly, burrow systems have a significant ability to buffer animals against temperature fluctuations, but instantaneous differences between ambient shade temperatures and internal burrow temperatures rarely exceed 10°C in the literature [50–55]. Rock-ice features in pika habitat in the Sierra Nevada range can also cool the average warm-season temperatures in talus interstices by 3.8°C, compared with the surface. In contrast, we observed differences of up to 11°C between average surface and interstitial microclimates in the warm season, and instantaneous disparities up to 31.5°C between talus interstitial temperatures and ambient temperature measured on site. For comparison, this disparity is roughly equivalent to 4500 m of elevation change (according to lapse rates for this region) and far exceeds the magnitude of observed and predicted directional climate change.

CRG microhabitats were also highly heterogeneous. A lack of spatial autocorrelation between temperature measurements (even between loggers at the same site, Fig. 2) suggests that spatially uncorrelated noise exceeds deterministic trends in temperature at scales over ~10 m in these microhabitats. This result may be due to the patchy nature of moss and vegetation cover on the talus or to the movement of shade cover across the talus throughout the day. Interestingly, strong correlations were observed in the Great Basin between interstitial temperatures in montane pika habitat and ambient temperatures measured at valley weather stations 40–60 km away [29]. The temperature disparities and spatial heterogeneity that we report here highlight the uniqueness of habitats with high vegetation cover, and more generally, the refugial value of places where surface conditions can partially decouple ambient temperatures from sub-surface microclimates. Importantly, high spatial heterogeneity in a habitat may also allow animals to behaviorally mitigate the impacts of directional climate change by shuttling between different areas within a habitat [12].

Taken together, these results suggest that the low elevation habitat in the CRG may actually be a better refuge from stresses caused by climate change than typical, high elevation habitat and that CRG pika populations have the potential to be relatively resilient to future environmental change. These results are a notable example of an unexpected or counter-intuitive thermal refuge, since the CRG was predicted to be unsuitable based on SDM results alone [e.g., 20,32,40]. Our results also suggest tractable conservation priorities for pikas and other thermally sensitive species in this region. Specifically, land managers could focus on protecting habitat features that contribute to the unique microclimates, including moss (e.g., from trampling by hikers or unsustainable harvest for the horticulture trade) and forest canopy cover (e.g., from logging or severe wildfires).

An ecologically similar species that is likely to be directly affected by talus microclimates in this region is the bushy-tailed woodrat, *Neotoma cinerea*. Like pikas, these woodrats inhabit exclusively rocky habitats in montane forests and are sensitive to high ambient temperatures [56]. This species has experienced localized extinctions in response to past climate change [57] and appears to be suffering a range collapse in response to contemporary climate change [1]. However, we have observed fresh woodrat sign at many talus patches in the CRG and speculate that cool microclimates and high vegetation availability also facilitate woodrat persistence at this unusually low elevation.

Course-scale SDMs should not be discarded as a predictive tool, but climate data and their resolution must be selected with great care. We echo calls from Potter et al. [13] that the ideal spatial resolution for climate data in SDMs must be related to the body size of the species of interest. However, we also suggest that heterogeneity of microclimate features and the spatial or temporal extent to which an organism samples these features should also inform selection of appropriate climate data in SDMs. For example, larger, less-mobile organisms are more likely to experience macroclimates observed at greater spatial grid lengths. Conversely, our results demonstrate that ambient temperature data collected at distant locations may have relatively little relevance for small, mobile species that can rapidly shuttle between complex microhabitats, particularly in environments or seasons when surface features such as snow or vegetation decouple macro- and microclimates. Finally, the timing, frequency and amount of precipitation are rarely part of habitat suitability analyses, though they should be considered.

Microhabitat data will be critical for developing and validating spatial statistics to understand the relationship between climate change and microclimate stability [13]. Relatively little is known about the degree to which microclimates will respond to macroclimatic changes or the time-scale on which these changes will occur. Collecting organism-relevant microhabitat data at finer spatial resolutions and longer time scales will be critical for bridging this gap and for determining the resiliency of microclimate refugia to continued changes in broad-scale temperature and precipitation patterns. Such studies are becoming increasingly feasible, given the availability of relatively inexpensive and small temperature dataloggers with long battery lives.

Our results highlight the need for fine-scale temperature measurements in microhabitats that are relevant to a focal species. When identifying refugia from climate change for conservation action, we can learn much from fine-scale observations of populations in seemingly marginal climates. Indeed, the conservation value of refugia lies in promoting species' ability to persist even under unfavorable climates [16].

Supporting Information

Figure S1 Comparison of moss cover at four sites in the Columbia River Gorge. For reference, pikas are also shown at sites of high (E) and low moss cover (F). *Photo credits*: (A–E) J. Varner, (F) J. J. Horns.

Figure S2 Moss transplant experiment. (A) In an area of the site where moss did not naturally grow, temperatures were measured under rocks and transplanted moss. (B) In an area of the site naturally covered with moss, temperatures were measured under rocks where moss was removed and a patch of unmanipulated moss. Arrows indicate datalogger locations.

Figure S3 Relative humidity compared to ambient temperature. Relative humidity measurements were collected every 2 hours in the talus interstices at sites 2 and 3 during June 2012–August 2013. Ambient temperatures (2 m height, shaded) were collected at site 3 at the same temperature intervals. Lines represent mean temperatures and shaded areas represent 95% confidence intervals.

Acknowledgments

We gratefully acknowledge E. Beever for many helpful discussions about temperature datalogger placement and analysis. M. Lambert, J. J. Horns, E. Westberg, K. Wolfenberger, E. Self, B. Scott and R. Connors helped to place and collect dataloggers in the field. D. Nesdill assisted with data deposition and accessibility.

Author Contributions

Conceived and designed the experiments: JV MDD. Performed the experiments: JV. Analyzed the data: JV. Wrote the paper: JV MDD.

References

1. Moritz C, Patton JL, Conroy CJ, Parra JL, White GC, et al. (2008) Impact of a century of climate change on small-mammal communities in Yosemite National Park, USA. Science 322: 261–264.
2. Rowe RJ, Terry RC, Rickart EA (2011) Environmental change and declining resource availability for small-mammal communities in the Great Basin. Ecology 92: 1366–1375.
3. Sheldon KS, Yang S, Tewksbury JJ (2011) Climate change and community disassembly: impacts of warming on tropical and temperate montane community structure. Ecol Lett 14: 1191–1200.
4. Chen I-C, Hill JK, Ohlemüller R, Roy DB, Thomas CD (2011) Rapid Range Shifts of Species Associated with High Levels of Climate Warming. Science 333: 1024–1026.
5. Şekercioğlu ÇH, Primack RB, Wormworth J (2012) The effects of climate change on tropical birds. Biol Cons 148: 1–18.
6. Austin MP, Van Niel KP (2011) Improving species distribution models for climate change studies: variable selection and scale. J Biogeogr 38: 1–8.
7. Pearson RG, Dawson TP (2003) Predicting the impacts of climate change on the distribution of species: are bioclimate envelope models useful? Glob Ecol Biogeog 12: 361–371
8. Buckley LB, Kingsolver JG (2012) Functional and Phylogenetic Approaches to Forecasting Species' Responses to Climate Change. Annu Rev Ecol Evol Syst 43: 205–226.
9. Kearney MR, Wintle BA, Porter WP (2010) Correlative and mechanistic models of species distribution provide congruent forecasts under climate change. Cons Lett 3: 203–213.
10. Sinervo B, Méndez-de-la-Cruz F, Miles DB, Heulin B, Bastiaans E, et al. (2010) Erosion of Lizard Diversity by Climate Change and Altered Thermal Niches. Science 328: 894–899.
11. Suggitt AJ, Gillingham PK, Hill JK, Huntley B, Kunin WE, et al. (2011) Habitat microclimates drive fine-scale variation in extreme temperatures. Oikos 120: 1–8.
12. Graae BJ, De Frenne P, Kolb A, Brunet J, Chabrerie O, et al. (2012) On the use of weather data in ecological studies along altitudinal and latitudinal gradients. Oikos 121: 3–19.
13. Potter KA, Woods HA, Pincebourde S (2013) Microclimatic challenges in global change biology. Glob Chang Biol 19: 2932–2939.
14. Channell R, Lomolino MV (2000) Dynamic biogeography and conservation of endangered species. Nature 403: 84–86.
15. Keppel G, Wardell-Johnson GW (2012) Refugia: keys to climate change management. Glob Chang Biol 18: 2389–2391.
16. Ashcroft MB (2010) Identifying refugia from climate change. J Biogeogr 37: 1407–1413.
17. Ashcroft MB, Gollan JR, Warton DI, Ramp D (2012) A novel approach to quantify and locate potential microrefugia using topoclimate, climate stability, and isolation from the matrix. Glob Chang Biol 18: 1866–1879.
18. Dobrowski SZ (2011) A climatic basis for microrefugia: the influence of terrain on climate. Glob Chang Biol 17: 1022–1035.
19. Smith A, Weidong L, Hik D (2004) Pikas as harbingers of global warming. Species 41: 4–5.
20. Hafner DJ (1993) North American pika (*Ochotona princeps*) as a Late Quaternary biogeographic indicator species. Quat Res 39: 373–380.
21. MacArthur RA, Wang LCH (1974) Behavioral thermoregulation in the pika, *Ochotona princeps*: a field study using radiotelemetry. Can J Zool 52: 353–358.
22. Smith AT (1974) The distribution and dispersal of pikas: influences of behavior and climate. Ecology 55: 1368–1376.
23. Smith AT (1974) The distribution and dispersal of pikas: consequences of insular population structure. Ecology 55: 1112–1119.
24. Smith AT, Weston ML (1990) *Ochotona princeps*. Mammal Species 352: 1–8.
25. Dearing MD (1997) The function of haypiles of pikas (*Ochotona princeps*). J Mammal 78: 1156–1163.
26. Ray C, Wilkening JL, Sweazea K (2009) Signs of demograpic change and physiological stress in Rocky Mountain pikas. North American Pika Conference, Jackson, WY.
27. Beever EA, Ray C, Wilkening JL, Brussard PF, Mote PW (2011) Contemporary climate change alters the pace and drivers of extinction. Glob Chang Biol 17: 2054–2070.
28. Beever EA, Dobrowski SZ, Long J, Mynsberge AR, Piekielek NB (2013) Understanding relationships among abundance, extirpation, and climate at ecoregional scales. Ecology 94: 1563–1571.
29. Beever EA, Ray C, Mote PW, Wilkening JL (2010) Testing alternative models of climate-mediated extirpations. Ecol Appl 20: 164–178.

30. Erb LP, Ray C, Guralnick R (2011) On the generality of a climate-mediated shift in the distribution of the American pika (*Ochotona princeps*). Ecology 92: 1730–1735.
31. Stewart JAE, Wright DH (2012) Assessing persistence of the American pika at historic localities in California's northern Sierra Nevada. Wild Soc Bull 36: 759–764.
32. Simpson WG (2009) American pikas inhabit low-elevation sites outside the species' previously described bioclimatic envelope. WNAN 69: 243–250.
33. Jeffress MR, Rodhouse TJ, Ray C, Wolff S, Epps C (2013) The idiosyncrasies of place: geographic variation in the climate-distribution relationships of the American pika. In press. Ecol Appl.
34. Varner J, Dearing MD (2013) Dietary plasticity in pikas as a strategy for atypical resource landscapes. J Mammal 94: 72–81.
35. Janis MJ, Robeson SM (2004) Determining the spatial representativeness of air-temperature records using variogram-nugget time series. Phys Geog 25: 513–530.
36. Pebesma E (2004) Multivariable geostatistics in S: the gstat package. Comput Geosci 30: 683–691.
37. R Core Development Team (2011) R: A language and environment for statistical computing. R Foundation for Statistical Computing. Vienna, Austria.
38. Stone PH, Carlson JH (1979) Atmospheric lapse rate regimes and their parameterization. J Atmos Sci 36: 415–423.
39. Mote PW, Lundquist JD, Minder J (2009) Surface and free-air lapse rates in the Cascade Mountains of Washington. 21st Conference on Climate Variability and Change. Phoenix, AZ: American Meteorological Society.
40. Galbreath KE, Hafner DJ, Zamudio KR (2009) When Cold is Better: Climate-Driven Elevation Shifts Yield Complex Patterns of Diversification and Demography in an Alpine Specialist (American Pika, *Ochotona princeps*). Evolution 63: 2848–2863.
41. Zechmeister HG (1995) Growth rates of five pleurocarpous moss species under various climatic conditions. J Bryol 18: 455–468.
42. Qiu G, Li H, Zhang Q, Chen W, Liang X, et al. (2013) Effects of Evapotranspiration on Mitigation of Urban Temperature by Vegetation and Urban Agriculture. J Integr Agric 12: 1307–1315.
43. Yang X, Zhao L, Bruse M, Meng Q (2013) Evaluation of a microclimate model for predicting the thermal behavior of different ground surfaces. Build Environ 60: 93–104.
44. Beringer J, Lynch AH, Chapin FS, Mack M, Bonan GB (2001) The representation of arctic soils in the land surface model: the importance of mosses. J Clim 14: 3324–3335.
45. Li H, Harvey J, Kendall A (2013) Field measurement of albedo for different land cover materials and effects on thermal performance. Build Environ 59: 536–546.
46. Erb LP, Ray C, Guralnick R (2013) Determinants of pika population density vs. occupancy in the Southern Rocky Mountains. Ecol Appl 24: 429–435.
47. Millar CI, Westfall RD (2010) Distribution and climatic relationships of the American pika (*Ochotona princeps*) in the Sierra Nevada and western Great Basin, USA; periglacial landforms as refugia in warming climates. Arct Antarct Alp Res 42: 76–88.
48. Parker GG, Lowman M, Nadkarni N (1995) Structure and microclimate of forest canopies. In: Lowman M, Nadkarni N, editors. Forest canopies. San Diego, CA: Academic Press. 73–106.
49. Rambo TR, North MP (2008) Spatial and Temporal Variability of Canopy Microclimate in a Sierra Nevada Riparian Forest. Northwest Sci 82: 259–268.
50. Johnson SR (1970) The thermal regulation, microclimate, and distribution of the mountain beaver, *Aplodontia rufa pacifica* Merriam: Oregon State University. Available: http://hdl.handle.net/1957/9420. Accessed 20 June 2014.
51. Shenbrot G, Krasnov B, Khokhlova I, Demidova T, Fielden L (2002) Habitat-dependent differences in architecture and microclimate of the burrows of Sundevall's jird (*Meriones crassus*) (*Rodentia: Gerbillinae*) in the Negev Desert, Israel. J Arid Environ 51: 265–279.
52. Šumbera R, Chitaukali W, Elichová M, Kubová J, Burda H (2004) Microclimatic stability in burrows of an Afrotropical solitary bathyergid rodent, the silvery mole-rat (*Heliophobius argenteocinereus*). J Zool 263: 409–416.
53. Du Plessis A, Kerley GI, Winter PD (1992) Refuge microclimates of rodents: a surface nesting *Otomys unisulcatus* and a burrowing *Parotomys brantsii*. Acta Theriol 37: 351–351.
54. Pike D, Mitchell J (2013) Burrow-dwelling ecosystem engineers provide thermal refugia throughout the landscape. Anim Conserv 16: 694–703.
55. Stark HE (1963) Nesting habits of the California vole, *Microtus californicus*, and microclimatic factors affecting its nests. Ecology: 663–669.
56. Smith FA (1997) *Neotoma cinerea*. Mammal Species 564: 1–8.
57. Grayson DK, Madsen DB (2000) Biogeographic implications of recent low-elevation recolonization by *Neotoma cinerea* in the Great Basin. J Mammal 81: 1100–1105.

15

Why Do Some People Do "More" to Mitigate Climate Change than Others? Exploring Heterogeneity in Psycho-Social Associations

José Manuel Ortega-Egea*, Nieves García-de-Frutos, Raquel Antolín-López

Department of Economics and Business, University of Almería (ceiA3), Almería, Spain

Abstract

The urgency of climate change mitigation calls for a profound shift in personal behavior. This paper investigates psycho-social correlates of *extra* mitigation behavior in response to climate change, while also testing for potential (unobserved) heterogeneity in European citizens' decision-making. A person's extra mitigation behavior in response to climate change is conceptualized—and differentiated from common mitigation behavior—as some people's broader and greater levels of behavioral engagement (compared to others) across specific self-reported mitigation actions and behavioral domains. Regression analyses highlight the importance of environmental psychographics (i.e., attitudes, motivations, and knowledge about climate change) and socio-demographics (especially country-level variables) in understanding extra mitigation behavior. By looking at the data through the lens of segmentation, significant heterogeneity is uncovered in the associations of attitudes and knowledge about climate change—but not in motivational or socio-demographic links—with extra mitigation behavior in response to climate change, across two groups of environmentally active respondents. The study has implications for promoting more ambitious behavioral responses to climate change, both at the individual level and across countries.

Editor: Bayden D. Russell, The University of Adelaide, Australia

Funding: The authors acknowledge financial support from the Spanish Ministry of Economy and Science (National R&D Project ECO2011-24921 and predoctoral grant program) and the European Regional Development Fund (ERDF/FEDER), from the University of Almería's (UAL, ceiA3) predoctoral grant program, and CySOC. The funders had no role in study design, data collection and analysis, decision to publish, or preparation of the manuscript.

Competing Interests: The authors have declared that no competing interests exist.

* Email: jmortega@ual.es

Introduction

Climate change is recognized as a major, anthropogenically-induced environmental threat, with potentially severe and far-reaching consequences for human and natural systems [1–2]. It is now beyond dispute that, in industrialized countries, people contribute to climate change through unsustainable high-carbon lifestyles [3]. To illustrate, in the European Union (EU), private households are directly responsible for as much as 20% of total greenhouse gas (GHG) emissions and over one-fourth of the final energy consumed [4–5]; substantially larger shares of EU energy use and GHG emissions are attributable in an indirect way—via consumption expenditures—to household/consumer activities [5–6].

Clearly, then, individuals have a central role to play in addressing climate change risks [7]. People can respond to climate change through adaptation—to potential and unavoidable climate impacts—and mitigation efforts—focused on reducing GHG emissions to prevent (or delay) further damages [3,8]. With climate change being primarily rooted in excessive energy consumption, the public's engagement in mitigation activities is recognized as critical to achieving a sustainable future—that is, shifting towards a new, low-carbon paradigm [1,3,7]. Over the past two decades, increased media coverage—coupled with economic incentives, subsidies, and related interventions—has substantially raised citizens' awareness and concern about climate change, but has typically failed to induce persistent behavioral changes [1,9]. In Europe, estimates of (some form of) personal action to mitigate climate change have ranged from 53% to 63% of the population, according to four surveys conducted between 2008 and 2011 in all EU-27 countries [10–11]; however, as shown elsewhere [12], there is limited engagement of most EU citizens in mitigation efforts beyond recycling [10–11].

Given the urgency of climate change mitigation, a profound shift is needed in personal behavior—from inaction or limited action levels—towards broader and greater levels of behavioral engagement [12–14]. Such *extra* behavioral responses—comprising additional mitigation actions and specific behavioral levels that go beyond what most people do—hold promise for a further incremental impact in addressing climate change. Hence, an important question arises as to what makes some people make an extra commitment (i.e., do "more") to mitigate climate change through personal action, as compared to others.

Considerable attention has been directed at the correlates of individual pro-environmental behavior, in general, or at specific types (or subsets) of personal action—both in private and public spheres [15–16]. For example, past research has investigated recycling [17–18], reducing car use and choosing an environmentally friendly mode of transport [19–20], engaging in environmental citizenship [21], and many other forms of personal pro-environmental behavior. Such green behaviors have been studied through a variety of social-psychological variables and factors, including psychographics—such as environmental attitudes [15,22], concern [23], values and motivations [24], or knowledge [25]—and socio-demographics—such as gender [26], age [27], education [28], income [29], or nationality [30–31].

Corresponding advances are being rapidly made in the field of climate change research, in relation to the correlates of personal mitigation (and adaptation) behavior [1–3,8]. The present article aims to enhance understanding of the psycho-social correlates of extra (i.e., broader and greater) mitigation behavior in response to climate change, among the EU public. It contributes to ongoing climate change research in the following three ways. First, the authors investigate factors that relate to a person's extra behavioral engagement to address climate change, as opposed to more common (or typical) types and levels of mitigation behavior. This is important because, as advocated in previous research [14], "we must do a lot" to effectively respond to climate change. For example, if altruism is the most important motivational process associated with extra mitigation behavior, then this suggests that reinforcing altruistic values—rather than egoistic ones—is a more powerful way to support this increasingly critical action pattern. Second, the study tests for potential unobserved heterogeneity—across respondents—in the relations of psychographics and socio-demographics to extra mitigation behavior in response to climate change. For instance, by looking at the data through the lens of segmentation (i.e., through latent class regression), the analyses will clarify whether specific types of knowledge/information about climate change issues (i.e., the causes, consequences, and ways of fighting this threat) relate to similarly or differently to the breath and level of self-reported behavioral engagement (i.e., extra vs. common) in climate change mitigation. Third, through the analysis of a large-scale survey dataset of European citizens, the study increases current understanding of cross-national variations in personal mitigation behavior in response to climate change. Overall, the findings reported here should help researchers and policy-makers to promote broader and greater levels of personal response to climate change, both at the individual level and across countries.

Extra Mitigation Behavior (in Response to Climate Change)

Pro-environmental behavior—and, more specifically, mitigation (and adaptation) responses to climate change—can be operationalized at multiple levels of analysis, such as individual, group, organizational, or regional/national levels. The focus here is on individual-level, personal mitigation behavior that, according to previous literature, can be broadly described as comprising voluntary and future-oriented behavioral responses to climate change (e.g., consumer reduction in energy consumption with mid- to long-term positive impacts on climate change) [3,32]. Given the multi-faceted nature of personal mitigation behavior, it potentially encompasses a broad range of actions in private and public spheres of life, one-off and regular decisions, simple and more difficult steps, as well as low and high impact actions—as regards their effectiveness in mitigating climate change [1,12,14].

Extra mitigation behavior—the focal variable of interest in this study—is viewed as comprising additional mitigation actions (i.e., broader) and enactment levels of specific mitigation behaviors (i.e., greater) that go beyond what most people do to address climate change. A person's extra behavioral engagement to mitigate climate change is captured—and differentiated from common mitigation behavior—on a set of self-reported past/present mitigation actions spanning in-home and out-of-home settings (e.g., saving energy at home vs. reducing car use), high- and low-impact mitigation practices (e.g., energy conservation vs. recycling), one-off and frequent choices (e.g., installing solar panels in the home vs. buying seasonal, locally produced food), relating to four specific, environmentally significant domains of personal action (i.e., domestic energy/water conservation, waste reduction, eco-friendly transportation, and eco-shopping) [12]. Thus, several different types of personal action are covered so as to profile (and measure) the behavior of people who report extra behavioral engagement to mitigate climate change, as compared to others; this approach allows for sufficient variability in the difficulty, impact, frequency, and context of the specific mitigation actions being examined.

Proposed Models of Extra Mitigation Behavior

The models tested in this study propose that environmental psychographics and socio-demographics help to explain why some people make an extra commitment to mitigate climate change through personal action, as compared to others, while testing for potential (unobserved) heterogeneity among respondent groups.

Psychographics and extra mitigation behavior

To date, numerous studies have assessed how psychological traits relate to ecological behavior (e.g., [33]). The importance of psychographics (for pro-environmental behavior) has remained fairly stable over time [15,34], suggesting their association with different types of climate change-motivated behavior. This study focuses first on three psychographic processes—i.e., attitudes, motivations, and knowledge about climate change (model set 1)—that likely relate to an individual's self-reported extra behavioral engagement to mitigate climate change, as compared to others.

Attitude is defined as the positive or negative feeling that an individual holds about a psychological object [25,35]. The potential targets of attitudes cover a broad spectrum of discriminable aspects of the physical world, such as a physical entity, a person or group of people, an abstract concept or issue, or a behavior [36]. In the environmental literature, attitude is acknowledged as a major proximal factor for ecological intention and behavior [15,33]. Two meta-analyses have confirmed a significant, moderate association between attitude and pro-environmental behavior, with estimated mean correlations of approximately 0.4 [15,34]. Nonetheless, the empirical evidence has been mixed for attitudinal associations with behavior—in line with the widely reported attitude–action gap [1]. Attitude–behavior links across environmental studies are contingent on various factors, including the consideration or omission of intention as a possible mediator of this relationship [15,35], attitude strength [37], attitude certainty and ambivalence [13], situational constraints to pro-environmental action [33,38], and importantly, congruence in the level of specificity/generality (i.e., the scope) and time interval (e.g., simultaneous or lagged) between attitude and behavior measures [33,39]. On the latter point, expectancy-value models recommend that attitudes be measured at the level of behavior—e.g., by measuring attitudes toward pro-environmental action, rather than attitudes toward the environment—, lagged but close in time to the measure of behavior, in order to optimize congruence in attitude–behavior measurement and to reinforce attitude associations with behavior [25,40–41]. Importantly also, the analysis of attitude relations to behavior is likely to be affected by the focus on self-reports or objective outcome measures of actual behavior [41].

It is worth stressing here that investigating the attitude–action gap in climate change-motivated behavior is beyond the scope of this study, given the focus here on self-reported (instead of actual or objectively measured) mitigation behaviors, the simultaneous measurement of attitudes and behavior, and not measuring important factors such as intentions, attitudinal strength, certainty, or ambivalence. Nonetheless, attitude variables should be significantly related to EU citizens' breath and level of self-reported behavioral engagement (i.e., extra vs. common) in response to climate change, owing to: (1) the matching level of specificity/generality in attitudes (towards the threat of climate change and the role of mitigation efforts) and mitigation behavior (in response to climate change)—in a way that measurement congruence exists at the level of climate change as an environmental problem and, at least partly, at the level of behavior; (2) the close time correspondence between (present) attitudes and (past/present) self-reported behaviors; and, (3) the high degree of volitional control over performance of the behavior, which is modeled as an aggregate of specific (and highly voluntary) mitigation actions [40,42].

Respondents with more positive/desirable attitudes, understood and measured here as positive (desirable) feelings that climate change is harmful and that mitigation actions are important, should be more likely to display extra behavioral engagement to mitigate climate change than others [12,14–15]. Likewise, respondents with negative (undesirable) attitudes towards the threat of climate change (i.e., that climate change is not harmful) and the role of mitigation efforts (i.e., that mitigation behavior is not important or ineffective) should not be disposed towards extra mitigation behavior in response to climate change, as compared to others [12,14].

H1a. *Positive/desirable attitudes* (towards the threat of climate change and the role of mitigation efforts) *will be positively associated with extra mitigation behavior.*

H1b. *Negative/undesirable attitudes* (towards the threat of climate change and the role of mitigation efforts) *will be negatively associated with extra mitigation behavior.*

Motivation is usually described as the driving force of behavior [25] or the "reason why a given behavior occurs" [43]. Motives can be both overt and hidden, depending on people's awareness (or not) of their motives for behavior [43]. Researchers also distinguish between primary (general) motives for a whole class of behaviors—e.g., acting in environmentally responsible ways—and selective (domain specific) motives for particular actions, such as recycling or reducing car use [25,43]. The present study examines the role of primary/general environmental motivations (for mitigating climate change) because of the assessment of aggregate, self-reported pro-environmental actions—i.e., people's breath and level of behavioral engagement in climate change mitigation—at a comparable level of generality [42,44].

People may be concerned about environmental issues for several reasons [45–46]. Thus, previous research has explored the different types of value orientations underlying motivations for environmentally significant behavior (see [47]). Owing to the prominence of Schwartz's norm-activation model [48], most studies have differentiated between self-transcendent (altruistic) and self-enhancement (egoistic) values [47–49]. Stern et al. [24] further subdivided altruistic values into social-altruistic and biospheric value orientations. Similarly, Gagnon Thompson and Barton [45] drew a distinction between ecocentric and anthropocentric motives and values. Ecocentric individuals attach importance to the environment for itself and will engage in pro-environmental behavior, even if it involves some sort of sacrifice on their part [45]; this behavior pattern is largely rooted in biospheric values [24]. Anthropocentrics' actions are more deeply grounded in social-altruistic and egoistic values [24,47]; that is, these individuals will engage in pro-environmental behavior, such as climate change mitigation behavior, only if it has positive consequences for mankind and does not diminish their quality of life or wealth [45].

Previous research suggests that, in general, pro-environmental behavior is more closely linked to biospheric values than to social-altruistic or egoistic ones [24,29,45–46]. However, not only ecocentric (biospheric) motivations, but also anthropocentric (social or egoistic) ones can relate to environmentally significant behavior [24,29]; as shown in previous research [8,12], non-altruistic motives—such as financial motivations—often underpin mitigation actions (e.g., energy conservation practices). There is in fact evidence of multiple motivations—i.e., altruistic and egoistic ones—for mitigation behavior [8], suggesting that both self-transcendent (altruistic) and self-enhancement (egoistic) motives may independently relate to, conflict, or converge for extra (i.e., broader and greater) behavioral engagement to mitigate climate change (among the EU public) [7,49–50].

Given the mixed evidence from previous research regarding the link between self-transcendent and self-enhancement motives for and personal mitigation behavior, the issue is posed here as an explorative research question rather than as a hypothesis:

RQ1. *How do self-transcendent (altruistic) and self-enhancement (egoistic) motives relate to extra mitigation behavior?*

Knowledge of environmental issues and problems has been a significant correlate of pro-environmental awareness, moral norms, attitude, intention, and behavior [15,25,33]. In particular, recent meta-analytic evidence suggests the close association of environmental knowledge with pro-environmental behavior [15]. Regardless of the assumed importance of environmental knowledge (and information) as a major, but not sufficient, rational precondition for ecological action [51], its specific role in pro-environmental decision-making has long been debated [15,34]. Information deficit approaches to behavior change—depicted largely as linear-sequential models [environmental knowledge → awareness and concern (environmental attitude) → pro-environmental behavior]—have been criticized as being too simplistic or ineffective [25]. Consistent with this, informational efforts to encourage voluntary, public engagement in climate change mitigation actions—mostly through the provision of scientifically sound information—appear to have had little success [1,9,13].

Multiple factors may work to strengthen knowledge/information associations with climate change mitigation action (and general pro-environmental behavior), or to cause the widely reported knowledge–action gap. First, the distinction between objective and subjective (self-reported) environmental knowledge is important; most past research implicitly assumes or explicitly states that self-assessments—used here to measure subjective knowledge about three climate change issues—serve as valid proxy measures of objective environmental knowledge (e.g., [52–53]), although each of these two knowledge types (objective and subjective) can be differently associated with specific pro-environmental behaviors [54]. Second, besides structural and situational constraints (see, e.g., [7]), the level and type of environmental knowledge and information have been shown to affect the strength of knowledge–behavior links [25,40,55]. Basic information provision is necessary for people to recognize environmental problems—e.g., to overcome the public's lack of knowledge about climate change—and consciously engage in mitigation behavior [1,3,25]. In contrast, an excessive amount of environmental information or very detailed technical data, concerning complex and far-reaching environmental issues such as climate change and global warming, can lead to public confusion and frustration [25,56]. Following the distinction between declarative, procedural, and effectiveness environmental knowledge [55], knowledge about the nature and causes of environmental problems (declarative knowledge) and knowledge about ecological action strategies or "how-to" knowledge (procedural knowledge) have been especially linked to individual pro-environmental behavior [33–34,55] and—importantly here—to mitigation behavior in response to climate change [8,57]. Arguably, extra (i.e., broader and greater) personal engagement in mitigation actions should depend on the public's knowledge about the causes of climate change (and global warming)—e.g., knowledge about human contributions to climate change—and knowledge about available courses of action [8,12]. The role of effectiveness environmental knowledge—which can be approached from two related, but distinct, angles—is somewhat more controversial. First, effectiveness environmental knowledge, when understood as knowledge about the relative ecological consequences (i.e., effectiveness) of different behavioral alternatives [55], has been shown to relate to individual pro-environmental behavior [52] and thus could be expected to be associated with extra mitigation behavior in response to climate change [8]. A second view of effectiveness environmental knowledge, referring to knowledge about the consequences of environmental problems [52]—and examined in this study in relation to self-reported climate change-motivated behavior—, has been criticized for eliciting feelings of frustration, owing to people's increased awareness of the limited impact of their actions on environmental protection [25]; thus, improving knowledge about the serious consequences of climate change—e.g., emphasizing the consequences of not engaging in mitigation behavior [52], or using fear appeals in climate change communication [9]—is not likely to result in an extra (i.e., broader and greater) level of mitigation behavior in response to climate change.

H2. *Both knowledge about the causes (declarative knowledge) and knowledge about the ways of fighting climate change (procedural knowledge) will be positively associated with extra mitigation behavior.*

Socio-demographics and extra mitigation behavior

Socio-demographic variables have generally shown modest or equivocal associations with pro-environmental behavior [16,26,51]. In fact, some authors have argued that socio-demographics (e.g., age, sex, race, or political orientation) will be less associated with environmental concern and ecological behavior over time, due to widespread green concerns across many demographic groups [58–59]—particularly in Western countries [26]. This contention contrasts with recent climate change studies showing that socio-demographic variables can be significant correlates of general or specific types of personal action to mitigate climate change [3,8,12,60].

Part of the between-study variation in socio-demographic associations with ecological behavior can be ascribed to methodological problems and differences across studies [61], analysis of direct vs. indirect relationships [29,62], country-specific factors [26], or the type of behavior studied [63]. An important argument here is that some socio-demographic (background) variables may be proxies for personal capabilities—that is, the knowledge and skills necessary for particular behaviors [16,36]. Thus, demographic variables like age, education, and income should be related to climate change mitigation efforts that depend strongly on personal capabilities [16]—i.e., mitigation actions potentially influenced by objective or subjective constraints [1,64]. This is particularly the case of high-impact mitigation actions—i.e., energy conservation practices—which appear to be significantly associated with an individual's age (see [8]). This study evaluates the association of external socio-demographic factors with self-reported extra mitigation behavior, both at the person level—i.e., gender, age, education, and political ideology—and at the country level—i.e., country values and country wealth.

Gender. The evidence from prior environmental and climate change research—though far from conclusive (see [61])—suggests that women typically report greater environmental concern and involvement in environmentally significant behaviors, relative to men [3,12,26,36,65]. Specifically, women appear more likely than men to engage in private-sphere and regular pro-environmental activities in response to climate change, such as reducing waste (e.g., recycling) [12,66] and conserving energy in the course of daily routines [12,66].

Three theoretical explanations have been offered for gender distinctions in general environmentalism and climate change behavior. The first rationale is that traditional gender roles and socialization patterns largely underlie women's greater environmental involvement [65]. Traditional female socialization has been linked to pro-environmental behavior, owing to women's other and ecocentric value orientations [65] and caretaker role [66]. Women tend to be more attentive to the interconnections between the natural environment and things they value—e.g., other people [24]; as a result, women will be more sensitive than men to the environmental consequences of their actions [51]. The second rationale lies in the fact that, overall, women tend to judge the world as more risky [67], perceive higher levels of environmental risk [68], and thus are likely to take more pro-environmental actions than men [62]. Finally, women appear to perceive fewer (subjective and objective) constraints on personal engagement with climate change mitigation, relative to men [64].

H3. *Female gender will be positively associated with extra mitigation behavior.*

Age. There is much controversy surrounding age relations to environmental knowledge, attitudes, and behavior [26,61]. Most studies have reported negative associations of age with environmental attitudes and concern, indicating that younger people tend to be more concerned with environmental problems such as climate change [2,38,61]. Less clear is the relationship between age and environmental knowledge, with age showing either non-significant or weak negative associations with knowledge about various environmental issues [26]. Also, researchers have studied the linkage between age and pro-environmental behavior with differing results—that is, age has been reported to be negatively, positively, or non-significantly related to environmentally-significant behavior (see [26]). This mixed evidence is also reflected in the study of personal mitigation behavior in response to climate change [3,8]. In fact, asymmetric age associations with different types of action have been observed [8,12]; to illustrate, as regards energy conservation actions, older individuals appear more likely to engage in less painful or simple energy conservation activities (e.g., buying energy-saving light bulbs or turning off unused lights), but less likely than younger people to engage in more difficult transport-related energy conservation [8,12].

Consistent with the common negative links between age and environmental concern (i.e., attitudinal measures), Diamantopoulos et al. [26] argued that age tends to negatively correlate with *intended* ecological behavior—i.e., with intentional measures of behavior [12]; conversely, positive age–behavior linkages are typically found in studies using measures of *current* pro-environmental behavior [8,12,26]. Environmental attitudes and intentions may not translate into climate change mitigation behavior in younger people, partly because of their lack of necessary resources (e.g., financial means) for environmentally significant actions [26,64]. Life-cycle and cohort effects may account as well for age differences in pro-environmental decisions [38,61]. The life-cycle age effect points to a non-linear (inverted U-shaped) relationship between age and climate change concern; that is, highest levels of environmental concern during middle-age [38,61]. In addition, researchers generally agree on a cohort effect, resulting from greater exposure of birth cohorts from the 1950s (or 1960s) to public discussion and concern about environmental problems—such as climate change and global warming—, compared to previous cohorts [38,63]. Finally, "differences in time horizons in relation to climate change" [13] would be suggestive of negative age associations with personal mitigation behavior.

Taking together the available evidence—particularly the life-cycle and cohort rationales—, middle-aged European citizens are more likely to report extra mitigation behavior in response to climate change, compared to their younger and older counterparts.

H4. *There will be a curvilinear (inverted U-shaped) association between age and extra mitigation behavior.*

Education. Several studies have examined the potential role of education level as an indirect correlate—e.g., through environmental knowledge, attitudes, or concern—and direct correlate of pro-environmental behavior [26,69]. With few exceptions (see [12,69]), findings have been fairly consistent across studies: better-educated individuals tend to be more knowledgeable, concerned, and involved in pro-environmental activities [26,61,63]—including climate change mitigation actions [2–3,12,57]. Much like age and income, educational attainment may be a good proxy for personal capabilities involved in environmentally significant behavior [16,62]. In this regard, people with more years of formal education have shown greater concern and behavioral commitment to environmental protection [26]; such individuals have access to more sources and types of information [70], and can be expected to understand highly-complex environmental issues, such as climate change, more fully than less educated citizens [26,57,61]. As a result, the following hypothesis is proposed:

H5. *Education level will be positively associated with extra mitigation behavior.*

Political ideology has often been employed, along with other psychological and demographic variables, to gain a deeper understanding of individual green behaviors (e.g., [8,47,69]). Studies including political ideology have reported very consistent results, on a left to right (liberalism–conservatism) continuum, people with left-of-center political views tend to show higher levels of concern, verbal commitment, attitudes, and environmentally significant behavior, compared to conservatives [22,51,63,69]; moreover, conservative political values appear to be strongly associated with skepticism about climate change [8]. Only a few studies have not found significant associations of political orientation with pro-environmental behavior [60,71] or climate change mitigation behavior [12]. Thus, political ideology is considered one of the most robust and stable socio-demographic correlates of environmental concern and behavior [51,61].

H6. *Left-of-center political ideology will be positively associated with extra mitigation behavior.*

Country variations. In accordance with calls for more international research [25], cross-cultural analyses of pro-environmental and climate change-motivated behavior have garnered increased attention over the past decade [21,31,66,72–79]. The far-reaching consequences of environmental degradation, coupled with increased (societal and governmental) environmental activism in wealthy and developing countries [80–81], led some authors to conclude the emergence of global environmentalism [30,79,82]. However, the *globalization hypothesis* has been disputed by several authors (e.g., [31,83]); in this respect, substantial empirical evidence has accumulated in support of cross-national variations in public environmental concern and protection [46,66,74–75,81]—including climate change mitigation efforts [78]—, both within Europe and across continents.

Post-materialism hypothesis. It is widely believed that international variations in pro-environmental attitudes and behavior are a consequence of different values and primary goals held across cultures [46,75]. Particularly, Inglehart's theory of post-materialism [80] provides a prevalent value priorities approach to understanding country differences in public environmental concern (see [81]). Post-materialist theory posits that environmental concern emerges only once basic individual needs are fulfilled [80]. According to this view, people from countries with a predominant post-materialist orientation tend to be more concerned about the environment and climate change and, consequently, can be expected to make and report an extra commitment to pro-environmental and climate change mitigation behaviors [31,80], compared to people from non-post-materialist countries. Although challenged on important points [30,79,82], the post-materialism hypothesis has received strong support from recent cross-cultural environmental studies [21,31,73,81,83].

H7a. *Post-materialist EU-27 countries will be positively associated with extra mitigation behavior.*

Wealth hypothesis. Cross-national variations in pro-environmental attitudes can also be explained by national differences in wealth [76,81,83]. Environmental concern may be an indirect consequence of wealth—i.e., mediated through post-materialist values, as asserted by Inglehart [80]—; in contrast, the prosperity/affluence hypothesis posits a direct link from wealth to environmental concern [31,76,83]. Regardless of small differences between the direct and indirect influence paths of wealth [31], sufficient evidence exists to suggest that citizens of wealthier nations tend to give higher priority to global environmental protection goals, compared to individuals in poorer nations [76,80–81,83]. Strong correlations have been obtained between wealth (GDP per capita) and priority/global indexes of environmental concern in the works of Franzen and colleagues—i.e., correlations of approximately 0.8, accounting for more than 50% of the cross-national variance in environmental concern (see [81]).The opposite relationship—i.e., negative correlations—has also been observed between wealth and measures of local environmental concern [30,83]. In poorer nations, lower environmental quality and pressing ecological problems are more likely to be sources of public concern and support for local environmental protection than in rich countries [31,81,83]. In the present study, given the global scope of climate change, respondents from wealthier EU-27 countries are expected to report more pro-environmental attitudes and extra behavioral engagement in climate change mitigation, compared to citizens of less-wealthy nations.

H7b. *Wealthier EU-27 countries will be positively associated with extra mitigation behavior.*

Unobserved heterogeneity in psycho-social associations

The implicit assumption in most environmental and climate change studies that data are collected from a single homogeneous population is, in general, unrealistic [84]. Individuals are often heterogeneous with regard to environmental psychographics—i.e., people hold different views and have different information levels of environmental problems, such as climate change—and relevant socio-demographic characteristics that can positively (or negatively) relate to pro-environmental action [85]. Such heterogeneity in public conceptualizations and preferences for mitigating climate change impacts is acknowledged as a central question in the climate change research literature [1–2,9,60]. Imposing the assumption of homogeneity—when, in fact, there is substantial (psychographic or socio-demographic) heterogeneity within the population—is likely to produce misleading inferences and biased results [84,86]. In particular, if heterogeneity across individuals is present but ignored in regression-based studies, researchers run the risk of obtaining inconsistent model parameters and probability estimates [86].

It is important to clarify that the term heterogeneity—as used in this article—refers to both *distinct subpopulations* and *variation across individuals* [87]. In general, two forms of heterogeneity are present in data sampled from a heterogeneous population: observed and unobserved to an analyst [86–87]. Observed heterogeneity has been frequently dealt with—in the study of pro-environmental and climate change behavior—by the use of observed socio-economic variables (e.g., demographics like gender) that define a priori subgroups [85,87–88]. However, few studies have incorporated—or attempted to uncover—unobserved heterogeneity in models of individual pro-environmental behavior or (even less so) of climate change mitigation behavior [85–86,88]. With long tradition in the marketing and management literature, unobserved heterogeneity is commonly given precedence over observed heterogeneity in uncovering subpopulations (i.e., for segmentation purposes) [84,87], probably for two main reasons. First, in uncovering unobserved heterogeneity researchers "let the data speak for itself" [87]; that is, subpopulations are unobserved by the analyst (not predefined) and have to be inferred from the data [87]. Second, unobserved heterogeneity is arguably the preferred approach for uncovering subpopulations on psychographic constructs, such as environmental attitudes and motivations [88].

Put simply, the issue of unobserved heterogeneity revolves around uncovering subgroups or segments with distinctive path model estimates [84]. In this study, potential unobserved heterogeneity is accounted for in both psychographic and socio-demographic correlates of extra personal mitigation behavior in response to climate change. By looking at the data through the lens of segmentation (i.e., response-based segmentation) [89], the findings will clarify if psycho-social associations with people's breath and level of self-reported behavioral engagement (i.e., extra vs. common) in climate change mitigation is affected by heterogeneity, or the extent to which heterogeneity exists [84].

Methods

Data source

The empirical analyses are performed on the cross-national dataset "Eurobarometer 69.2—Europeans' attitudes towards climate change". A primary goal of this EU-wide survey was to investigate European citizens' climate change-related attitudes and behavior. Data were collected between March 25th and May 4th 2008 by TNS Opinion & Social, at the request of the European Commission, Directorate-General for Communication, Research and Political Analysis Unit. The Eurobarometer survey covers the population—aged 15 and over—of the 27 EU member states, three candidate countries (Croatia, Turkey, and the Former Yugoslav Republic of Macedonia), and the Turkish Cypriot Community. In each country, a stratified, multistage probability sampling design was used to guarantee the reliability of national and European estimates. A total of 30,170 individuals were interviewed face-to-face in their homes, and in the appropriate national language. The questionnaire addressed European citizens' self-reported attitudes, motivations, level of knowledge, and personal enactment of specific mitigation activities in response to climate change; other relevant measures were available, including materialist/post-materialist values and socio-demographic indicators, such as gender, age, education, political ideology, and country. Access to the Eurobarometer data was provided by the GESIS Data Archive for the Social Sciences (Cologne, Germany). A detailed description of the Eurobarometer dataset used here is available as electronic supplementary information (Appendix S1).

Measurement items

Indicators of mixed scale types (i.e., categorical and continuous items) were used to measure the outcome and independent variables of the study—detailed in Tables 1 to 4.

Measures of personal mitigation behavior. Respondents already engaged in some form of climate change-motivated activity were asked to report, on a binary nominal scale (1 = *yes*; 0 = *no*), whether they had undertaken each of 11 types of actions aimed at fighting climate change (see Table 2 for the list of behaviors); these activities entail different levels of mitigation difficulty, impact, and frequency in the behavioral domains of domestic energy/water conservation, waste reduction, eco-friendly transportation, and eco-shopping. An important behavioral domain not covered in this study—and reflective of high-impact mitigation behavior—is that of public sphere (public/political) environmental activism and citizenship. The focal outcome of interest, a categorical aggregate score of people's behavioral engagement (i.e., extra vs. common) in climate change mitigation, was created through segmentation analysis on all self-reported mitigation activities.

Measures of attitudes, motivations, and knowledge (see Table 3). Environmental attitudes (five items) and self-reported knowledge about climate change issues (three items) were both rated on four-point scales from 1 to 4; in the attitude measures, 1 indicates *totally* disagree and 4 indicates *totally agree*; in the self-reported knowledge items, 1 denotes *not at all informed* and 4 denotes *very well informed*. Ecological motivations were measured through five possible reasons for fighting climate change; all motivation items were rated on a binary scale (1 = *yes*; 0 = *no*).

Individual-level demographics (see Tables 1 and 4). Gender, measured as biological sex, was coded with 1 designating *male* and 2 designating *female*. Age, initially measured as a continuous variable, was divided into six age categories (coded from 1 to 6): *15–24, 25–34, 35–44, 45–54, 55–64*, and *65 years and over*. Education was measured by the age at which respondents stopped full-time education, and recoded into a five-category variable ranging from 0 to 4: *no full-time education*, *up to 15 years of age*, *16–19 years old*, *20 years and over*, and *still studying*. Political ideology was assessed through respondents' self-placement on a 10-point, left-to-right continuum; scores 1–4 were combined into 1 = *left/liberal*; categories 5–6 into 2 = *moderate*; and scores 7–10 into 3 = *right/conservative*.

Country-level variables (see Table 4). Participants' country is a nominal variable, with categories ranging from 1 to 33. Post-materialism, the first hypothesized explanation for country variations in citizens' self-reported behavioral engagement (i.e., extra vs. common) in climate change mitigation, was measured at the individual level through Inglehart's four-item materialist/post-materialist value battery [80]; latent class segmentation performed on these value priorities, and profiled by country, led to the classification of European countries into three materialist vs. post-materialist groups; the grouping variable was coded 1 = *materialist countries*, 2 = *countries with mixed values*, and 3 = *post-materialist*

Table 1. Demographic profile of respondents.

Demographic variables	**Total sample (n = 30,170)**	*Subsample 1*: **environmentally active citizens (n = 17,233)**	*Subsample 2*: **environmentally inactive citizens (n = 12,937)**	χ^2 **tests:** *subsample 1 vs. 2*			
				χ^2	**d.f.**	*p*	**Phi/Cramer's V**
Gender				7.72	1	0.005	0.016
Male	45.6%	44.9%	46.5%				
Female	54.4%	55.1%	53.5%				
Age				[illegible]50.39	5	<0.001	0.071
15–24	12.6%	10.8%	15.1%				
25–34	15.1%	15.0%	15.2%				
35–44	17.2%	18.0%	16.2%				
45–54	17.3%	17.9%	16.5%				
55–64	16.9%	17.7%	15.8%				
65+ years	20.9%	20.7%	21.2%				
Education (*age at which respondents stopped full-time education*)				528.37	5	<0.001	0.132
No full-time education	0.5%	0.2%	0.8%				
15– years	2[illegible].9%	19.7%	24.8%				
16–19	4[illegible].7%	41.3%	42.3%				
20+ years	25.3%	29.5%	19.7%				
Still studying	8.4%	7.7%	9.3%				
Refusal/DK	2.2%	1.6%	3.1%				
Political ideology				659.16	3	<0.001	0.148
Left/liberal	24.1%	27.[illegible]%	20.3%				
Moderate	30.2%	3[illegible].7%	28.3%				
Right/conservative	24.6%	25.3%	23.8%				
Refusal/DK	21.0%	16.0%	27.7%				

Table 2. Behavioral characterization of "extra vs. common" mitigation behavior.

Relative sizes/Mitigation actions		Segment membership probabilities [a]	
		Segment 1: ***"common" mitigation behavior***	**Segment 2:** ***"extra" mitigation behavior***
Relative size of segments		0.7706	0.2294
qe6	*Which of the following actions aimed at fighting climate change have you personally taken?*		
qe6.1	You have purchased a car that consumes less fuel, or is more environmentally friendly	0.1345	0.2973
qe6.2	You are reducing the use of your car, for example by car-sharing or using your car more efficiently	0.1651	0.4316
qe6.3	You have chosen an environmentally friendly way of transportation (by foot, bicycle, public transport)	0.2654	0.4769
qe6.4	You are reducing your consumption of energy at home (for example by turning down air conditioning or heating, not leaving appliances on stand-by buying energy efficient products such as low-energy bulbs or appliances)	0.5842	0.9254
qe6.5	You are reducing your consumption of water at home (for example not leaving water running when washing the dishes, etc.)	0.5116	0.7926
qe6.6	Where possible you avoid taking short-haul flights	0.0570	0.2958
qe6.7	You have switched to an energy supplier or tariff supplying a greater share of energy from renewable sources than your previous one	0.0537	0.1558
qe6.8	You are separating most of your waste for recycling	0.6284	0.9305
qe6.9	You are reducing the consumption of disposable items (for example plastic bags, certain kind of packaging, etc.)	0.2618	0.8318
qe6.10	You buy seasonal and local products to avoid products that come from far away, and thus contribute to CO_2 emissions (because of the transport)	0.1500	0.6398
qe6.11	You have installed equipment in your own home that generates renewable energy (for example, a wind turbine, solar panels)	0.0411	0.1051

[a]Conditional (marginal) probabilities clarifying how segment-membership relates to each climate change mitigation action.

countries, which closely resembles the three-class classification used in Inglehart's short post-materialism index [80,90]. Eurostat data on GDP per capita (year 2008) was used as a proxy for EU countries' wealth; based on terciles of GDP per capita, countries were divided into three wealth groups (coded from 1 to 3): countries with *low*, *intermediate*, and *high* wealth levels.

Statistical methodology

This paper applies latent class models (LC cluster and regression), as implemented in the Latent Gold v4.5 software, to synthesize the outcome variable and test the hypotheses linking psychographic and socio-demographic variables to the breath and level of personal mitigation behavior (i.e., extra vs. common) in response to climate change. Latent class analysis provides a powerful probabilistic approach for capturing *unobserved* heterogeneity in survey responses, and is especially useful for modeling (dependent and independent) categorical variables with varying numbers of categories [91–92], as in the present study. Alternative methods such as multi-group SEM allow researchers to account for *observed* heterogeneity—instead of unobserved heterogeneity—, where both the source of variation and subpopulations are known and defined a priori by the analyst. A more detailed description of statistical analysis and procedures is available as electronic supplementary information (Appendix S2).

Results

Descriptive profile of respondents

The sample is well-balanced in terms of gender, age, education, and political ideology (see Table 1 for results); yet, there was greater participation of female (54.4%), middle-aged (mean age = 47.6 years), and moderately educated individuals (41.7%), with a center political orientation (30.2%). All participants were asked to report whether they had "personally taken actions aimed at helping to fight climate change"; more than half of the sample (57.1%; n = 17,233) *totally agreed* or *tended to agree* with this statement. Demographically, the subsamples of environmentally active and inactive EU citizens differed significantly (based on χ^2 tests), but weakly (based on association measures such as Phi and Cramer's V), in gender, age, education, and political ideology. As expected, the subsample of EU citizens already engaged in some form of climate change-motivated activity is an older, better-educated, leftist/liberal, female group, compared to environmentally inactive respondents. Subsequent analyses focused only on environmentally (i.e., climate change) active EU citizens in the year 2008 (subsample 1 in Table 1). The decision to restrict the analyses to the subsample of climate change-active citizens is consistent with this study's investigation of the correlates of "extra vs. common" personal engagement in mitigation behavior—i.e., why some people go beyond what most other environmentally active people do to mitigate climate change through extra personal action. However, it is worth raising a cautionary note about

Table 3. Psychographic associations with extra mitigation behavior (model set 1).

Independent variables		Parameter estimates (*z*-values) [a]				Wald	*p*-value	Wald (=)	*p*-value
		Class 1		**Class 2**					
Model 1a: **Environmental attitudes**									
qe5…	*Agreement with the following statements:*								
qe5.1	Climate change is an unstoppable process, we cannot do anything about it	−0.04	(−2.56)	−0.04	(−2.56)	6.5	0.011	*class-independent*	
qe5.2	The seriousness of climate change has been exaggerated	n.s.		−0.30	(−2.52)	6.4	0.012	-	-
qe5.3	Emission of CO_2 (carbon dioxide) has only a marginal impact on climate change	−0.10	(−5.51)	n.s.		30.4	3.5e-08	-	-
qe5.4	Fighting climate change can have a positive impact on the European economy	n.s.		0.26	(1.80)	3.3	*0.072*	-	-
qe5.5	Alternative fuels, such as 'bio fuels', should be used to reduce GHG emissions	0.19	(2.55)	−1.07	(−3.24)	71.2	3.5e-16	20.5	6.1e-06
Model 1b: **Ecological motivations**									
qe7…	*Reasons for taking mitigation actions:*								
qe7.1	You think that if everybody changed their behavior, it will have a real impact on CC	0.14	(12.3)			152.0	6.4e-35		
qe7.2	You think that it is your duty as a citizen to protect the environment	0.08	(7.36)			54.2	1.8e-13		
qe7.3	You are very concerned about the world that you will leave for the future and young generations	0.18	(16.9)			286.6	2.8e-64		
qe7.4	You think that taking these actions will save your money	0.09	(7.78)			60.5	7.5e-15		
qe7.5	You have been directly exposed to the consequences	n.s.				0.9	*0.34*		
Model 1c: **Environmental knowledge**									
qe3…	*Well informed about:*								
qe3.1	The different causes of climate change	−0.20	(−1.36)	0.67	(2.35)	9.8	0.007	9.5	0.002
qe3.2	The different consequences of climate change	0.40	(3.06)	−0.40	(−1.33)	17.4	0.000	8.4	0.004
qe3.3	Ways in which we can fight climate change	−0.16	(−1.28)	0.41	(2.27)	7.2	0.027	7.2	0.007

[a]Parameter estimates represent class-specific associations, of each independent variable, with extra mitigation behavior; *z*-values in brackets.

Table 4. Socio-demographic associations with extra mitigation behavior (model set 2).

Independent variables	*Model 2a*				*Model 2b*			
	Parameter estimates (z-values) [a]		**Wald**	***p*-value**	**Parameter estimates (z-values) [a]**		**Wald**	***p*-value**
Individual-level demographics:								
Gender			57.2	3.9e-14			61.1	5.3e-15
Male	−0.083	(−7.566)			−0.086	(−7.819)		
Female	0.083	(7.566)			0.086	(7.819)		
Age			82.4	2.6e-16			87.3	2.5e-17
15–24	−0.370	(−6.924)			−0.380	(−7.132)		
25–34	−0.073	(−2.635)			−0.084	(−3.013)		
35–44	0.095	(3.848)			0.097	(3.906)		
45–54	0.163	(6.688)			0.157	(6.427)		
55–64	0.120	(4.804)			0.132	(5.276)		
65+ years	0.066	(2.595)			0.078	(3.079)		
Education			147.7	8.2e-32			157.8	5.6e-34
15– years	−0.285	(−9.638)			−0.300	(−10.097)		
16–19	−0.038	(−1.682)			−0.051	(−2.276)		
20+ years	0.125	(5.469)			0.119	(5.182)		
Still studying	0.198	(3.827)			0.232	(4.490)		
Political ideology			19.7	5.4e-05			20.3	3.9e-05
Left/liberal	0.058	(3.774)			0.064	(4.138)		
Moderate	0.008	(0.539)			−0.002	(−0.108)		
Right/conservative	−0.066	(−4.030)			−0.062	(−3.781)		
Country-level variables:								
Materialism/post-materialism (country groups)			542.8	1.3e-118				
Materialist countries	−0.395	(−21.925)						
Countries with mixed values	0.102	(6.610)						
Post-materialist countries	0.294	(19.231)						
Wealth (country groups)							684.6	2.1e-149
Low					−0.296	(−12.530)		
Intermediate					−0.125	(−6.311)		
High					0.421	(26.050)		

Note: *Models 2a and 2b* separately include each of the two country-level variables, along with the four individual-level demographics.
[a]Parameter estimates represent category-specific associations, of each independent variable, with extra mitigation behavior; z-values in brackets.

generalizing the main study's findings to environmentally inactive people; that is, what drives extra mitigation behavior may differ greatly between environmentally active vs. inactive population segments.

Capturing extra mitigation behavior

As suggested in previous sections, LC cluster analysis was conducted on a set of 11 self-reported climate change mitigation actions (see Table 2), so as to capture and define the levels of the focal outcome variable of the study; thus, respondents were classified as showing different breath and level of personal engagement in mitigation behavior in response to climate change. The values of three segment retention criteria (BIC, AIC3, CAIC) and their percent reductions were examined for each of 10 potential cluster-solutions. These reached a minimum value for either 8 or 9 clusters; however, cluster-solutions with more than two segments were not regarded as appropriate, owing to practically insignificant percent reductions in all segment retention criteria (less than 1%), and excessive classification errors (over 10%) for the use of segment-membership as the outcome variable in subsequent regression analyses (see supplementary information—Appendix S3). The 2-cluster solution was also preferred over more complex models in terms of interpretability of the segment profiles, thus providing a higher theoretical and practical value. In sum, the results yielded an optimal solution of two segments (i.e., two differentiated behavior patterns) of EU citizens currently engaged in some form of climate change-motivated activity. This finding confirmed the distinction of two levels of "extra vs. common" personal engagement in mitigation behaviors.

Latent class sizes and levels of engagement in climate change mitigation behavior were substantially unbalanced between the two differentiated segments; *segment 1* (extra mitigation behavior) and *segment 2* (common mitigation behavior) respectively accounted for 77.1% and 22.9% of environmentally active respondents. The observations were reweighted to correct for potential biases in the latent class (segment) sizes (see Appendix S1 for details); nonetheless, the comparison of reweighted and unweighted LC cluster analyses revealed very similar results. As detailed in Table 2, respondents in segment 2 reported greater participation in mitigation efforts, compared to EU citizens showing more common types and levels of mitigation behavior (and classified into segment 1), in areas such as separating garbage for recycling (93.1% vs. 62.8%), and reducing the consumption of energy (92.5% vs. 58.4%) and water (79.3% vs. 51.2%) at home. Segment profiles differed even more in anti-shopping actions; a large majority of respondents in segment 2 claimed to be reducing their consumption of disposable items (83.2%), and avoiding products that come from far away-places (64.0%), compared to much lower shares in segment 1. The less common pro-environmental behaviors, both in the segments of people reporting extra and common mitigation behavior, refer to installing renewable energy systems in the household, switching to a greener energy supplier or tariff, and (where possible) avoiding taking short-haul flights. Overall, the profile of segment 2—and its comparison with that of segment 1—is largely in accordance with the definition of extra mitigation behavior in response to climate change given here: broader and greater engagement levels across all the specific mitigation actions and behavioral domains being examined (i.e., domestic energy/water conservation, waste reduction, eco-friendly transportation, and eco-shopping), compared to people showing more common mitigation behavior to address climate change.

Regression results

Following LC cluster (segmentation) analysis, respondents' segment membership—i.e., a two-level categorical measure differentiating extra from common behavioral engagement in climate change mitigation—was treated as the outcome variable in two sets of LC regression models. Respectively, model sets 1 and 2 examine how environmental psychographics and socio-demographics relate to the breath and level of self-reported personal mitigation behavior in response to climate change, while exploring the extent of unobserved heterogeneity.

Psychographic correlates (model set 1). The psychographic variables assessed in *model set 1* include respondents' self-reported knowledge (about the causes, ways of fighting, and consequences of climate change), positive/desirable and negative/undesirable attitudes (towards the threat of climate change and the role of mitigation efforts), and altruistic and egoistic ecological motivations (for mitigating climate change). Knowledge, attitude, and motivation associations were first examined in separate models, so as to ascertain the level of heterogeneity (if any) in the relationship of each subset of psychographic variables to the breath and level of self-reported personal mitigation behavior. Within the range of 1 to 3 latent classes, *2-class* solutions were deemed optimal in "attitude" and "knowledge" models (models 1a and 1c, respectively)—thus suggesting that unobserved heterogeneity exists in how knowledge and attitude variables relate to extra mitigation behavior, across two subgroups of environmentally active EU citizens; conversely, the *1-class* solution of homogeneity was preferable in the "motivational" model (model 1b). For a detailed description of the criteria supporting the model comparison and selection process, please see supplementary information (Appendix S3).

As an important matter of clarification at this point, the following should be noted: (1) *1-class* models entail the existence of a similar (perfectly homogeneous) pattern of association between the examined correlates with extra mitigation behavior for all environmentally active respondents; (2) *2-class* models imply that the correlates show a different pattern of association with extra mitigation behavior between two subgroups of environmentally active respondents.

All variables analyzed in the "attitude model" (model 1a) had significant links to the breath and level of self-reported personal mitigation behavior (see Table 3). For the most part, attitude associations were class dependent—that is, four of the five attitude variables under investigation (items qe5.2 to qe5.5) behaved differently across two subgroups of environmentally active respondents. In the largest subgroup (class 1), as expected, positive/desirable attitude items (qe5.5) linked positively, and negative/undesirable attitude items (qe5.1 and qe5.3) negatively, to extra mitigation behavior. In the smallest subgroup (class 2), negative/undesirable attitude items (qe5.1 and qe5.2) were also, as hypothesized, negatively linked to extra mitigation behavior; the findings were mixed for the positive/desirable attitude items in class 2—that is, showing positive and negative associations (respectively, items qe5.4 and qe5.5) with extra mitigation behavior. In summary, the trends observed here for attitude items—despite heterogeneity across two groups environmentally active EU citizens—provide substantial support for hypotheses H1a and H1b positing that positive/desirable attitudes (towards the threat of climate change and the role of mitigation efforts) would be associated positively, and negative/undesirable attitudes negatively, with some people's extra mitigation behavior in response to climate change, as compared to others.

In the 1-*class* "motivational model" (model 1b), almost all tested variables were significantly associated with the breath and level of self-reported personal mitigation behavior—except for one motivation item: qe7.5, *you have been directly exposed to the consequences of climate change* (Wald $= 0.92$; $p = 0.34$)—a finding that relates to the literature on personal experience of and behavioral responses to climate change (see the Discussion section). The other four altruistic and egoistic motivational variables (items qe7.1 to qe7.4) were significant positive correlates of extra mitigation behavior. The findings stressed the greater importance of respondents' social-altruistic motivations in understanding why some people engage in extra mitigation behavior to address climate change, as compared to others; two such social-altruistic items (qe7.3 and qe7.1) ranked first and second, respectively, in order of statistical significance (see Table 3). These results shed light on RQ1, by showing the existence of both self-transcendent (altruistic) and self-enhancement (egoistic) motives for some people's extra mitigation behavior, as compared to others, and that self-transcendent (altruistic) motives can be expected to show greater positive (desirable) associations than egoistic ones.

The three knowledge variables (about climate change issues) tested in the "knowledge model" (model 1c) were significant, class-dependent correlates of the breath and level of self-reported personal mitigation behavior (see Table 3). Interestingly, the associations of environmental knowledge items were all in the opposite direction across two classes of environmentally active respondents. In the largest subgroup (class 1), only item qe3.2, *knowledge about the consequences of climate change*, was positively associated with extra behavioral engagement to mitigate climate change; conversely, better *knowledge about the causes* (item qe3.1) and *ways of fighting climate change* (item qe3.3) was linked to extra mitigation behavior in the smallest segment (class 2). These findings are considered to partially support H2, in that both knowledge about the causes and ways of fighting climate change positively relate to extra mitigation behavior in response to climate change, but only in the comparatively small segment of more environmentally engaged EU citizens. Contrary to the authors' expectations, *knowledge about the consequences* of climate change may relate to extra mitigation efforts (in response to climate change) for the majority of environmentally active EU citizens.

Overall, the findings of the three related psychographic models (model set 1) provide evidence of unobserved heterogeneity in attitude and knowledge correlates—across two subgroups of environmentally active citizens—, and homogeneity in motivational correlates of people's breath and level of behavioral engagement (i.e., extra vs. common) in mitigation behavior to address climate change. Because of different number of *latent classes* in the "motivational model" (1-*class* model)—compared to the "attitude model" and the "knowledge model" (2-*class* models), these three types of psychographic variables were not entered into a single model simultaneously.

Socio-demographic correlates (model set 2). Preliminary inspection of separate socio-demographic associations with the breath and level of personal mitigation behavior suggested the consideration of *1-class* models of homogeneity as optimal solutions across the variables under investigation. Thus—unlike in the regression model set 1—, individual and country-level socio-demographics could be entered jointly into LC regression models. Two different specifications of *model set 2* (models 2a and 2b) were then tested to examine the relationships of four individual-level demographics (gender, age, education, and political ideology) and two country-level variables (materialism/post-materialism and wealth) with EU citizens' extra (vs. common) behavioral engagement in climate change mitigation. In models 2a and 2b, each country-level variable was separately investigated, along with individual-level demographics. Within the range of 1 to 3 latent classes, *1-class* solutions were deemed optimal in the socio-demographic models 2a and 2b, thus revealing homogeneity in socio-demographic associations with extra mitigation behavior (see Appendix S3 for details).

All tested socio-demographic variables were significantly associated with the breath and level (extra vs. common) of climate change mitigation behavior in the final, 1-*class* versions of models 2a and 2b (see Table 4). Country-level variables were most significantly associated with extra mitigation behavior, with wealth ranking first and materialism/post-materialism second in statistical significance. The post-materialism hypothesis for explaining country variations in EU citizens' extra (vs. common) mitigation behavior was tested and strongly supported in model 2a; accordingly, the post-materialist country group showed the strongest positive association with extra personal engagement in mitigation behavior. Further, the results of model 2b confirmed the importance of the wealth hypothesis; only the wealthiest country group showed a positive association with extra behavioral engagement to mitigate climate change. In summary, the previous findings yielded full support for the two hypotheses involving country-level variables (support for H7a and H7b).

The parameter estimates for individual-level demographics were almost identical in the alternative model specifications 2a and 2b (all significant at $p<0.001$). As expected, female gender was positively, although somewhat weakly, related to extra mitigation behavior to address climate change (support for H3). As suggested by the visual inspection of findings in Table 4, a curvilinear-like or "plateau" relationship appeared to exist between age and the breath and level of personal mitigation behavior, with the 45–56 age group showing the highest association with extra mitigation action. A complementary test of non-linearity was performed to clarify the precise form of age associations; for this purpose, the age variable—with linear ranges and of equal span—was converted into a quasi-interval scale and the authors tested whether a linear, curvilinear, or quadrilateral relationship was the best way to describe age relations to extra mitigation behavior. As expected, and detailed in the supplementary information (Appendix S4), a curvilinear function provided the optimal fit to the age data. Thus, there was considerable support for hypothesis H4 of a curvilinear (inverted U-shaped) relationship between age and EU citizens' extra behavioral engagement in mitigation behavior—mostly grounded in the interplay of cohort and life-cycle age effects on pro-environmental (climate change) concern and behavior. Regarding education, a positive association was identified between the age at which respondents stopped full-time education and extra mitigation behavior—which is fully supportive of H5. Finally, the results supported the hypothesis (H6) that people with a leftist/liberal political orientation would display extra (i.e., broader and greater) behavioral engagement in climate change mitigation, compared to right-wing/conservative environmentally active respondents.

Discussion

A fundamental shift is needed towards broader and greater levels of behavioral engagement in the population that provide further incremental benefits in addressing climate change. The cross-national research reported here aimed to enhance understanding of what makes some people make an extra commitment (i.e., do "more") to mitigate climate change through personal action, as compared to others. The authors assessed the role of psychographics, individual and country-level socio-demographics, and the extent of (unobserved) heterogeneity in an individual's extra behavioral engagement to address climate change.

In line with past environmental research (e.g., [93–94]), segmentation analysis revealed two heterogeneous (and differentiated) action patterns in response to climate change, among environmentally active EU citizens. These findings validated the a priori, intuitive two-level distinction of "extra vs. common" mitigation behavior in the focal variable of interest. As expected, engaging extra to mitigate climate change through personal behavior, compared to more common (or typical) mitigation behavior, was reflected in broader and greater self-reported engagement in all specific mitigation endeavors—entailing different levels of difficulty, impact, and frequency—and behavioral domains being examined [8,14].

The definition and measurement in aggregate of extra mitigation behavior allowed the analysis of factors that could influence broader and greater levels of mitigation behavior in a variety of settings. Overall, the findings reinforce earlier evidence that environmental psychographics (i.e., attitudes, motivations, and knowledge about climate change) are more associated with personal mitigation efforts in response to climate change than socio-demographics [12].

A central question in this study was: Is there heterogeneity in psycho-social correlates of extra mitigation behavior? The findings revealed that unobserved heterogeneity significantly affects how attitude and knowledge variables relate to extra (vs. common) personal behavior to mitigate climate change, but does not affect the links with motivations or socio-demographics. These results warn of the risk of ignoring the potential presence of unobserved heterogeneity in the analysis of attitudinal and knowledge associations with climate change-motivated behavior—i.e., researchers run the risk of obtaining biased or inconsistent results.

Positive/desirable and negative/undesirable attitude variables were significantly associated with extra mitigation behavior. Unlike most expectancy-value models [15,34,40], attitudes were not measured here only at the level of behavior; however, the congruent level of specificity/generality in attitudes (towards the impact of climate change and mitigation efforts) and self-reported mitigation behavior (in response to climate change), the close time correspondence between attitudes and behaviors, coupled with the assessment of aggregate mitigation behavior—i.e., offsetting differences in behavioral control across 11 types of mitigation actions—, is likely to have reinforced the significance attitude relations to behavior [33,39–40]. Overall, the findings emphasize the importance of building citizens' positive/desirable attitudes toward climate change issues (i.e., feelings that climate change is harmful and that mitigation actions are important), and reducing negative/undesirable ones (i.e., feelings that climate change is not harmful and that mitigation behavior is not important or ineffective)—but differently for two subgroups of environmentally active EU citizens—, so as to effectively promote broader and greater levels of engagement in mitigation action in response to climate change.

An important finding concerned the role of self-reported knowledge about climate change issues. Respondents' level of knowledge about the causes and ways of fighting climate change—on the one side—, and knowledge about the consequences of climate change—on the other side—were all significantly, but inversely, related to extra mitigation behavior, across two (environmentally active) respondent subgroups. In contrast with most previous studies [8,33,55], informing the public about the consequences of climate change appears to be more useful, than informing about the causes and ways of fighting climate change, in promoting extra mitigation behavior for the majority of environmentally active citizens—i.e., people that currently display limited action levels in response to climate change. Thus, not all types of climate change messages and information can be assumed to be equally effective in promoting extra mitigation efforts in the population.

In line with previous studies (e.g., [50]), both altruistic and egoistic motivations were positively associated with extra (vs. common) personal behavior to mitigate climate change for all environmentally active respondents. Thus, interventions aimed at encouraging citizens to undertake more ambitious mitigation efforts should appeal to both altruistic and egoistic ecological motives and values. Nonetheless, as suggested in earlier work (e.g., [95]), the results warn that motivational influences on pro-environmental behavior tend to vary in strength (and importance). The only motivational variable not significantly associated with the extra mitigation behavior concerned respondents' direct exposure to the consequences of climate change. Such a finding (as noted earlier) adds to the inconclusive evidence in the literature on the association of first-hand experience of climate change consequences and personal engagement in mitigation behavior [96]. There is evidence to suggest that personal experience of different types of climate change consequences (e.g., flooding vs. air pollution) are likely to elicit different personal behavioral responses [97]—an issue that could not be addressed here and requires further investigation.

This study further suggests that—despite equivocal evidence from previous environmental studies (e.g., [16,26,51])—, a variety of individual and country-level socio-demographics are significant correlates of extra mitigation behavior in response to climate change. Country-level variables were most significantly related to respondents' level of climate change mitigation behavior. In line with Franzen's work (e.g., [31,81]), strong support was obtained for the post-materialism and wealth hypotheses—i.e., broader and greater levels of behavioral engagement to mitigate climate change in EU countries with predominant post-materialist values and in wealthier nations. These findings exemplify the cultural and economic underpinnings of personal mitigation behavior in response to climate change. Accordingly, climate change campaigns encouraging people to do "more" or "a lot" through environmentally responsible behaviors should be tailored to each cultural and wealth EU country group.

The results for individual-level demographics also support the contention that demographic variables continue to be significant correlates of pro-environmental behavior [22,57,60,63]. As hypothesized, the findings confirm the need to encourage broader and greater mitigation responses to climate change, especially among men, conservatives, less educated individuals, and both the youngest and oldest population groups. The curvilinear (inverted U-shaped) link between age and extra mitigation behavior warn environmental researchers of potential non-linear associations of socio-demographics with pro-environmental (climate change) behavior. Thus, researchers are advised to use statistical methods which can detect the presence of potential non-linearities (and heterogeneity), such as latent class analysis or neural networks.

Limitations and Recommendations for Further Study

The present study has a number of limitations that should be addressed in future work. First, the use of secondary survey data—i.e., the Eurobarometer dataset—limited the scope of the research questions and the operationalization of the study variables. Despite the advantages of Eurobarometer data, such as providing a rich, cross-national source of information—with minimum time and financial investment—, its use inhibited the analysis of other relevant correlates of extra mitigation behavior, such as behavioral intentions, subjective norm, or perceived behavioral control [40]. Thus, future studies should consider including unmodeled factors such as citizens' perceptions of self-relevance or involvement with climate change and global warming issues, using alternative and refined measures of psychographics (e.g., attitudes toward the behavior) and behavior (e.g., lagged or future mitigation action), and assessing situational deterrents (and enhancers) of broader and greater levels of behavioral engagement to mitigate climate change. In addition, the important behavioral domain of public sphere (public/political) environmental activism and citizenship, and reflective of high-impact mitigation behavior—that could not be covered in this study—should be accounted for in the future.

Second, this study is based on self-reported (instead of actual or objectively measured) mitigation behavior and the correlates of interest. On the one hand, the focus on self-reported mitigation behavior, coupled to the simultaneous measurement of attitudes and behavior, did not allow the analysis of attitude–action gap in climate change-motivated behavior—an important issue that warrants further investigation through comparison of self-reports and objective measures of actual behavior. On the other hand, it is possible that social desirability biases may have played a role, leading some respondents to overstate their knowledge about climate change issues and their engagement in mitigation actions. Past research has shown the correlation and redundancy between measures of perceived and objective environmental knowledge, but also their potentially differential associations with pro-environmental behaviors [54]. The disparity between perceived and objective knowledge (and their relationships) deserves future attention in studies on climate change-motivated behavior. As regards mitigation behavior, social desirability arguably represents a minor problem in light of the analysis of a representative, non-student sample of environmentally active EU citizens, and the satisfactory correspondence of the self-report measures used in this study to *past/present* behavior—measured as *yes/no* present enactment of specific mitigation actions. Non-student samples, people with high scores of ecological behavior, and measures of past/present (rather than *intended/future*) pro-environmental behavior are likely to be less affected by social desirability biases [52].

Third, the analyses reported here were restricted only to participants already engaged in some form of climate change-motivated activity (i.e., environmentally active citizens). This decision was consistent with this study's investigation of the correlates of "extra vs. common" to mitigate climate change through personal behavior. In pursuing this objective, respondents not reporting any action on climate change did not provide relevant information on the focal variable (extra mitigation behavior) and its underlying motivations. Yet, the exclusion from the analyses of environmentally inactive respondents should be acknowledged as a significant methodological and practical limitation that cautions against generalizing the present study's findings to environmentally inactive people—around 40% of EU-27 citizens [10–11]; this is important because environmentally active vs. inactive population segments are likely to differ in what drives extra (i.e., broader and greater levels of) mitigation behavior. The shortcomings of the current approach could be overcome in future studies that extend understanding of what makes people do "a lot" to mitigate climate change beyond the more receptive (and arguably less challenging) population segment of environmentally active citizens. Such broader investigations would make additional progress toward ambitious goals in the public's behavioral engagement to mitigate climate change.

Fourth, this study examined only direct (psychographic and socio-demographic) correlates of extra climate change mitigation behavior. However, full understanding of the complex and dynamic mechanisms involved in such environmentally significant behavior requires that direct, indirect, and moderating influences be considered. Statistical methodologies such as structural equation modeling (SEM) or partial least squares (PLS) are most appropriate to unravel the interplay between internal and external correlates of broader and greater levels of personal response to climate change [69]—for instance, by placing the variables under the nomological structure of expectancy-value models [40]. If possible, future research should explore lagged effects of knowledge, motivations, and attitudes on intentions and future behavior.

Future research should continue to address the important topic of unobserved heterogeneity in pro-environmental decisions [93–94]. Particularly, additional empirical evidence is needed to verify the heterogeneous (class-dependent), psychographic links to personal mitigation behavior in response to climate change, observed in this study. The significance and strength of country-level associations warrant additional international analyses of pro-environmental behavior in response to climate change. Also, the interaction between individual-level psychographics (e.g., knowledge and attitudes) and countries' wealth and post-materialism levels deserves closer attention in relation to personal mitigation behavior. Preliminary analyses (not reported here but available upon request) show that, amongst post-materialist and wealthier EU countries, the findings tend to be more consistent with previous literature [8,33,55]—e.g., there is greater presence of the segment of environmentally active people for whom knowledge about the causes and ways of fighting climate change relates to extra behavioral engagement to mitigate climate change, and positive/desirable and negative/undesirable attitudes are significantly associated (positively and negative, respectively) with extra mitigation behavior. Certainly, wealth and post-materialism are not the only valid approaches to explaining country differences in climate change mitigation behavior. Other potentially relevant cultural dimensions for cross-national environmental studies include harmony [21]; individualism, long-term orientation, and locus of control [77]; and traditional and altruistic values [72]. Country and regional heterogeneity in environmental policy and legislations are also likely to account for national differences in public mitigation behavior. Multilevel analysis seems to be most appropriate for environmental research questions involving variables from different levels—e.g., individual and country-level influences on climate change-motivated behavior.

Concluding Remarks

As a result of the variety of variables analyzed in relation to extra (vs. common) personal behavior to mitigate climate change, the following can be concluded:

1. A profound shift is needed in personal behavior—from inaction or limited action levels—toward extra (i.e., broader and greater levels of) behavioral engagement to mitigate climate change.
2. The population segments of environmentally active and inactive EU citizens differed significantly, but weakly, in demographic terms.
3. Environmentally active citizens—the population segment under study—can be differentiated in two intuitive categories: people reporting "extra" vs. "common" behavioral engagement to mitigate climate change.
4. Both psychographics and (individual and country-level) socio-demographics help to explain why some people make an extra commitment to mitigate climate change through personal action, as compared to others.

5. Psychographics tend to show greater association than socio-demographics with extra mitigation behavior in response to climate change.
6. There is heterogeneity in the associations involving attitude and knowledge variables, whereas homogeneity exists in the links of motivations and socio-demographics with extra mitigation behavior.
7. The findings draw attention to the importance of potential non-linearities in socio-demographic correlates.
8. The study has implications for promoting more ambitious behavioral responses to climate change, both at the individual level and across countries.

Supporting Information

Appendix S1 Detailed description of Eurobarometer 69.2.

Appendix S2 Supplementary description of statistical analysis.

Appendix S3 Criteria supporting the model comparison and selection process: LC cluster and regression models.

Appendix S4 Complementary test of non-linear age associations.

Acknowledgments

The authors gratefully thank the GESIS Data Archive for the Social Sciences (Cologne, Germany) for their continued support and providing access to the dataset used in this study. We are also grateful to Dr. Bayden D. Russell and two anonymous reviewers for their insightful and constructive feedback on earlier versions of this manuscript.

Author Contributions

Conceived and designed the experiments: JMO NGF RAL. Performed the experiments: JMO NGF RAL. Analyzed the data: JMO NGF RAL. Contributed reagents/materials/analysis tools: JMO NGF RAL. Wrote the paper: JMO NGF RAL.

References

1. Lorenzoni I, Nicholson-Cole S, Whitmarsh L (2007) Barriers perceived to engaging with climate change among the UK public and their policy implications. Global Environ Change 17: 445–459. doi: 10.1016/j.gloenvcha.2007.01.004
2. Whitmarsh L (2011) Scepticism and uncertainty about climate change: Dimensions, determinants and change over time. Global Environ Change 21: 690–700. doi: 10.1016/j.gloenvcha.2011.01.016
3. Semenza JC, Hall DE, Wilson DJ, Bontempo BD, Sailor DJ, George LA (2008) Public perception of climate change: Voluntary mitigation and barriers to behavior change. Am J Prev Med 35: 479–487. doi: 10.1016/j.amepre.2008.08.020
4. Junta de AndalucûÙa (2010) Andalucía Innova. Especial cambio climático. Available: http://www.andaluciainvestiga.com/revista/pdf/100PreguntasMedioAmbiente/100medioambiente.pdf. Accessed 2013 Jun 3.
5. European Environment Agency (2013) Final energy consumption by sector. Available: http://www.eea.europa.eu/data-and-maps/indicators/final-energy-consumption-by-sector-5/assessment#toc-1. Accessed 2013 Jun 3.
6. Bin S, Dowlatabadi H (2005) Consumer lifestyle approach to US energy use and the related CO_2 emissions. Energy Policy 33: 197–208. doi: 10.1016/S0301-4215(03)00210-6
7. Whitmarsh L, Seyfang G, O'Neill S (2011) Public engagement with carbon and climate change: to what extent is the public 'carbon capable'? Global Environ Change 21: 56–65. doi: 10.1016/j.gloenvcha.2010.07.011
8. Whitmarsh L (2009) Behavioural responses to climate change: Asymmetry of intentions and impacts. J Environ Psychol 29: 13–23. doi: 10.1016/j.jenvp.2008.05.003
9. Howell RA (2011) Lights, camera . . . action? Altered attitudes and behaviour in response to the climate change film *The Age of Stupid*. Global Environ Change 21: 177–187. doi: 10.1016/j.gloenvcha.2010.09.004
10. European Commission (2008) Europeans' attitudes towards climate change. Directorate-General for Communication. Available: http://ec.europa.eu/public_opinion/archives/ebs/ebs_300_full_en.pdf. Accessed 2009 Nov 30.
11. European Commission (2011) Climate change. Directorate-General for Communication. Available: http://ec.europa.eu/public_opinion/archives/ebs/ebs_372_en.pdf. Accessed 2013 Jun 3.
12. Whitmarsh L, O'Neill S (2010) Green identity, green living? The role of pro-environmental self-identity in determining consistency across diverse pro-environmental behaviours. J Environ Psychol 30: 305–314. doi: 10.1016/j.jenvp.2010.01.003
13. Poortinga W, Spence A, Whitmarsh L, Capstick S, Pidgeon NF (2011) Uncertain climate: An investigation into public scepticism about anthropogenic climate change. Global Environ Change 21: 1015–1024. doi: 10.1016/j.gloenvcha.2011.03.001
14. Thøgersen J, Crompton T (2009) Simple and painless? The limitations of spillover in environmental campaigning. J Consum Pol 32: 141–163. doi: 10.1007/s10603-009-9101-1
15. Bamberg S, Moser G (2007). Twenty years after Hines, Hungerford and Tomera: A new meta-analysis of psychosocial determinants of pro-environmental behavior. J Environ Psychol 27: 14–25. doi: 10.1016/j.jenvp.2006.12.002
16. Stern PC (2000) Toward a coherent theory of environmentally significant behavior. J Soc Issues 56: 407–424. doi: 10.1111/0022-4537.00175
17. Cheung SF, Chan DK-S, Wong ZS-Y (1999) Reexamining the theory of planned behavior in understanding wastepaper recycling. Environ Behav 31: 587–612. doi: 10.1177/00139169921972254
18. Hopper JR, Nielsen JM (1991) Recycling as altruistic behavior: Normative and behavioral strategies to expand participation in a community recycling program. Environ Behav 23: 195–220. doi: 10.1177/0013916591232004
19. Abrahamse A, Steg L, Gifford R, Vlek C (2009) Factors influencing car use for commuting and the intention to reduce it: A question of self-interest or morality? Transport Res F 12: 317–324. doi: 10.1016/j.trf.2009.04.004
20. Bamberg S, Schmidt P (2003) Incentives, morality or habit? Predicting students' car use for university routes with the models of Ajzen, Schwartz, and Triandis. Environ Behav 35: 264–285. doi: 10.1177/0013916502250134
21. Oreg S, Katz-Gerro T (2006) Predicting proenvironmental behavior cross-nationally: values, the theory of planned behavior, and value-belief-norm theory. Environ Behav 38: 462–483. doi: 10.1177/0013916505286012
22. Scott D, Willits FK (1994) Environmental attitudes and behavior: A Pennsylvania survey. Environ Behav 26: 239–260. doi: 10.1177/0013916594026002006
23. Stern PC, Dietz T, Guagnano GA (1995) The new ecological paradigm in social-psychological context. Environ Behav 27: 723–743. doi: 10.1177/0013916595276001
24. Stern PC, Dietz T, Kalof L (1993) Value orientations, gender and environmental concern. Environ Behav 25: 322–348. doi: 10.1177/0013916593255002
25. Kollmuss A, Agyeman J (2002) Mind the gap: Why do people act environmentally and what are the barriers to pro-environmental behavior? Environ Educ Res 8: 239–260. doi: 10.1080/13504620220145401
26. Diamantopoulos A, Schlegelmilch BB, Sinkovics RR, Bohlen GM (2003) Can socio-demographics still play a role in profiling green consumers? A review of the evidence and an empirical investigation. J Bus Res 56: 465–480. doi: 10.1016/S0148-2963(01)00241-7
27. Mohai P, Twight BW (1987) Age and environmentalism: An elaboration of the Buttel model using national survey evidence. Soc Sci Q 68: 798–815.
28. Iyer ES, Kashyap RK (2007) Consumer recycling: Role of incentives, information, and social class. J Consum Behav 6: 32–47. doi: 10.1002/cb.206
29. Clark CF, Kotchen MJ, Moore RM (2003) Internal and external influences on pro-environmental behavior: Participation in a green electricity program. J Environ Psychol 23: 237–246. doi: 10.1016/S0272-4944(02)00105-6
30. Dunlap RE, Mertig AG (1995) Global concern for the environment: Is affluence a prerequisite? J Soc Issues 51: 121–137. doi: 10.1111/j.1540-4560.1995.tb01351.x
31. Franzen A (2003) Environmental attitudes in international comparison: An analysis of the ISSP surveys 1993 and 2000. Soc Sci Q 84: 297–308. doi: 10.1111/1540-6237.8402005
32. Parker SK, Williams HM, Turner N (2006) Modeling the antecedents of proactive behavior at work. J Appl Psychol 91: 636–652. doi: 10.1037/0021-9010.91.3.636
33. Kaiser FG, Wölfing S, Fuhrer U (1999) Environmental attitude and ecological behaviour. J Environ Psychol 19: 1–19. doi: 10.1006/jevp.1998.0107
34. Hines JM, Hungerford HP, Tomera AN (1986/87) Analysis and synthesis of research on responsible environmental behavior: A meta-analysis. J Environ Educ 18: 1–8.
35. Fishbein M, Ajzen I (1975) Belief, attitude, intention, and behavior: An introduction to theory and research. Reading, MA: Addison-Wesley.
36. Ajzen I, Gilbert N (2008) Attitudes and the prediction of behavior. In: Crano WD, Prislin R, editors. Attitudes and attitude change. New York, NY: Psychology Press. pp. 289–311.

37. Smith SM, Haugtvedt CP, Petty RE (1994) Attitudes and recycling: Does the measurement of affect enhance behavioral prediction? Psychol Market 11: 359–374. doi: 10.1002/mar.4220110405
38. Kaiser FG, Keller C (2001) Disclosing situational constraints to ecological behavior: a confirmatory application of the mixed Rasch model. Eur J Psychol Assess 17: 212–221. doi: 10.1027//1015-5759.17.3.212
39. Fazio RH (1990) Multiple processes by which attitudes guide behavior: The MODE model as an integrative framework. In: Zanna MP, editor. Advances in experimental social psychology, Vol. 23. New York, NY: Academic Press. pp. 75–109.
40. Ajzen I (1991) The theory of planned behavior. Organ Behav Hum Dec 50: 179–211. doi: 10.1016/0749-5978(91)90020-T
41. Webb TL, Sheeran P (2006) Does changing behavioral intentions engender behavior change? A meta-analysis of the experimental evidence. Psychol Bull 132: 249–268. doi: 10.1037/0033-2909.132.2.249
42. Epstein S (1983) Aggregation and beyond: Some basic issues on the prediction of behavior. J Pers 51: 360–392. doi: 10.1111/j.1467-6494.1983.tb00338.x
43. Moisander J (2007) Motivational complexity of green consumerism. Int J Consum Stud 31: 404–409. doi: 10.1111/j.1470-6431.2007.00586.x
44. Booth-Kewley S, Vickers RRJr (1994) Associations between major domains of personality and health behavior. J Pers 62: 281–298. doi: 10.1111/j.1467-6494.1994.tb00298.x
45. Gagnon Thompson S, Barton M (1994) Ecocentric and anthropocentric attitudes toward the environment. J Environ Psychol 14: 149–157. doi: 10.1016/S0272-4944(05)80168-9
46. Schultz PW, Zelezny L (1999) Values as predictors of environmental attitudes: Evidence for consistency across 14 countries. J Environ Psychol 19: 255–265. doi: 10.1006/jevp.1999.0129
47. De Groot JIM, Steg L (2008) Value orientations to explain beliefs related to environmental significant behavior: How to measure egoistic, altruistic, and biospheric value orientations. Environ Behav 40: 330–354. doi: 10.1177/0013916506297831
48. Schwartz SH (1977) Normative influences on altruism. In: Berkowitz L, editor, Advances in experimental social psychology: Vol. 10. New York, NY: Academic Press. pp.221–279. doi: 10.1016/S0065-2601(08)60358-5
49. Nordlund AM, Garvill J (2002). Value structures behind proenvironmental behavior. Environ Behav 34: 740–756. doi: 10.1177/001391602237244
50. De Young R (2000) Expanding and evaluating motives for environmentally responsible behavior. J Soc Issues 56: 509–526. doi: 10.1111/0022-4537.00181
51. Mobley C, Vagias WM, DeWard SL (2010) Exploring additional determinants of environmentally responsible behavior: The influence of environmental literature and environmental attitudes. Environ Behav 42: 420–447. doi: 10.1177/0013916508325002
52. Frick J, Kaiser FG, Wilson M (2004) Environmental knowledge and conservation behavior: Exploring prevalence and structure in a representative sample. Pers Indiv Differ 37: 1597–1613. doi: 10.1016/j.paid.2004.02.015
53. McCright AM, Dunlap RE (2011) Cool dudes: The denial of climate change among conservative white males in the United States. Global Environ Change 21: 1163–1172. doi: 10.1016/j.gloenvcha.2011.06.003
54. Ellen PS (1994) Do we know what we need to know? Objective and subjective knowledge effects on pro-ecological behaviors. J Bus Res 30: 43–52. doi: 10.1016/0148-2963(94)90067-1
55. Kaiser FG, Fuhrer U (2003) Ecological behavior's dependency on different forms of knowledge. Appl Psychol-Int Rev 52: 598–613. doi: 10.1111/1464-0597.00153
56. Immerwahr J (1999) Waiting for a signal: Public attitudes towards global warming, the environment and geophysical research. Public Agenda Foundation. Available: http://www.policyarchive.org/handle/10207/bitstreams/5662.pdf. Accessed 2011 Feb 15.
57. O'Connor RE, Bord RJ, Fisher A (1999) Risk perceptions, general environmental beliefs, and willingness to address climate change. Risk Anal 19: 461–471. doi: 10.1023/A:1007004813446
58. Jones RE, Dunlap RE (1992) The social bases of environmental concern: Have they changed over time? Rural Sociol 57: 28–47. doi: 10.1111/j.1549-0831.1992.tb00455.x
59. Roberts JA (1996) Will the real socially responsible consumer please step forward? Bus Horizons 39: 79–83. doi: 10.1016/S0007-6813(96)90087-7
60. Leiserowitz A (2006) Climate change risk perception and policy preferences: The role of affect, imagery, and values. Climatic Change 77: 45–72. doi: 10.1007/s10584-006-9059-9
61. Xiao C, McCright A (2007) Environmental concern and socio-demographic variables: A case study of statistical models. J Environ Educ 38: 3–13. doi: 10.3200/JOEE.38.1.3-14
62. Slimak MW, Dietz T (2006) Personal values, beliefs, and ecological risk perception. Risk Anal 26: 1689–1705. doi: 10.1111/j.1539-6924.2006.00832.x
63. Dietz T, Stern PC, Guagnano GA (1998) Social structural and social psychological bases of environmental concern. Environ Behav 30: 450–471. doi: 10.1177/001391659803000402
64. Sutton SG, Tobin RC (2011) Constraints on community engagement with Great Barrier Reef climate change reduction and mitigation. Global Environ Change 21: 894–905. doi: 10.1016/j.gloenvcha.2011.05.006
65. Zelezny LC, Chua P, Aldrich C (2000) Elaborating on gender differences in environmentalism. J Soc Issues 56: 443–457. doi: 10.1111/0022-4537.00177
66. Hunter LM, Hatch A, Johnson A (2004) Cross-national gender variation in environmental behaviors. Soc Sci Q 85: 677–694. doi: 10.1111/j.0038-4941.2004.00239.x
67. Slovic P (1999) Trust, emotion, sex, politics, and science: Surveying the risk-assessment battlefield. Risk Anal 19: 689–701. doi: 10.1111/j.1539-6924.1999.tb00439.x
68. Davidson DJ, Freudenburg WR (1996) Gender and environmental risk concerns: A review and analysis of available research. Environ Behav 28: 302–339. doi: 10.1177/0013916596283003
69. Cottrell SP (2003) Influence of sociodemographics and environmental attitudes on general responsible environmental behavior among recreational boaters. Environ Behav 35: 347–375. doi: 10.1177/0013916503035003003
70. Vining J, Ebreo A (1990) What makes a recycler? A comparison of recyclers and nonrecyclers. Environ Behav 22: 55–73. doi: 10.1177/0013916590221003
71. Newman K (1986) Personal values and commitment to energy conservation. Environ Behav 18: 53–74. doi: 10.1177/0013916586181003
72. Aoyagi-Usui M, Vinken H, Kuriyabashi A (2003) Pro-environmental attitudes and behaviors: An international comparison. Hum Ecol Rev 10: 23–31.
73. Gelissen J (2007) Explaining popular support for environmental protection: A multilevel analysis of 50 nations. Environ Behav 39: 392–415. doi: 10.1177/0013916506292014
74. Guerin D, Crete J, Mercier J (2001) A multilevel analysis of the determinants of recycling behavior in the European countries. Soc Sci Res 30: 195–218. doi: 10.1006/ssre.2000.0694
75. Gifford R, Scannell L, Kormos C, Smolova L, Biel A, et al. (2009) Temporal pessimism and spatial optimism in environmental assessments: An 18-nation study. J Environ Psychol 29: 1–12. doi: 10.1016/j.jenvp.2008.06.001
76. Kemmelmeier M, Król G, Kim YH (2002) Values, economics and proenvironmental attitudes in 22 societies. Cross-Cult Res 36: 256–285. doi: 10.1177/10697102036003004
77. Sarigöllü E (2009) A cross-country exploration of environmental attitudes. Environ Behav 41: 365–386. doi: 10.1177/0013916507313920
78. Bord RJ, Fisher A, O'Connor RE (1998) Public perceptions of global warming: United States and international perspectives. Climate Res 11: 75–84. doi: 10.3354/cr011075
79. Brechin SR (2003) Comparative public opinion and knowledge on global climatic change and the Kyoto protocol: The U.S. against the world? Int J Sociol Soc Pol 23: 106–134. doi: 10.1108/01443330310790318
80. Inglehart R (1997) Modernization and postmodernization: Cultural, economic, and political change in 43 societies. Princeton, NJ: Princeton University Press.
81. Franzen A, Meyer R (2010) Environmental attitudes in cross-national perspective: A multilevel analysis of the ISSP 1993 and 2000. Eur Sociol Rev 26: 219–234. doi: 10.1093/esr/jcp018
82. Brechin SR, Kempton W (1994) Global environmentalism: A challenge to the postmaterialism thesis? Soc Sci Q 75: 245–269.
83. Diekmann A, Franzen A (1999) The wealth of nations and environmental concern. Environ Behav 31: 540–549. doi: 10.1177/00139169921972227
84. Vinzi VE, Ringle CM, Squillacciotti S, Trinchera L (2007) Capturing and treating unobserved heterogeneity by response based segmentation in PLS path modeling. A comparison of alternative methods by computational experiments. ESSEC Research Center Working Paper 07019. Paris, France: ESSEC Business School.
85. Ojea E, Loureiro ML (2007) Altruistic, egoistic and biospheric values in willingness to pay (WTP) for wildlife. Ecol Econ 63: 807–814. doi: 10.1016/j.ecolecon.2007.02.003
86. Bhat CR (2000) Incorporating observed and unobserved heterogeneity in urban work travel mode choice modeling. Transport Sci 34: 228–238. doi: 10.1287/trsc.34.2.228.12306
87. Tay L, Diener E, Drasgow F, Vermunt JK (2011) Multilevel mixed-measurement IRT analysis: An explication and application to self-reported emotions across the world. Organ Res Methods 14: 177–207. doi: 10.1177/1094428110372674
88. Aldrich GA, Grimsrud KM, Thacher JA, Kotchen MJ (2007) Relating environmental attitudes and contingent values: How robust are methods for identifying preference heterogeneity? Environ Resour Econ 37: 757–775. doi: 10.1007/s10640-006-9054-7
89. Barr S, Shaw G, Coles T (2011) Times for (Un)sustainability? Challenges and opportunities for developing behaviour change policy. A case-study of consumers at home and away. Global Environ Change 21: 1234–1244. doi: 10.1016/j.gloenvcha.2011.07.011
90. Moors G (2007) Testing the internal validity of the Inglehart thesis by means of a latent class choice model. Acta Sociol 50: 147–160. doi: 10.1177/0001699307077656
91. Magidson J, Vermunt JK (2004) Latent class models. In: Kaplan D, editor. The Sage handbook of quantitative methodology for the social sciences. Thousand Oaks, CA: Sage Publications. pp. 175–198.
92. Vermunt JK, Magidson J (2002) Latent class cluster analysis. In: Hagenaars JA, McCutcheon AL, editors. Applied latent class analysis. Cambridge, UK: Cambridge University Press. pp. 89–106.
93. Bodur M, Sarigöllu E (2005) Environmental sensitivity in a developing country: Consumer classification and implications. Environ Behav 37: 487–510. doi: 10.1177/0013916504269666
94. Dolnicar S, Grün B (2009) Environmentally friendly behavior: Can heterogeneity among individuals and contexts/environments be harvested for improved

sustainable management? Environ Behav 41: 693–714. doi: 10.1177/0013916508319448

95. Thøgersen J (2009) The motivational roots of norms for environmentally responsible behavior. Basic Appl Soc Psychol 31: 348–362. doi: 10.1080/01973530903317144

96. Spence A, Poortinga W, Butler C, Pidgeon NF (2011) Perceptions of climate change and willingness to save energy related to flood experience. Nature Clim Change 1: 46–49. doi: 10.1038/nclimate1059

97. Whitmarsh L (2008) Are flood victims more concerned about climate change than other people? The role of direct experience in risk perception and behavioural response. J Risk Res 11: 351–374. doi: 10.1080/13669870701552235

Combining XCO_2 Measurements Derived from SCIAMACHY and GOSAT for Potentially Generating Global CO_2 Maps with High Spatiotemporal Resolution

Tianxing Wang*, Jiancheng Shi, Yingying Jing, Tianjie Zhao, Dabin Ji, Chuan Xiong

State Key Laboratory of Remote Sensing Science, Institute of Remote Sensing and Digital Earth, Chinese Academy of Sciences. Beijing, China

Abstract

Global warming induced by atmospheric CO_2 has attracted increasing attention of researchers all over the world. Although space-based technology provides the ability to map atmospheric CO_2 globally, the number of valid CO_2 measurements is generally limited for certain instruments owing to the presence of clouds, which in turn constrain the studies of global CO_2 sources and sinks. Thus, it is a potentially promising work to combine the currently available CO_2 measurements. In this study, a strategy for fusing SCIAMACHY and GOSAT CO_2 measurements is proposed by fully considering the CO_2 global bias, averaging kernel, and spatiotemporal variations as well as the CO_2 retrieval errors. Based on this method, a global CO_2 map with certain UTC time can also be generated by employing the pattern of the CO_2 daily cycle reflected by Carbon Tracker (CT) data. The results reveal that relative to GOSAT, the global spatial coverage of the combined CO_2 map increased by 41.3% and 47.7% on a daily and monthly scale, respectively, and even higher when compared with that relative to SCIAMACHY. The findings in this paper prove the effectiveness of the combination method in supporting the generation of global full-coverage XCO_2 maps with higher temporal and spatial sampling by jointly using these two space-based XCO_2 datasets.

Editor: Juan A. Añel, University of Oxford, United Kingdom

Funding: The work described in this paper has been jointly supported by project of "Climate Change: Carbon Budget and Related Issues" (Grant nr. XDA05040402) from Chinese Academy of Sciences (CAS), the CAS/SAFEA International Partnership Program for Creative Research Teams (Grant nr. KZZD-EW-TZ-09) and National Natural Science foundation of China (Grant nr. 41301177). The funders had no role in study design, data collection and analysis, decision to publish, or preparation of the manuscript.

Competing Interests: The authors have declared that no competing interests exist.

* Email: wangtx@radi.ac.cn

Introduction

In recent years, global warming caused by emission of CO_2 has attracted considerable attention from the public. During the past decade, although tremendous efforts have been made toward improving the understandings of the mechanism between CO_2 increase in the atmosphere and global warming, some uncertainties still exist in the spatiotemporal characteristics of CO_2 sinks/sources on regional and global scales due to the lack of high-density measurements of such variables with good accuracy [1,2]. To date, the estimates of CO_2 flux from inverse methods rely mainly on ground-based measurements [3,4]. Although providing highly accurate atmospheric CO_2 records, the traditional ground-based networks intrinsically suffer from sparse spatial coverage [2,5]. Satellite-based measurements with various spatial and temporal resolutions provide a unique opportunity to accurately map atmospheric CO_2 in both daytime and nighttime over large areas, thus having the potential to bridge this gap. As a result, various satellite-based platforms have been equipped in recent years for deriving the CO_2 concentrations.

Generally, methods for retrieving CO_2 from space can be grouped into two categories: (1) inferring CO_2 concentrations by measuring shortwave infrared (SWIR) reflected solar radiation around 1.6 and 2.0 μm with sufficient spectral resolution. This includes the Greenhouse gases Observing SATellite (GOSAT), operating since 2009 [6], the Scanning Imaging Absorption spectrometer for Atmospheric CartograpHY (SCIAMACHY), in orbit since 2002 [7], and the second Orbiting Carbon Observatory (OCO-2), which, as a rebuild of OCO [8,9], is planned to be launched in July 2014. In addition, CarbonSat will also be scheduled to be launched in 2018 (http://www.iup.uni-bremen.de/carbonsat/). These measurements have a nearly uniform sensitivity to CO_2 from the surface up through the middle troposphere, and thus are frequently used to derive the column-average dry air mole fraction of atmosphere CO_2 (XCO_2) during the daytime; (2) retrieving CO_2 concentrations by interpreting the recorded spectra of the Earth-atmosphere system in thermal infrared (TIR) bands (around 15 μm). Instruments that work in such a way include AIRS [10,11], IASI [12,13], and FTS (Band 4) of GOSAT [6]. These measurements bring the advantage that they can detect CO_2 during both day and night time, while the lack of sensitivity in the lower troposphere makes them inappropriate to estimate CO_2 near the surface where the largest signals of CO_2 sources and sinks occur [1]. The complementarities of these platforms allow us to combine the SWIR and TIR measurements for obtaining enhanced understanding of CO_2 spatiotemporal variations globally. Since XCO_2 is much less affected by vertical transport of CO_2, it is particularly useful for investigation of CO_2 sources and sinks using inversion modeling [14,15]. On the other hand, the spatial and temporal variations in XCO_2 are even smaller than that in the surface CO_2; therefore, unprecedented measurement precision and accuracy are highly required for such column measurements [16–19]. SCIAMACHY (operation stopped in April 2012) and GOSAT are two typical instruments that can be used to derive XCO_2 from space, and a variety of retrieval algorithms have been developed for SCIAMACHY [1,20–27] and GOSAT [2,4,5,28–30] with eyes on improving XCO_2 retrieval accuracy to a great extent. At present, a number of XCO_2 products have been released. These will definitely enhance our understanding of the global carbon cycle.

Unfortunately, almost all typical instruments currently used to derive atmospheric CO_2 concentration are working in the infrared spectral range (less than 16 μm). Thus, except for the instrument's observation mode (for example, GOSAT observes in lattice points), the spatial coverage of the derived CO_2 is severely restricted by the presence of clouds. In addition, the lower signal-to-noise level over ice/snow covered surfaces and ocean for SWIR instruments (e.g., SCIAMACHY) also contributes to the CO_2 sparse coverage. For instance, it has been pointed out that only about 10% of GOSAT data can be used for retrieval of XCO_2 due to the cloud contaminations [4]. The amount of CO_2 measurements will be even smaller if additional screening criteria such as quality of spectral fit, aerosol loadings, etc. are further applied. Although the amount of remaining CO_2 measurements from certain space-based instruments may largely surpass that of ground-based sites, it is still not sufficient enough for accurately quantifying the spatiotemporal distribution of CO_2 over the global scale. As a result, it is greatly desired to jointly use these available CO_2 measurements derived from various space-based data. Recently, a novel method has been proposed for combining CO_2 values from seven different algorithms, and a new Level-2 CO_2 database (EMMA) from one algorithm is composed according to the median of monthly average of seven CO_2 products in each 10°×10° latitude/longitude grid box [31]. In fact, this method cannot increase the number of CO_2 observations but chooses a product with moderate oscillation among the available products. Despite the usefulness of the XCO_2 measurements (Level 2) in their own right, further spatiotemporal analysis for interpreting their scientific merit is essentially necessary due to the retrieval uncertainties and sparse coverage of such Level-2 observations [32]. For this point, many works have attempted to generate global full-coverage (i.e., Level 3) maps from XCO_2 values derived from single satellite observations using a geospatial statistics approach [32–34]. However, as reflected in these studies (for instance, Fig. 1 in the work of [33]), a compromise has to be made between the interpolated accuracy and the spatiotemporal resolution of Level-3 product because of the limited amount of Level-2 XCO_2 observations being used. For this point, instead of using Level-2 XCO_2 from a single dataset (e.g., GOSAT or OCO-2) as performed in the existing literature, we attempt to explore the potential of combining two CO_2 datasets (GOSAT and SCIAMACHY) in assisting in global Level-3 generation, aiming to: (1) propose a general strategy for combining (fusing) various CO_2 datasets with different instruments, algorithms, averaging kernels, etc.; and 2) increase the number of daily CO_2 points (utilized in Level-3 map interpolations) through the combination of two datasets, so that potentially improved Level-3 maps with higher accuracy and shorter time scale can be generated. The better the interpretation of the satellite-based CO_2 observations one can make, the higher the resolution (both temporal and spatial) of the generated global CO_2 maps.

Datasets

For GOSAT, the Fourier transform spectrometer (FTS) on GOSAT is the fundamental unit to retrieve atmospheric CO_2 and CH_4. It observes sunlight reflected from the earth's surface, and light emitted from the atmosphere and the surface. It is composed of three narrow bands in the SWIR region (0.76, 1.6, and 2.0 μm) and a wide TIR band (5.5–14.3 μm) at a spectral and spatial resolution of 0.2 cm^{-1} and 10.5 km, respectively [35]. Specifically, four CO_2 products from GOSAT have currently been released to the public: University of Leicester product [9,36], the RemoTeC product [28], NIES GOSAT product [35] and the product generated by NASA's Atmospheric CO_2 Observations from Space (ACOS) team (hereafter called ACOS product) [2,30]. The difference between some of the above mentioned products with various versions have been investigated in a recent study [37]. In the present paper, the ACOS product of 2009–2010 with version v2.9 has been employed.

SCIAMACHY was successfully launched on board Environmental Satellite (ENVISAT) in 2002 (unfortunately ceased in April 2012), which is a detector elements satellite spectrometer covering the spectral range 0.24–2.38 μm with a moderate spectral resolution of about 0.2–1.6 nm, and spatial resolution at nadir of 60×30 km [7]. It has eight spectral channels, with 1024 individual detector diodes for each band, observing the spectral regions 0.24–1.75 μm (band 1–6), 1.94–2.04 μm (band 7), and 2.26–2.38 μm (band 8) simultaneously in nadir and limb and solar and lunar occultation viewing geometries [22]. As mentioned in Section 1, till today, a number of CO_2-retrieval algorithms have been developed for SCIAMACHY. The IUP/IFE of University of Bremen has released two XCO_2 products, i.e., WFM-DOAS product [21,22] and the Bremen Optimal Estimation DOAS (BESD) product [1,26]. In this study, the BESD product with the versions of v02.00.08 for 2009–2010 is used.

In addition, CO_2 profiles of CT [38] are also collected here to allow the data mentioned above to be properly fused. CT is a NOAA data assimilation system, which provides the 3D profiles of CO_2 mole fractions in the atmosphere over the globe. For this study, CT data with version CT2011 is collected. This dataset provides global CO_2 profiles with 3°×2° latitude/longitude grid and 3 hours temporal resolution (a total 8 times from 01 to 22 in UTC) spanning the time period from January 2000 to December 2010. The CT dataset is used here mainly to assist in adjusting and time-shifting of the two CO_2 products being combined.

Methodologies

For combining the different space-based CO_2 measurements, three steps are adapted in this study. First, taking the global ground measurements of CO_2 as reference, remove the bias of the individual CO_2 retrievals for ensuring the accuracy of the fused CO_2 product; then make some adjustment for both the ACOS and BESD products, so that they can be physically comparable and thus combined; finally fuse the ACOS and BESD CO_2 products considering their retrieval uncertainties, spatial scales, differences in averaging kernels and overpass times, etc.

3.1 Global bias corrections

Removal of any global bias of the retrieved CO_2 when compared with the ground *in situ* measurements is essential before performing joint use. Many researches [4,25] frequently pointed out that CO_2 retrievals from GOSAT are low biased with different levels due to the uncertainties in pressure, radiometric calibration, line shape model, cloud and aerosol scattering, etc.

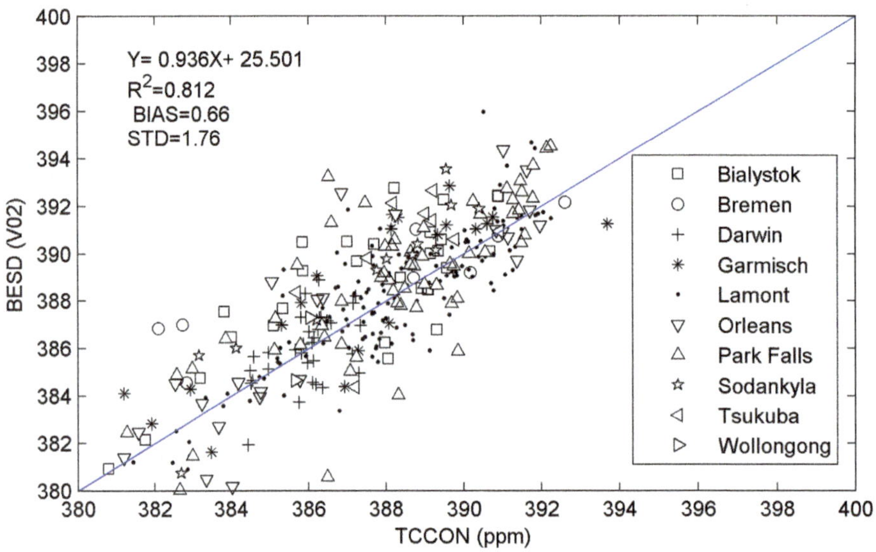

Figure 1. Validation of the BESD products against *in situ* TCCON CO_2 measurements over globe for 2009–2010.

Fortunately, a recent study has proposed a method for evaluating systematic errors in CO_2 and showed that the new version of ACOS product (v2.9) has a low global bias (<0.5 ppm) [39]. Thus, there is no global bias correction for the ACOS product being conducted here, but only the ACOS retrievals that pass the filter of table B1 in the work of [38] and marked as "good" in the quality flag are used. For the BESD product, we select Total Carbon Column Observing Network (TCCON) [15] measurements for 2009–2010 as the ground truth to determine its global bias. Specifically, BESD retrievals within ±2.5° and ±2.5° latitude/longitude box centered at each TCCON site and the mean FTS value (within ±1 h time window of satellite overpass time) are

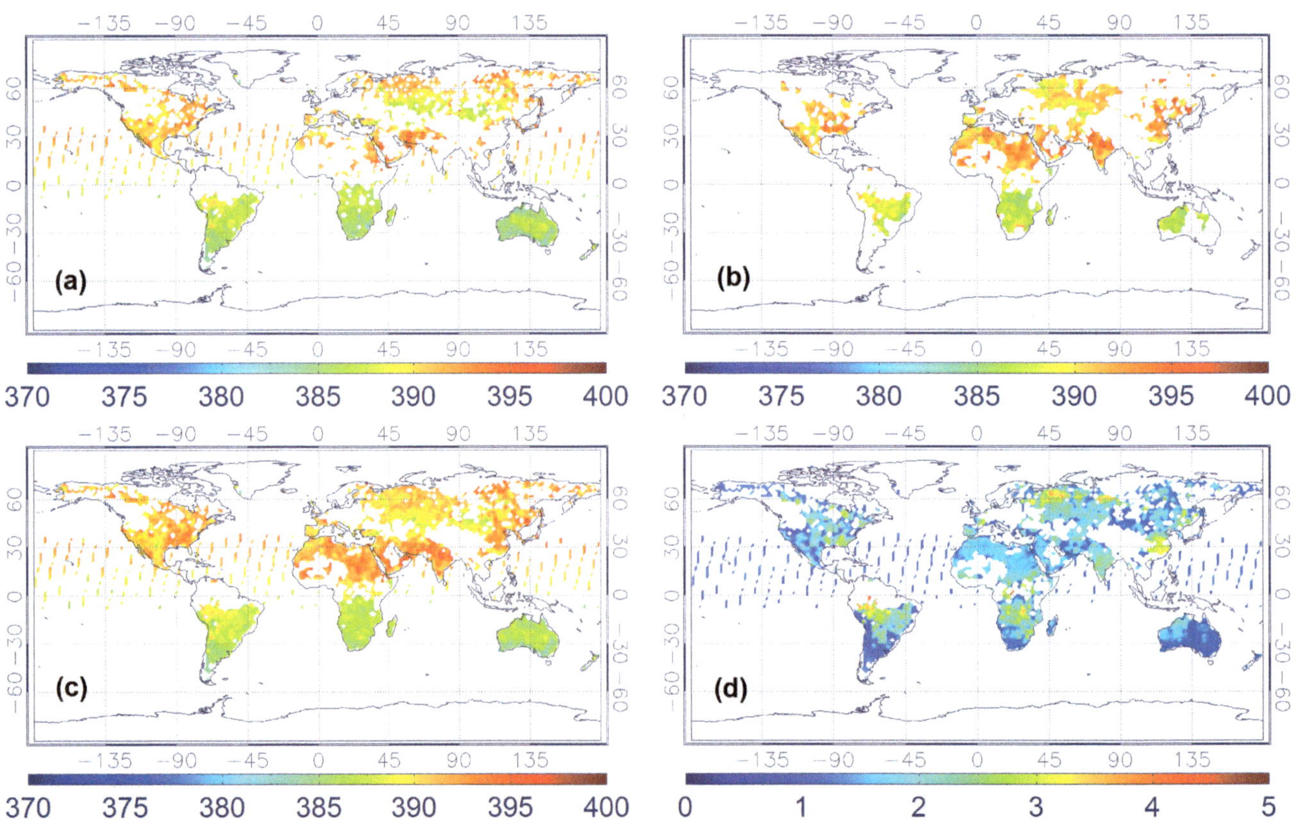

Figure 2. XCO_2 monthly mean maps in May of 2010. ((a) ACOS XCO_2, (b) BESD XCO_2, (c) combined product, and (d) XCO_2 uncertainties of the combined product).

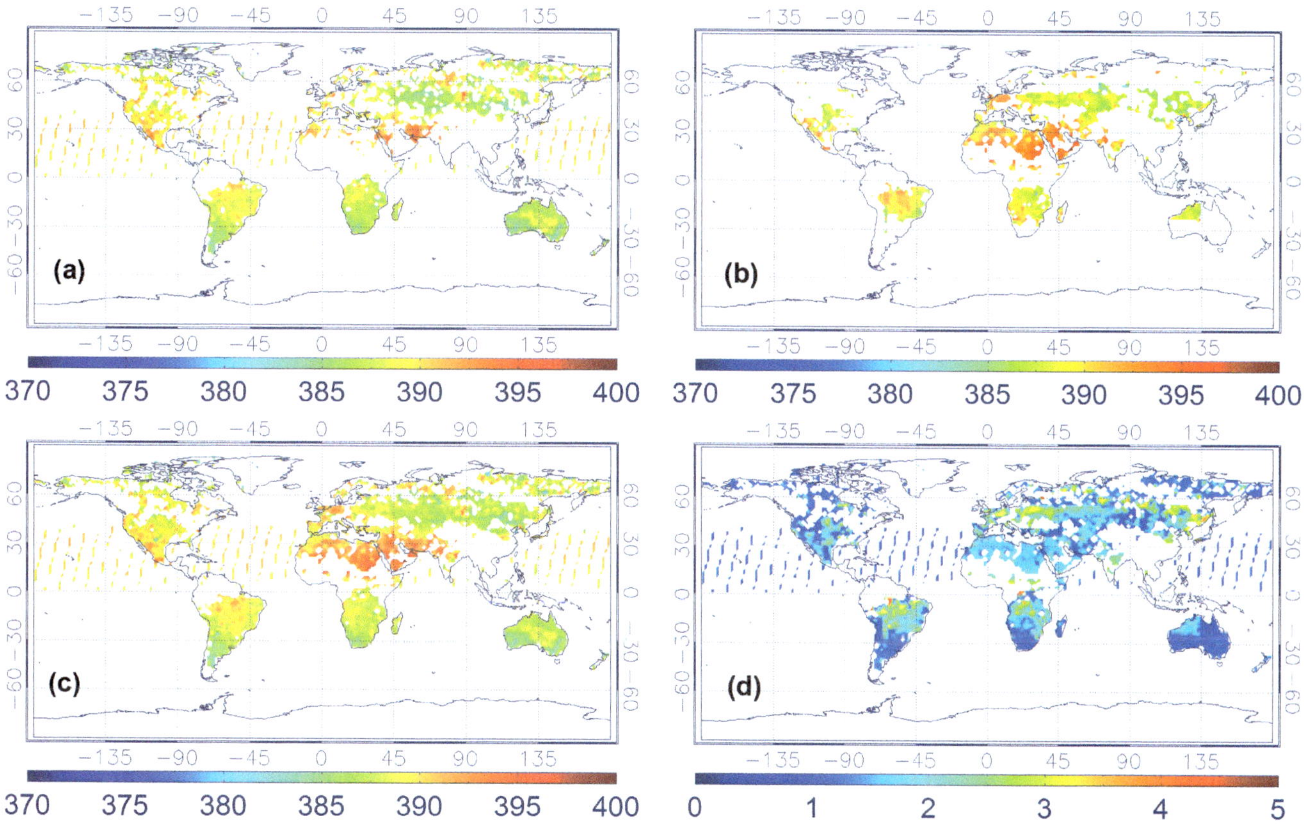

Figure 3. XCO_2 monthly mean maps in June of 2010. ((a) ACOS XCO_2, (b) BESD XCO_2, (c) combined product, and (d) XCO_2 uncertainties of the combined product).

extracted and compared (totally ten TCCON sites are utilized). The coincidence criteria mentioned above ultimately yield a total of 338 pairs of CO_2 measurements. The comparison result is shown in Fig. 1.

3.2 Retrieval adjustments

As pointed out by most researchers, it is not reasonable to directly compare or use two XCO_2 measurements. A suitable way to do that is to take the a priori profiles and variations in averaging kernel into account during the comparison [26,40]. To tackle the a priori issue, after correcting their global biases, both BESD and ACOS products are adjusted for a common a priori profile, which we assume to be the CT profile interpolated at the middle of the two overpass times (Equation (1)). Specifically, the a priori CO_2 profile of both the ACOS and BESD are first interpolated or extrapolated to the level of the CT CO_2 profile according to their pressure layers. After interpolation, the a priori profiles for both ACOS and BESD have the same dimension as the CT profile. Here the reason we take the CT profile at the middle of the two overpass times is that the time difference for GOSAT (1:00 pm) and SCIAMACHY (10:00 am) is relative large (3 hours), if we take one satellite time as reference, the induced error would be large for the other satellite measurements considering the CO_2 natural diurnal variation. So a middle time between these two satellite overpass times is selected for minimizing the CO_2 uncertainties during the adjustment.

$$XCO_2\_adj = XCO_2\_ret + (h^T I - a)(xCT - xa) \qquad (1)$$

Here, XCO_2_adj is the adjusted XCO_2 for ACOS or BESD; XCO_2_ret corresponds to retrieved XCO_2 of ACOS or BESD; a is the column-averaging kernel (row vector) of ACOS or BESD; h is pressure-weighting function (column vector); I is an identity matrix; xCT and xa (column vectors) are the common CT CO_2 profile and the corresponding a priori CO_2 profile for ACOS or BESD, respectively.

While it is not trivial to accurately consider the smoothing error without an estimate of the true atmospheric variability which is generally not readily available for most cases [39]. Fortunately, some works revealed that the smoothing error is generally small [26,39]. Consequently, for the remainder of this paper, only the adjustment in Equation (1) is applied for both the ACOS and BESD CO_2 products (after bias corrections).

3.3 Combination and time shifting

Based on the processes described above, the world is divided into a number of 0.5°×0.5°latitude/longitude grid box (totally 720×360). For each grid cell, Equation (2) is used to combine the corresponding CO_2 measurements within that grid.

$$XCO_2\_Fued = \sum_{i=1}^{m} \left(XCO_2\_i \times \frac{1 - Uncert\_ratio^i}{\sum_{i=1}^{m} (1 - Uncert\_ratio^i)} \right) \qquad (2)$$

where XCO_2_Fued is the combined XCO_2; m is the total number of space-based CO_2 retrievals (ACOS and/or BESD) within a certain grid; XCO_2_i is the ith XCO_2 retrieval in a grid for which

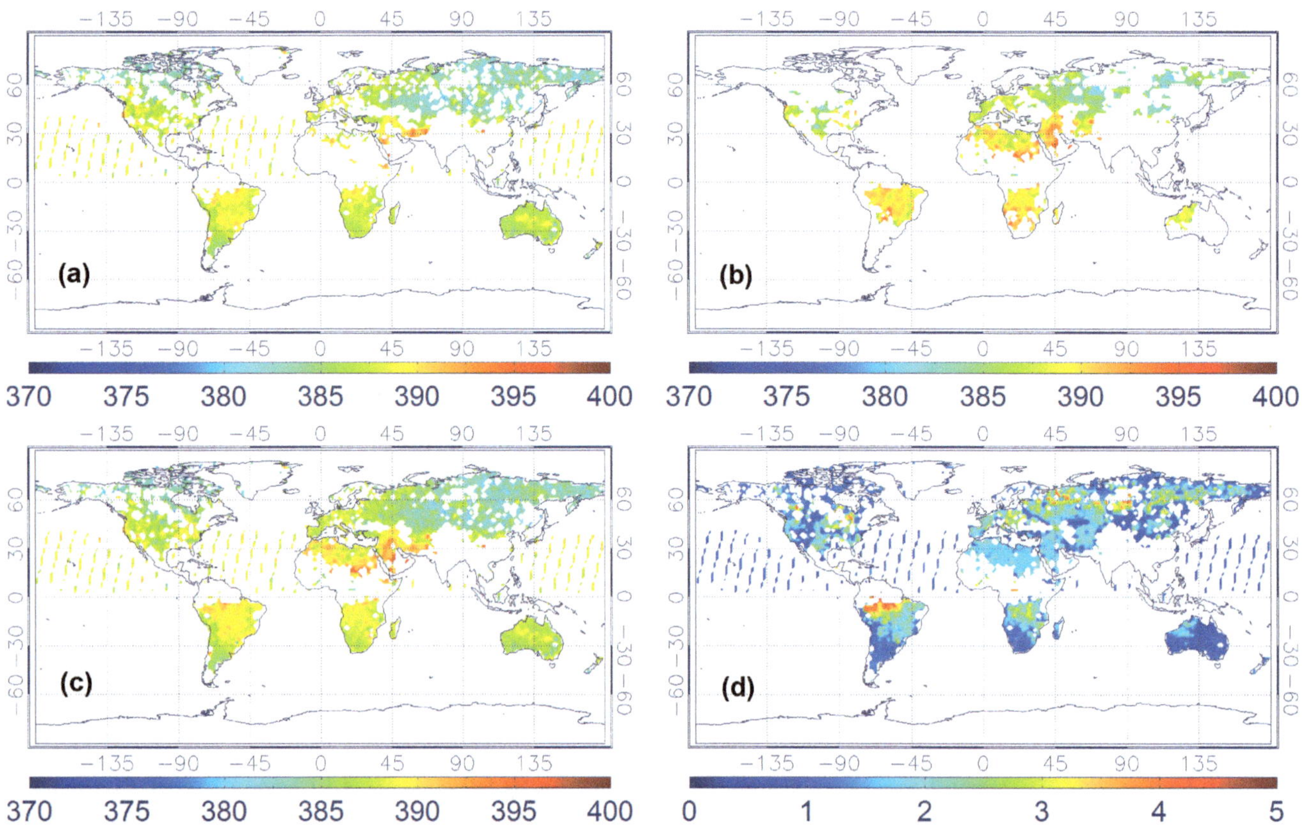

Figure 4. XCO_2 monthly mean maps in July of 2010. ((a) ACOS XCO_2, (b) BESD XCO_2, (c) combined product, and (d) XCO_2 uncertainties of the combined product).

the global bias and Equation (1) are supposed to be applied; $Uncert\_ratio^i$ is the ratio of uncertainty of the ith XCO_2 retrieval to its XCO_2 value.

Please note that since different CO_2 retrievals have distinct overpass times, it is necessary to unify them to avoid uncertainties induced from the time discrepancy before fusion. To this end, a method for considering the CO_2 shifting along time has been developed (Equation (3)). First, designate a specific time or select one overpass time as reference, then transfer CO_2 measurements at various overpass times to that of the reference time by interpolating the CT CO_2 at temporal scale. Here, it should be pointed out that despite the CO_2 absolute values of CT not being accurate enough, the daily cycle pattern of atmospheric CO_2 it reflects is assumed to be correct.

$$XCO_2\_ref = \frac{\omega^T X\_ref^{CT}}{\omega^T X\_t^{CT}} \times XCO_2\_t \tag{3}$$

Here, XCO_2_ref is the transformed XCO_2 (ACOS or BESD) at the reference time; XCO_2_t is the retrieved XCO_2 from ACOS or BESD at overpass time t; X_ref^{CT} and X_t^{CT} are CO_2 profiles of CT at times of reference and t, respectively; ω is the pressure-weighting vector (column vector).

Based on the time-shifting strategy proposed here, a global CO_2 map at any specific time can be theoretically produced by employing the pattern of the CO_2 daily cycle reflected by CT data. For instance, we can unify all XCO_2 retrievals being combined with various overpass times to that of UTC = 1.

Results

Evaluation analysis showed that the global bias for the BESD product is generally small. In this study, the bias of the BESD product is corrected by subtracting 0.6 ppm from all XCO_2 values according to the results in Fig. 1. Although the systematic bias of the XCO_2 retrievals is removed, it is supposed that the error characteristics (random error) within the data are still unchanged. The bias-corrected XCO_2 retrievals of both ACOS and BEDS are used as fundamental data for the combination algorithm.

By applying the series of processes shown in Section 3, daily, weekly, as well as monthly maps of combined XCO_2 for 2009 and 2010 are generated. Here, as an example, only four maps (from May to August) of monthly mean XCO_2 of 2010 are shown here (Fig. 2–Fig. 5). In addition, the total XCO_2 uncertainties of the combined product which mainly depend on the uncertainties of the original ACOS or BESD XCO_2 retrievals are also illustrated.

From Fig. 2–Fig. 5, it is not difficult to observe that the combined data realize the physical complementary of the two products in terms of spatial coverage. The number of valid CO_2 measurements in the fused product is the union of the CO_2 data from both the ACOS and BESD at the same geographical location. In addition, the combined XCO_2 demonstrates similar spatiotemporal characteristics with that of ACOS and BESD over the globe, which implies that all processes associated with the combination do not distort the essential information of the original XCO_2 products (ACOS or BESD). Similar findings can also be observed in the daily mean and weekly mean XCO_2 maps. To quantitatively investigate the improvement of fused XCO_2 in spatial coverage, the fractional coverage of all three variables

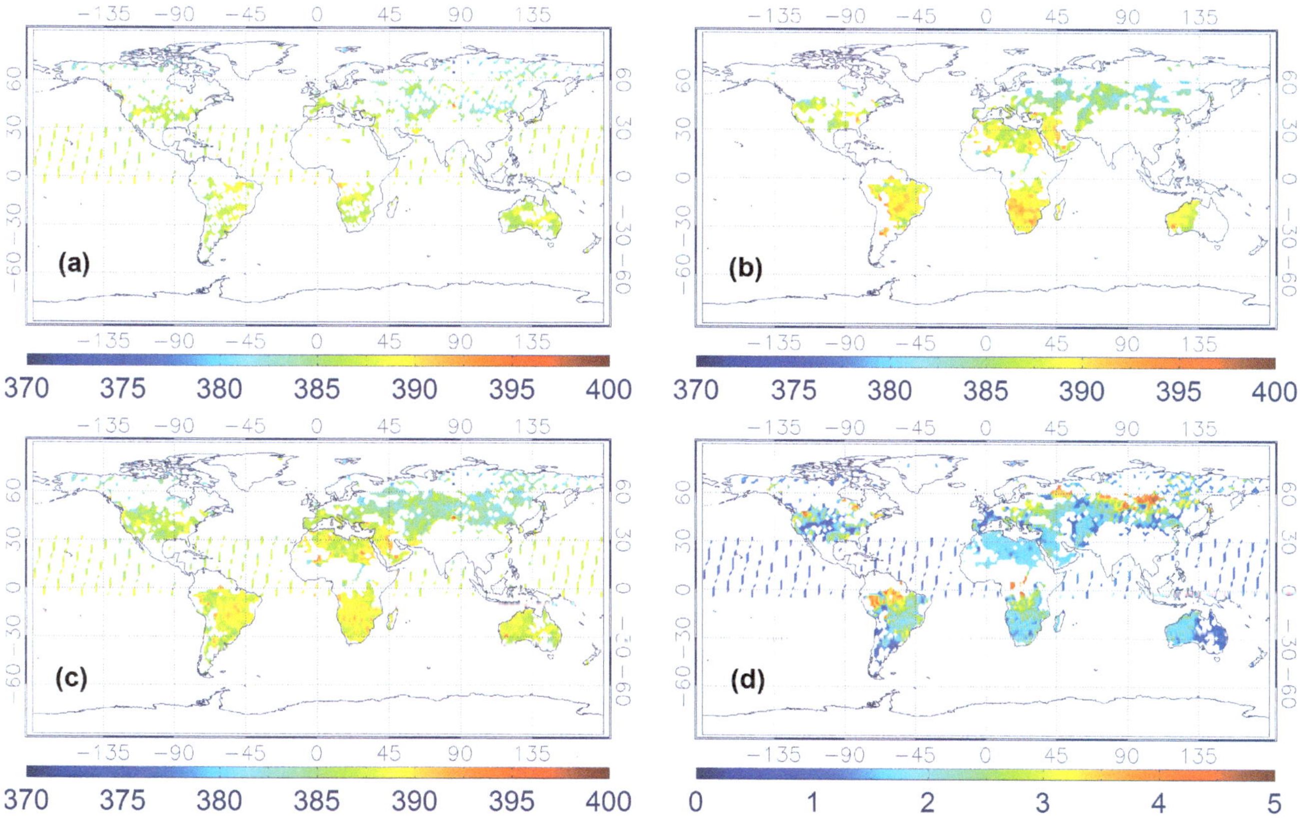

Figure 5. XCO_2 monthly mean maps in August of 2010. ((a) ACOS XCO_2, (b) BESD XCO_2, (c) combined product, and (d) XCO_2 uncertainties of the combined product).

(ACOS, BESD, and combined XCO_2) on both daily and monthly scales is calculated (Fig. 6). From Fig. 6, it can be seen that the average global coverage of ACOS and BESD is around 0.46% and 0.21%, respectively, on a daily scale. The monthly mean coverage of such products accounts for about 5.70% and 3.75%, respectively. While spatial coverage of combined XCO_2 can reach up to 0.65% and 8.42% on daily and monthly scales, respectively, it accounts for increments of 41.3% and 47.7% on the daily and

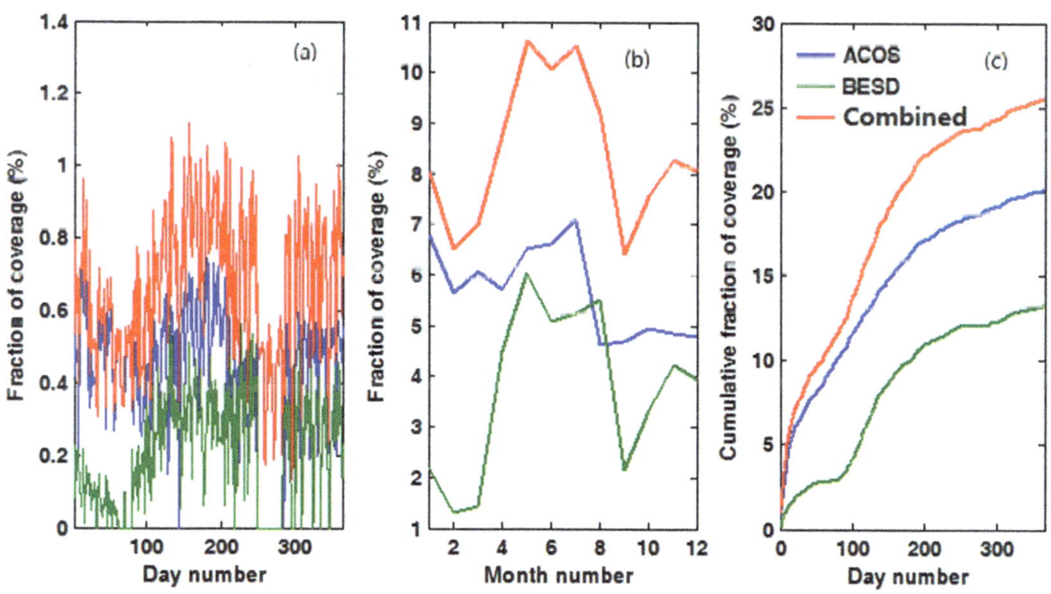

Figure 6. XCO_2 fraction of coverage of ACOS, BESD, and combined products. (a) Daily coverage. (b) Monthly coverage. (c) Cumulative coverage.

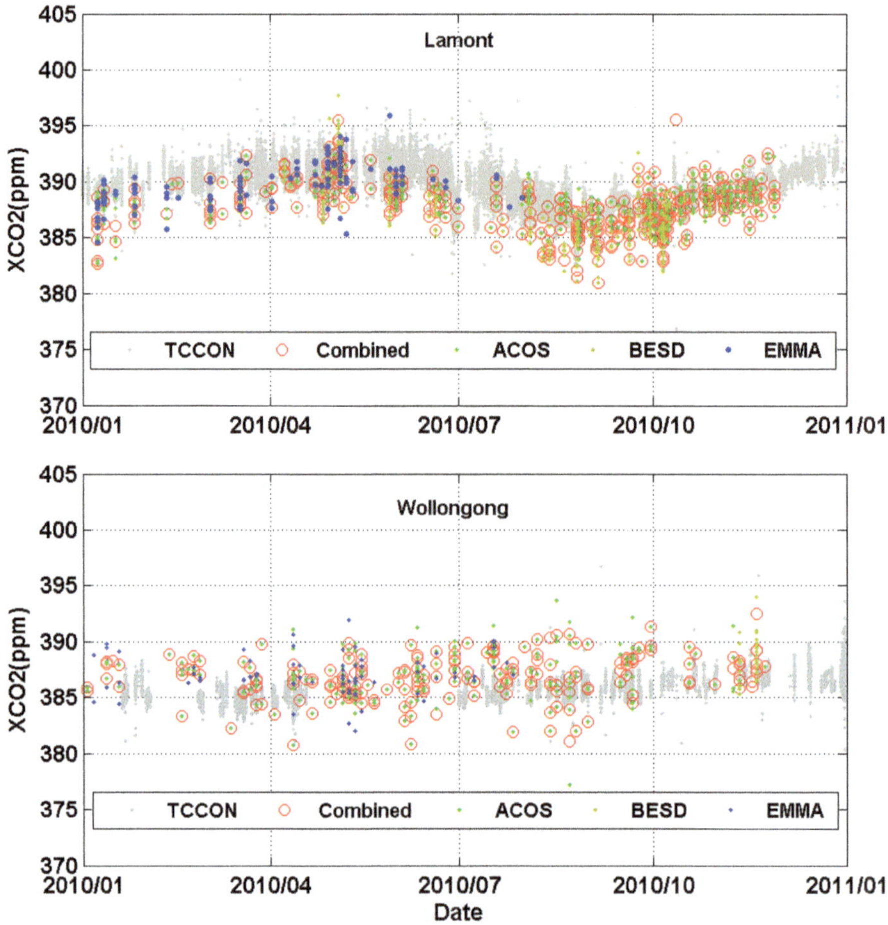

Figure 7. Comparison of XCO_2 measurements from TCCON, ACOS, BESD, EMMA, and our new combination method over Wollongong and Lamont sites (distance<0.25 degree, temporal difference<1 hour).

monthly scales with respect to that of GOSAT and it is even higher relative to the coverage of SCIAMACHY. Likewise, the cumulative fraction of coverage of the combined XCO_2 has risen to 25% when compared with 20% and 13% for ACOS and BESD, respectively. The increase in the XCO_2 spatial coverage indicates the potential advantage of the combined XCO_2 observations in generating global Level-3 XCO_2 maps when compared with any single dataset by providing more satellite-based XCO_2 retrievals used for optimal interpolating.

For evaluating the performance of our combination strategy, the combined XCO_2 values are compared with that retrieved from ACOS and BESD as well as XCO_2 in the EMMA database at two TCCON sites (Fig. 7). The results reveal that the XCO_2 values from the combination method show generally consistent variation in time with TCCON measurements except for a small overall bias (especially for the Lamont site). On the whole, the new combined XCO_2 product shows good consistency with the EMMA data, and they are comparable in terms of CO_2 magnitude, while the combined XCO_2 are shown with a longer time period, which is in line with the satellite observations, and possess more data points even over the same period.

Discussions and Conclusions

Despite the fact that space-based measurements can provide a unique opportunity to map atmospheric CO_2 over large areas, the number of valid CO_2 measurements from a single space-based instrument is generally limited for a certain day over a specific region due to the presence of clouds. In addition, although these Level-2 XCO_2 retrievals themselves are very important for inversion modeling of surface carbon sources/sinks, further comprehensive analysis by investigating the spatiotemporal full-coverage XCO_2 (Level 3) distribution is needed for interpreting their significant scientific merit [32]. While the limited satellite observations restrict the generation of Level-3 XCO_2 maps with high spatial and temporal resolutions when only a single satellite-based XCO_2 dataset is considered. This is our main motivation in this paper.

In this study, a strategy for combining SCIAMACHY and GOSAT CO_2 measurements has been proposed by fully accounting for the CO_2 global bias, differences in averaging kernels and overpass times, and the Level-2 retrieval errors of the CO_2 measurements being used. The results indicated that the average global coverage of both ACOS and BESD is less than 0.5% on a daily scale, and less than 6% on a monthly scale. While spatial coverage of combined XCO_2 can reach up to 0.65% and 8.42% on daily and monthly scales, respectively, the comparison analysis reveals that the combined XCO_2 product is consistent with TCCON and EMMA in both temporal variation and magnitude except for a small bias when compared with the TCCON measurements. All these findings herein prove the effectiveness of the combination method in supporting generation global full-coverage XCO_2 maps with higher temporal and spatial sampling by jointly using two space-based XCO_2 datasets. Similar to the existing studies (e.g. [32–34]), although these combined XCO_2 are not intended to be used in inverse modeling studies, they deliver a key complement for such research, and can be deemed as an independent dataset for comparison with model predictions. Similar to the existing study [31], an improved fusion approach (based on multiple XCO_2 datasets) to create Level-2 XCO_2 measurements that can be directly used for inverse modeling is also attempted and will be presented in another paper.

A last point that needs to be addressed is that although we employed CO_2 data of GOSAT and SCIAMACHY in this study, the proposed strategies are not restricted to such data. As a general strategy, it can be refined and adapted to further combine other XCO_2 products, such as OCO-2, CarbonSat, etc. in the future, and even to be applied to the fusion of other trace gases, such as O_3, CH_4.

Acknowledgments

The authors would like to thank the SCIAMACHY team at University of Bremen IUP/IFE as well as the ACOS scientific teams for providing us the CO_2 products. The authors also thank the anonymous reviewers for their helpful and valuable comments to improve this work.

Author Contributions

Conceived and designed the experiments: TW JS YJ TZ DJ CX. Performed the experiments: TW JS YJ TZ DJ CX. Analyzed the data: TW JS YJ TZ DJ CX. Contributed reagents/materials/analysis tools: TW JS YJ TZ DJ CX. Wrote the paper: TW JS YJ TZ DJ CX.

References

1. Reuter M, Bovensmann H, Buchwitz M, Burrows JP (2010) A method for improved SCIAMACHY CO_2 retrieval in the presence of optically thin clouds. Atmos. Meas. Tech 3: 209–232.
2. O'Dell CW, Connor B, Bösch H, O'Brien D, Frankenberg C, et al. (2012) The ACOS CO_2 retrieval algorithm – Part 1: Description and validation against synthetic observations. Atmos Meas Tech 5(1): 99–121.
3. Baker DF, Law RM, Gurney KR, Rayner P, Peylin P, et al (2006) TransCom 3 inversion intercomparison: Impact of transport model errors on the interannual variability of regional CO_2 fluxes, 1988–2003. Global Biogeochem Cy 20: GB1002, doi:10.1029/2004GB002439.
4. Morino I, Uchino O, Inoue M, Yoshida Y, Yokota T, et al. (2011) Preliminary validation of column-averaged volume mixing ratios of carbon dioxide and methane retrieved from GOSAT short-wavelength infrared spectra. Atmos Meas Tech 4: 1061–1076.
5. Butz A, Hasekamp OP, Frankenberg C, Aben I (2009) Retrievals of atmospheric CO_2 from simulated space-borne measurements of backscattered near-infrared sunlight: accounting for aerosol effects. Appl Opt 18: 3322–3336.
6. Kuze A, Suto H, Nakajima M, Hamazaki T (2009) Thermal and near infrared sensor for carbon observation Fourier-transform spectrometer on the Greenhouse Gases Observing Satellite for greenhouse gases monitoring, Appl Opt 35: 6716–6733.
7. Bovensmann H, Burrows JP, Buchwitz M, Frerick J, Noel S, et al (1999) SCIAMACHY –Mission objectives and measurement modes. J Atmos Sci 56: 127–150.
8. Crisp D, Atlas RM, Breon FM, Brown LR, Burrows JP, et al (2004) The Orbiting Carbon Observatory (OCO) mission. Adv Space Res 34: 700–709.
9. Boesch H, Baker D, Connor B, Crisp D, Miller C (2011) Global Characterization of CO_2 Column Retrievals from Shortwave-Infrared Satellite Observations of the Orbiting Carbon Observatory-2 Mission. Remote Sens 3: 270–304.
10. Aumann HH, Chahine MT, Gautier C, Goldberg MD, Kalnay E, et al. (2003) AIRS/AMSU/HSB on the Aqua Mission: Design, science objectives, data products, and processing systems. IEEE Trans Geosci Remote Sens 41: 253–264.
11. Chahine M, Barnet C, Olsen ET, Chen L, Maddy E (2005) On the determination of atmospheric minor gases by the method of vanishing partial derivatives with application to CO_2. Geophys Res Lett 32: L22803, doi:10.1029/2005GL024165.
12. Phulpin T, Cayla F, Chalon G, Diebel D, Schlussel P (2002) IASI on board Metop: Project status and scientific preparation. Proceedings of the 12th International TOVS Study Conference. Lorne, Vi ctoria, Australia.
13. Turquety S, Hadji-Lazaro J, Clerbaux C, Hauglustaine DA, Clough SA, et al. (2004) Operational trace gas retrieval algorithm for the Infrared Atmospheric Sounding Interferometer. J Geophys Res 109: D21301, doi:10.1029/2004JD004821.
14. Yang Z, Washenfelder RA, Keppel-Aleks G, Krakauer NY, Randerson JT, et al. (2007) New constraints on Northern Hemisphere growing season net flux. Geophys Res Lett 34: L12807. doi:10.1029/2007GL029742.
15. Wunch D, Toon GC, Blavier JFL, Washenfelder R, Notholt J, et al. (2011) The Total Carbon Column Observing Network. Philos T Roy Soc A 369: 2087–2112.
16. Rayner PJ, O'Brien DM (2011) The utility of remotely sensed CO_2 concentration data in surface source inversions. Geophys Res Lett 28: 175–178.
17. Houweling S, Breon FM, Aben I, Rodenbeck C, Gloor M, et al. (2004) Inverse modeling of CO_2 sources and sinks using satellite data: a synthetic inter-comparison of measurement techniques and their performance as a function of space and time. Atmos Chem Phys 4: 523–538.
18. Olsen SC, Randerson JT (2004) Differences between surface and column atmospheric CO_2 and implications for carbon cycle research. J Geophys Res 109; D02301. doi:10.1029/2003JD003968.
19. Miller CE, Crisp D, DeCola PL, Olsen SC, Randerson JT, et al. (2007) Precision requirements for space-based XCO_2 data. J Geophys Res 112: D10314, doi:10.1029/2006JD007659.
20. Buchwitz M, Beek R, Burrows JP, Bovensmann H, Warneke T, et al. (2005) Atmospheric methane and carbon dioxide from SCIAMACHY satellite data: initial comparison with chemistry and transport models, Atmos Chem Phys 5: 941–962.
21. Buchwitz M, Beek R, Nol S, Burrows JP, Bovensmann H, et al. (2005) Carbon monoxide, methane and carbon dioxide columns retrieved from SCIAMACHY by WFM-DOAS: year 2003 initial data set. Atmos Chem Phys 5: 3313–3329.
22. Buchwitz M, Rozanov VV, Burrows JP (2000) A near-infrared optimized DOAS method for the fast global retrieval of atmospheric CH_4, CO, CO_2, H_2O, and N_2O total column amounts from SCIAMACHY Envisat-1 nadir radiances. J Geophys Res 105: 15231–15245, doi:10.1029/2000JD900191.
23. Houweling S, Hartmann W, Aben I, Schrijver H, Skidmore J, et al. (2005) Evidence of systematic errors in SCIAMACHY-observed CO_2 due to aerosols. Atmos Chem Phys 5: 3003–3013.
24. Barkley MP, Frieβ U, Monks PS (2006) Measuring atmospheric CO_2 from space using Full Spectral Initiation (FSI) WFM-DOAS, Atmos Chem Phys 6: 3517–3534.
25. Schneising O, Buchwitz M, Burrows JP, Bovensmann H, Reuter M, et al. (2008) Three years of greenhouse gas column-averaged dry air mole fractions retrieved from satellite – Part 1: Carbon dioxide. Atmos Chem Phys 8: 3827–3853.
26. Reuter M, Bovensmann H, Buchwitz M, Burrows JP, Connor BJ, et al. (2011) Retrieval of atmospheric CO_2 with enhanced accuracy and precision from SCIAMACHY: Validation with FTS measurements and com-parison with model results. J Geophys Res 116: D04301, doi:10.1029/2010JD015047.
27. Bösch H, Toon GC, Sen B, Washenfelder RA, Wennberg PO, et al. (2006) Space-based near-infrared CO_2 measurements: Testing the Orbiting Carbon Observatory retrieval algorithm and validation concept using SCIAMACHY observations over Park Falls, Wisconsin. J Geophys Res 111: D23302, doi:10.1029/2006JD007080.
28. Butz A, Guerlet S, Hasekamp O, Schepers D, Galli A, et al. (2011) Toward accurate CO_2 and CH_4 observations from GOSAT. Geophys Res Lett 14: L14812. DOI:10.1029/2011GL047888.
29. Yoshida Y, Ota Y, Eguchi N, Kikuchi N, Nobuta K, et al. (2011) Retrieval algorithm for CO_2 and CH_4 column abundances from short-wavelength infrared spectral observations by the Greenhouse Gases Observing Satellite. Atmos Meas Tech 4: 717–734.
30. Crisp D, Fisher B, O'Dell C, Frankenberg C, Basilio R, et al. (2012)The ACOS CO_2 retrieval algorithm - Part II: Global XCO_2 data characterization. Atmos Meas Tech 5: 687–707.
31. Reuter M, Bösch H, Bovensmann H, Bril A, Buchwitz M, et al. (2013) A joint effort to deliver satellite retrieved atmospheric CO_2 concentrations for surface flux inversions: the ensemble median algorithm EMMA. Atmos Chem Phys 13: 1771–1780.
32. Hammerling DM, Michalak AM, O'Dell C, Kawa SR (2012) Global CO_2 distributions over land from the Greenhouse Gases Observing Satellite(GOSAT). Geophys Res Lett 39: L08804, doi:10.1029/2012GL051203.
33. Hammerling DM, Michalak AM, Kawa SR (2012) Mapping of CO_2 at high spatiotemporal resolution using satellite observations: Global distributions from OCO-2. J Geophys Res 117: D06306, doi:10.1029/2011JD017015.
34. Zeng Z, Lei L, Hou S, Ru F, Guan X, et al. (2014) A Regional Gap-Filling Method Based on Spatiotemporal Variogram Model of CO_2 Columns. IEEE Trans Geosci Remote Sens 6: 3594–3603.
35. Yokota T, Yoshida Y, Eguchi N, Ota Y, Tanaka T, et al. (2009) Global Concentrations of CO_2 and CH_4 Retrieved from GOSAT: First Preliminary Results. SOLA 5: 160–163.
36. Parker R, Boesch H, Cogan A, Fraser A, Feng L, et al. (2011) Methane observations from the Greenhouse Gases Observing SATellite: Comparison to

ground-based TCCON data and model calculations. Geophys Res Lett 15: L15807, doi:10.1029/2011GL047871.
37. Wang TX, Shi JC, Jing YY, Xie YH (2013) Investigation of the consistency of atmospheric CO_2 retrievals from different space-based sensors: Intercomparison and spatio-temporal analysis, Chin Sci Bull, 33: 4161–4170.
38. Peters W, Jacobson AR, Sweeney C, Andrews AE, Conway TJ, et al. (2007) An atmospheric perspective on North American carbon dioxide exchange: Carbon Tracker. Proc Natl Acad Sci U.S.A. 48: 18925–18930.
39. Wunch D, Wennberg PO, Toon GC, Connor BJ, Fisher B, et al. (2011) A method for evaluating bias in global measurements of CO_2 total columns from space. Atmos Chem Phys 11: 12317–12337.
40. Rodgers CD (2000) Inverse Methods for Atmospheric Sounding: Theory and Practice. World Scientific Publishing Co. Ltd.193 p.

17

Global Decline in Suitable Habitat for *Angiostrongylus* (= *Parastrongylus*) *cantonensis*: The Role of Climate Change

Emily M. York[1,2]*, Christopher J. Butler[2], Wayne D. Lord[1,2]

1 W. Roger Webb Forensic Science Institute, University of Central Oklahoma, Edmond, Oklahoma, United States of America, **2** Department of Biology, University of Central Oklahoma, Edmond, Oklahoma, United States of America

Abstract

Climate change is implicated in the alteration of the ranges of species worldwide. Such shifts in species distributions may introduce parasites/pathogens, hosts, and vectors associated with disease to new areas. The parasite *Angiostrongylus* (=*Parastrongylus*) *cantonensis* is an invasive species that causes eosinophilic meningitis in humans and neurological abnormalities in domestic/wild animals. Although native to southeastern Asia, *A. cantonensis* has now been reported from more than 30 countries worldwide. Given the health risks, it is important to describe areas with potentially favorable climate for the establishment of *A. cantonensis*, as well as areas where this pathogen might become established in the future. We used the program Maxent to develop an ecological niche model for *A. cantonensis* based on 86 localities obtained from published literature. We then modeled areas of potential *A. cantonensis* distribution as well as areas projected to have suitable climatic conditions under four Representative Concentration Pathways (RCP) scenarios by the 2050s and the 2070s. The best model contained three bioclimatic variables: mean diurnal temperature range, minimum temperature of coldest month and precipitation of warmest quarter. Potentially suitable habitat for *A. cantonensis* was located worldwide in tropical and subtropical regions. Under all climate change RCP scenarios, the center of the projected distribution shifted away from the equator at a rate of 68–152 km per decade. However, the extent of areas with highly suitable habitat (>50%) declined by 10.66–15.66% by the 2050s and 13.11–16.11% by the 2070s. These results conflict with previous studies, which have generally found that the prevalence of tropical pathogens will increase during the 21st century. Moreover, it is likely that *A. cantonensis* will continue to expand its current range in the near future due to introductions and host expansion, whereas climate change will reduce the total geographic area of most suitable climatic conditions during the coming decades.

Editor: Carles Lalueza-Fox, Institut de Biologia Evolutiva - Universitat Pompeu Fabra, Spain

Funding: Funding for this project was provided by the University of Central Oklahoma Office of Research and Grants. The funder had no role in study design, data collection and analysis, decision to publish, or preparation of the manuscript.

Competing Interests: The authors have declared that no competing interests exist.

* Email: eyork@ou.edu

Introduction

Changes in the distribution and phenology of many organisms were observed as the earth warmed by 0.6±0.2°C during the 20th century [1]–[4]. Since 1945, warming of the earth has been greater than any other time during the past 1,000 years [5]. Changing climate is predicted to drive 11% to 58% of vertebrate, invertebrate, and plant species to extinction by 2050 [6], and is also expected to promote expansion and/or geographic shift of tropical diseases into temperate areas [7]. Consequently, there is an urgent need to examine and model how climate change might alter infectious disease emergence within human, domestic, and wild animal populations worldwide [8].

Ecological niche modeling (ENM) predicts the fundamental and realized niche of species by relating point occurrence data of species to environmental factors [9], [10]. These models are useful in predicting the geographic range in which a species might be found, but are limited by the exclusion of detailed environmental characteristics (e.g. biotic interactions, heterogeneous landscapes). Maximum Entropy (Maxent) modeling uses environmental conditions and species presence only data to accurately estimate the distribution of a species [11]. By predicting the entire geographic range in which a species might occur, the fundamental niche of an organism is not limited by its realized niche. This approach can assess the relative importance of specific environmental factors to a species distribution, locate areas of current suitable habitat, and project changes in its distribution over time [11].

Epidemiology uses a multifaceted approach to monitor, predict and prevent disease outbreaks. ENM is a valuable epidemiological tool because it determines the functional geographic responses of parasites and pathogens to climate change, both proximate and future. Recent studies have incorporated ENM to assess the potential impacts of climate change on infectious diseases vectors, reservoirs and/or pathogens (e.g., leishmaniasis, monkeypox, Chagas' disease, malaria and blastomycosis) [12]–[18].

Angiostrongylus (= Parastrongylus) cantonensis is a parasitic nematode and a cause of the reemerging zoonotic disease, human

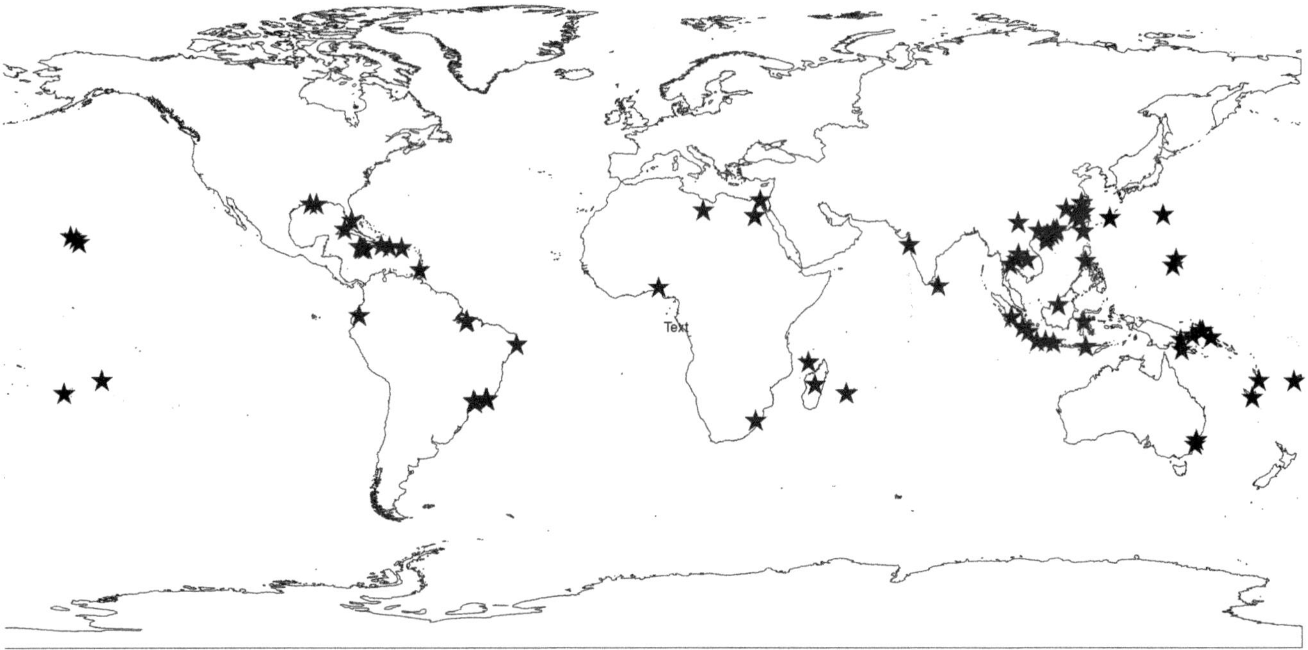

Figure 1. Current distribution of *A. cantonensis* based on locality data (n = 86) obtained from the literature.

Table 1. Comparison of the top three possible models.

Variables in Model	Log Likelihood*	AICc scores	ΔAICc	wAICc	Mean AUC
BIO 2, BIO 6, BIO 18	−1054.82	2152.646731	0	0.9987	0.945
BIO 2, BIO 6, BIO 15, BIO 18	−1045.22	2166.226719	13.579988	0.00112	0.946
BIO 2, BIO 6, BIO 12, BIO 18	−1047.25	2170.297934	17.6512034	0.00015	0.946

*Log-likelihood is the natural log of the probability of the data given in the model. AICc is a corrected AIC score, used for a small sample size by increasing the cost for each parameter. Delta AICc is the difference between the model with the lowest score (the "best" model) and the AICc score for each model. The model weight (wAICc) is the relative likelihood for each model, divided by the total relative likelihood for all models that were considered. AUC (area under the curve) is a measure of the accuracy of the model.

Table 2. Total area predicted to have >50% probability of suitable habitat conditions for *A. cantonensis* under each climate change scenario.

Scenario	Area (km^2)	% change in area	Area common to current (km^2)	% of current distribution retained
Current	6160942.17			
2050s-26	5357536.56	−13.04%	4665876.03	86.96%
2050s-45	5387894.55	−12.55%	4432958.46	87.45%
2050s-60	5504137.11	−10.66%	4517113.23	89.34%
2050s-85	5196405.69	−15.66%	4023687.78	84.34%
2070s-26	5353298.55	−13.11%	4639496.58	86.89%
2070s-45	5235326.19	−15.02%	4176861.57	84.98%
2070s-60	5339287.17	−13.34%	4108620.96	86.63%
2070s-85	5168469.42	−16.11%	3387207.87	83.89%

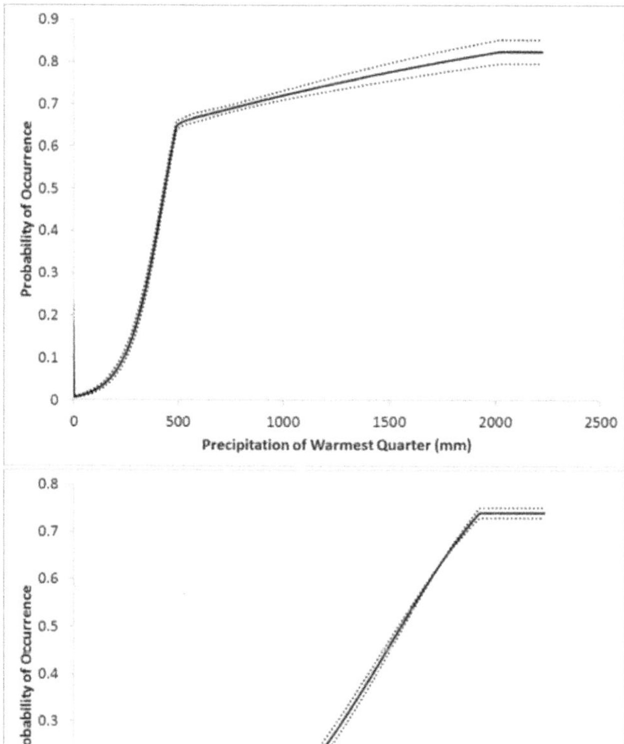

Figure 2. Probability of *A. cantonensis* presence in response to ecogeographical variables in the best-fit model.

eosinophilic meningitis, as well as neurological abnormalities in wildlife and domestic animals [19], [20]. Definitive and intermediate hosts for the parasite include rats and mollusks, respectively [21], [22]. Humans and other mammals are incidental hosts that become infected upon consumption of the third-stage larvae. Infection primarily occurs by consuming raw or undercooked mollusks or other infected paratenic hosts (e.g., freshwater prawns, frogs, monitor lizards) [21], [23], [24].

Angiostrongylus cantonensis was first documented in Guangzhou (Canton), China in 1935 [25]. During the past 50 years, the parasite has spread from Southeast Asia to over 30 countries worldwide [26], [27]. There have been more than 2,800 cases of *A. cantonensis* infection in humans worldwide with 116 cases involving U.S. citizens [27], as well as numerous infections in other animals. Given the rapid dispersal of the parasite and the health implications for humans and wildlife, there is a need to determine the potential distribution of *A. cantonensis*.

To our knowledge, no global model for the current and potential distribution of *A. cantonensis* has been published. Although Lv et al. [22] published a comprehensive distribution of *A. cantonensis* within China, their model did not examine the potential distribution worldwide. The aim of this study was to use Maxent modeling to determine the maximum range distribution for the parasite globally and predict the potential future distribution of *A. cantonensis* under Intergovernmental Panel on Climate Change (IPCC) climate change scenarios.

Methods

Maxent was used to model the current and projected distribution of *A. cantonensis*. Documented occurrences of *A. cantonensis* were collected from published records. Records that met one or more of the following criteria were incorporated into our models: 1) documentation of the parasite in accepted endemic areas; 2) multiple cases of human infection (3 or more) within an area; 3) reports of the parasite found in intermediate or definitive hosts (with ≥ 3% prevalence) (Table S1). A total of 86 locations were included (Figure 1). Elevation and 19 climate variables were downloaded from WorldClim [28] with a resolution of 5 arc-minutes (100 km^2; Table S2). All variables were included in the model initially. However, only the variables with the highest gain independent of others (Fig. S1) were retained, as these variables accounted for the greatest amount of the observed variation. In addition, the environmental variables that lowered the training gain the greatest when omitted were retained (Fig. S2), as these variables contained the most unique information. These variables were then retained for high multicollinearity ($|r|>0.8$) [29]. Additionally, Akaike's information criterion for small sample correction (AICc) was used to evaluate the regularization of the models and to avoid overfitting [30]. All possible combinations of the variables that did not exhibit high multicollinearity were examined. Ten-fold cross-validation was used and receiver operating characteristic (ROC) curves were created by plotting sensitivity vs. 1–specificity to evaluate the accuracy of the resulting model. The area under the curve (AUC) was used to evaluate models. Models with an AUC score of 0.5 indicated a model preforming no better than random, while models with AUC score of 1 indicated a perfect model [11], [31]. However, AUC scores are not without limitations [32], [33] and should be used in conjunction with other model evaluation methods [34]. Consequently, we used AICc scores and model weights along with AUC scores to determine the model that best describes the current distribution of *A. cantonensis*.

IPCC 5 data for future climate conditions for the 2050s and 2070s were obtained from the International Centre for Tropical Agriculture (CIAT) [35] to project the potential future distribution of *A. cantonensis* at 5 arc-minutes (100 km^2). Eleven IPCC models under four Representative Concentration Pathways (RCPs) (2.6, 4.5, 6.0, and 8.5) for the 2050s and 2070s were evaluated: BCC-CSM1-1, CCSM4, GISS-E2-R, HadGEM2-AO, HadGEM2-ES, IPSL-CM5A-LR, MIROC-ESM-CHEM, MIROC-ESM,

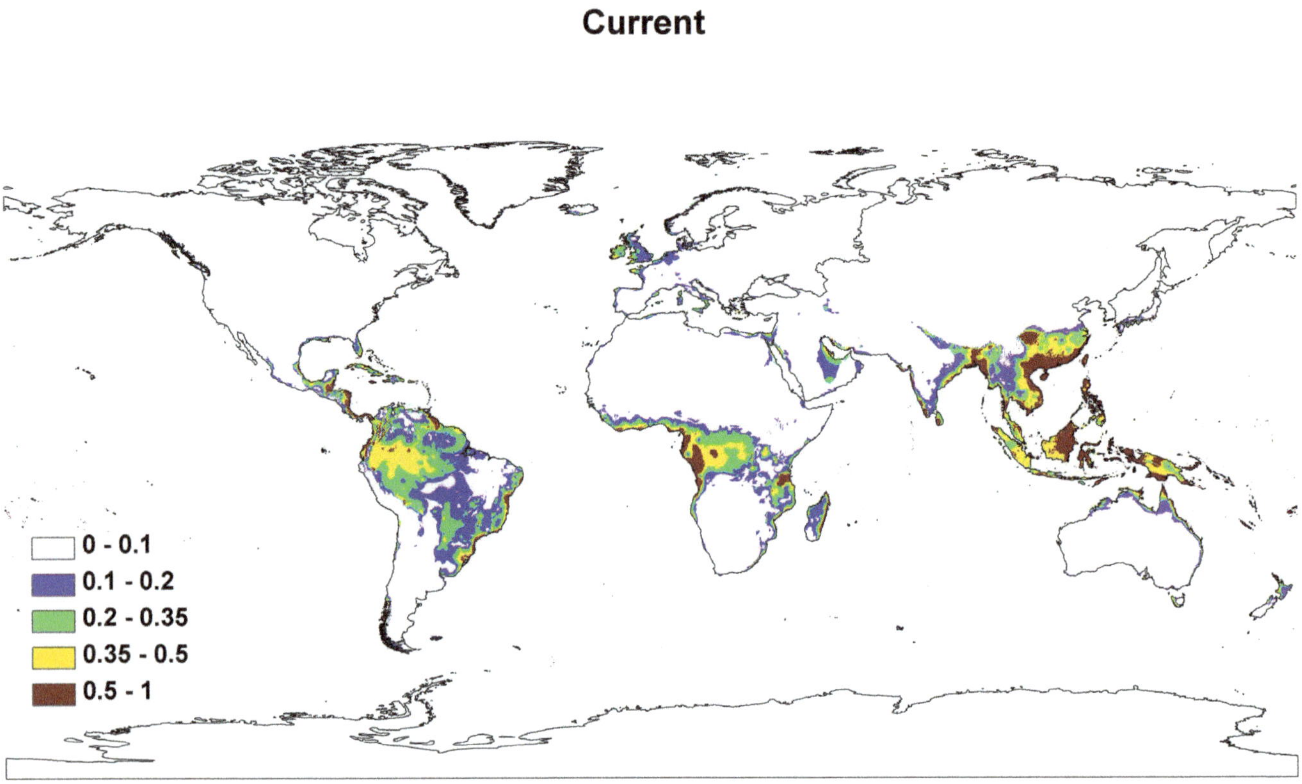

Figure 3. The Maxent model of the projected current distribution for *A. cantonensis.*

MIROC5, MRI-CGCM3, and NorESM1-M [35]. The eleven models for each RCP scenario were then averaged to produce a total of 8 future models (four for the 2050s and four for the 2070s). We present our results as mean ± standard deviation. The change in area for each RCP during the 2050s and 2070s was calculated for highly suitable habitat (>50%) and total suitable habitat (> 11.8%). We determined total suitable habitat using a 5% omission rate.

Results

The best model (i.e., the model with the lowest small sample corrected variant of the AICc score) included three environmental variables: mean diurnal temperature range (BIO 2), minimum temperature of coldest month (BIO 6), and precipitation of warmest quarter (BIO 18; Table 1). The AUC was 0.945±0.029 for this model. Figure 2 displays suitability in response to the three variables. Areas that were predicted to have suitability >50% had a mean diurnal temperature range of 5.13–8.76°C, a minimum temperature of the coldest month of 14.79–32.17°C, and precipitation of the warmest quarter of 438.37–2,224.20 mm (Fig. 2). Areas with >50% suitability were found primarily in tropical areas (Fig. 3), including the reported native range in southeast Asia. When examining total suitable habitat of >11.8%, we found an increase in area futher north and south of the equator, extending into Europe and to New Zealand (Fig. 3).

Areas with suitable climatic conditions for *A. cantonensis* are predicted to decline by the 2050s and the 2070s under all four RCP scenarios (Fig. 4). Currently, 6,160,942.17 km^2 are highly suitable (i.e. >50% chance of suitability) and 33,432,536.52 km^2

Table 3. Total area predicted to have >11.8% probability of suitable habitat conditions for *A. cantonensis* under each climate change scenario.

Scenario	Area (km^2)	% change in area	Area common to current (km^2)	% of current distribution retained
Current	33432536.52			
2050s-26	32577236.91	−2.56%	29954427.66	89.60%
2050s-45	33078100.5	−1.06%	30045069.18	99.50%
2050s-60	33265264.86	−0.50%	29835763.38	97.63%
2050s-85	32658191.55	−2.32	28675327.05	97.85%
2070s-26	32713026.21	−2.15%	29997067.23	98.25%
2070s-45	32847518.16	−1.75%	29113052.94	98.45%
2070s-60	32914374.93	−1.55	28863269.82	98.45%
2070s-85	32428906.56	−3.00%	26600172.48	97.00%

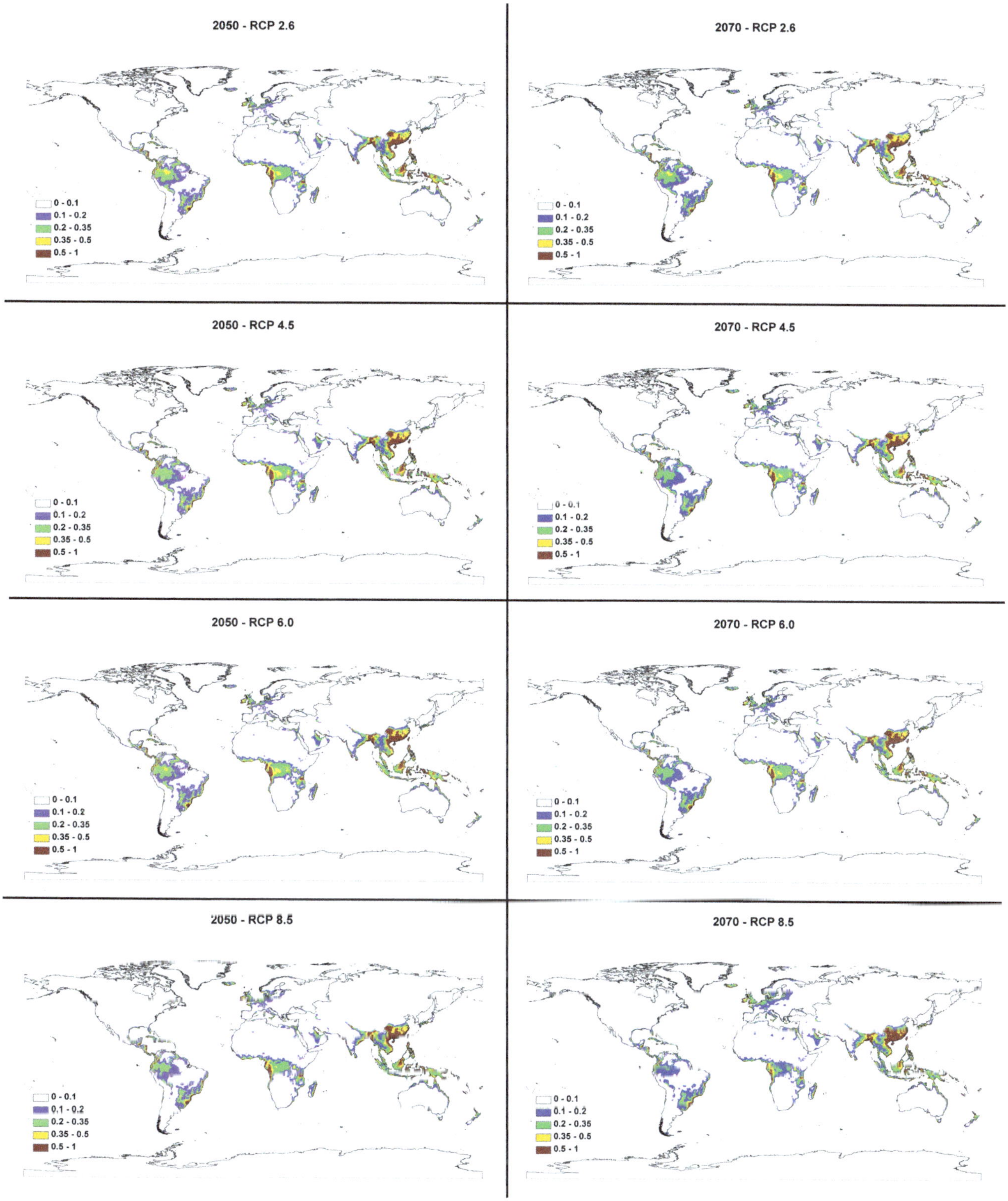

Figure 4. Comparison of model runs for *A. cantonensis*.

may be considered suitable (i.e. >11.8% chance of suitability). However, by the 2050s the amount of highly suitable habitat is expected to decrease to 5,196,405.69–5,504,137.11 km^2, with 84.34–89.34% of the area in common with the current model (Table 2). The total habitat suitability also indicates an overal decrease to 32,577,236.91–33,265,264.86 km^2 with 89.60–99.50% of the area in common with the current model (Table 3). By the 2070s, the area of highly suitable habitat will further decline in all four RCP scenarios, ranging from 5,168,469.42–5,353,298.55 km^2, (Table 2). The four models based on RCP

Table 4. Summary of the distance from each projected centroid for each RCP scenario (2.6, 4.5, 6.0, and 8.5) to the current centroid as well as the rate per decade.

Scenario	Distance (km) to current centroid	Rate of km per decade
Current North		
2050s–2.6	440.27 (NE)	88.05 km/decade
2050s–4.5	583.41 (NE)	116.68 km/decade
2050s–6.0	580.02 (NE)	116.00 km/decade
2050s–8.5	762.00 (NE)	152.40 km/decade
2070s–2.6	499.82 (NNE)	71.40 km/decade
2070s–4.5	695.15 (NE)	99.31 km/decade
2070s–6.0	679.03 (NNE)	97.00 km/decade
2070s–8.5	1027.87 (NNE)	146.84 km/decade
Current South		
2050s–2.6	479.80 (ESE)	95.96 km/decade
2050s–4.5	462.34 (SE)	92.47 km/decade
2050s–6.0	369.34 (SE)	73.87 km/decade
2050s–8.5	610.53 (ESE)	122.11 km/decade
2070s–2.6	547.55 (ESE)	78.22 km/decade
2070s–4.5	529.11 (ESE)	75.59 km/decade
2070s–6.0	566.90 (ESE)	80.99 km/decade
2070s–8.5	479.80 (ESE)	68.54 km/decade

scenarios had 83.89–86.89% of the area shared with the current model (Table 2). When examining total suitable habitat, the future models had 97.00–98.45% of the area in common with the current model (Table 3). The centroid (geometric center of the species range) for the northern hemisphere is predicted to shift northeast 440.27–762.00 km by the 2050s, a rate of 88.05–152.40 km per decade (Table 4, Fig. 5). By the 2070s, the centroid is expected to continue to shift to the north or northeast by 499.82–1,027.87 km, a rate of 71.40–146.84 km per decade. In the southern hemisphere, the centroid is expected to shift southeast and east-southeast 369.34–610.53 km by the 2050s for all scenarios at a rate of 73.87–122.11 km per decade (Table 4, Fig. 5). By the 2070s, the centroid is predicted to shift east-southeast 479.80–566.90 km, a rate of 68.54–80.99 km per decade (Table 4). The only exception is that the centroid for the RCP 2.6 scenario is expected to shift northward by the 2070s from the 2050s centroid but remain east-southeast of the current centroid (Fig. 5).

Discussion

The effects of global climate change are hypothesized to influence disease range expansions (via pathogen spread) and indirect expansions (via reservoirs, hosts, or vector range expansions). This will increase the frequency of disease outbreaks and expand the pool of at-risk populations [16], [36], [37]. Hales et al. [38], predict an increase in land area compatible for Dengue fever transmission by 2085, with 50–60% of the world's population at risk. Within North America, leishmaniasis reservoirs and vectors are predicted to undergo a range expansion northward, leading to greater human exposure [18]. However, an emerging picture of the effects of global climate change on disease is that an increase in habitat suitability in one area will be balanced by decreased suitability elsewhere, leading to a range shift or reduction [4], [39]. Although the proximate expansion of *A. cantonensis* into new suitable regions continues via introduction of definitive and intermediate hosts, our findings predict an ultimate decline of up to 16% in area of highly suitable bioclimatic habitat and up to a 3% decline in total suitable bioclimatic habitat by the 2070s.

The global model for the present distribution of *A. cantonensis* predicts that the most suitable habitat is located near the equator in tropical to subtropical regions. Three bioclimatic variables were found to contribute the most to predicting the potential distribution of the parasite: minimum temperature of the coldest month, minimum diurnal temperature range, and precipitation of the warmest quarter. Under all IPCC climatic scenarios, our models predict a shift in the distribution of suitable habitat for *A. cantonensis* in the Northern and Southern hemispheres. Under all RCP climatic scenarios, a shift in the distribution of the parasite is expected to occur north and east in the Northern hemisphere by the 2050s (range = 88.05–152.40 km per decade), continuing through the 2070s (range = 71.40–146.84 km per decade). The shift in the distribution of the parasite in the Southern hemisphere under all climatic scenarios is predicted to occur southeastward by the 2050s (range = 73.87–122.11 km per decade) and through the 2070s (range = 68.54–80.99 km per decade). Although there have been no endemic reports of *A. cantonensis* within Europe, all four models suggest an increase in suitable habitat for *A. cantonensis* within Europe while simultaneously showing an overall decline in global suitability. This potential range shift into Europe is most likely due to a predicted increase in the minimum temperature of the coldest month, which demonstrates the need for additional monitoring programs within Europe. These programs should include long-term surveying or screening for the parasite within definitive, paratenic, and intermediate hosts. Furthermore, an increase in public health programs targeted at awareness of the

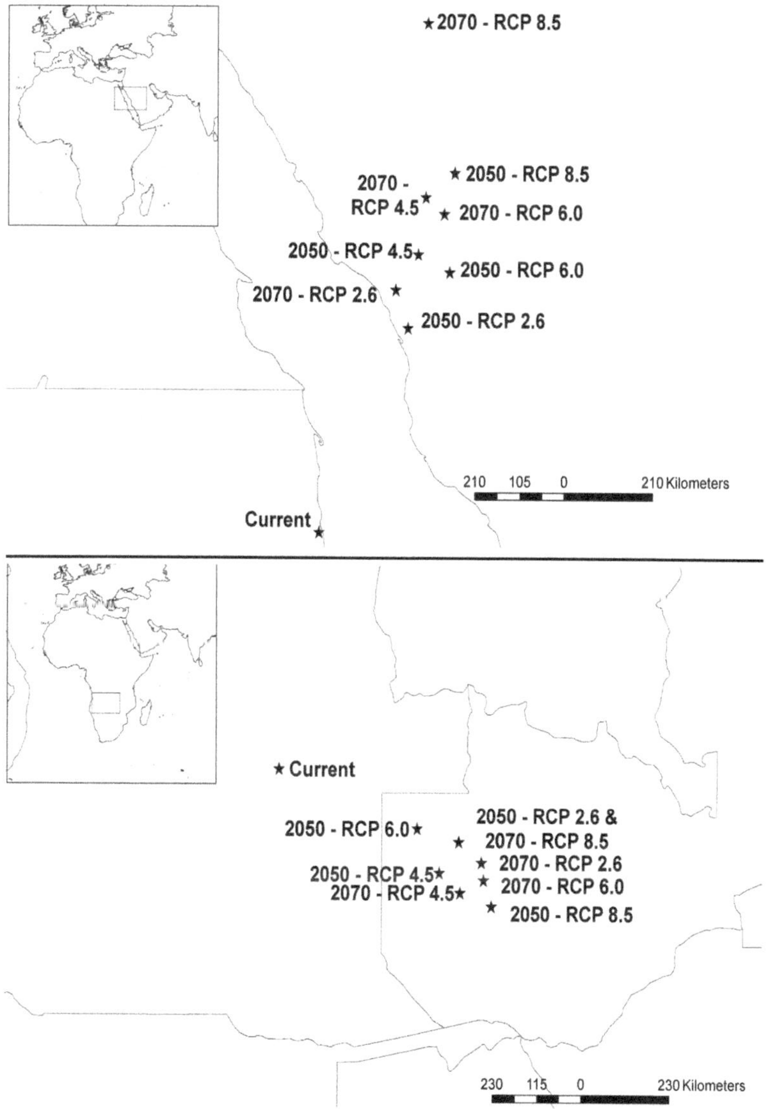

Figure 5. Northern and Southern Hemisphere centroids (indicated by stars) for all scenarios.

parasite and its transmission will be essential in deterring an increase in human infection.

Temperature and precipitation are environmental variables that significantly influence the distribution of *A. cantonensis*. Because temperature plays a critical role in influencing biological processes [4] it will likely have a significant impact on pathogens, infectious disease hosts, vectors and reservoirs. As global temperatures rise (IPCC), there is increased potential for vector-borne diseases and pathogens to spread and/or increase in severity [40]–[43]. Increases in temperature can speed the rate of development for some malarial protozoa, increasing the risk of transmission from mosquito to host [44]. However, the positive association between temperature and pathogen transmission might be offset by a pathogen's total bioclimatic requirements for survival. If such requirements are not met, host, vector, and/or pathogen mortality might increase. Similarly, increased temperatures might initially promote the spread and occurrence of *A. cantonensis*. However, with an expected temperature increase of 1.4°C–5.8°C from 1990 to 2100 [43] in areas where the bio-climatic norm exceeds an ecologically critical threshold temperature, resources needed to support parasitic growth and reproduction may become increasingly limited [4], [45]. Such demands could restrict the distribution of the parasite to areas with sufficient resources, potentially limiting disease incidence.

Global climate change is expected to increase the risk of intense precipitation and increased humidity in some regions, whereas other regions will experience extreme drought [35]. The effects of climatic variability in precipitation might induce the emergence of diseases in new areas or intensify infection rates of endemic pathogens. In several cases, disease occurrence has been demonstrated to be positively [41], [46]–[49] associated with rainfall. Alternatively, regions experiencing drought might negatively impact pathogen viability. Many parasites having intermediate hosts, such as *A. cantonensis,* require moist or wet environments for development and survival. Without sufficient precipitation, the distribution of the parasite might become more restricted, thereby decreasing the risk of transmission.

As *A. cantonensis* continues to spread, health complications in both humans and wildlife are expected to increase. Following introduction into a new area, *A. cantonensis* quickly infects and causes illness in humans, domestic animals and wildlife [19], [50]–[52]. Infected humans are often hospitalized with eosinophilic meningitis, and might also experience extraocular muscular paralysis [53]. In wildlife, *A. cantonensis* can cause a variety of symptoms (e.g. lethargy, limb paralysis) due to neurological invasion and might result in death [19], [50], [54]–[58]. To deter future outbreaks, *A. cantonensis* monitoring programs should be established worldwide, evaluating known definitive, intermediate and paratenic hosts and other wildlife. In addition, increased public awareness of the parasite, and the means by which it is transmitted, may lead to a lower incidence of infection.

These results provide the first global perspective of higher-risk areas for *A. cantonensis* colonization and overall changes in habitat suitability. The methodology employed here has been applied broadly to other studies on global climate change. More recently the application of ENM in evaluating disease distribution, risks and spread [12], [13], [17], [18], [59], [60] has proven useful. By identifying and documenting the distributions and ontogenetic requirements of known hosts (e.g., *Rattus* sp., molluscs) and conducting field surveys for the parasite, future studies might provide a much improved and conservative representation of the current and future range of *A. cantonensis*.

Acknowledgments

We thank William Caire, Joshua York, and Gloria Caddell for helpful comments on earlier drafts of the manuscript and technical support. We also thank the University of Central Oklahoma Department of Biology and W. Roger Webb Forensic Science Institute for logistic support.

Author Contributions

Conceived and designed the experiments: EMY CJB. Performed the experiments: EMY CJB. Analyzed the data: EMY CJB. Contributed reagents/materials/analysis tools: EMY CJB. Wrote the paper: EMY. Reviewed and edited manuscript: EMY CJB WDL.

References

1. Hughes L (2000) Biological consequences of global warming: is the signal already apparent?. Trends Ecol Evol 15: 56–61.
2. Butler CJ (2003) The disproportionate effect of climate change on the arrival dates of short distance migrant birds. Ibis 145: 484–495.
3. Parmesan C, Yohe G (2003) A globally coherent fingerprint of climate change impacts across natural systems. Nature 421: 37–42. doi:10.1038/nature01286.
4. Lafferty KD (2009) The ecology of climate change and infectious diseases. Ecology 90: 888–900.
5. Intergovernmental Panel on Climate Change (2007). Fourth Assessment Report of the Intergovernmental Panel on Climate Change. Cambridge: Cambridge University Press.
6. Thomas CD, Cameron A, Green RE, Bakkenes M, Beaumont LJ, et al. (2004) Extinction risk from climate change. Nature 427: 145–148. doi:10.1038/nature02121.
7. Epstein PR (2000) Is global warming harmful to health? Sci Am 283: 50–57.
8. Daszak P, Cunningham AA, Hyatt AD (2000) Emerging infectious diseases of wildlife–threats to biodiversity and human health. Science 287: 443–499.
9. Peterson AT, Kluza DA (2005) Ecological niche modeling as a new paradigm for large-scale investigations of diversity and distribution of birds. USDA Forest Service General Technical Reports PSW-GTR-191: 1201–1204.
10. Pearson RG, Dawson TP (2003) Predicting the impacts of climate change on the distribution of species: are bioclimate envelope models useful? Glob Ecol Biogeogr 12: 361–371.
11. Phillips SJ, Anderson RP, Schapire RE (2006) Maximum entropy modeling of species geographic distributions. Ecol Model 190: 231–259.
12. Peterson AT, Shaw J (2003) *Lutzomyia* vectors for cutaneous leishmaniasis in Southern Brazil: ecological niche models, predicted geographic distributions, and climate change effects. Int J Parasitol 33: 919–931.
13. Levine RS, Peterson AT, Yorita KL, Carroll D, Damon IK, Reynolds MG (2007) Ecological niche and geographic distribution of human monkeypox in Africa. PLoS One 2: e176.
14. Costa J, Peterson AT, Beard CB (2002) Ecological niche modeling and differentiation of populations of *Triatoma brasiliensis* Neiva, 1911, the most important Chagas disease vector in northeastern Brazil (Hemiptera, Reduviidae, Triatominae). Am J Trop Med Hyg 67: 516–520.
15. Tanser FC, Sharp B, Le Sueur D (2003) Potential effect of climate change on malaria transmission in Africa. The Lancet 362: 1792–1798.
16. Martens WJ, Niessen LW, Rotmans J, Jetten TH, McMichael AJ (1995) Potential impact of global climate change on malaria risk. Environ Health Perspect 103: 458–464.
17. Reed KD, Meece JK, Archer JR, Peterson AT (2008) Ecological niche modeling of *Blastomyces dermatitidis* in Wisconsin. PLoS ONE 3: e2034. Doi:10.1371/journal.pone.0002034.
18. González C, Wang O, Strutz SE, González-Salazar C, Sánchez-Cordero V, et al. (2010) Climate change and risk of Leishmaniasis in North America: predictions from ecological niche models of vector and reservoir species. PLoS Negl Trop Dis 4: e585. doi:10.1371/journal.pntd.0000585.
19. Kim DY, Stewart TB, Bauer RW, Mitchell M (2002) *Parastrongylus* (=*Angiostrongylus*) *cantonensis* now endemic in Louisiana. J Parasitol 88: 1024–1026.
20. Qvarnstrom Y, Sullivan JJ, Bishop HS, Hollingsworth R, da Silva AJ (2007) PCR-based detection of *Angiostrongylus cantonensis* in tissue and mucus secretions from molluscan hosts. Ap Appl Environ Microbiol 73: 1415–1419.
21. Chikweto A, Bhaiyai M, Maepherson CNL, DeAllie C, Pickney RD, et al. (2009) Existence of *Angiostrongylus cantonensis* in rats (*Rattus norvegicus*) in Grenada, West Indies. Vet Parasitol 162: 160–162.
22. Lv S, Zhang Y, Liu H-X, Hu L, Yang K, et al. (2009) Invasive snails and an emerging infectious disease: results from the first national survey on *Angiostrongylus cantonensis* in China. PLoS Negl Trop Dis 3: e368.
23. Lai C-H, Yen C-M, Chin C, Chung H-C, Kuo H-C, et al. (2007) Eosinophilic meningitis caused by *Angiostrongylus cantonensis* after ingestion of raw frogs. Am J Trop Med Hyg 76: 399–402.
24. Radomyos P, Tungtrongchitr A, Praewanich R (1992) Experimental infection of yellow tree monitor (*Varanus bengalensis*) and related species with *Angiostrongylus cantonensis*. Southeast Asian J Trop Med Public Health 23: 167–168.
25. Chen HT (1935) Un nouveau nematode pulmonaire, *Pulmonema cantonensis* n. g., n. sp., des rats de Canton. Ann Parasitol Hum Comp 13: 312–317.
26. Kliks MM, Palumbo NE (1992) Eosinophilic meningitis beyond the Pacific Basin: the global dispersal of a peridomestic zoonosis by *Angiostrongylus cantonensis*, the nematode lungworm of rats. Soc Sci Med 34: 199–212.
27. Wang QP, Lai D-H, Zhu X-Q, Chen X-G, Lun Z-R (2008) Human angiostrongyliasis. Lancet Infect Dis 8: 621–630.
28. Hijmans RJ, Cameron SE, Parra JL, Jones PG, Jarvis A (2005) Very high resolution interpolated climate surfaces for global land areas. Int J Climatol 25: 1965–1978.
29. Jones CC, Acker SA, Halperin CB (2010) Combining local- and large-scale models to predict the distributions of invasive plant species. Ecol Appl 20: 311–326.
30. Warren DL, Seifert SM (2011) Ecological niche modeling in Maxent: the importance of model complexity and the performance of model selection criteria. Ecol Appl 21: 335–342.
31. Phillips SJ, Dudik M, Schapire RE (2004) A maximum entropy approach to species distribution modeling. Pgs. 655–662 In Proceedings of the Twenty-First International Conference on Machine Learning.
32. Lobo JM, Jiménez-Valverde A, Real R (2008) AUC: a misleading measure of the performance of predictive distribution models. Glob Ecol Biogeogr 17: 145–151. doi:10.1111/j.1466-8238.2007.00358.x.
33. Elith J, Phillips SJ, Hastie T, Miroslav D, Chee YE, et al. (2011) A statistical explanation of Maxent for ecologists. Divers Distrib 17: 43–57.

34. So H-C, Sham PC (2010) A unifying framework for evaluating the predictive power of genetic variants based on the level of heritability explained. PLoS Genet 6: e1001230. doi:10.1371/journal.pgen.1001230.
35. Ramirez J, Jarvis A (2008) High resolution statistically downscaled future climate surfaces. Available: http://www.ccafs-climate.org/. Accessed 2014 April 27.
36. Patz JA, Martens WJ, Focks DA, Jetten TH (1998) Dengue fever epidemic potential as projected by general circulation models of global climate change. Environ Health Perspect 106: 147–153.
37. Harvell CD, Mitchell CE, Ward JR, Altizer S, Dobson AP, et al. (2002) Climate warming and disease risks for terrestrial and marine biota. Science 296: 2158–2162.
38. Hales S, de Wet N, Maindonald J, Woodward (2002) Potential effect of population and climate changes on global distribution of dengue fever: an empirical model. Lancet 360: 830–834.
39. Kutz SJ, Jenkins EJ, Veitch AM, Ducrocq J, Polley L, et al. (2009) The Arctic as a model for anticipating, preventing, and mitigating climate change impacts on host–parasite interactions. Vet Parasitol 163: 217–228.
40. Kutz SJ, Hogberg EP, Polley L, Jenkins EJ (2005) Global warming is changing the dynamics of Arctic host–parasite systems. Proc R Soc B 272: 2571–2576.
41. Lal A, Baker MG, Hales S, French NP (2012) Potential effects of global environmental changes on cryptosporidiosis and giardiasis transmission. Trends Parasitol 29: 83–90.
42. Poulin R (2006) Global warming and temperature-mediated increases in cercarial emergence in trematode parasites. Parasitology 132: 143–151.
43. Houghton JT, Ding Y, Griggs DJ, Noguer M, van der Linden PJ, et al. (2001) Climate Change 2001: The Scientific Basis. Cambridge: Cambridge University Press.
44. McArthur RH (1972) Geographical Ecology: Patterns in the distribution of a species. New Jersey: Princeton University Press.
45. Smith G (1990) The population biology of the free living phase of *Haemonchus contortus*. Parasitology 101: 309–316.
46. Bouma MJ, Hugo JK (1996) The El Niño Southern Oscillation and the historic malaria epidemics on the Indian subcontinent and Sri Lanka: an early warning system for future epidemics?. Trop Med Int Health 1: 86–96.
47. Bouma MJ, Poveda G, Rojas W, Chavasse D, Quinones M, et al. (1997) Predicting high-risk years for malaria in Colombia using parameters of El Niño Southern Oscillation. Trop Med Int Health 2: 1122–1127.
48. Woodruff RE, Guest CS, Garner MG, Becker N, Lindesay J, et al. (2002) Predicting Ross River virus epidemics from regional weather data. Epidemiology 13: 384–393.
49. Curriero FC, Patz JA, Rose JB, Lele S (2001) The association between extreme precipitation and waterborne disease outbreaks in the United States, 1948–1994. Am J Public Health 91: 8 1194–1199.
50. Duffy MS, Miller CL, Kinsella JM, de Lahunta A (2004) *Parastrongylus cantonensis* in a nonhuman primate, Florida. Am J Trop Med Hyg 10: 2207–2210.
51. New D, Little MD, Cross J (1995) *Angiostrongylus cantonensis* infection from eating raw snails. N Engl J Med 332: 1105–1106.
52. Panackel C, Cherian G, Vijayakumar K, Sharma RN (2006) Eosinophilic meningitis due to *Angiostrongylus cantonensis*. Indian J Med Microbiol 24: 220–221.
53. Morgan U (2000) Detection and characterization of parasites causing emerging zoonoses. Int J Parasitol 30: 1407–1421.
54. Higgins D, Carlisle-Novack M, Mackie J (1997) Neural angiostrongylosis in three captive rufous bettongs (*Aepyprymnus rufescens*). Aust Vet J 75: 564–566.
55. Mason K (1987) Canine neural angiostrongylosis: The clinical and therapeutic features of 55 natural cases. Aust Vet J 64: 201–203.
56. McKenzie R, Green P, Wood A (1978) *Angiostrongylus cantonensis* infection of the brain of a captive Bennett's wallaby (*Macropus rufogrieseus*). Aust Vet J 54: 86–88.
57. Wright J, Kelly W, Wadell A, Hamilton J (1991) Equine neural angiostrongylosis. Aust Vet J 68: 58–60.
58. Monks DJ, Carlisle MS, Carrigan M, Rose K, Spratt D, et al. (2005) *Angiostrongylus cantonensis* as a cause of cerebrospinal disease in a yellow-tailed black cockatoo (*Calyptorhynchus funereus*) and two tawny frogmouths (*Podargus strigoides*). J Avian Med Surg 19: 289–293.
59. Peterson AT, Bauer JT, Mills JN (2004) Ecologic and geographic distribution of filovirus disease. Emerg Infect Dis 10: 40–47.
60. Levine RS, Peterson AT, Benedict MQ (2004) Geographic and ecologic distributions of the *Anopheles gambiae* complex predicted using a genetic algorithm. Am J Trop Med Hyg 70: 105–109.

18

DNA Barcodes and Species Distribution Models Evaluate Threats of Global Climate Changes to Genetic Diversity

Wei-wei Zhou[1], Bao-lin Zhang[1,2], Hong-man Chen[1], Jie-qiong Jin[1], Jun-xiao Yang[1], Yun-yu Wang[1], Ke Jiang[1], Robert W. Murphy[1,3], Ya-ping Zhang[2]*, Jing Che[1]*

1 State Key Laboratory of Genetic Resources and Evolution, and Yunnan Laboratory of Molecular Biology of Domestic Animals, Kunming Institute of Zoology, Chinese Academy of Sciences, Kunming, China, **2** Laboratory for Conservation and Utilization of Bio-resources, Yunnan University, Kunming, China, **3** Centre for Biodiversity and Conservation Biology, Royal Ontario Museum, Toronto, Ontario, Canada

Abstract

Anthropogenic global climate changes are one of the greatest threats to biodiversity. Distribution modeling can predict the effects of climate changes and potentially their effects on genetic diversity. DNA barcoding quickly identifies patterns of genetic diversity. As a case study, we use DNA barcodes and distribution models to predict threats under climate changes in the frog *Nanorana parkeri*, which is endemic to the Qinghai-Tibetan Plateau. Barcoding identifies major lineages W and E. Lineage W has a single origin in a refugium and Lineage E derives from three refugia. All refugia locate in river valleys and each greatly contributes to the current level of intraspecific genetic diversity. Species distribution models suggest that global climate changes will greatly influence *N. parkeri*, especially in the level of genetic diversity, because two former refugia will fail to provide suitable habitat. Our pipeline provides a novel application of DNA barcoding and has important implications for the conservation of biodiversity in southern areas of the Qinghai-Tibetan Plateau.

Editor: Dirk Steinke, Biodiversity Insitute of Ontario - University of Guelph, Canada

Funding: This work was supported by grants of the Strategic Priority Research Program (B) (XDB03030113), the Ministry of Science and Technology of China (MOST) (2014FY210200), and the "Western Light" Talents Training Program of the Chinese Academy of Sciences (CAS) to ZWW; the Key Research Program (KJZD-EW-L07) of CAS, the National Natural Science Foundation of China (31090250), MOST (2011FY120200), and the Bureau of Science and Technology of Yunnan, China (2010CI045) to JC; and the Major Innovation Program of CAS (KSCX2-EW-Z-2) to ZYP. Research also was supported by Discovery Grant 3148 from the Natural Sciences and Engineering Research Council of Canada to RWM. The funders had no role in study design, data collection and analysis, decision to publish, or preparation of the manuscript.

Competing Interests: The authors have declared that no competing interests exist.

* Email: chej@mail.kiz.ac.cn (JC); zhangyp@mail.kiz.ac.cn (YPZ)

Introduction

Climatic changes influence organisms and an understanding of how this occurs is important for conservation. More than one line of evidence documents the impact anthropogenic global climate change (GCC) exerts on organisms [1]. Explorations into how past climate changes influenced organisms may serve to predict future impacts of GCC. Genetic diversity is important in conservation because higher levels maintain the evolutionary potential of species. However, the distribution of genetic diversity is often uneven across the range of a species and many factors may contribute to this. Environmental changes during glacial-interglacial cycling in the Quaternary is one of the most important historical drivers of genetic patterns [2,3]. For example, by retaining suitable habitat over several glacial cycles, refugia hold higher levels of genetic diversity compared with recently occupied areas [2,3]. Refugia also drive genetic distinctiveness within species owing to providing long-term geographic isolation. A clear understanding a species' evolutionary history and its drivers is important for planning conservation [4].

Genetic analyses form the cornerstone of conservation planning, especially in defining objective evolutionarily significant units (ESUs) and management units (MUs) [5]. DNA barcoding [6], which uses a short, universal genetic marker (*COI* in eukaryotes) to identify matrilines and species, may serve to efficiently identify ESUs and MUs.

Species distribution models (SDMs) also provide information useful for conservation planning [7–9]. Comparisons between SDMs for current and the Last Glacial Maximum (LGM) may provide a chance to explore the impacts of past climate changes to organisms. As an extension of this application, SDMs can compare the distributions of current and future habitats. This allows assessments of the risk of local extirpation and extinction caused by future habitat degradation. Whereas both DNA barcoding and SDMs provide valuable insights, their synthesis may serve to evaluate threats of GCC to organisms.

The Qinghai-Tibetan Plateau (QTP), which covers more than 2.5 million km^2 and has an average elevation of about 4000 m above sea level [10], is the largest and highest plateau on Earth. Unlike North America and Europe, no unified ice sheet formed on the QTP during the LGM, yet its environment changed substantially. The average temperature was from 6°C to 9°C lower than today and precipitation decreased by from 30% to 70% [11].

In addition to climatic drivers, geographic features also play a role in the formation of patterns of genetic diversity. In the southern QTP, two mountains stretch from east to west and the Yarlung Zangpo River (YZR; Brahmaputra River) (Fig. S2) occurs between them. The Nianqingtanggula Mountains (NM; Nyainqentanglha Mountains) in the east and Gangdisi Mountains (GM; Kailas Range) in the west form the northern mountains and the Himalayas form the southern boundary. Several rivers flow southward across the latter. The complex geography may have offered refugia during dramatic Quaternary climatic changes. Analyses of matrilineal dispersal are likely to reveal such history. Driven by the GCC, future severe temperature changes of the QTP may exceed those of lower elevations [12]. Thus, the GCC may impose enormous impacts on organisms living in the QTP. We assess this possibility by synthesizing DNA barcoding and SDMs analyses.

Because of their relatively low mobility and physiological requirements, amphibians are sensitive to environmental changes, especially temperature and precipitation. Further, amphibians retain high-resolution signals of historical responses to environmental perturbations [13]. The environment on the QTP is harsh for amphibians; suitable environment is rare except in the eastern and southern edges. A previous study on *Rana kukunoris*, a common frog on eastern QTP, revealed a unique genetic pattern compared to other taxa in the region [14]. Comparative amphibian studies from the southern QTP will yield insights into how environmental changes affect the biota fauna of the QTP.

Nanorana parkeri, a median-sized frog belonging to the family Dicroglossidae, is endemic to southern and southeastern Tibet. It occurs at elevations ranging from 2800 to 5000 m in valleys of the YZR and north of the NM [15]. Climate changes during LGM greatly affected this area. Our study evaluates samples from across the entire range of *N. parkeri*. Using the species as model system, we test the effectiveness of a pipeline that combines DNA barcoding and SDMs to identify future threats to genetic diversity in the face of GCC.

Materials and Methods

Ethics Statement

All the species included in our study (Table S1) are not endangered or protected species according to the "Law of the People's Republic of China on the Protection of Wildlife" and "Regulations for the Implementation of the People's Republic of China on the Protection of terrestrial Wildlife" (State Council Decree [1992] No. 13). The permission for field work in four major management areas (39 localities) including Xigaze, Lhasa, Nyingchi and Qamdo was issued by the Forestry Department of Xizang [Tibet] Autonomous Region, China. Local people of the Forestry Bureaus involved and helped during the whole survey. Totally, 549 tissue samples including toe tips, muscle, livers, tadpoles and egg masses were obtained, following the Animal Use Protocols approved by the Animal Care and Ethics Committee of the Kunming Institute of Zoology, Chinese Academy of Sciences. For egg masses, we separated 5–10 eggs from egg mass and stored them in 95% alcohol after removing egg jelly. For adult, within each locality only five individuals were euthanized using clove oil firstly. Following euthanization, tissues dissected from adult specimens were preserved in 95% ethanol. More other adults were just cut two toe tips and then released. Tadpoles were also euthanized using clove oil firstly, then stored in 95% alcohol after removing gut. Table S1 lists our samplings information, including species name, locality, GPS coordinates, and accession nos. in Genbank etc.

Population sampling and laboratory techniques

Our samplings covered the entire documented distribution range of *N. parkeri* (Fig. 1, Table S1). One individual of *N. pleskei*, *N. ventripunctata* and *N. liebigii* was used as an outgroup representative [16,17].

We extracted genomic DNA using standard phenol–chloroform extraction protocol [18]. Partial sequences of cytochrome c oxidase subunit I (*COI*) were sequenced for all individuals using universal primers [19]. PCR products were purified and used as the template DNA for cycle sequencing reactions performed using BigDye Terminator Cycle Sequencing Kit (v.2.0, Applied Biosystems, Foster City, USA) and an ABI PRISM 3730 DNA Analyzer.

Sequence alignment and phylogenetic analyses

Nucleotide sequences were checked by eye using LASERGENE 7.0 and aligned using CLUSTALX 1.81 [20] with default parameters. Subsequently, the aligned sequences were checked and optimized in MEGA 4.0 [21]. Identical haplotypes for mtDNA were collapsed using DNASP 5.10 [22]. The overall value of nucleotide diversity (π) and haplotype diversity (H) were also estimated using DNASP.

Phylogenetic analyses of the *COI* data were conducted using Bayesian inference (BI), maximum likelihood (ML) and maximum parsimony (MP). BI analyses were performed using MrBayes 3.1.2 [23]. We tested three different partition strategies based on codon positions (no partition; 1+2, 3; and 1, 2, 3). The best strategy was chosen based on the Bayes factor test [24] (Table 1), as it represented a robust method testing partitioning strategies [25]. Nucleotide substitution models were selected for each data partition using the Akaike information criterion in MrModeltest v2.3 [26]. BI analyses for each partition strategy were run 3 million generations while sampling trees every 1000 generations. The first 50% of the sampled trees were discarded as burn-in. The final analyses employing the best partition strategy were run for 10 million generations. AWTY [27] was used to confirm satisfactory convergence of the topological split frequencies. MP analyses were implemented using PAUP* 4.0b10 [28]. Heuristic searches with TBR were executed for 1000 random pseudoreplicates with all characters treated as unordered and equally weighted. Bootstrap analyses were conducted using 1000 replicates to assess nodal reliabilities. ML analyses were conducted using RaxML 7.0.4 [29] and the GTR+I+G model was implemented for each data partition.

We built a median-joining network (MJN) using NETWORK 4.5 [30] to visualize the frequencies of the haplotypes. To remove excessive links and median vectors, we used the MP option [31].

Divergence time estimation

Different demographic and molecular clock models were compared using path sampling and stepping-stone sampling (Table 2) [32]. Employing the best models, time to most recent common ancestor was estimated using BEAST 1.7.5 [33]. Due to the absence of a reliable fossil record and an established substitution rate of *COI* for *N. parkeri*, we used the divergence time between *N. parkeri* and *N. pleskei* (8.9±2.7 Ma) [16] as a secondary calibration point. Model selection and partition scheme were the same as used in the BI analyses. The final analyses involved two independent runs of 30 million generations each, while sampling trees every 1000th generation. Burn-in and convergence of the chains were determined with Tracer 1.5

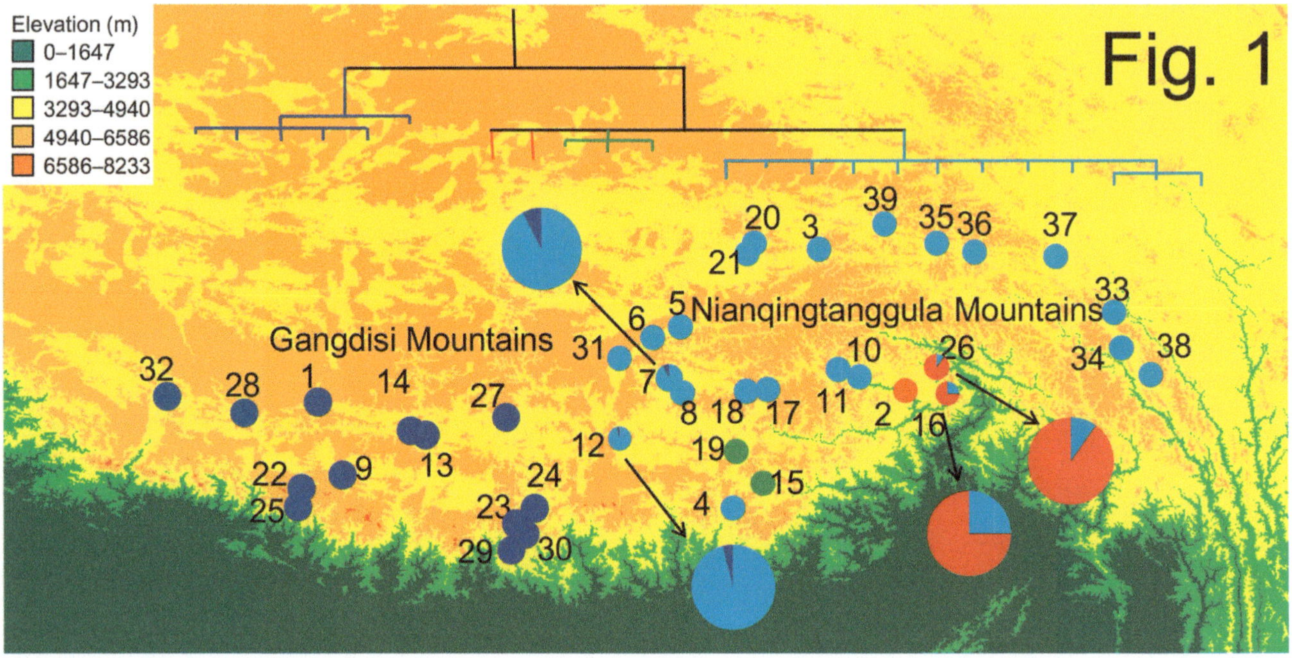

Figure 1. Sampling sites and the maternal genealogy based on mtDNA sequence data. Site numbers refer to Table S1. Colors of the matrilines match those of samplings localities on the map. Four localities containing two matrilines each are show in larger pie figures.

[34]. The measures of effective sample sizes were used to determine the Bayesian statistical significance of each parameter.

Population structure and demographic analyses

We explored population structure and genetic diversity landscape using SPADS 1.0 [35]. Groups of populations were defined as for SAMOVA [36]. We explored values of K (number of groups) ranging from 2 to 10 with 100 simulated annealing processes. The optimum value of K was identified by exploring the behavior of the indices F_{CT} and F_{SC}. Spatial patterns of genetic diversity based on allelic richness (Ar) and π across the landscape were explored using the GDivPAL function in SPADS.

To test for the influences of mountains and rivers on the genetic structure of *N. parkeri*, we measured the population structure by three independent analyses of molecular variance (AMOVA) [37]: populations north and south of the YZR; populations north of NM, west of the boundary of GM and NM, and the remaining ones; and four groups according the results of the SAMOVA. Analyses were performed using Arlequin 3.5 [38] and significance was assessed by 10000 permutations.

We investigated past changes of effective population size using the neutral test and mismatch distributions. Tajima's *D* [39] and Fu's *Fs* statistics [40] with 10000 coalescent simulations were calculated using Arlequin. Pairwise mismatch distributions [41] were calculated for each group in Arlequin. The expected distribution under a model of sudden demographical expansion was generated using 10000 permutations. The significance of deviations from this model was tested using the sum of squared deviation (SSD) and raggedness index (Rag). All analyses were performed for each lineage separately because population subdivisions could have masked the effects of expansions.

Species distribution modeling

We inferred the potential geographic range of *N. parkeri* using the maximum entropy model implemented in MAXENT 3.3.3 [42,43]. Environmental variables from the WorldClim database with resolutions of 2.5 arc-minutes [44] were downloaded as environmental layers. Because of the controversy about whether correlated variables should be removed or not, we used all of the 19 bioclimatic layers [45]. All layers were cropped to span from 83°E to 99°E and from 26°N to 33°N.

Random null distributions were built to test for the significance of our SDM. For this test, we built a new SDM using 39 random points (as same number as our sampling localities). This process was repeated 100 times and the areas under the curves (AUCs) were used as null distributions [46].

Assuming niche conservatism over time [47–49], we predicted the former distribution of *N. parkeri* by projecting our model on

Table 1. Bayes factor of each partition strategy.

Partition Strategy	Ln(BF)	Standard Deviation
no partition	−1885.375	0.134
1+2, 3	−1844.969	0.14
1, 2, 3	−1885.375	0.12

2lnBF (H_1–H_0) >10 was treated as decisive support for each hypothesis.

Table 2. Results of demographic model selection.

	strick		lognormal relaxed clock	
	PS	SS	PS	SS
constant	−1553.29	−1553.31	−1553.33	−1553.34
exponential	−1556.44	−1556.53	−1556.24	−1556.27
logistic	−1557.29	−1557.36	−1555.73	−1555.77
expansion	−1554.42	−1554.45	−1555.05	−1555.07
bayesian skyline	−1544.51	−1544.56	−1545.42	−1545.46

PS: path sampling; SS: stepping-stone sampling.

LGM climatic layers. Predicted distributions during the LGM were generated by downloading both the community climate system model (CCSM) [50] and model for interdisciplinary research on climate (MIROC) [51] from the WorldClim database. To predict the influences of GCC to this species, we also projected the model to climate data of the 2080 s based on MIROC model under the A1b scenario. In this way, we predicted the distribution changes in future.

Results

Sequence information

We obtained 549 partial *COI* sequences from 39 localities of *N. parkeri*, plus one sequence each from *N. pleskei*, *N. ventripunctata* and *N. liebigii*. The fragment consisted of 539 base pairs, of which 35 positions exhibited variation and 27 were potentially parsimony informative, resulting in 23 haplotypes in *N. parkeri*. All sequences were deposited in GenBank (Table S1). The overall value of π was 0.01746±0.00055 and H was 0.678±0.016.

Genealogical analyses of *COI*

Bayes factor test showed a preference for the partition strategy of 1+2, 3 (Table 1). The best fit substitution model for the first and second codon was HKY and GTR+G model was the best fit for third codon.

Matrilineal genealogies obtained from BI, ML and MP analyses were nearly identical (Fig. 1 & Fig. S1) and all methods recovered lineages East (E) and West (W). The boundary of GM and NM separated these two lineages. Lineage W was comprised of 6 haplotypes, which had no clear structure. Lineage E was comprised 17 haplotypes of which 12 formed highly supported sublineage E1. The MJN depicted patterns similar to those of the gene tree (Fig. 2). Haplotypes H1 and H11, which were more common than the other haplotypes, occupied central positions.

Divergence time estimates

Results of model comparisons were shown in Table 2. Both path sampling and stepping-stone sampling suggested that the BSP demographic model under strict clock outperformed the alternative models. The average divergence time estimations were shown in Fig. 3. Lineages E and W diverged about 1.4 Ma (95% CI: 0.6–2.4 Ma). Lineage E diverged about 0.41 Ma (95% CI: 0.15–0.81 Ma) and the radiation of Lineage W happened about 0.15 Ma (95% CI: 0.05–0.32 Ma).

Genetic structure and demographic history

Results of population clustering were illustrated in Figure 4. F_{CT} plateaued when K was 4 (Table S2). Samples from Lineage W grouped together (Fig. 4, dark blue). Lineage E contained 3 groups, of which populations in the southern edges of the QTP formed two small groups (localities 2, 16 and 26; and locality 15 and 19); the remaining localities comprised the third group (Fig. 4, light blue). Spatial patterns of genetic diversity based on Ar and π were shown in Figure 5 and listed in Table S3. Several peaks indicated high levels of genetic diversity across the distribution of *N. parkeri*. This was congruent with the pattern of lineage divergence.

When forming two groups based on the YZR, among-group diversity accounted for 16.04% of the overall variation and among-populations within groups accounted for 80.76%. Dividing populations according to the mountains, among-group diversity accounted for 87.90% of the overall variation and among-populations within

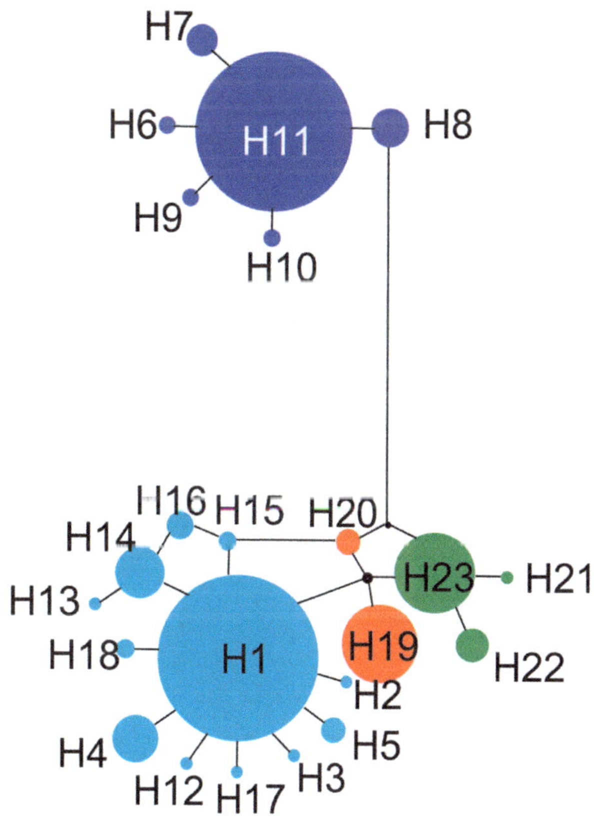

Figure 2. The median-joining network of *COI* haplotypes. Colors correspond to maternal lineages in Figure 1.

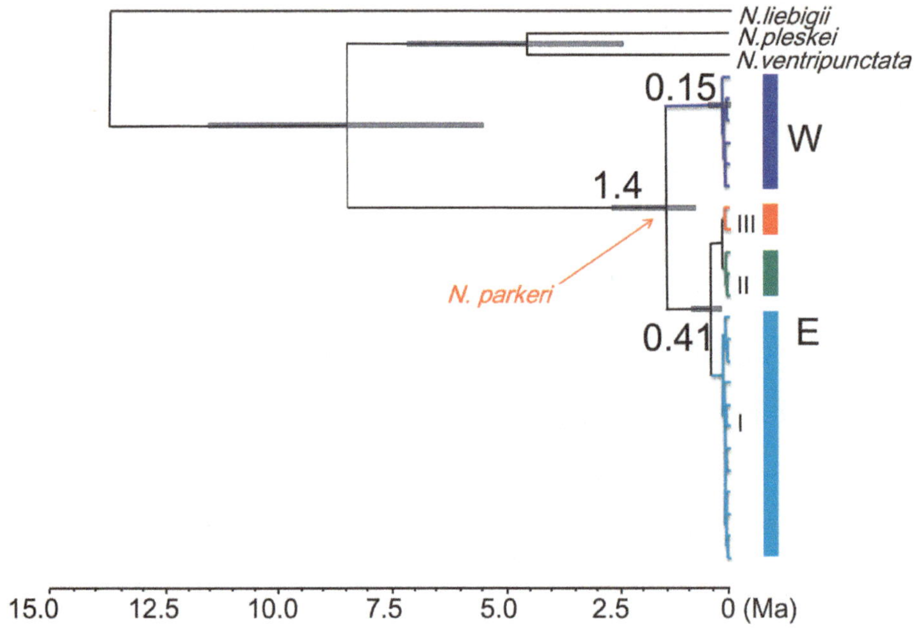

Figure 3. Estimates of divergence times obtained using BEAST 1.7.5.

groups accounted for 9.67%. Finally, after dividing populations based on results of the SAMOVA, among-group diversity accounted for 96.88% of the overall variation and among-populations within groups accounted for 0.97% (Table 3).

For Lineage W, the values of Tajima's *D* and Fu's *Fs* were significantly negative (Table 4), which indicated population expansions. The hypothesis of sudden expansion was not rejected by mismatch distribution analyses (Fig. 6) as the SSD and Rag were insignificant ($P_{SSD} = 0.381$ and $P_{RAG} = 0.658$). In sublineage E1, significantly negative values of both Tajima's *D* and Fu's *Fs* supported population expansions. The mismatch distribution analyses did not reject the sudden expansion model ($P_{SSD} = 0.545$ and $P_{RAG} = 0.609$). In sublineages E2, Tajima's *D* and Fu's *Fs* were negative, but insignificant. In sublineage E3, Tajima's *D* was negative but Fu's *Fs* was positive, yet all values were not significant. The null hypothesis of sudden expansion model was not rejected by mismatch distribution analyses.

Species distribution modeling

The AUCs from random data ranged from 0.605 to 0.800 (0.733±0.033; Table S4). For the SDMs of *N. parkeri*, the AUC of our data was 0.904 and this was significant better than that of a random model ($P<0.001$).

The MIROC model predicted that the distribution of *N. parkeri* at the LGM was much smaller than that of today (Fig. 7). The results based on CCSM model suggested a similar pattern, but with larger areas than those from MIROC model. Both models suggested range fluctuations after the LGM in northern, high altitude areas. The boundary area between GM and NM, where lineages W and E divided, has been unsuitable for *N. parkeri* since the LGM. The predicted distribution under GCC assuming the A1b scenario was shown in Figure 7. *Nanorana parkeri* is predicted to have a larger distribution in the northwestern QTP. However, the species' distribution will contract in the southeastern QTP.

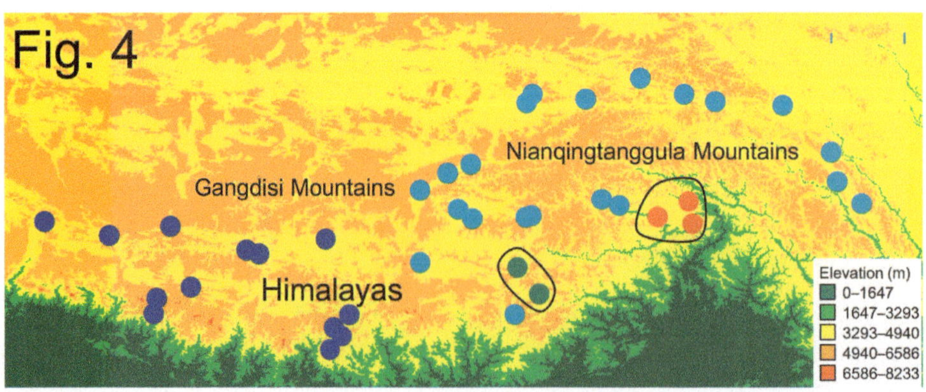

Figure 4. Results of SAMOVA. Different colors represent different groups. A solid line encircles two groups of Lineage E from microrefugia.

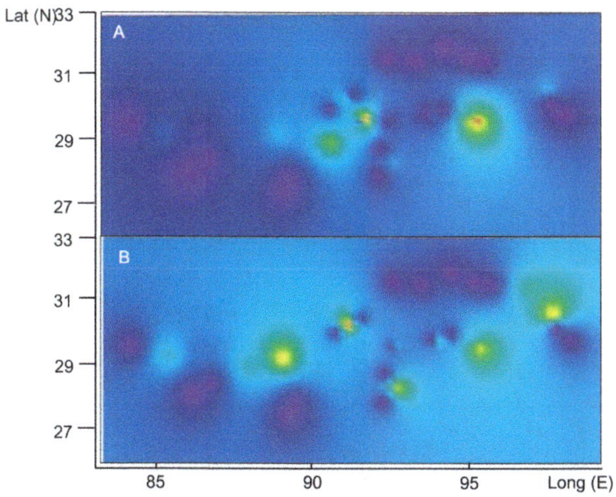

Figure 5. Genetic diversity across the range of *Nanorana parkeri*. X axis represents longitude and the Y axis latitude. A: π, pattern of nucleotide diversity; B: Ar, pattern of allelic richness.

Discussion

Lineage divergence and ESUs

ESUs, which can be determined by genetic differentiation at neutral markers caused by isolation, are important when considering conservation actions. Our study builds on the framework for delineating ESUs of *N. parkeri* by reconstructing the patterns of lineage divergence and its drivers. Our analyses suggest the recognition of ESUs that correspond to lineages W, E1, E2 and E3. Our research demonstrates that *COI* can quickly delineate ESUs.

The gene trees and MJN depict a clear east-west split for *N. parkeri*. SDMs suggest that unsuitable habitat between the GM and NM, both now and during the LGM, drives this pattern. The boundary is the demarcation line of two QTP climatic zones that are largely congruent with the 400 mm precipitation line. East of this line, the region is semi-humid, and to the west semi-arid [52]. Similar pattern was also found in *Hippophae tibetana* [53,54].

Geographic features, such as rivers or mountains, appear to contribute little to the patterns of genetic divergence. The YZR, which would impose a north-south split, does not drive population divergence. The AMOVA explains only 16.04% of the genetic variation in this scenario. Elevations of NM range from 5000 m to 6000 m, but no substantial genetic divergence distinguishes frogs in south and north of NM. Lineages E and W are separated by the boundary of GM and NM. However, habitat barriers and environmental differences more likely generate this pattern than geographical features. Two observations support this hypothesis. First, populations of lineages W and E both occur south of the GM and NM. Second, *N. parkeri* occurs along the YZR valley, which flows across the boundary and connects lineages W and E.

The pattern of genetic divergence within each major lineage differs. The 6 haplotypes in Lineage W do not yield a clear pattern (Fig. 1). The network depicts a star-like haplogroup (Fig. 2). SAMOVA also suggests populations from W comprise a single group. In contrast, Lineage E differs in having three groups (Figs. 1, 2). Results from SAMOVA present the same pattern. Most northern populations form a group as do frogs from localities 15, 19 (E2) and localities 2, 16, 26 (E3). The difference in patterns between E and W may owe to topography. The intricate topography of the East Himalayas, which is a global biodiversity hotspot, provides suitable habitats and generates barriers that limit dispersal for Lineage E. Other species exhibit a similar pattern; populations in the fringes or southern slope of Himalayas are genetically different from those on the QTP [54,55]. Expansion and contraction of distribution ranges caused by climate changes provided opportunity for populations divergence in QTP [54–57].

The genetic pattern within *N. parkeri* suggests recognition of four ESUs, W, E1, E2 and E3. Certainly gene flow occurs within and between lineages E and W, as lineages or sublineages mix at localities 8, 12, 16 and 26. However, matrilineal uniqueness occurs in populations far from the boundary.

Glacial refugia and genetic diversity of *N. parkeri*

Refugia in the QTP prevented extinction during the LGM. In doing so, they conserved high levels of genetic diversity. In contrast, genetic diversity in recently occupied areas is much lower because of founder effects [2,3]. The identification of refugia is an important part of conservation because these areas preserve genetic diversity. Synthesizing DNA barcoding and SDMs, we quickly and effectively identify refugia for *N. parkeri*. Populations in refugia retain higher levels of diversity than population in newly occupied places (Fig. 5 and Table S3).

The multiple refugia hypothesis corresponds with patterns of mtDNA divergence. The matrilineal genealogy, MJN and SAMOVA analyses indicate the absence of genetic structure within Lineage W. Thus, Lineage W likely originates from a single refugium. The SDMs indicate the presence of suitable habitat in the river valley near locality 27 (Fig. 7) and the area retains a much higher level of genetic diversity (Fig. 5) than other sites. The neutral test and mismatch distribution analyses clearly detect a population expansion in Lineage W.

Our analyses identify several refugia in Lineage E. Sublineage E1 may originate from a northern refugium near localities 7 and 8, which occur in the valley at the confluence of the Lhasa River and the YZR (Fig. 7). During the LGM the suitable environment harbored a high level of genetic diversity. The absence of genealogical structure in sublineage E1 is congruent with a sudden population expansion, as are significantly negative values of Tajima's D and Fu's Fs, and the mismatch distribution analyses. Further, our analyses identify two microrefugia in the river valleys along southern edges of the QTP. Localities 2, 16 and 26, which

Table 3. Results of AMOVA.

Group compositions	Among group	Among populations	Within populations	ΦSC	ΦST	ΦCT
By River	16.04%	80.76%	3.20%	0.96183	0.96795	0.16039
By Mountain	87.90%	9.67%	2.42%	0.79973	0.97578	0.87905
By SAMOVA	96.88%	0.97%	2.16%	0.30944	0.97844	0.96877

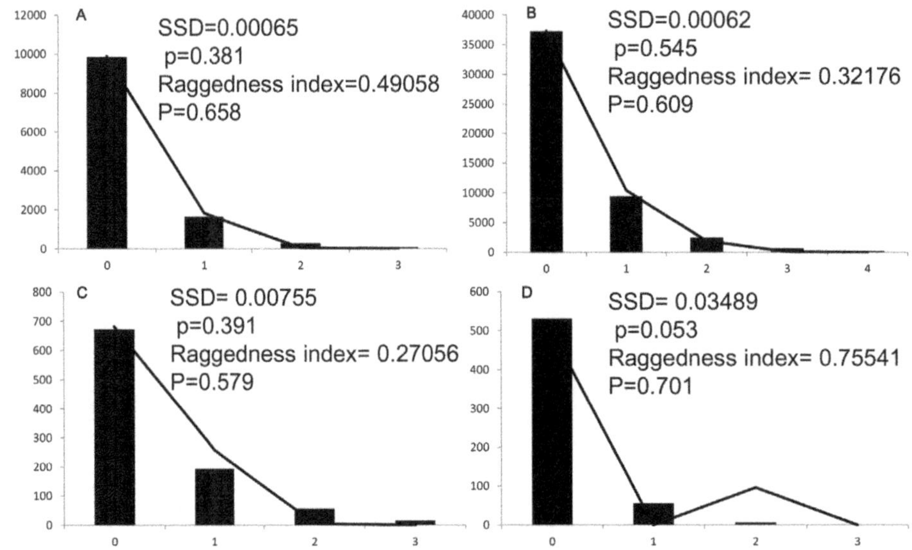

Figure 6. Results of the mismatch distribution for *Nanorana parkeri*. Bars indicate simulated mismatch distributions and lines denote the observed mismatch distributions. A: lineage W; B: sublineage E1; C: sublineage E2; D: sublineage E3.

are in the YZR Valley near Nyingchi and Medog, seemingly constitute a microrefugium. The other microrefugium consists of localities 15 and 19, which are also in a river valley. This river flows southwards across the Himalayas. The lower elevations of the river valleys make them less susceptible to the influences montane glaciers. River valleys in the Himalayas, especially those connected with southern slopes, appear to offer microrefugia.

Populations from sampling localities north of the NM also retain a high level of genetic diversity (Fig. 5). Private haplotypes H4 (locality 38), H2 and H3 (locality 33), and H5 (locality 37) occur here (Table S1). The SDMs suggest that suitable habitat existed in northern areas of NM. This area appears to harbor another microrefugium. However, these private haplotypes fail to cluster together and the most common haplotype in the area is H1 (Fig. 2 & Table S1). Thus, we cannot reject the hypothesis that these private haplotypes originated during population expansions after the LGM.

Potential threats of GCC to *N. parkeri*

Our study suggests that the combination of DNA barcoding and SDMs can detect threats of GCC to the survival of species. This pipeline facilitates conservation. SDMs predict that suitable habitat for *N. parkeri* may experience great shifts in the near future. Whereas the Northwest QTP will offer suitable habitat for *N. parkeri*, the Southeast QTP will become unsuitable. Although suitable habitat will experience an overall expansion, populations in the Southeast QTP may experience sharp decreases in population size or become extirpated. Given that amphibians have poor dispersal abilities, the latter scenario may be an unfortunate consequence of GCC.

Our pipeline suggests that many populations of Lineage E may suffer from developing unsuitable habitats. The microrefugium near Nyingchi and Medog will become unsuitable under this prediction (Fig. 7). Fortunately, suitable habitat may persist in the two other refugia. Thus, genetic variation will decrease yet all may not be lost. For Lineage W, suitable habitat will disappear near the refugium at locality 27. Loss of this refugium will greatly decrease the amount of genetic diversity.

Our analyses suggest that genetic diversity of *N. parkeri* may greatly decrease. Although this facilitates effective conservation planning, we urge caution. Our SDMs for the future assume A1b scenarios, which involve rapid global economic growth with a balance of fossil and non-fossil energy sources. Each of possible scenarios yields different predictions about the future climate. All scenarios require constant adjustment according to global economic conditions. The complex landscape in the southern QTP may supply suitable microhabitats for *N. parkeri* which cannot be detected by our SDMs. Accordingly, we suggest the urgent development of an effective monitoring program, especially for populations in refugia that may lose suitable habitats.

Conclusion

DNA barcoding detects the genetic structure of *N. parkeri* and serves to define ESUs. Our analyses recover major lineages E and

Table 4. Statistics of neutrality tests for each lineage and sublineage.

	Tajima's D	P value	Fu's Fs	P value
W	−1.62115	0.0044	−6.41652	0.0005
E1	−1.80395	0.0033	−14.10103	0
E2	−0.69425	0.2176	−0.77698	0.1858
E3	−0.65715	0.2279	0.92064	0.5177

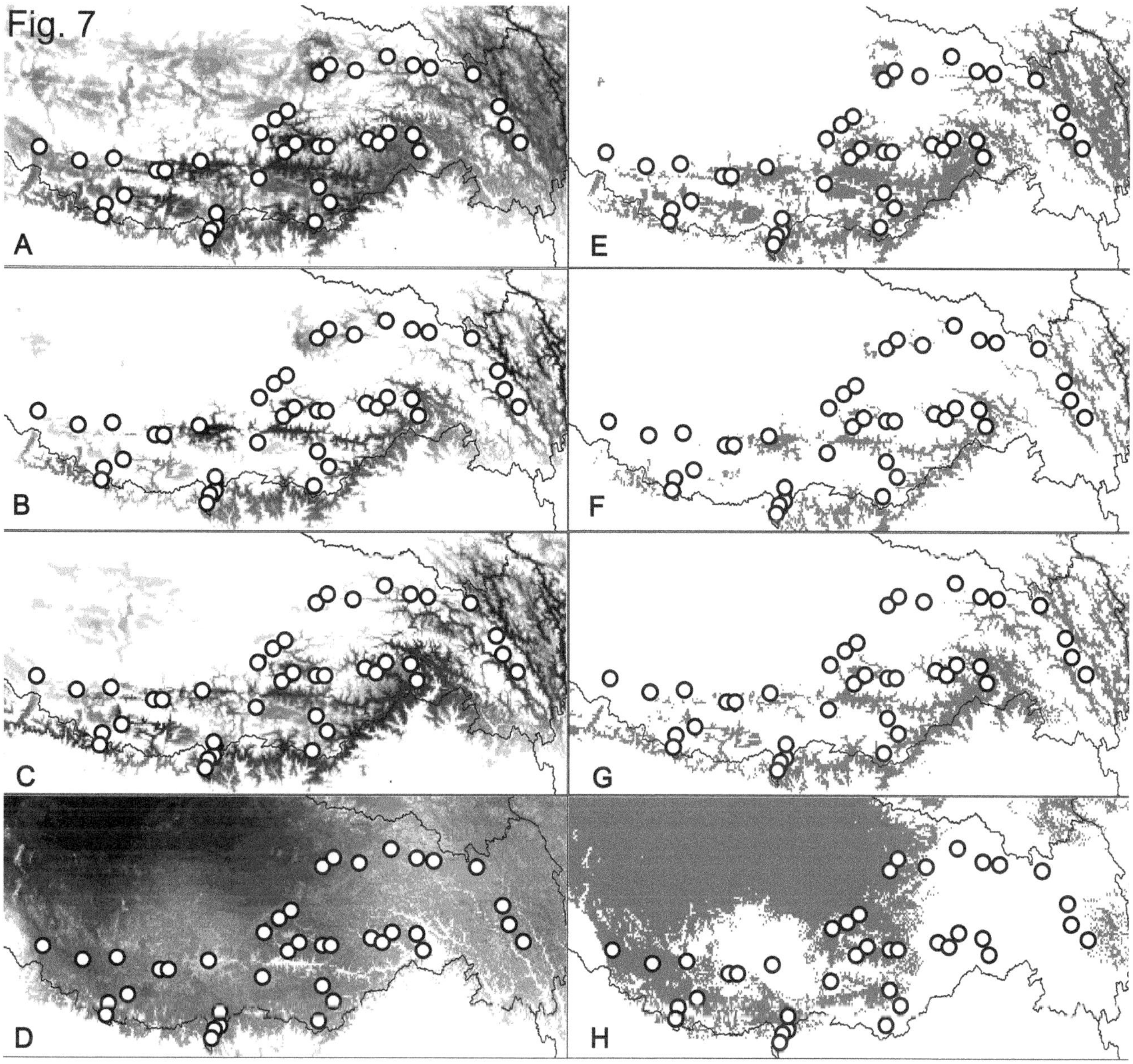

Figure 7. The predicted distribution of *Nanorana parkeri*. A: predicted distribution based on current data; B: distribution during the LGM based on MIROC model; C: distribution during the LGM based on CCSM model; D: the predicted distribution in the 2080 s under GCC; E–H: predicted distributions above the threshold for models A–D, respectively.

W, which separate at the boundary of the GM and NM. Habitat barriers and environmental differences, combined with geographic features, are the drivers of genetic divergences. Lineage E contains three parts: E1, E2 and E3. Thus our analyses define four ESUs that correspond to matrilines W, E1, E2 and E3. Genetic and environmental data identify four historical refugia, each of which corresponds to a lineage or sublineage. Lineage W originates from a refugium in river valley near locality 27. Most populations of Lineage E originate from a refugium in the river valley near Lhasa. Two microrefugia occur in river valleys along the Himalayas. Our analyses detect a population mixture after the LGM in some localities but these occurrences do not influence our designations of ESUs. Our study highlights the importance of valleys along the Himalayas for biodiversity conservation. Based on climate models under GCC, we predict the potential distribution changes and threats to genetic diversity of *N. parkeri*. Our pipeline, which combines DNA barcoding and SDMs, is an effective approach in conservation.

Supporting Information

Figure S1 Matrilineal genealogy of *Nanorana parkeri* based on BI analyses of *COI* sequence data. Bootstrap proportions ≥70% and Bayesian posterior probabilities ≥95% were treated as strongly supported (▼) and bootstrap proportions ≥70% and Bayesian posterior probabilities ≥90% were treated as being moderately supported. Bootstrap proportions<70% and Bayesian posterior probabilities <90% were treated as being unsupported (*).

Acknowledgments

Many thanks are owed to help from Mian Hou, Peng Guo, Kai Wang, Duan You and Yong-biao Xu in the field. We thank the local conservation departments for their help in field work. We also thank for the technical support from the laboratory of YPZ. DNA barcoding experiments were partly performed in the Southern China DNA Barcoding Center (SCDBC) in KIZ, CAS.

Author Contributions

Conceived and designed the experiments: WWZ JC YPZ. Performed the experiments: JQJ HMC BLZ. Analyzed the data: WWZ BLZ. Contributed reagents/materials/analysis tools: KJ JXY YYW. Wrote the paper: WWZ JC RWM.

References

1. Rosenzweig C, Karoly D, Vicarelli M, Neofotis P, Wu Q, et al. (2008) Attributing physical and biological impacts to anthropogenic climate change. Nature 453: 353–357.
2. Hewitt GM (2000) The genetic legacy of the Quaternary ice ages. Nature 405: 907–913.
3. Hewitt GM (2004) Genetic consequences of climatic oscillations in the Quaternary. Phil Trans R Soc B 359: 183–195.
4. Crandall KA, Bininda-Emonds OR, Mace GM, Wayne RK (2000) Considering evolutionary processes in conservation biology. Trends Ecol Evol 15: 290–295.
5. Moritz C (1994) Defining 'evolutionarily significant units' for conservation. Trends Ecol Evol 9: 373–375.
6. Hebert PD, Cywinska A, Ball SL (2003) Biological identifications through DNA barcodes. Proc R Soc Lond B Biol Sci 270: 313–321.
7. Guisan A, Thuiller W (2005) Predicting species distribution: offering more than simple habitat models. Ecol Lett 8: 993–1009.
8. Thomas CD, Cameron A, Green RE, Bakkenes M, Beaumont LJ, et al. (2004) Extinction risk from climate change. Nature 427: 145–148.
9. Guisan A, Tingley R, Baumgartner JB, Naujokaitis-Lewis I, Sutcliffe PR, et al. (2013) Predicting species distributions for conservation decisions. Ecol Lett 16: 1424–1435.
10. Zheng D (1996) The system of physico-geographical regions of the Qinghai-Xizang (Tibet) Plateau. Science in China (series D) 39: 410–417.
11. Zhao J, Shi Y, Wang J (2011) Comparison between Quaternary Glaciations in China and the Marine Oxygen Isotope Stage (MIS): An Improved Schema. Acta Geographica Sinica 66: 867–884.
12. Yao T, Liu X, Wang N (2000) The question of the range of climatic change on the Qinghai-Tibetan Plateau. Chinese Science Bulletin 45: 98–106.
13. Beebee T (2005) Conservation genetics of amphibians. Heredity 95: 423–427.
14. Zhou W, Yan F, Fu J, Wu S, Murphy RW, et al. (2013) River islands, refugia and genetic structuring in the endemic brown frog *Rana kukunoris* (Anura, Ranidae) of the Qinghai-Tibetan Plateau. Mol Ecol 22: 130–142.
15. Fei L, Ye C, Jiang J, Xie F, Huang Y (2005) An Illustrated Key to Chinese Amphibians. Sichuan Publishing House of Science and Technology, Chengdu.
16. Che J, Zhou W, Hu J, Yan F, Papenfuss TJ, et al. (2010) Spiny frogs (Paini) illuminate the history of the Himalayan region and Southeast Asia. Proc Natl Acad Sci USA 107: 13765–13770.
17. Che J, Hu J, Zhou W, Murphy RW, Papenfuss TJ, et al. (2009) Phylogeny of the Asian spiny frog tribe Paini (Family Dicroglossidae) sensu Dubois. Mol Phylogenet Evol 50: 59–73.
18. Sambrook J, Russell DW (2001) Molecular cloning: a laboratory manual. Cold Spring Harbor Laboratory Press, New York.
19. Che J, Chen H, Yang J, Jin J, Jiang K, et al. (2012) Universal *COI* primers for DNA barcoding amphibians. Mol Ecol Resour 12: 247–258.
20. Thompson JD, Gibson TJ, Plewniak F, Jeanmougin F, Higgins DG (1997) The CLUSTAL_X windows interface: flexible strategies for multiple sequence alignment aided by quality analysis tools. Nucleic Acids Res 25: 4876–4882.
21. Tamura K, Dudley J, Nei M, Kumar S (2007) MEGA4: molecular evolutionary genetics analysis (MEGA) software version 4.0. Mol Biol Evol 24: 1596–1599.
22. Librado P, Rozas J (2009) DnaSP v5: a software for comprehensive analysis of DNA polymorphism data. Bioinformatics 25: 1451–1452.
23. Ronquist F, Huelsenbeck JP (2003) MrBayes 3: Bayesian phylogenetic inference under mixed models. Bioinformatics 19: 1572–1574.
24. Brandley MC, Schmitz A, Reeder TW (2005) Partitioned Bayesian Analyses, Partition Choice, and the Phylogenetic Relationships of Scincid Lizards. Syst Biol 54: 373–390.
25. Brown JM, Lemmon AR (2007) The Importance of Data Partitioning and the Utility of Bayes Factors in Bayesian Phylogenetics. Syst Biol 56: 643–655.
26. Nylander JAA (2004) MrModeltest v2. Program distributed by the author. Evolutionary Biology Centre, Uppsala University.
27. Nylander JAA, Wilgenbusch JC, Warren DL, Swofford DL (2008) AWTY (are we there yet?): a system for graphical exploration of MCMC convergence in Bayesian phylogenetics. Bioinformatics 24: 581–583.
28. Swofford DL (2003) PAUP*: Phylogenetic Analysis Using Parsimony (* and Other Methods) Version 40b10. Sunderland, MA.: Sinauer Associates.
29. Stamatakis A, Hoover P, Rougemont J (2008) A rapid bootstrap algorithm for the RAxML Web servers. Syst Biol 57: 758–771.
30. Bandelt HJ, Forster P, Rohl A (1999) Median-joining networks for inferring intraspecific phylogenies. Mol Biol Evol 16: 37–48.
31. Polzin T, Daneshmand SV (2003) On Steiner trees and minimum spanning trees in hypergraphs. Operations Research Letters 31: 12–20.
32. Baele G, Lemey P, Bedford T, Rambaut A, Suchard MA, et al. (2012) Improving the Accuracy of Demographic and Molecular Clock Model Comparison While Accommodating Phylogenetic Uncertainty. Mol Biol Evol 29: 2157–2167.
33. Drummond AJ, Suchard MA, Xie D, Rambaut A (2012) Bayesian phylogenetics with BEAUti and the BEAST 1.7. Mol Biol Evol 29: 1969–1973.
34. Rambaut A, Drummond A (2007) Tracer v1.4. Available from http://beast.bio.ed.ac.uk/Tracer.
35. Dellicour S, Mardulyn P (2013) SPADS 1.0: a toolbox to perform spatial analyses on DNA sequence data sets. Mol Ecol Resour: in press.
36. Dupanloup I, Schneider S, Excoffier L (2002) A simulated annealing approach to define the genetic structure of populations. Mol Ecol 11: 2571–2581.
37. Excoffier L, Smouse PE, Quattro JM (1992) Analysis of molecular variance inferred from metric distances among DNA haplotypes: application to human mitochondrial DNA restriction data. Genetics 131: 479–491.
38. Excoffier L, Lischer H (2010) Arlequin suite ver 3.5: a new series of programs to perform population genetics analyses under Linux and Windows. Mol Ecol Resour 10: 564–567.
39. Tajima F (1989) Statistical method for testing the neutral mutation hypothesis by DNA polymorphism. Genetics 123: 585–595.
40. Fu Y (1997) Statistical tests of neutrality of mutations against population growth, hitchhiking and background selection. Genetics 147: 915–925.
41. Rogers AR, Harpending H (1992) Population growth makes waves in the distribution of pairwise genetic differences. Mol Biol Evol 9: 552–569.
42. Phillips SJ, Anderson RP, Schapire RE (2006) Maximum entropy modeling of species geographic distributions. Ecol Model 190: 231–259.
43. Phillips SJ, Dudík M (2008) Modeling of species distributions with Maxent: new extensions and a comprehensive evaluation. Ecography 31: 161–175.
44. Hijmans RJ, Cameron SE, Parra JL, Jones PG, Jarvis A (2005) Very high resolution interpolated climate surfaces for global land areas. Int J Climatol 25: 1965–1978.
45. Merow C, Smith MJ, Silander JA (2013) A practical guide to MaxEnt for modeling species' distributions: what it does, and why inputs and settings matter. Ecography 36: 1058–1069.
46. Smith AB (2013) On evaluating species distribution models with random background sites in place of absences when test presences disproportionately sample suitable habitat. Divers Distrib 19: 867–872.
47. Holt RD (2003) On the evolutionary ecology of species' ranges. Evol Ecol Res 5: 159–178.
48. Peterson A, Soberón J, Sánchez-Cordero V (1999) Conservatism of ecological niches in evolutionary time. Science 285: 1265–1267.
49. Wiens JJ, Graham CH (2005) Niche conservatism: integrating evolution, ecology, and conservation biology. Annu Rev Ecol Evol Syst 36: 519–539.
50. Collins WD, Bitz CM, Blackmon ML, Bonan GB, Bretherton CS, et al. (2006) The community climate system model version 3 (CCSM3). J Clim 19: 2122–2143.
51. Hasumi H, Emori S (2004) K-1 coupled GCM (MIROC) description. Center for Climate System Research, University of Tokyo, National Institute for

Environmental Studies, Frontier Research Center for Global Change, Tokyo, Japan.
52. Li S, Xu L, Guo Y, Qian W, Zhang G, et al. (2007) Change of annual precipitation over Qinghai-Xizang Plateau and sub-regions in recent 34 Years. Journal of Desert Research 27: 307–314.
53. Wang H, Laqiong,Sun K, Lu L, Wang Y, et al. (2010) Phylogeographic structure of *Hippophae tibetana* (Elaeagnaceae) highlights the highest microrefugia and the rapid uplift of the Qinghai-Tibetan Plateau. Mol Ecol 19: 2964–2979.
54. Jia D, Liu T, Wang L, Zhou D, Liu J (2011) Evolutionary history of an alpine shrub *Hippophae tibetana* (Elaeagnaceae): allopatric divergence and regional expansion. Biol J Linnean Soc 102: 37–50.
55. Wang L, Wu Z, Bystriakova N, Ansell SW, Xiang Q, et al. (2011) Phylogeography of the Sino-Himalayan Fern *Lepisorus clathratus* on the Roof of the World? PloS one 6: e25896.
56. Wang L, Abbott R, Zheng W, Chen P, Wang Y, et al. (2009) History and evolution of alpine plants endemic to the Qinghai-Tibetan Plateau: *Aconitum gymnandrum* (Ranunculaceae). Mol Ecol 18: 707–721.
57. Yang F, Li Y, Ding X, Wang X (2008) Extensive population expansion of *Pedicularis longiflora* (Orobanchaceae) on the Qinghai-Tibetan Plateau and its correlation with the Quaternary climate change. Mol Ecol 17: 5135–5145.

19

Response of Two Dominant Boreal Freshwater Wetland Plants to Manipulated Warming and Altered Precipitation

Yuanchun Zou[1], Guoping Wang[1], Michael Grace[2], Xiaonan Lou[1], Xiaofei Yu[1], Xianguo Lu[1]*

1 Key Lab of Wetland Ecology and Environment, Northeast Institute of Geography and Agroecology, Chinese Academy of Sciences, Changchun, China, **2** Water Studies Centre and School of Chemistry, Monash University, Clayton, Australia

Abstract

This study characterized the morphological and photosynthetic responses of two wetland plant species when they were subject to 2–6°C fluctuations in growth temperature and ±50% of precipitation, in order to predict the evolution of natural wetlands in Sanjiang Plain of North-eastern China. We investigated the morphological and photosynthetic responses of two dominant and competitive boreal freshwater wetland plants in Northeastern China to manipulation of warming (ambient, +2.0°C, +4.0°C, +6.0°C) and altered precipitation (−50%, ambient, +50%) simultaneously by incubating the plants from seedling to senescence within climate-controlled environmental chambers. Post-harvest, secondary growth of *C. angustifolia* was observed to explore intergenerational effects. The results indicated that *C. angustifolia* demonstrated a greater acclimated capacity than *G. spiculosa* to respond to climate change due to higher resistance to temperature and precipitation manipulations. The accumulated effect on aboveground biomass of post-harvest secondary growth of *C. angustifolia* was significant. These results explain the expansion of *C. angustifolia* during last 40 years and indicate the further expansion in natural boreal wetlands under a warmer and wetter future. Stability of the natural surface water table is critical for the conservation and restoration of *G. spiculosa* populations reacting to encroachment stress from *C. angustifolia* expansion.

Editor: Eric Gordon Lamb, University of Saskatchewan, Canada

Funding: The research was supported by the National Basic Research Program of China (2010CB951301-2), the National Key Technology R & D Program (2012BAC19B05), and the National Natural Science Foundation of China (41271107, 41101195) in the study design, data collection and analysis and preparation of the manuscript.

Competing Interests: The authors have declared that no competing interests exist.

* Email: luxg@iga.ac.cn

Introduction

Wetlands are among the most important terrestrial carbon pools and play an important role in global carbon cycling [1,2]. Boreal wetlands comprise about half of the total global wetland area, store about one third of the world's organic carbon in the form of living or partially decayed vegetation and soil organic carbon and contribute 34% of global atmospheric methane flux [3,4]. Due to their large and dynamic carbon pool, boreal wetlands may not be only significantly impacted by climate change, but may also provide negative (e.g. increasing organic carbon sink) or positive feedback (e.g. increasing greenhouse gas emissions) mechanisms to anthropogenic climate change [5,6] and mitigate climate warming overall.

Boreal wetlands are considered particularly vulnerable to climate change because they depend on specific climatic conditions with low temperature and high water availability [7,8]. To quantify anticipated climate effects on boreal wetland ecosystem processes, field experiments simulating climate change have been undertaken in North America and northern Europe [2,5,9], and these studies suggested that both the performance of a specific plant and the compositions of plant communities will change in different directions and at different magnitudes as a response to warming and/or changes in the depth of the surface water table that usually connected with temperature and precipitation change. Despite these efforts, species-specific eco-physiological responses of wetland plants to changes in temperature and precipitation, and especially the interactive effects of these factors, remain unclear. Simulations provide a mechanism for exploring the importance of these temperature-precipitation interactions on wetland plants under future climate conditions [10]. Multifactor experimental manipulations of temperature and precipitation in the laboratory can complement field studies by enabling controlled adjustment of covarying climate variables based on the historical data to generate relatively realistic climate change experimental scenarios [11]. The responses then could be scaled up to the level of the ecosystem when the initial structure of the wetland and the environmental driving variables are consequently well coupled.

As typical dominant boreal freshwater wetland plants in Northeast China, Siberia and Far East Russia, *Calamagrostis angustifolia* and *Glyceria spiculosa* both belong to the family Poaceae, and are often distributed as large, dense clumps in palustrine, riverine and lacustrine wetlands [12,13]. Generally, *C. angustifolia* covers saturated microhabitats while *G. spiculosa* covers the flooded microhabitats with greater surface water depth. Both plants play important ecological functions in the wetlands as primary producers for food webs, shelter for some invertebrates and major organic carbon sources for carbon sequestration. In recent years, with climate warming and surface water deficiency in some wetlands of the Sanjiang Plain, the microhabitats once covered by *G. spiculosa* have been almost replaced by *C. angustifolia* due to the drawdown of surface water depth, caused either by climate change or anthropogenic activities such as drainage.

In this paper, we simultaneously considered two major factors (temperature and precipitation) involved in climate change effects on boreal wetland ecosystem functioning. The two plant species, *C. angustifolia* and *G. spiculosa*, were incubated from seedling to senescence in environmental chambers under manipulated warming and altered precipitation conditions. The objectives of this study were: (1) to characterize the morphological and photosynthetic responses of the two plants to 2–6°C warming and ±50% of precipitation changes; and (2) to analyze the resistance and acclimation of the two plants and predict their potential distribution in the future under the projected changed climate conditions in the Sanjiang Plain wetlands of Northeast China.

Materials and Methods

Study site and target species

The study site is located in Xingkai Lake (Khanka) lacustrine wetlands of Sanjiang Plain, the largest expanse of freshwater wetlands in China. Xingkai Lake is the largest lake in Northeast Asia and a transboundary lake shared by China and Russia. According to calculations based on the 54 years of local meteorological records, the annual mean precipitation is 561 mm and the temperature ranges from −39°C in January to 36°C in July with the annual mean of 3.5°C for 1957–2010. According to a linear fitting of local meteorological records in Xingkai Lake lacustrine wetlands (Annual temperature = −52.55 + 0.028*year, r>0.99, p<0.0001; unpublished), there is a significant warming trend of 0.28°C per decade in Xingkai Lake over the past half century. This is even greater than the average level of 0.22°C per decade for the whole country [14].

The target plant species for the three-month incubation experiments were collected *in situ* from *Calamagrostis angustifolia* (45°20′59″ N, 132°18′52″ E) and *Glyceria spiculosa* (45°21′49″ N, 132°19′26″ E) communities within Xingkai Lake lacustrine wetlands. Representative and homogeneous individuals of *C. angustifolia* and *G. spiculosa* were excavated with intact roots and soils using a planting shovel in May of 2011 and 2012, respectively, when the first new shoots were apparent in springtime. The soil types were silty clay meadow marsh soil and humus marsh soil, respectively. The seedlings and soils with the depth of approximate 20 cm were cut into small blocks with rhizome based, transported to the laboratory in Northeast Institute of Geography and Agroecology within two days, and transplanted into incubation pots with 10-cm diameter and 15-cm depth with 4–5 seedlings in each pot. To minimize loss of vitality during transplantation, these potted seedlings were first incubated in a greenhouse for one month to germinate new ramets before placement into environmental chambers. The temperature in the greenhouse was set as the average air temperature of June in Xingkai Lake lacustrine wetlands. All the field studies got the permission from administrative authority of Xingkai Lake National Nature Reserve and did not involve any endangered or protected species.

Experimental design

The design between the warming and the precipitation was a two-way randomized block experiment. Four environmental chambers were deployed to control the temperature. Within each chamber, three plastic container (35 cm × 25 cm × 10 cm) were used to control water level fluctuations caused by precipitation manipulation. Each container contained three replicated pots. Before incubation, any pot growing approximate ten vigorous seedlings with similar performance (about 20 cm and 15 cm heights for *C. angustifolia* and *G. spiculosa*, respectively) was selected and randomly separated into 4 groups corresponding to the 4 temperature regimes. The ambient temperature was set as the local monthly average air temperatures in July, August and September, respectively fitted. The ambient temperature was adjusted five times per day based on the local monthly (July, August and September) average air temperatures over the time periods 1:00–4:00, 4:00–10:00, 10:00–16:00, 16:00–19:00 and 19:00–1:00. Table 1 provides the temperature and illuminance values for the plant incubations in the environmental chambers. The illuminance values were based on the field observation. These pots were exchange their positions in the chambers daily to keep the temperature and light intensity similar for a given pot. The warming scenarios were then set according to the ambient temperature regime with +2°C, +4°C and +6°C added for each interval. The lights were plant growth fluorescent lamps that are recognized as suitable for plants' growth [15].

The ambient water depths were set to mimic the natural conditions in Xingkai Lake wetlands, with no net water for *C. angustifolia* and 10 cm of water above the soil surface for *G. spiculosa*. The ambient precipitation regime was set according to the local monthly average precipitation. These seedlings were gently watered over one hour using sprinkler nozzles to simulate natural precipitation every 3 days with one tenth of the amount of monthly precipitation. The altered precipitations were set as -50% and +50% based on the ambient precipitation rates.

After the three month incubation in the environmental chambers and subsequent harvesting of all above-ground biomass, one *C. angustifolia* pot was randomly selected from the 3 replicates of each precipitation treatment container within each environmental chamber. Twelve pots were consequently selected

Table 1. Temperature and illuminance control for plants' incubation in the environmental chambers.

Time period		0:00–7:00	7:00–11:00	11:00–14:00	14:00–16:00	16:00–24:00
Illuminance (lux)		0	10000	20000	15000	5000
Temperature	July	16	20	28	24	17
(°C)	August	17	19	26	23	18
	September	11	14	19	17	12

for a secondary growth test to analyze the potential effect of climate change on the next generation of this plant species. These pots were incubated for one month in the same environmental chamber with the average temperature and precipitation in June of 17.4°C and 74.4 mm, respectively in Xingkai Lake wetlands.

Response measurements

The morphological indicators (height, stem diameter and leaf area) were observed once per month using a meter stick, vernier caliper and portable area meter (LI-3000, Li-Cor Inc., USA) from seedling to senescence, with each indicator measured by 4 replicates for *C. angustifolia* and 3 replicates for *G. spiculosa*. The photosynthetic indicators, including leaf net photosynthetic rate (Pn), leaf stomatal conductance (Gs) and leaf transpiration rate (Ts) were measured by a portable photosynthesis system (LI-6400, Li-Cor, Inc., USA) during the peak growth phase (around August), with each indicator measured by 4 replicates for *C. angustifolia* and 3 replicates for *G. spiculosa*. The chlorophyll content was determined by homogenizing and extracting the leaf tissue in 80% acetone until the green colour turn white completely, then measuring the optical absorbance at 663 and 645 nm for chlorophyll a and b, respectively using a spectrophotometer (UV-2500, Shimadzu, Japan), and calculating concentrations using the specific absorption coefficients [16,17]. The reported leaf chlorophyll content was the sum of chlorophyll a and b, with 3 replicates both for *C. angustifolia* and *G. spiculosa*.

Above-ground biomass was measured after plant senescence by drying to constant weight at 80°C. The total nitrogen content of harvested leaves was determined using a continuous flow analyzer (SAN^{++}, SKALAR, the Netherlands) after digesting 0.3000 g of homogeneous, milled material with H_2SO_4/H_2O_2, with 3 replicates both for *C. angustifolia* and *G. spiculosa*. The below-ground biomass was not measured as it is a mixture of newly developed rhizomes/roots within that growing season and the rhizomes/roots from previous growing seasons. Older material usually accounts for most of the total below-ground biomass [18].

The response measurements for the secondary growth test of *C. angustifolia* were conducted as described above. The measured indicators were height, stem diameter above the soil, leaf area of each leaf on the plant based on measuring a suitable number of leaves, Pn, Gs, Ts and above-ground biomass, with each indicator measured by 3 replicates.

Statistical analyses

Statistical analyses were performed using SPSS Statistics 21.0 (SPSS Inc., USA). The main and the interaction effects of temperature and precipitation on the two plants' morphological indicators were compared by repeated measures analysis of variance (ANOVA). For repeated measures ANOVA, Mauchly's Test of Sphericity was first examined and the Greenhouse-Geisser adjustment was adopted when the null hypothesis was rejected. The main and the interaction effects of temperature and precipitation on the two plants' photosynthetic indicators, leaf chlorophyll contents during the phase of peak growth, the above-ground biomasses and leaf nitrogen contents after senescence were compared by two-way ANOVA. One-way ANOVA was performed to compare the differences of above-ground biomasses of *C. angustifolia* and *G. spiculosa*. Differences in morphology, photosynthesis and above-ground biomass between different secondary grown *C. angustifolia* after the temperature and precipitation manipulation for one month were also compared using one-way ANOVA. The least significant difference (LSD) was used respectively to perform *post hoc* multiple comparisons when

Table 2. Repeated-measures analysis of variance (ANOVA) of morphological indicators for *Calamagrostis angustifolia* and *Glyceria spiculosa*.

Variable		Month		Month × Temperature		Month × Precipitation		Month × Temperature × Precipitation	
		F	*p*	*F*	*p*	*F*	*p*	*F*	*p*
C. angustifolia	Height	30.864	**<0.001**	0.910	0.482	2.611	0.052	2.025	**0.044**
	Basal stem diameter	0.064	0.860	0.538	0.702	0.343	0.764	0.513	0.835
	Leaf area	151.032	**<0.001**	2.449	**0.047**	0.271	0.857	0.922	0.517
G. spiculosa	Height	111.065	**<0.001**	1.038	0.412	2.203	0.116	2.520	**0.033**
	Basal stem diameter	3.165	0.051	0.412	0.867	1.426	0.240	0.839	0.612
	Leaf area	61.079	**<0.001**	1.739	0.155	3.271	**0.031**	1.363	0.239

Significant effects ($p<0.05$) are shown in bold. When the highest order of interaction effect is significant, the significant main effect is not in bold.

equal variances were assumed; otherwise, Tamhane's T2 method was used.

Results

Morphological indicators

Repeated-measures ANOVA of morphological indicators showed that the interaction effects of temperature and precipitation on the height of *C. angustifolia* and *G. spiculosa* were both significant ($F = 2.025$, $p = 0.044$; $F = 2.520$, $p = 0.033$, respectively). For both species, main effects of temperature or precipitation, and any interaction effects, on basal stem diameter were all insignificant. The main effects of temperature on leaf area of *C. angustifolia* and the effect of precipitation on that of *G. spiculosa* were both significant ($F = 2.449$, $p = 0.047$; $F = 3.271$, $p = 0.031$, respectively) (Table 2).

C. angustifolia's height increased from July to September. The estimated marginal means (as the average of the results for each month) of the height suggested that the greatest heights were observed in the ambient precipitation for each warming manipulation (Fig. 1A). Under ambient precipitation conditions (Fig. 1B), the greatest height occurred at ambient temperature (except for the early growth of July), while the lowest height was found in the +2°C manipulation (except in September). When the precipitation decreased by 50% from ambient (Fig. 1C), the greatest height occurred in the +4°C manipulation (except for the early growth of July), while the lowest height was in the +6°C manipulation. When the precipitation increased by 50% from ambient, the greatest height occurred in the +4°C manipulation, while the other three treatments produced plants of comparable height (Fig. 1D).

G. spiculosa's height also increased throughout the incubation (Fig. 2B–D). The estimated marginal means of the height suggested that ambient precipitation and 2–4°C of warming could support greater plant height (Fig. 2A). In the ambient precipitation conditions, there were no significant differences in height between the different temperature treatments (Fig. 2B). For both the −50% and +50% precipitation treatments, the greatest height occurred in the +4°C manipulation (Fig. 2C, D).

The leaf area of the two species increased first and then decreased throughout the incubation. For *C. angustifolia*, the leaf area with the +6°C manipulation was smaller than other manipulations in July. There was no significant difference in leaf area between the ambient and the +2°C manipulation in July. In

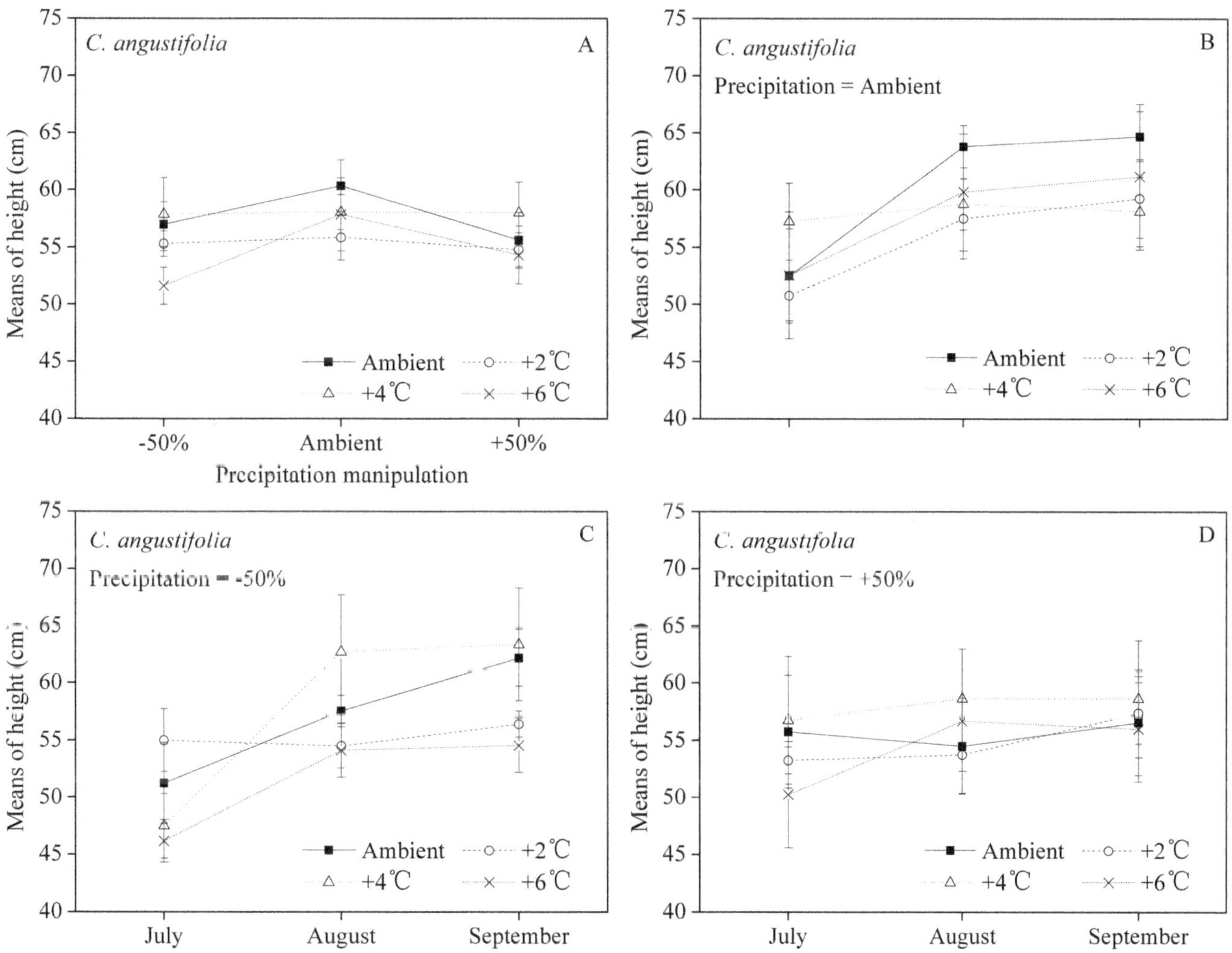

Figure 1. Interaction of temperature and precipitation manipulations on plant height (A) of *Calamagrostis angustifolia*, and the monthly estimated marginal means of height of *C. angustifolia* for different temperature treatments when precipitation was manipulated at ambient (B), −50% (C) and +50% (D) levels, respectively. The error bars represent means ± 1 standard error.

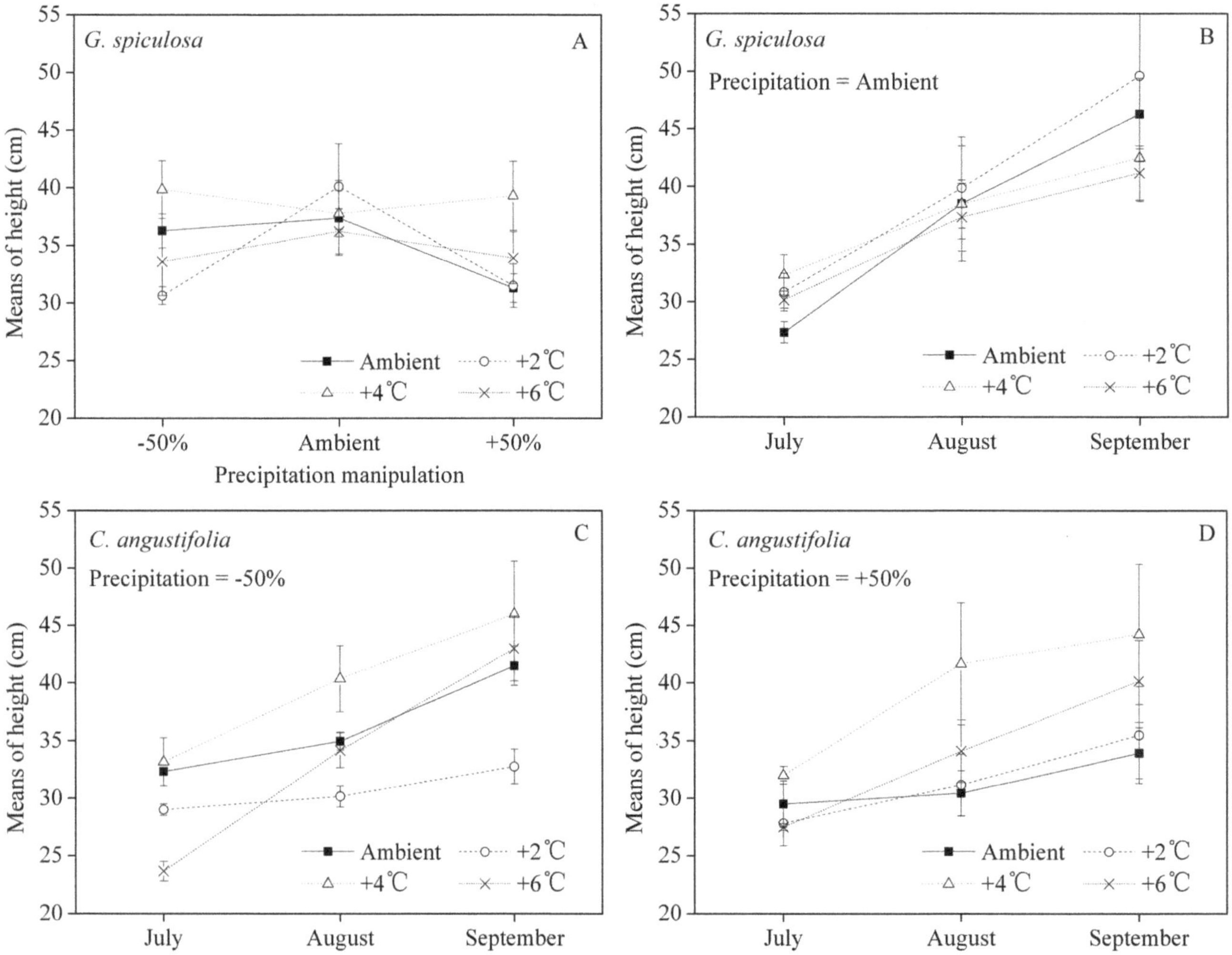

Figure 2. Interaction of temperature and precipitation manipulations on plant height (A) of *Glyceria spiculosa*, and the monthly estimated marginal means of height of *G. spiculosa* for different temperature treatments when the precipitation was manipulated at ambient (B), −50% (C) and +50% (D) levels, respectively. The error bars represent means ± 1 standard error.

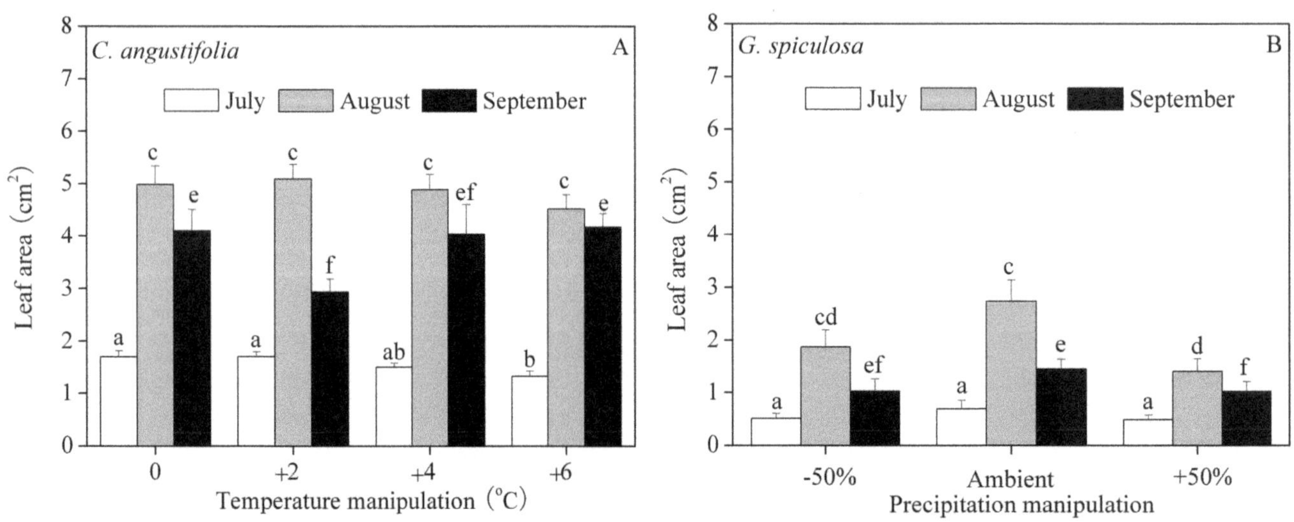

Figure 3. Leaf area of *C. angustifolia* (A) and *G. spiculosa* (B) monthly. For each month, different letters over the bars indicate significant differences ($p<0.05$) between temperature manipulations for *C. angustifolia* and precipitation manipulations for *G. spiculosa*. The letters 'a' and 'b' are for the July results, 'c' and 'd' for the August results, and 'e' and 'f' for the September results. The error bars represent means ± 1 standard error.

Table 3. Two-way ANOVA of photosynthetic indicators and leaf chlorophyll content for *C. angustifolia* and *G spiculosa* during the phase of peak growth, and above-ground biomass and leaf total nitrogen content after harvest.

Variable		Temperature		Precipitation		Temperature × Precipitation	
		F	*p*	*F*	*p*	*F*	*p*
C. angustifolia	Pn	9.878	**<0.001**	7.516	**0.002**	4.974	**0.001**
	Gs	3.691	**0.021**	8.639	**0.001**	6.825	**<0.001**
	Ts	4.410	**0.010**	3.013	0.062	9.208	**<0.001**
	Chlorophyll	0.966	0.419	0.765	0.473	2.545	**0.037**
	Aboveground biomass	4.512	**0.012**	2.349	0.117	0.842	0.550
	Leaf total nitrogen	7.125	**0.001**	3.642	0.052	0.524	0.784
G. spiculosa	Pn	1.653	0.201	11.361	**<0.001**	2.244	0.073
	Gs	17.389	**<0.001**	1.966	0.162	3.400	**0.014**
	Ts	5.091	**0.007**	2.883	0.075	3.170	**0.020**
	Chlorophyll	4.920	**0.008**	7.153	**0.004**	2.322	**0.046**
	Aboveground biomass	4.156	**0.017**	13.637	**<0.001**	1.002	0.447
	Leaf total nitrogen	2.737	0.066	1.134	0.338	2.394	0.059

Significant effects ($p<0.05$) are shown in bold. When the highest order of interaction effect is significant, the significant main effect is not in bold.

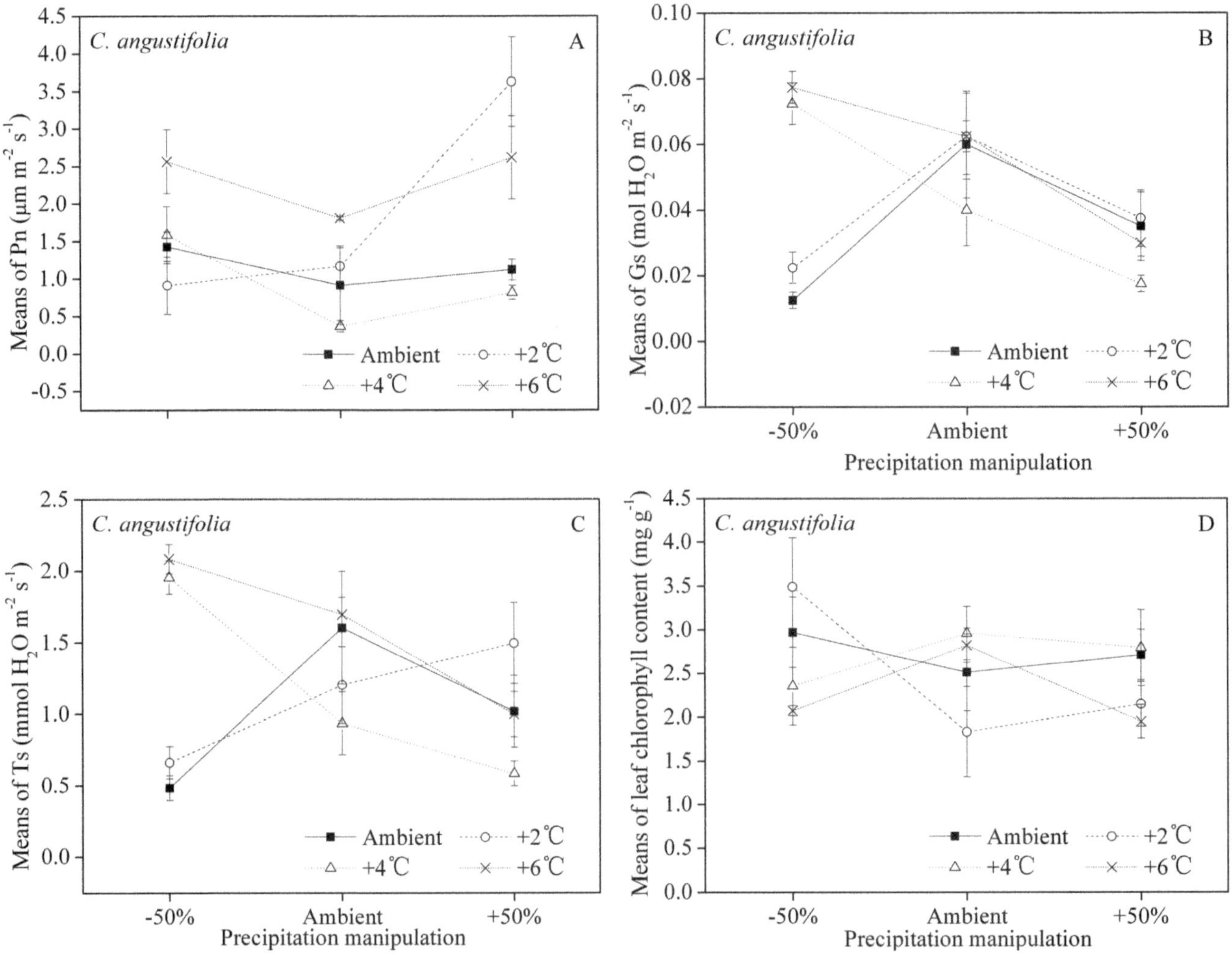

Figure 4. Interactions of temperature and precipitation manipulations on *C. angustifolia*'s leaf net photosynthetic rate (Pn, A), leaf stomatal conductance (Gs, B), leaf transpiration rate (Ts, C) and leaf chlorophyll content (D), respectively. The error bars represent means ± 1 standard error.

September, the leaf area with the +4°C manipulation was not significantly different from either the ambient or the +6°C manipulation. There was no significant difference in leaf area with different temperature manipulations in August (Fig. 3A). For *G. spiculosa*, there was no significant difference between ambient and the −50% precipitation treatments in any of the three months, while the leaf area with the 50% precipitation treatment was smaller than other treatments in August and September (Fig. 3B).

Photosynthetic indicators and leaf chlorophyll content

Two-way ANOVA of photosynthetic indicators and leaf chlorophyll content showed that the interaction effects of temperature and precipitation on these two species were both significant, except for the net photosynthetic rate (Pn) for *G. spiculosa*, which was only significantly affected by precipitation (Table 3).

For *C. angustifolia*, smaller Pn occurred under ambient precipitation except for the +2°C manipulation, and greater Pn was measured in the +6°C manipulation with the exception of the +50% precipitation treatment (Fig. 4A). The leaf stomatal conductance, Gs, decreased with the enhanced precipitation for the +4°C and +6°C manipulations (Fig. 4B). The greatest leaf transpiration rates, Ts, were found in the +4°C and +6°C manipulations under −50% precipitation (Fig. 4C). Ts rates decreased in both these temperature treatments with increasing precipitation. There was no systematic pattern in Ts at ambient temperature but the +2°C manipulation showed Ts increased with increasing precipitation. There were no consistent trends in leaf chlorophyll content with temperature or precipitation treatments (Fig. 4C) but all concentrations fell within the range 1.83–3.49 mg/g.

For *G. spiculosa*, Pn decreased with precipitation (Fig. 5A). With the exception of the +2°C treatment under the −50% precipitation regimen, the highest Gs results were found under ambient temperature (Fig. 5B). There was no consistent response in Gs to increased precipitation over the temperature treatments. There was no consistent trend in Ts across either temperature or precipitation treatments (Fig. 5C). All temperature treatments showed a decrease in leaf chlorophyll content between −50% and ambient precipitation (Fig. 5D). Increasing precipitation to 50% above ambient resulted in no consistent trend in chlorophyll with increasing temperature regime.

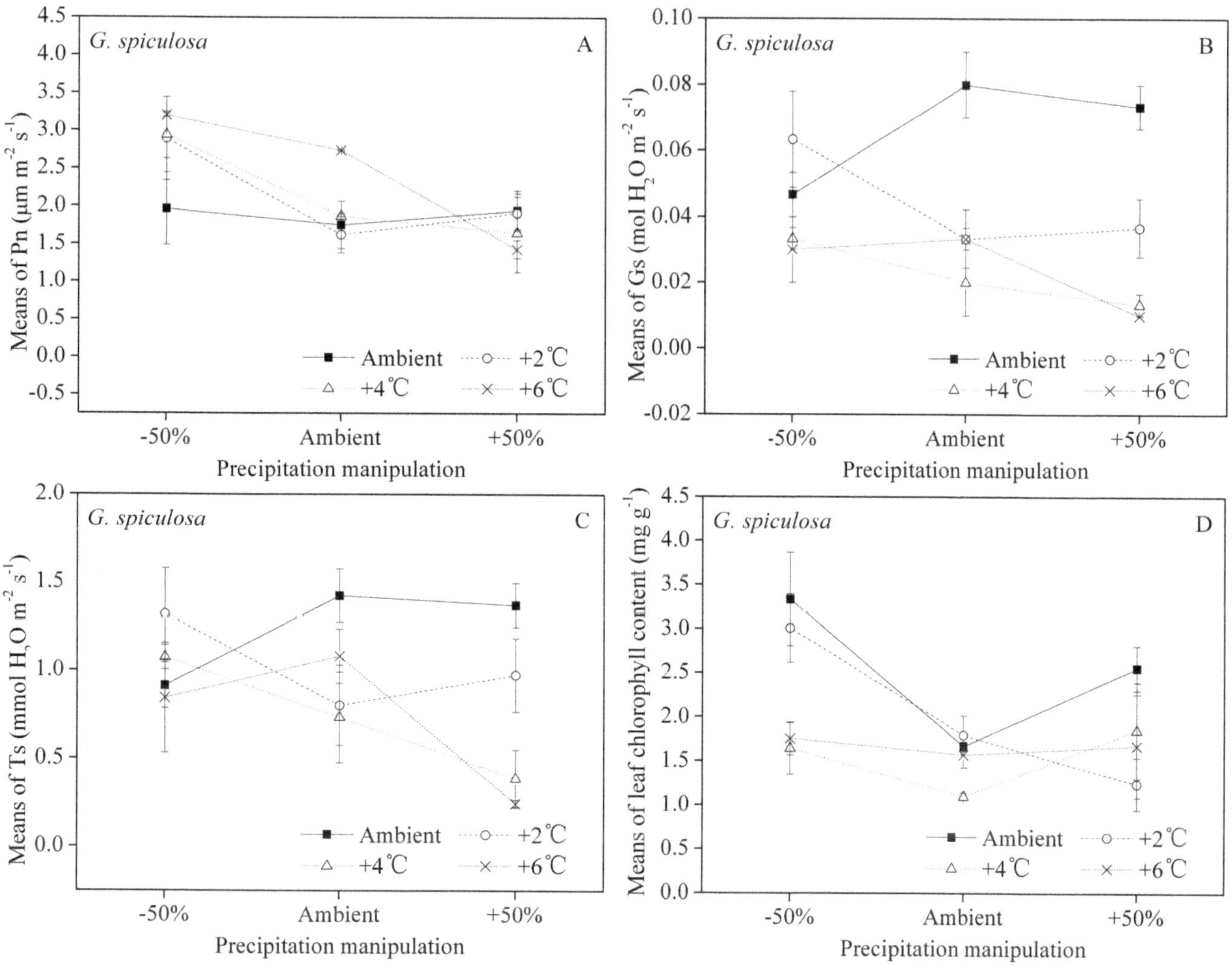

Figure 5. Main effect of precipitation manipulation on *G. spiculosa*'s leaf net photosynthetic rate (Pn, A), interactions of temperature and precipitation manipulations on leaf stomatal conductance (Gs, B), leaf transpiration rate (Ts, C) and leaf chlorophyll content (D), respectively. Different letters shared by the bars indicate significant differences ($p<0.05$) between precipitation treatments. The error bars represent means ± 1 standard error.

Above-ground biomass and leaf nitrogen content

Above-ground biomass of *C. angustifolia* was only significantly affected by temperature ($F=4.512$, $p=0.012$), while the effect of precipitation and the interaction effect were insignificant (Table 3). Biomass was at a maximum in the +4°C manipulation, increasing by 58% above that at ambient temperature (Fig. 6A). Leaf total nitrogen of *C. angustifolia* was also only significantly affected by temperature ($F=7.125$, $p=0.001$; Table 3). All three temperature increments resulted in statistically significant increases above ambient temperature, with a maximum value at +6°C, representing a 39% increase above ambient temperature (Fig. 6B).

Above-ground biomass of *G. spiculosa* was significantly affected both by temperature and precipitation ($F=4.156$, $p=0.017$; $F=13.637$, $p<0.0001$, respectively), while the interaction effect was insignificant (Table 3). As noted with *C. angustifolia*, biomass was highest in the +4°C manipulation, with an increase of 37% above ambient temperature (Fig. 6C). Both increasing and decreasing precipitation had negative effects on above-ground biomass, with reductions of 29% and 32%, respectively compared to ambient precipitation (Fig. 6D). Any effects of temperature or precipitation, as well as their interaction effect, on leaf total nitrogen of *G. spiculosa* were insignificant (Table 3).

Secondary growth of *C. angustifolia*

One-way ANOVA results (Table 4) showed that only above-ground biomass displayed a significant difference with the incubation temperature treatment after one month's secondary growth ($F=4.096$, $p=0.019$); no other morphological or photosynthetic indicator showed significant differences (Table 4). Above-ground biomass of secondary growth was greatest in the +4°C manipulation, 1.11 times higher than the ambient temperature (Fig. 7).

Discussion

Interaction effects of warming and precipitation alteration

Boreal wetlands are becoming increasingly vulnerable in a warming world [8]. Measuring responses of dominant plant species to warming and precipitation alteration is fundamental to assess and project the vulnerability of boreal wetlands under

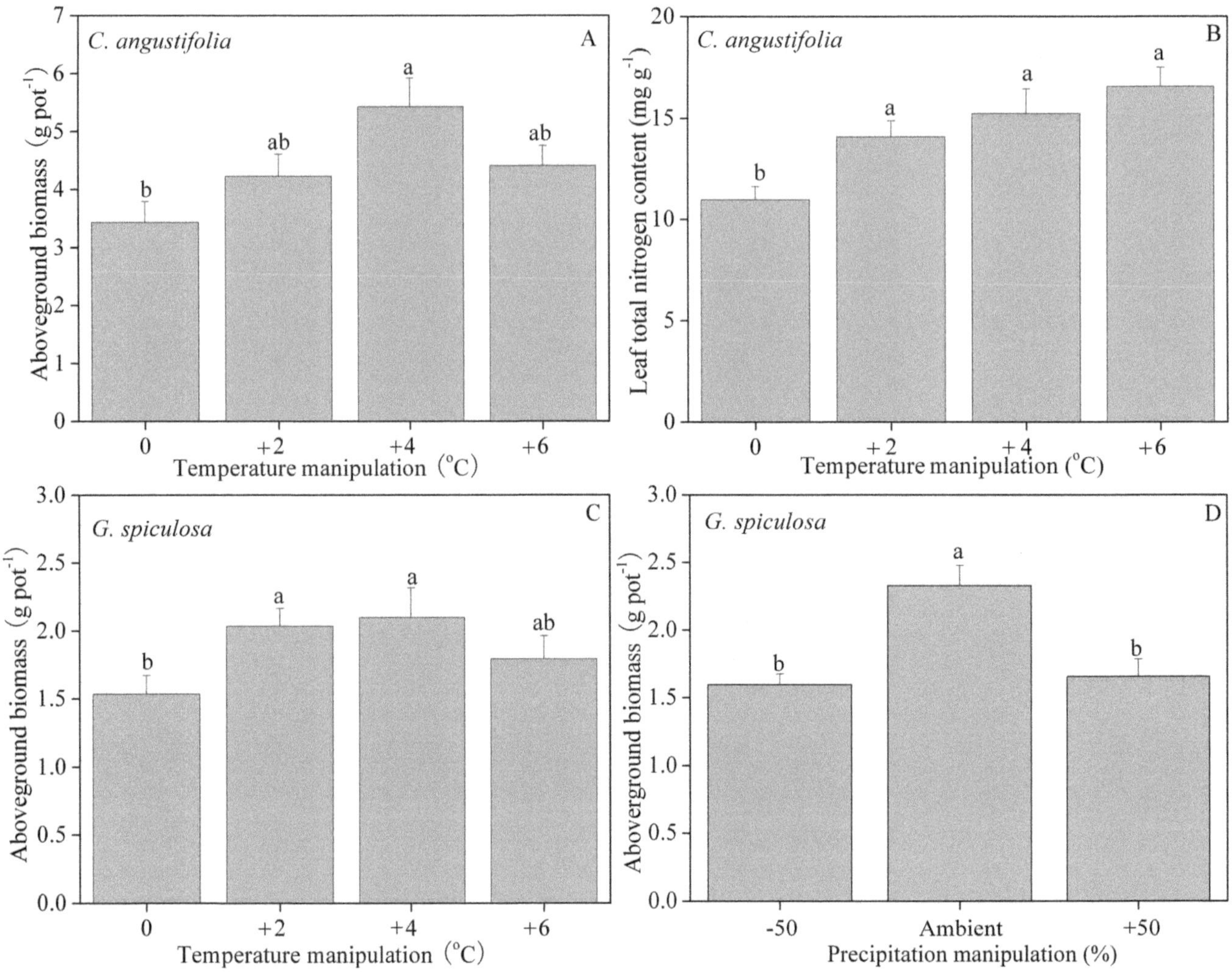

Figure 6. Aboveground biomass (A), leaf total nitrogen (B) of *C. angustifolia*, and aboveground biomass of *G. spiculosa* (C, D) after harvest. Different letters shared by the bars indicate significant differences ($p<0.05$) between temperature or precipitation manipulations. The error bars represent means ± 1 standard error.

climate change. The interactions among multiple climate-change factors may cause responses in wetlands that may not be detected when focusing on a single factor, because climate change will include simultaneous changes in temperature and precipitation regimes [19,20]. The above-ground biomass of *C. angustifolia* and *G. spiculosa* initially showed an increasing trend and then decreased with continued increments in mean monthly temperature. The +4°C temperature treatment produced the highest biomass (Fig. 6A, C), which is consistent with the single factor study by Breeuwer et al. [10] and multi-factor meta-analysis results from Lin et al. [21].

The monthly heights in different warming manipulations show intersecting trend lines (Fig. 1A, 2A), indicating that interaction effects of temperature and precipitation should be considered, especially for the +4°C manipulation. In the decreased precipitation treatment (−50%), greater plant heights were at ambient temperature and in the +4°C manipulation for *C. angustifolia* and *G. spiculosa* (Fig. 1C, 2C, respectively). With the exclusion of injury, height is one of the best indicators of an herbaceous plant's growth and competitiveness in light [22,23]. Our results show that any study on species-specific or interspecific plant response to climate change should also include consideration of the accompanying precipitation regime.

A limited increase of temperature within the optimal range often enhances photosynthesis, depending on the species and its environmental acclimation [24,25]. However, this expected enhancement of Pn can be moderated by concurrent water or drought stress on plants [26]. Our results show that Pn of *C. angustifolia* increased with temperature only in the driest (−50% precipitation) treatment, while ambient precipitation resulted in smaller Pn values in two of the three elevated temperatures (Fig. 4A), indicating that a change in precipitation would counter-intuitively limit the photosynthetic capacity of *C. angustifolia*. The Pn of *G. spiculosa* decreases with increasing precipitation (Fig. 5A), suggesting that soil water availability affects photosynthesis more than temperature variation [27]. Photosynthetic capacity is closely associated with, and approximately proportional to, leaf chlorophyll and nitrogen contents, and all three are favored by increased temperature [28,29,30]. Only the leaf nitrogen content of *C. angustifolia* increased with temperature (Table 3, Fig. 6B), while that of *G. spiculosa* was not significantly affected either by temperature or precipitation (Table 3). Although the leaf chlorophyll contents of two species were altered both by

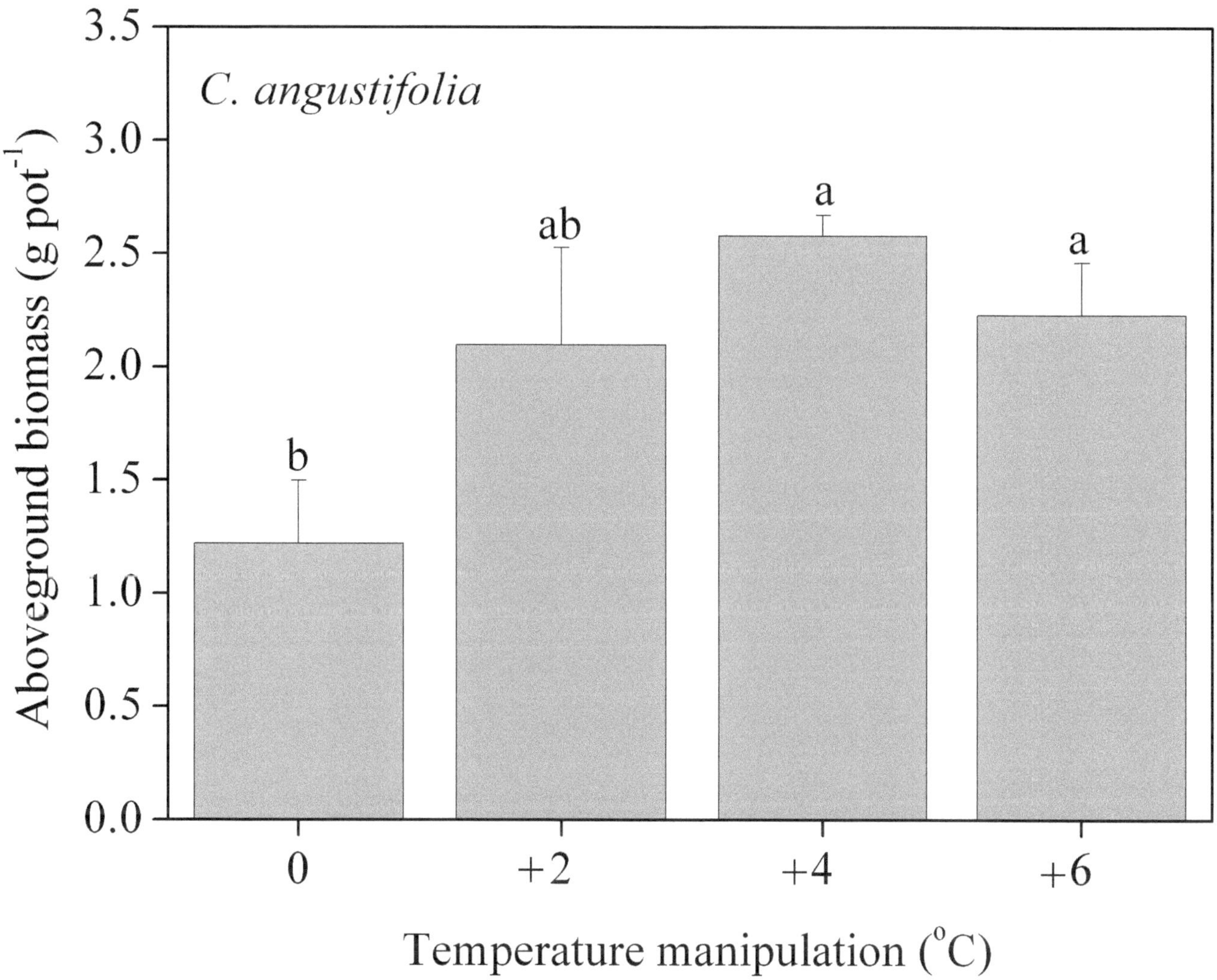

Figure 7. Aboveground biomass of *C. angustifolia* after secondary growth for one month. Different letters shared by the bars indicate significant differences ($p<0.05$) between the temperature manipulations. The error bars represent means ± 1 standard error.

temperature and precipitation (Table 3), the temperature responses under different precipitation treatments differed between species (4D, 5D).

These differential, and even opposite, responses of two species to our manipulated temperature and precipitation treatments are consistent with the widely supported observations and modeling that show species respond individualistically to changes in environmental conditions [5].

Species-specific responses to temperature and precipitation change

The effects of climate change on boreal wetland ecosystems ranges from regional, ecosystem, vegetation type, functional type to species [21,31,32,33]. Our results suggest that within the same functional plant type, species-specific response was a significant determinant in height (Table 3, Fig. 1, 2), leaf area (Fig. 3), photosynthetic indicators, leaf chlorophyll contents (Table 3, Fig. 4, 5), and above-ground biomass (Table 3, Fig. 6).

The most marked species-specific difference was leaf area and above-ground biomass (Table 2, 3). Leaf area is only affected by temperature for *C. angustifolia* (Fig. 3A) and precipitation for *G. spiculosa* (Fig. 3B). Above-ground biomass of *C. angustifolia* is only affected by temperature (Fig. 6A), while both temperature and precipitation influence *G. spiculosa* (Fig. 6C, D). Given the direct relationship with net primary productivity, above-ground biomass is considered as one of the plant characteristics responding strongly to climate change [34]. The +2°C manipulation increased above-ground biomass 23% and 31% compared with ambient conditions for *C. angustifolia* (Fig. 6A) and *G. spiculosa* (Fig. 6C), respectively. According to the linear fitting of local meteorological records in Xingkai Lake lacustrine wetlands, a 2°C warming is likely to take place by the 2080s. Consequently, this species-specific difference in predicted biomass would affect carbon and nutrient budgets in boreal wetland ecosystems responding to future temperature change [31].

Previous species-specific comparisons under simulated climate change are mostly based on only one growth season, and the intergenerational response is not readily apparent [35,36]. The significant difference in above-ground biomass of secondary growth (Fig. 7) with the different temperature treatments indicates that effects of climate change would continue to the next generation, because *C. angustifolia* mainly reproduces clonally by rhizomes [12]. These rhizomes would store a large amount of

Table 4. One-way ANOVA of aboveground biomass, morphological and photosynthetic indicators of *C. angustifolia* after secondary growing for one month.

ANOVA	Height	Basal stem diameter	Leaf area	Pn	Gs	Ts	Aboveground biomass
F	0.281	0.874	1.849	2.963	2.650	2.802	4.096
p	0.838	0.494	0.217	0.097	0.120	0.108	**0.049**

Significant effects ($p<0.05$) are shown in bold.

biomass below-ground, translocated from above-ground when the shoot is harvested. Greater biomass results in better performance during secondary growth [37]. Considering the biomass accumulation of biomass year by year, *C. angustifolia* is more acclimated to temperature and precipitation change. This acclimated capacity supports the continued presence of *C. angustifolia* under climate change and perhaps even expansion, because a competitor is not as evolutionarily successful in adapting to change and hence will decline leaving an ecological niche to be filled [31].

Inter-specific interaction to temperature and precipitation change

Interspecies interactions play a key role in regulating the distribution and composition of plant communities. The impacts of inter-species interactions can be altered by external drivers such as climatic conditions or anthropogenic activities [38]. For well conserved boreal wetlands, climate change may affect wetland plant communities by directly limiting or fostering the performance of particular species or by altering species competition and abundance [10,39].

As the two dominant species in Sanjiang Plain wetlands of Northeast China, *C. angustifolia* competes with *G. spiculosa* both in mesic and hydric habitats. Based on sequentially documented surveys of plant communities in a mesic habitat from 172 sampling plots from 1973 to 2003, the "importance values", as the average of relative density, frequency and coverage which gives an overall estimate of the influence of importance of a plant species in the community, of *C. angustifolia* decreased from 0.55 in the 1970s to 0.50 in the 2000s and that of *G. spiculosa* from 0.30 in the 1970s to 0.13 in the 2000s, accompanied by an increase in mesophytes. Although importance values for both hygrophytes declined, the former decreased by 9%, while the latter decreased 57%. Another sequentially documented *Carex lasiocarpa* community in a hydric habitat showed that the importance value of *G. spiculosa* decreased from 0.19 in the 1970s to 0.13 in the 2000s, while that of *C. angustifolia* increased from 0.021 to 0.023 [40]. Consequently, *C. angustifolia* is more competitive than *G. spiculosa* and *G. spiculosa* is more environmentally sensitive than *C. angustifolia*, which might be partly attributed to the greater acclimated capacity of *C. angustifolia* responding to precipitation and/or temperature change (excluding anthropogenic activities). Firstly, the greatest above-ground biomass of *C. angustifolia* presents in the +4°C treatment (Fig. 6A), while that of *G. spiculosa* is in the +2°C and +4°C manipulations (Fig. 6C), suggesting that *C. angustifolia* might be more acclimated to greater warming, even occasional extreme high temperatures. Secondly, there is no significant difference in the precipitation manipulations for *C. angustifolia*'s leaf area (Table 3) or above-ground biomass (Table 3), while either -50% or +50% of precipitation significantly decreased the leaf area (Fig. 3B) in August and September and above-ground biomass (Fig. 6D) of *G. spiculosa*.

Considering the warming trend of 2.5–5.4°C in the Sanjiang Plain within this century [41], it is predicted that the distribution of *C. angustifolia* will expand while that of *G. spiculosa* will shrink. However, the composition dynamics of these communities will be determined both by the intrinsic biological characteristics of each species and the external environment including climate change [5,42]. The species-specific responses described in this research coupled with inter-species interactions demonstrate the complexity of multi-factor effects. In addition to climate, other factors, e.g. soil nutrient concentrations, will modify or attenuate temperature and precipitation effects [35]. Consequently, minimization of other external (anthropogenic) factors is required in order to assess climate change impacts and this can best be achieved using well-conserved boreal wetlands.

Implications for boreal wetland conservation and restoration

Boreal wetland ecosystems, especially those that are well conserved, have an innate resistance and acclimated self-regulation potential, allowing them to respond to climate change and variation in habitat conditions [7,8]. However, long-term stress caused by high temperature, extreme drought or flooding, or the degradation and eventual loss of dominant species or functional groups will alter fundamental ecosystem properties and processes, because changes in the abundance, production, or distribution of species are often sufficient to alter both structure and function of the ecosystem [43]. Therefore, investigating the thresholds of irreversible climate change effects is both urgent and critical to the conservation and restoration of boreal wetland ecosystems.

Stability and acclimation can by delineated by ecological thresholds. Within these thresholds, the species interactions of freshwater wetland plants tend to shift from competitive to facilitative with increased stress [44]; however, when the environment changes beyond these thresholds, perhaps through a combination of habitat alteration by climate change and direct damage by humans, boreal wetlands may shift to other ecosystems such as grassland or shrubland [45]. This may result in loss of the organic carbon sequestered in these wetlands for hundreds or thousands of years, ultimately reducing the net cooling effect of a carbon sink on global warming [7,45]. To avoid this consequence, conservation policies and restoration practices will need to be revised in the face of potential thresholds for irreversible change to key wetland plant species under climate change. For example, hydrologically damaged wetlands will experience greater vulnerability to climate change effects compared to wetlands with an intact hydrological regime [8]. According to our results (Fig. 6D), the stability of the natural surface water table is critical for the conservation and restoration of the *G. spiculosa* population, which indicates special actions may be required at both the management and policy level if this species is to be safeguarded in the future.

Acknowledgments

The authors express gratitude to the reviewers and editors for their critical comments on an earlier version of the manuscript. We gratefully acknowledge Jiawei Guo and Haiyang Zhao for seedling collection, and Yuxia Zhang and Bokun Lou for sample analyses.

Author Contributions

Conceived and designed the experiments: YZ GW. Performed the experiments: X. Lou. Analyzed the data: XY. Contributed reagents/materials/analysis tools: X. Lu. Contributed to the writing of the manuscript: YZ MG.

References

1. Gorham E (1991) Northern peatlands, role in the carbon cycle and probable responses to climatic warming. Ecol Appl 1:182–195.
2. Aerts R, Callaghan TV, Dorrepaal E, van Logtestijn RSP, Cornelissen JHC (2012) Seasonal climate manipulations have only minor effects on litter decomposition rates and N dynamics but strong effects on litter P dynamics of sub-arctic bog species. Oecologia 170:809–819.
3. Bartlett KB, Harriss RC (1993) Review and assessment of methane emissions from wetlands. Chemosphere 26:261–320.
4. Bridgham SD, Pastor J, Updegraff K, Malterer TJ, Johnson K, et al. (1999) Ecosystem control over temperature and energy flux in northern peatlands. Ecol Appl 9:1345–1358.
5. Weltzin JF, Bridgham SD, Pastor J, Chen J, Harth C (2003) Potential effects of warming and drying on peatland plant community composition. Global Change Biol 9:141–151.
6. Flanagan LB, Syed KH (2011) Stimulation of both photosynthesis and respiration in response to warmer and drier conditions in a boreal peatland ecosystem. Global Change Biol 17:2271–2287.
7. Dise N (2009) Peatland response to global change. Science 326:810–811.
8. Essl F, Dullinger S, Moser D, Rabitsch W, Kleinbauer I (2012) Vulnerability of mires under climate change, implications for nature conservation and climate change adaptation. Biodivers Conserv 21:655–669.
9. Charles H, Dukes JS (2009) Effects of warming and altered precipitation on plant and nutrient dynamics of a New England salt marsh. Ecol Appl 19:1758–1773.
10. Breeuwer A, Heijmans MMPD, Robroek BJM, Berendse F (2008) The effect of temperature on growth and competition between *Sphagnum* species. Oecologia 156:155–167.
11. Thompson RM, Beardall J, Beringer J, Grace M, Sardina P (2013) Means and extremes, building variability into community-level climate change experiments. Ecol Lett 16:799–806.
12. Editorial Committee of Flora Republicae Popularis Sinicae (1987) Flora Republicae Popularis Sinicae 9(3). Beijing, China: Science Press. 221–223 p.
13. Editorial Committee of Flora Republicae Popularis Sinicae (2002) Flora Republicae Popularis Sinicae 9(2). Beijing, China: Science Press. 327–328 p.
14. Ren G, Guo J, Xu M, Chu Z, Zhang L, et al. (2005) Climate changes of China's mainland over the past half century. Acta Meteorol Sin 63:942–956.
15. Guo R, Shi L, Ding X, Hu Y, Tian S, et al. (2010) Effects of saline and alkaline stress on germination, seedling growth, and ion balance in wheat. Agron J 102:1252–1260.
16. Arnon DI (1949) Copper enzymes in isolated chloroplasts Polyphenoxidase in *Beta vulgaris*. Plant Physiol 24:1–15.
17. Porra RJ (2002) The chequered history of the development and use of simultaneous equations for the accurate determination of chlorophylls a and b. Photosynth Res 73:149–156.
18. Zou YC, Jiang M, Yu XF, Lu XG, David JL, et al. (2011) Distribution and biological cycle of iron in freshwater wetlands of Sanjiang Plain, Northeast China. Geoderma 164:238–248.
19. Fenner N, Freeman C, Lock MA, Harmens H, Reynolds B, et al. (2007) Interactions between elevated CO_2 and warming could amplify DOC exports from peatland catchments. Environ Sci Technol 41:3146–3152.
20. Kardol P, Cregger MA, Campany CE, Classen AT (2010) Soil ecosystem functioning under climate change, plant species and community effects. Ecology 91:767–781.
21. Lin DL, Xia JY, Wan SQ (2010) Climate warming and biomass accumulation of terrestrial plants, a meta-analysis. New Phytol 188:187–198.
22. Wilson SD, Tilman D (1993) Plant competition and resource availability in response to disturbance and fertilization. Ecology 74:599–611.
23. Gough L, Gross KL, Cleland EE, Clark CM, Collins SL, et al. (2012) Incorporating clonal growth form clarify the role of plant height in response to nitrogen addition. Oecologia 169:1053–1062.
24. Berry J, Bjorkman O (1980) Photosynthetic response and adaptation to temperature in higher plants. Annu Rev Plant Physiol 31:491–543.
25. Rustad LE, Campbell JL, Marion GM, Norby RJ, Mitchell MJ, et al. (2001) A meta-analysis of the response of soil respiration, net nitrogen mineralization, and aboveground plant growth to experimental ecosystem warming. Oecologia 126: 543–562.
26. De Boeck HJ, Lemmens CMHM, Zavalloni C, Gielen B, Malchair S, et al. (2008) Biomass production in experimental grasslands of different species richness during three years of climate warming. Biogeosciences 5:585–594.
27. Ge Z, Zhou X, Kellomäki S, Zhang C, Peltola H, et al (2012) Acclimation of photosynthesis in a boreal grass (*Phalaris arundinacea* L) under different temperature, CO_2, and soil water regimes. Photosynthetica 50:141–151.
28. Farquhar GD, Sharkey TD (1982) Stomatal conductance and photosynthesis. Annu Rev Plant Physiol 33:317–345.
29. Evans JR (1989) Photosynthesis and nitrogen relationship in leaves of C_3 plants. Oecologia 78:9–19.
30. Mae T (1997) Physiological nitrogen efficiency in rice, nitrogen utilization, photosynthesis, and yield potential. Plant Soil 196:201–210.
31. Weltzin JF, Pastor J, Harth C, Bridgham SD, Updegraff K, et al. (2000) Response of bog and fen plant communities to warming and water-table manipulations. Ecology 81:3464–3478.
32. Erwin KL (2009) Wetlands and global climate change, the role of wetland restoration in a changing world. Wetl Ecol Manage 17:71–84.
33. Liancourt P, Spence LA, Song D, Lkhagva A, Nharkhuu A, et al. (2013) Plant response to climate change varies with topography, interactions with neighbors, and ecotype. Ecology 94:444–453.
34. Van Minnen JG, Onigkeit J, Alcamo J (2002) Critical climate change as an approach to assess climate change impacts in Europe, development and application. Environ Sci Policy 5:335–347.

35. Luo WB, Xie YH (2009) Growth and morphological responses to water level and nutrient supply in three emergent macrophyte species. Hydrobiologia 624:151–160.
36. Gedan KB, Bertness MD (2010) How will warming affect the salt marsh foundation species *Spartina patens* and its ecological role? Oecologia 164:479–487.
37. Qiu ZC, Wang M, Lai WL, He FH, Chen ZH (2011) Plant growth and nutrient removal in constructed monoculture and mixed wetlands related to stubble attributes. Hydrobiologia 661:251–260.
38. Brooker RW (2006) Plant-plant interactions and environmental change *New Phytol*, 171, 271–284.
39. Pastor J, Peckham B, Bridgham S, Weltzin J, Chen J (2002) Plant community dynamics, nutrient cycling and alternative stable equilibria in peatlands. Am Nat 160:553–568.
40. Lou YJ (2008) Species biodiversity spatial patterns of typical wetland plants and their changes within 30 years in the Sanjiang Plain. Graduate University of Chinese Academy of Sciences, Beijing, China.
41. Yin XM (2013) Simulation Study of Climate Change impacts on Wetlands Productivity in Sanjiang Plain. Graduate University of Chinese Academy of Sciences, Beijing, China.
42. Hoeppner SS, Dukes JS (2012) Interactive responses of old-field plant growth and composition to warming and precipitation. Global Change Biol 18:1754–1768.
43. Berry PM, Dawson TP, Harrison PA, Pearson RG (2002) Modelling potential impacts of climate change on the bioclimatic envelope of species in Britain and Ireland. Global Ecol Biogeogr 11:453–462.
44. He Q, Bertness MD, Altieri AH (2013) Global shifts towards positive species interactions with increasing environmental stress. Ecol Lett 16:695–706.
45. Hobbs RJ, Higgs E, Harris JA (2009) Novel ecosystems, implications for conservation and restoration. Trends Ecol Evol 24:599–605.

20

Germination Shifts of C_3 and C_4 Species under Simulated Global Warming Scenario

Hongxiang Zhang[1], Qiang Yu[2], Yingxin Huang[1], Wei Zheng[3], Yu Tian[4], Yantao Song[5], Guangdi Li[1,3], Daowei Zhou[1]*

1 Northeast Institute of Geography and Agroecology, Chinese Academy of Sciences, Changchun, China, **2** State Key Laboratory of Forest and Soil Ecology, Institute of Applied Ecology, Chinese Academy of Sciences, Shenyang, China, **3** Graham Centre for Agricultural Innovation, Wagga Wagga, New South Wales, Australia, **4** Animal Science and Technology College, Jilin Agricultural University, Changchun, China, **5** College of Environment and Resources, Dalian Nationalities University, Dalian, China

Abstract

Research efforts around the world have been increasingly devoted to investigating changes in C_3 and C_4 species' abundance or distribution with global warming, as they provide important insight into carbon fluxes and linked biogeochemical cycles. However, changes in the early life stage (e.g. germination) of C_3 and C_4 species in response to global warming, particularly with respect to asymmetric warming, have received less attention. We investigated germination percentage and rate of C_3 and C_4 species under asymmetric (+3/+6°C at day/night) and symmetric warming (+5/+5°C at day/night), simulated by alternating temperatures. A thermal time model was used to calculate germination base temperature and thermal time constant. Two additional alternating temperature regimes were used to test temperature metrics effect. The germination percentage and rate increased continuously for C_4 species, but increased and then decreased with temperature for C_3 species under both symmetric and asymmetric warming. Compared to asymmetric warming, symmetric warming significantly overestimated the speed of germination percentage change with temperature for C_4 species. Among the temperature metrics (minimum, maximum, diurnal temperature range and average temperature), maximum temperature was most correlated with germination of C_4 species. Our results indicate that global warming may favour germination of C_4 species, at least for the C_4 species studied in this work. The divergent effects of asymmetric and symmetric warming on plant germination also deserve more attention in future studies.

Editor: Wagner L. Araujo, Universidade Federal de Vicosa, Brazil

Funding: This research has been supported by the National Natural Science Foundation of China (41330640). The funder had no role in study design, data collection and analysis, decision to publish, or preparation of the manuscript.

Competing Interests: The authors have declared that no competing interests exist.

* Email: zhoudaowei@iga.ac.cn

Introduction

Climate change may strongly shift the distribution of C_3 and C_4 vegetation. C_3 and C_4 distribution helps to estimate global vegetation primary production and carbon uptake by the terrestrial biosphere [1]. Many previous examinations of the distribution of C_3 and C_4 grasses have attempted to predict changes due to the effect of rising atmospheric CO_2 concentration, and have suggested that the direct and indirect effects are likely to push C_3/C_4 relative abundances in opposite directions [2,3]. A recent study indicates that C_4 species tend to spread toward more northern latitudes and higher altitudes in the Inner Mongolia grassland. This was mainly triggered by increasing temperature, which overwhelmed the positive effect of rising CO_2 concentrations on C_3 species [4]. Others have found that temperature is the primary driver of C_4 grass species distribution [5] and the aboveground productivity ratios of C_3 and C_4 plants [6,7]. Long-term data sets from the shortgrass steppe have indicated that increased spring minimum temperature was correlated with decreased net primary production by the dominant C_4 grass (*Bouteloua gracilis*) and increased abundance and production by exotic and native C_3 forbs [8].

Global warming has advanced the yearly first-flowering times of plants [9]. This may change reproductive output and seed maturation time, which affects seed germination [10]. Germination percentage and rate of C_4 species increased with temperature until 35°C to 40°C, while germination of C_3 species was favored by cooler conditions [11]. Different germination responses of C_3 and C_4 species may affect ecosystem structure and functioning via species' altered relative competitive ability and/or net primary productivity. However, there has been little research on the consequences of global warming for plant seed germination [12], let alone for the effects on germination of C_3 and C_4 species.

Global warming has been found to be asymmetric [13], i.e. there are greater increases in daily minimum than maximum temperatures, resulting in declining diurnal temperature ranges [14,15]. This pattern has been empirically demonstrated in several regions [16,17]. To date, however, most modelling efforts and experimental manipulations investigating plant or ecosystem responses to climate change have assumed that future warming will occur primarily during the day or uniformly over the diurnal cycle [18,19]. Only a few researchers have studied the effect of nocturnal temperature elevation on ecosystem functions [20–22].

Table 1. The alternating temperature regimes used in the study.

Maximum temperature constant (TmaxC)	Average temperature constant (TaveC)	Asymmetric warming (AW)	Symmetric warming (SW)
TmaxC1: 5/35 (20, 30)*	TaveC1: 5/35 (20, 30)	AW1: 5/20 (12.5, 15)	SW1: 5/15 (10, 10)
TmaxC2: 8/35 (21.5, 27)	TaveC2: 8/32 (20, 24)	AW2: 11/23 (17, 12)	SW2: 10/20 (15, 10)
TmaxC3: 11/35 (23, 24)	TaveC3: 11/29 (20, 18)	AW3: 17/26 (21.5, 9)	SW3: 15/25 (20, 10)
TmaxC4: 14/35 (24.5, 21)	TaveC4: 14/26 (20, 12)	AW4: 23/29 (26, 6)	SW4: 20/30 (25, 10)
TmaxC5: 17/35 (26, 18)	TaveC5: 17/23 (20, 6)	AW5: 29/32 (30.5, 3)	SW5: 25/35 (30, 10)
TmaxC6: 20/35 (27.5, 15)	TaveC6: 20/20 (20, 0)	AW6: 35/35 (35, 0)	SW6: 30/40 (35, 10)
TmaxC7: 23/35 (29, 12)			
TmaxC8: 26/35 (30.5, 9)			
TmaxC9: 29/35 (32, 6)			
TmaxC10: 32/35 (33.5, 3)			
TmaxC11: 35/35 (35, 0)			

*The mean temperature and diurnal temperature range of each temperature regime were presented in parenthesis with comma between them.

To our knowledge, there is no study that has investigated the effect of asymmetric warming on plant regeneration from seed.

Predicting germination timing changes of C_3 and C_4 species under long term asymmetric warming compared with symmetric warming helps to elucidate C_3 and C_4 vegetation distribution changes. A thermal time approach can be used to model these aspects, though no one has attempted this before, except in modelling dormancy change [23] and seed germination [24], as well as seedling emergence [25] in current climate.

Previous research on germination response to alternating temperature has mainly focused on how seasonal changes in temperature regulate seed dormancy and germination [26–28] or on the effect of temperature fluctuations on germination [29–31]. In the current study, we alternated temperature regimes to simulate asymmetric and symmetric global warming and analyzed changes in germination patterns of C_4 and C_3 species using a thermal time model. The main objectives of this study were: (1) to determine the comparative effects of global warming on seed germination of C_3 and C_4 species; and (2) to compare the different effects on germination resulting from asymmetric versus symmetric warming.

Materials and Methods

Plant materials and seed collection

Six wild species were selected in this study, of which *Cynanchum chinense* R. Br., *Lappula myosotis* V. Worf., and *Saussurea amara* DC. were C_3 species, and *Amaranthus retroflexus* L., *Portulaca oleracea* L., and *Echinochloa crusgalli* (L.) Beauv. were C_4 species. These species are widely distributed in northern China. Mature seeds of the six target species were collected during autumn 2006 from wild populations in the Songnen grassland of China (sites near 44°40' N, 123°44' E), and stored in cloth bags at 4°C. The seeds were permitted to be collected around the land of the Grassland Ecosystem Experimental Station of Northeast Normal University. No protected species were sampled.

Table 2. The two-way ANOVA analysis of the effects of plant photosynthetic type (PPT) and temperature treatments (T) in four temperature regimes on germination percentage and germination rate.

	PPT (df = 1)	T (df = 10† or 5)	PPT * T (df = 10† or 5)
GP			
TmaxC	32.24***	1.24ns	2.38*
TaveC	0.13ns	0.30ns	3.23*
SW	0.43ns	0.88ns	3.18*
AW	0.05ns	0.19ns	3.40*
GR			
TmaxC	557.85***	5.37**	1.02ns
TaveC	<0.01ns	0.07ns	1.56ns
SW	1.05ns	1.11ns	14.16***
AW	0.84ns	1.18ns	4.97**

ns, $P>0.05$;
*, $P<0.05$;
**, $P<0.01$;
***, $P<0.001$
†The degree of freedom for TmaxC temperature regime is 10 and df of other temperature regimes is 5.

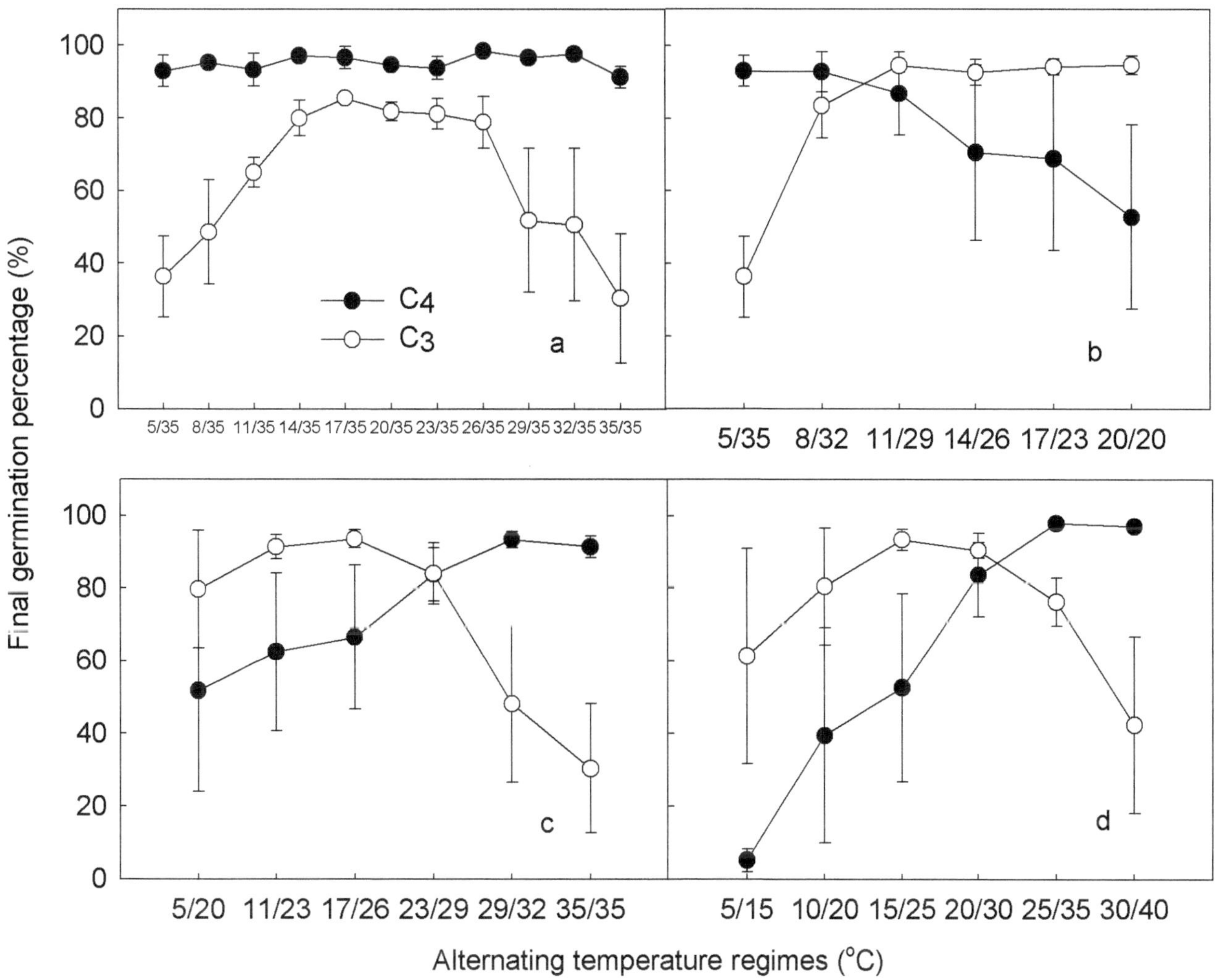

Figure 1. Germination percentages of C_4 (closed circles, average of *E. crusgalli, P. oleracea* and *A. retroflexus*) and C_3 species (open circles, average of *L. myosotis, S. amara* and *C. chinense*) under different alternating temperature regimes (a, TmaxC; b, TaveC; c, AW; d, SW). Mean ± S.E.

Temperature regimes

To simulate global asymmetric and symmetric warming, two alternating temperature regimes were established (Table 1): a) The differential rate of warming between maximum and minimum temperatures (asymmetric warming, AW). Under this scenario, minimum/maximum temperatures were incrementally increased by +6°C/+3°C for each treatment from 5°C/20°C to 35°C/35°C. As a result, the mean temperature increased by 4.5°C and diurnal temperature range (DTR) decreased by 3°C for each set of treatments, with 6 treatments total in this regime; b) The same rate of warming between maximum and minimum temperatures (symmetric warming, SW). Under this scenario, both minimum and maximum temperatures increased by 5°C for each treatment from 5°C/15°C to 30°C/40°C. As a result, the mean temperatures incrementally increased by 5°C while the DTR remained constant at 10°C, with 6 treatments total in this regime. To test temperature metrics effects, two additional temperature regimes were established (Table 1): c) Minimum temperatures increased from 5°C to 35°C with 3°C increments, but the maximum temperature remained constant at 35°C across treatments (TmaxC). Effectively, the mean temperature increased by 1.5°C and the DTR decreased by 3°C across treatments, with 11 treatments total in this regime; d) minimum temperatures increased from 5°C to 20°C in 3°C increments, but maximum temperatures decreased from 35°C to 20°C in 3°C decrements (TaveC). Effectively, the mean temperature remained constant at 20°C, while the DTR decreased by 6°C across treatments, with 6 treatments total in this regime. To complete statistical analysis of thermal time model parameters, germination at 5–40°C constant temperatures with 5°C interval was also tested, with 8 treatments total in this regime. Together, 37 temperature regimes were established in this study. The DTR is defined as the difference between the diurnal maximum and minimum temperature. Alternating temperatures can be dissected into minimum temperature, maximum temperature and the calculated DTR and average temperature.

Germination test

Germination studies were conducted in growth chambers in August 2007. Each chamber corresponded to each temperature treatment. A 12-h photoperiod (Sylvania cool white fluorescent lamps, 25 μmol m^{-2} s^{-1}, 400–700 nm) with 12-h dark period was maintained throughout the experiment. The temperatures cycled with half-hour linear transition periods between minimum and

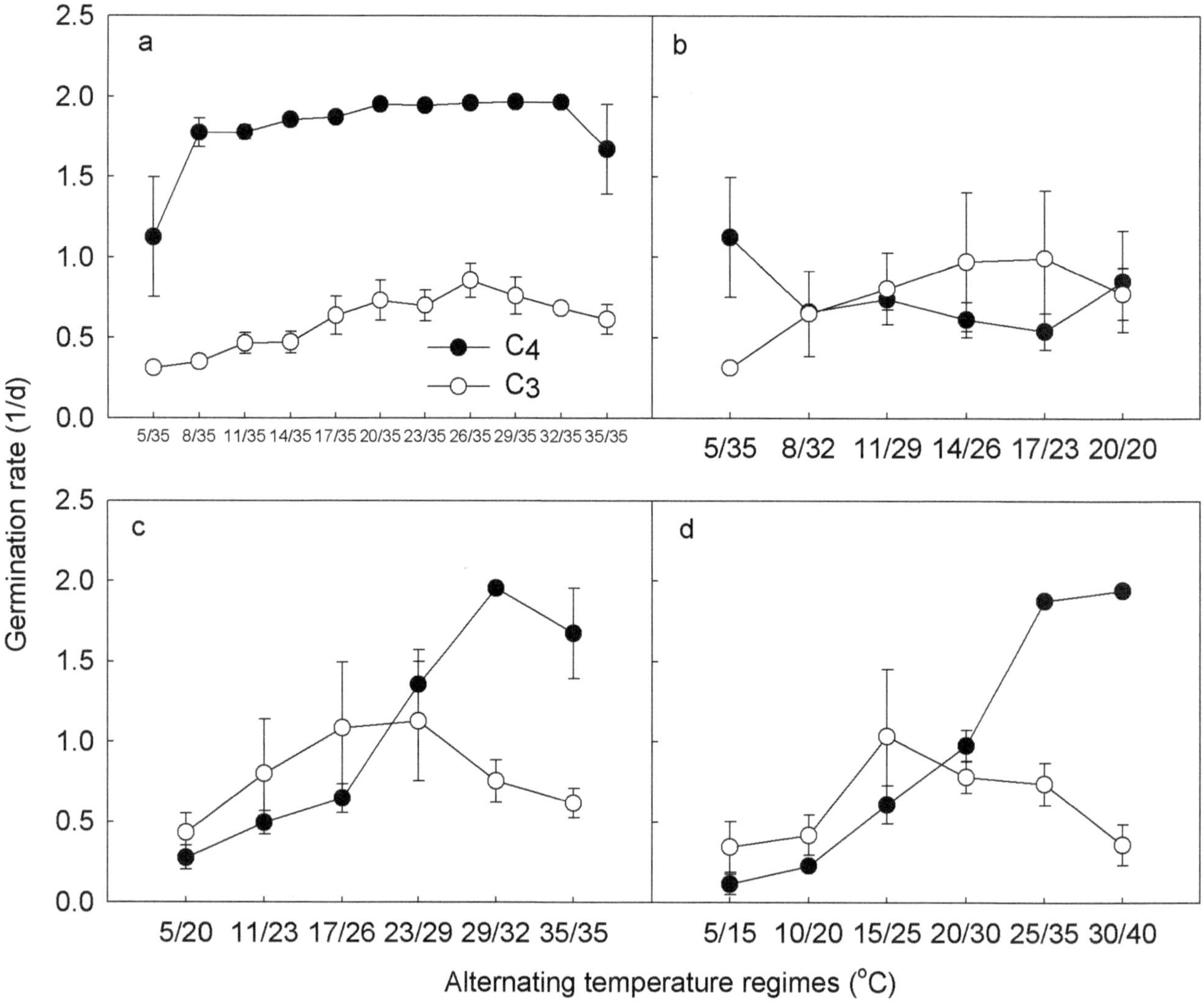

Figure 2. Germination rates of C_4 and C_3 species under different alternating temperature regimes. See Fig. 1 for symbols.

maximum temperatures (8:00–8:30 and 20:00–20:30, respectively).

Fifty seeds from each species were placed on two layers of filter paper in a grid (100 mm diameter) of plastic trays, and replicated 4 times within each temperature treatment. The filter paper was kept moistened with distilled water. Seeds were considered to have germinated when the radicle emerged. Germination status was recorded twice a day during the first 5 days, then once a day during the second 5 days. Germination was measured for 10 days because preliminary tests indicated that most of the germination events occurred within this period. The final germination percentage (GP) and germination rate (GR, the reciprocal of germination time) of each species were calculated with respect to the cumulative germination curves of each temperature treatment.

Thermal time model theory

Garcia-Huidobro et al. [32] presented a model with two equations relating thermal time (degree days above a base temperature) to germination rate (the reciprocal of the time taken for a given fraction of seed to germinate) at a constant temperature. The two equations of the thermal time model (TT model) are:

$$GR_g = 1/t_g = (T - T_b(g))/\theta_1(g) \quad T < T_o$$

$$GR_g = 1/t_g = (T_c(g) - T)/\theta_2(g) \quad T > T_o$$

For any given subpopulation g, germination rate GR can be described by two straight lines. The slopes of the two lines are $\theta_1(g)$ and $\theta_2(g)$ (thermal time constant at suboptimal and supra-optimal temperatures) with the intersection of the two lines defined as T_o (optimum temperature, at which maximum germination rate occurs). The two points where germination percentages equal zero are defined as $T_b(g)$ and $T_c(g)$ (the minimum and maximum temperature, below or above which no germination occurs). The parameters are useful for field predictions and can be used to compare germination in different species, climates, and locations.

If alternating temperatures were all above or all below the optimum temperature, then the formulas were no different to those for constant temperature. However, if the temperature fluctuated from below the optimum to above, then the equation changed [33]. In such cases, the predicted germination rate for

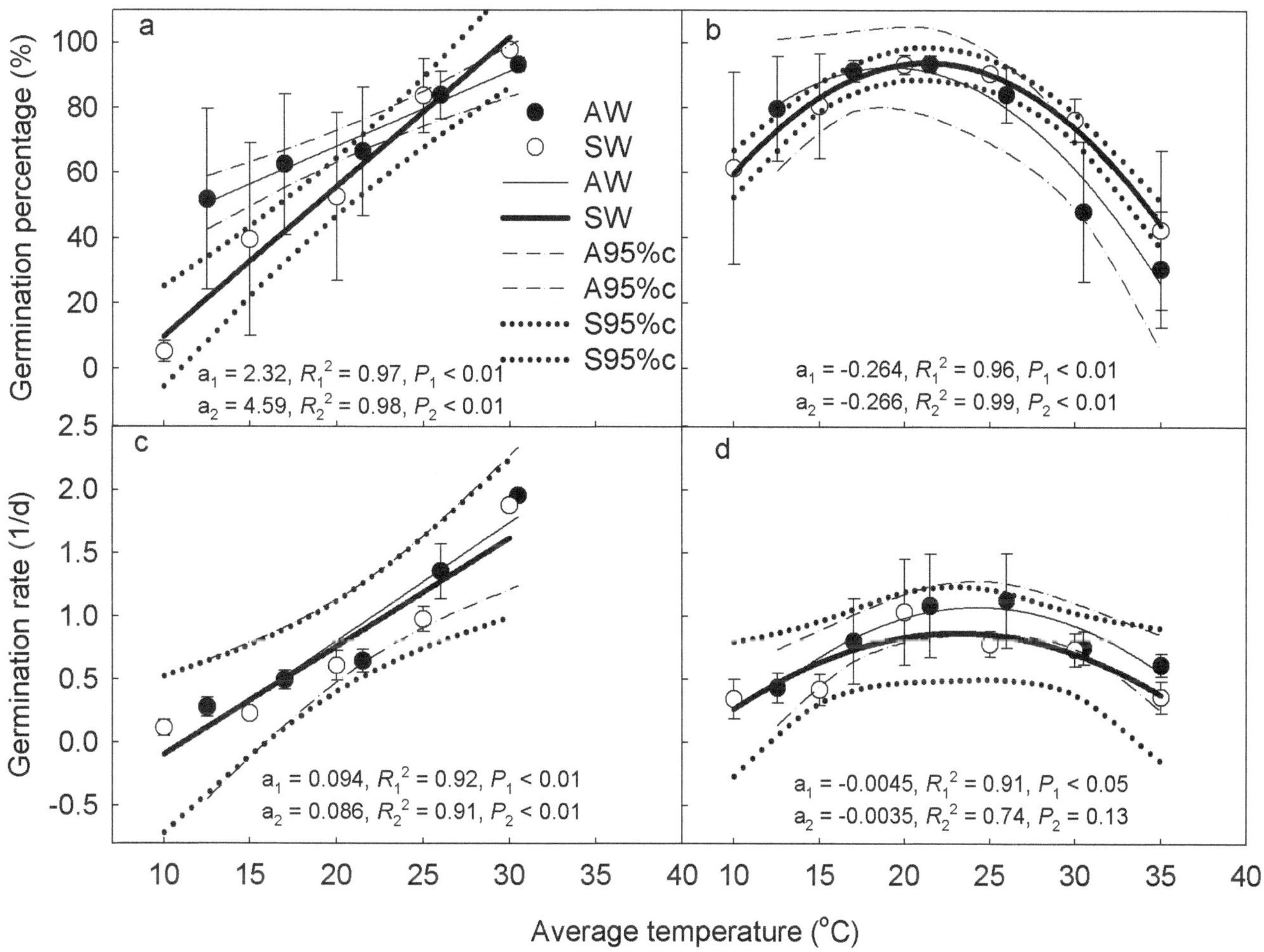

Figure 3. Germination percentage (a, b) and germination rate (c, d) changes of C_4 (a, c) and C_3 (b, d) species with increasing average temperature in the AW and SW temperature regimes. The relationship between C_4 species and the average temperature is fitted by linear equation; the relationship between C_3 species and the average temperature is fitted by polynomial equation. Dash-dot lines represent 95% confidence of AW temperature regime and dotted lines represent 95% confidence of SW temperature regime. a_1, R_1^2, P_1 represent the parameter, coefficient of determination and probability for the fitting in the AW regime and a_2, R_2^2, P_2 represent the parameter, coefficient of determination and probability for the fitting in the SW regime.

any subpopulation g is:

$$1/T_g = [(t_1/t)(T_1 - T_b(g))/\theta_1(g)] + [(t_2/t)(T_c(g) - T_2)/\theta_2(g)]$$

where t_1 is the time below T_o, with mean temperature of T_1, and t_2 is the time above T_o, with mean temperature T_2.

Data analysis

Data were analysed using SPSS (version 13.0, SPSS Inc., Chicago, Illinois, USA). Germination percentage and germination rate of C_3 or C_4 species were the average of the three species with the same photosynthetic type. The effects of different alternating temperature treatments in each regime and the plant photosynthesis type (C_3 and C_4) on germination percentage and germination rate were examined using two-way ANOVA. Germination times for 1 percent seed (1% germination) were calculated from the cumulative germination curves in AW and SW temperature regimes for the six species and converted to rates in this study, which represents the initiative germination time of each batch of seeds. This standard (1% germination) was used instead of 50% germination because seed germination of certain species was less than 50% for some temperature treatments (see Figure S1). Thermal time model (TT model) parameters were analyzed using repeated probit regression as described previously [33]. The average daily temperatures were used for determining alternating temperatures. The model parameter (T_b and θ_1) differences between C_3 and C_4 species, or between AW and SW temperature regimes within one photosynthetic type, were tested by Paired-Samples T-tests (2-tailed). The relationships between germination percentages/rates and the average temperatures in AW and SW regimes were fitted by either linear ($y = ax + y_0$) or quadratic regressions $y = ax^2 + bx + y_0$) using SigmaPlot (version 10.0, Systat Software Inc., Richmond, California, USA), with 95% confidence of the fitted lines given. The differences in the slope (a in the equation) of the regression between AW and SW regimes for germination of C_4 species were tested using the SMATR package in R software [34]. Pearson correlation analysis was carried out to test the correlation between GP/GR and temperature metrics (minimum, maximum, DTR, average) from the four alternating temperature regimes, totally 29 treatments.

Table 3. Thermal time model parameter estimates (T_b, minimum temperature; θ_1, thermal time constant) for C_4 and C_3 species under the symmetric warming (SW) and asymmetric warming (AW) alternating temperature regimes.

		T_b (°C)	θ_1 (°C·d)	R^2	P
C_4 species					
P. oleracea	SW	9.6	12.4	0.94	0.0013
	AW	10.4	11.3	0.94	0.0015
E. crusgalli	SW	10.4	12.3	0.95	0.0011
	AW	10.7	10.0	0.93	0.0079
A. retroflexus	SW	12.6	11.7	0.92	0.0028
	AW	13.3	10.5	0.87	0.0206
C_4 species average	SW	10.9	12.1		
	AW	11.5	10.6		
C_3 species					
L. myosotis	SW	6.3	8.2	0.76	0.3275
	AW	7.2	7.3	0.97	0.1123
S. amara	SW	1.7	34.6	0.93	0.0349
	AW	1.8	31.8	0.95	0.1372
C. chinense	SW	5.6	28.6	0.90	0.0133
	AW	4.3	25.5	0.94	0.0057
C_3 species average	SW	4.5	23.8		
	AW	4.4	21.5		

Results

Germination responses to different temperature regimes

The interaction of plant photosynthetic type and temperature treatment had significant effects on germination percentage (GP) and germination rate (GR), except in the case for GR under TmaxC and TaveC temperature regimes (Table 2).

GP of C_4 species were all above 90% in the TmaxC temperature regime, while GP of C_3 species increased linearly from 5/35°C to 17/35°C and remained around 80%–85% from 17/35°C to 26/35°C, then decreased sharply as the minimum temperature increased thereafter (Figure 1a). In the TaveC temperature regime, GP of C_4 species decreased as the minimum temperature increased, while GP of C_3 species increased, reached its highest value at 11/29°C, and remained subsequently constant (Figure 1b). In the AW and SW temperature regimes, GP of C_4 species increased, whereas GP of C_3 species increased and then decreased with the increasing temperatures (Figure 1c, d).

In the TmaxC temperature regime, GR of C4 species at 5/35°C was significantly lower than those at the other altered temperatures ($P<0.05$, Figure 2a). As the minimum temperature increased, GR of C_3 species increased gradually and reached peak at 26/35°C, and then decreased. In the TaveC temperature regime, GR of C_4 species tended to decrease with the increasing minimum temperature, while GR of C_3 species tended to increase until 17/23°C (Figure 2b). However, there were no significant differences in all values except GR at 5/35°C between C_4 and C_3 species ($P<0.05$). In the AW and SW temperature regimes, GR of C_4 species increased until 30°C of average temperature (29/32°C for AW or 25/35°C for SW), while GR of C_3 species increased and then decreased with the increasing temperatures (Figure 2c, d). GR of the three C_3 and C_4 species responded consistently to the temperature regimes, with only magnitude difference (Figure S2).

Comparison of TT model parameters and germination under asymmetric and symmetric warming

For all species in the AW and SW alternating temperature regimes (except C_3 species *L. myosotis* in the SW temperature regime), the linear regressions of the TT model explained more than 87% variation (Table 3). For the three C_4 species and the average, the estimated T_b of 1% germination in the AW temperature regime was higher than those in the SW temperature regime ($P = 0.059$). However, the estimated θ_1 in the AW regime were lower than those in the SW regime ($P = 0.057$). Therefore, the germination rates were difficult to compare between the two temperature regimes according to the TT model equation (Figure 3c). For *C. chinense* and the average of three C_3 species, T_b and θ_1 of 1% germination in the AW temperature regime were both lower than that in the SW temperature regime ($P = 0.891$ for T_b, $P = 0.081$ for θ_1), such that germination rates of the C_3 species in the AW temperature regime were higher than values in the SW temperature regime (Figure 3d). Compared with C_3 species, T_b of C_4 species were significantly higher in both the AW and SW temperature regimes ($P<0.05$). If C_3 species *L. myosotis* is excluded, θ_1 of C_4 species was also significantly lower than C_3 species in both temperature regimes ($P<0.05$).

For relationship between germination percentage and the average temperature, the parameter a in the AW regime was smaller than that in the SW regime, regardless of photosynthetic type (Figure 3a, b) and the difference was significant for C_4 species ($P<0.01$). This suggests that changes were more intense under symmetric warming (SW) than asymmetric warming (AW), especially for C_4 species. For relationship between germination rate and the average temperature, the parameter a in the AW regime was larger than that in SW regime for both C_4 and C_3 species. However, the parameter a between the two temperature regimes was not significantly different for C_4 species ($P = 0.74$, Figure 3c) and the linear regression between germination rate and

Table 4. Pearson correlation analysis of germination percentage (GP) and germination rate (GR) with temperature metrics (TM, minimum, maximum, average, diurnal temperature range (DTR)) from the four alternating temperature regimes for C_4 and C_3 species (*r*, correlation coefficient; *P*, probability for the correlation).

Species	TM	GP		GR	
		r	*P*	*r*	*P*
C_4 species					
P. oleracea	Minimum	0.388	0.037	0.590	0.001
	Maximum	0.870	<0.001	0.863	<0.001
	DTR	0.221	0.250	−0.005	0.981
	Average	0.678	<0.001	0.815	<0.001
E. crusgalli	Minimum	0.200	0.298	0.624	<0.001
	Maximum	0.434	0.019	0.822	<0.001
	DTR	0.103	0.594	−0.072	0.712
	Average	0.343	0.069	0.819	<0.001
A. retroflexus	Minimum	0.462	0.012	0.636	<0.001
	Maximum	0.958	<0.001	0.846	<0.001
	DTR	0.206	0.285	−0.067	0.728
	Average	0.771	<0.001	0.839	<0.001
C_4 species average	Minimum	0.439	0.017	0.640	<0.001
	Maximum	0.939	<0.001	0.876	<0.001
	DTR	0.216	0.260	−0.050	0.798
	Average	0.746	<0.001	0.855	<0.001
C_3 species					
L. myosotis	Minimum	−0.416	0.025	−0.002	0.992
	Maximum	−0.679	<0.001	−0.440	0.017
	DTR	−0.049	0.801	−0.324	0.087
	Average	−0.608	<0.001	−0.208	0.278
S. amara	Minimum	−0.576	0.001	0.438	0.017
	Maximum	−0.594	0.001	−0.071	0.713
	DTR	0.189	0.326	−0.531	0.003
	Average	−0.679	<0.001	0.270	0.157
C. chinense	Minimum	0.539	0.003	0.802	<0.001
	Maximum	0.312	0.100	0.385	0.039
	DTR	−0.358	0.057	−0.590	0.001
	Average	0.521	0.004	0.737	<0.001
C_3 species average	Minimum	−0.299	0.115	0.333	0.077
	Maximum	−0.514	0.004	−0.197	0.307
	DTR	−0.055	0.777	−0.509	0.005
	Average	−0.449	0.015	0.138	0.474

average temperature for C_3 species in the SW regime was not significant ($P = 0.13$, Figure 3d).

Correlation between GP/GR and temperature metrics

For C_4 species average, GP/GR were positively correlated with minimum, maximum and average temperature ($P<0.05$, Table 4), with the highest correlation coefficients between GP/GR and maximum temperature ($r = 0.939$ or $r = 0.876$). GRs of C_4 species were negatively correlated with DTR, though it was not significant ($P = 0.798$). All of the three C_4 species responded similarly. For C_3 species however, GPs were negatively correlated with maximum and average temperatures ($P<0.05$), and GRs were negatively correlated with DTR ($P<0.01$). Furthermore, the results of the three C_3 species were inconsistent.

Discussion

Germination percentage and germination rate of C_4 species increased linearly under the AW regime (Figure 1c and 2c), while those of C_3 species increased and then decreased with the increasing temperature. This suggests that long term global asymmetric warming may favor seed germination of C_4 species, at least for the tested species, which is consistent with the model prediction by Thorpe et al. [35] and inference from Sage and Kubien [36]. Spring events are changing more than autumn events as they are more sensitive to climate and are also undergoing the greatest alterations of climate relative to other seasons [37]. Germination and emergence are mainly spring events in most temperate regions [38], which are important research topics related to global warming. Though we just studied three C_3 and three C_4 species and more research is needed on this topic, our results somehow reflected the trends of C_3 and C_4 species germination and emergence shift under global warming.

In our study, symmetric and asymmetric warming had different impacts on germination of C_4 species. The symmetric warming treatments significantly overestimated the speed of germination percentage change with temperature ($P<0.05$), compared to the asymmetric warming (Figure 3a). Another evidence is that the average base temperature is higher and the average thermal time constant is lower in the AW alternating temperature regime compared to the SW regime (Table 3), which is near significant ($P=0.059$ for base temperature, $P=0.057$ for thermal time constant).

Ecosystem warming studies have been performed for more than 20 years using a variety of methods including heat-resistance cables, infrared (IR) lamps, field chambers (e.g. OTC), and night-time warming. Historic air temperature data and most models suggest that much of the global warming increase will occur during the night-time hours. Therefore, artificial night-time warming is ideally suited for replicating a potentially relevant form of climate change [39]. Field night warming has been achieved by heating with IR at night-time [21], by reflective curtains covering the vegetation at night [40], or by light-weight aluminum fabric shelters (mounted on rollers similar to a window shade) that are drawn across the warming plots at night [41]. In this study, we first used alternating temperature regimes controlled by different growth chambers to simulate consecutive global asymmetric (+3/+6°C at day/night continuously) and symmetric (+5/+5°C at day/night continuously) warming (Table 2). This method can be used to predict the trend of shifts in seed germination and seedling growth under global warming.

Phillips et al. [42] suggest that it is significant to understand the influence of both day and night-time warming on the carbon balance of plants and concluded that changes in daily mean temperatures, rather than changes in minimum or maximum temperature, are sufficient for predicting ecosystem carbon fluxes in a Mediterranean grassland system. Another report indicates that a changing daily temperature regime may be important in determining plant responses to warming temperatures and should be considered in predictions of plant and ecosystem responses to future climate change [43]. The correlation between germination and the temperature metrics (minimum, maximum, diurnal temperature range (DTR) and average temperature) from the four temperature regimes for C_4 and C_3 species were analyzed in this study (Table 4). For C_4 species, GP/GR was most correlated with the maximum temperature, while the results of the three C_3 species were inconsistent. Von Fischer et al. [7] suggest that daily maximum temperature better predicts percent of C_4 species than daily average or minimum temperature. Hattersley [44] also found summer (January) temperature in Australia had highest correlation with percent of C_4 species. DTR is an important index of climate change [45]. In our study, average and minimum temperature was significantly correlated with germination of C_4 species, but not C_3 species. However, DTR was not correlated with germination, except with the germination rates of the C_3 species *S. amara* and *C. chinense*.

Conclusions

This work indicates that global symmetric and asymmetric warming favors seed germination of C_4 species, rather than C_3 species, according to the results of the tested species in this study. Compared to asymmetric warming, symmetric warming overestimated the speed of germination percentage change with temperature for C_4 species. Asymmetric and symmetric warming had no significant effects on germination of C_3 species. Among the temperature metrics (minimum, maximum, diurnal temperature range and average of the alternating temperature), maximum temperature (day-time temperature) was found to be most correlated with germination for C_4 species. Germination of C_3 species responded inconsistently to the temperature metrics. Using alternating temperature regimes controlled by different growth chambers to simulate consecutive global asymmetric and symmetric warming is a good method, which can be used to predict the trend of shifts in plant early growth under global warming.

Supporting Information

Figure S1 Final germination percentages of six species under different alternating temperature regimes (TmaxC, a, b; TaveC, c, d; AW, e, f; SW, g, h). C_4 species: a, c, e, g; C_3 species: b, d, f, h.

Figure S2 Germination rates of six species under different alternating temperature regimes. See S1 for symbols.

Acknowledgments

We thank Dr Ping Wang, Wisdom Japhet, and Louis Irving for the comments on the previous version of the manuscript.

Author Contributions

Conceived and designed the experiments: HZ DZ. Performed the experiments: HZ WZ YT. Analyzed the data: HZ YH GL YS. Contributed to the writing of the manuscript: HZ QY.

References

1. Auerswald K, Wittmer MHOM, Männel TT, Bai YF, Schäufele R, et al. (2009) Large regional-scale variation in C_3/C_4 distribution pattern of Inner Mongolia steppe is revealed by grazer wool carbon isotope composition. Biogeosciences 6: 795–805.
2. Collatz GJ, Berry JA, Clark JS (1998) Effects of climate and atmospheric CO_2 partial pressure on the global distribution of C_4 grasses: present, past, and future. Oecologia 114: 441–454.
3. Winslow JC, Hunt ER, Piper SC (2003) The influence of seasonal water availability on global C_3 versus C_4 grassland biomass and its implications for climate change research. Ecol Model 163: 153–173.
4. Wittmer MHOM, Auerswald K, Bai YF, Schaufele R, Schnyder H (2010) Changes in the abundance of C_3/C_4 species of Inner Mongolia grassland: evidence from isotopic composition of soil and vegetation. Glob Change Biol 16: 605–616.
5. Bremond L, Boom A, Favier C (2012) Neotropical C_3/C_4 grass distributions-present, past and future. Glob Change Biol 18: 2324–2334.
6. Paruelo JM, Lauenroth WK (1996) Relative abundance of plant functional types in grasslands and shrublands of North America. Ecol Appl 6: 1212–1224.
7. Von Fischer JC, Tieszen LL, Schimel DS (2008) Climate controls on C_3 vs. C_4 productivity in North American grasslands from carbon isotope composition of soil organic matter. Glob Change Biol 14: 1141–1155.
8. Alward RD, Detling JK, Milchunas DG (1999) Grassland vegetation changes and nocturnal global warming. Science 283: 229–231.
9. Abu-Asab MS, Peterson PM, Shetler SG, Orli SS (2001) Earlier plant flowering in spring as a response to global warming in the Washington, DC, area, Biodivers Conserv 10: 597–612.
10. De Frenne P, Brunet J, Shevtsova A, Kolb A, Graae BJ, et al. (2011) Temperature effects on forest herbs assessed by warming and transplant experiments along a latitudinal gradient Glob Change Biol 17: 3240–3253.

11. Zhang H (2008) Research on seed germination ecology. Ph. D thesis.
12. Hovenden MJ, Wills KE, Chaplin RE, Schoor JKV, Williams AL, et al. (2008) Warming and elevated CO_2 affect the relationship between seed mass, germinability and seedling growth in *Austrodanthonia caespitosa*, a dominant Australian grass. Glob Change Biol 14: 1633–1641.
13. Xia J, Chen J, Piao S, Ciais P, Luo Y, et al. (2014) Terrestrial carbon cycle affected by non-uniform climate warming. Nat Geosci 7: 173–180.
14. Easterling DR, Horton B, Jones PD, Peterson TC, Karl TR, et al. (1997) Maximum and minimum temperature trends for the glob. Science 277: 364–367.
15. Vose RS, Easterling DR, Gleason B (2005) Maximum and minimum temperature trends for the globe: an update through 2004. Geophys Res Let 32: 5.
16. Karl TR, Jones PD, Knight RW, Kukla G, Plummer N, et al. (1993) A new perspective on recent global warming: asymmetric trends of daily maximum and minimum temperature. B Am Meteorol Soc 74: 1007–1023.
17. Liu B, Xu M, Henderson M, Qi Y, Li YQ (2004) Taking China's Temperature: daily range, warming trends, and regional variations, 1955–2000. J Climate 17: 4453–4462.
18. Zavaleta ES, Thomas BD, Chiariello NR, Asner GP, Shaw MR, et al. (2003) Plants reverse warming effect on ecosystem water balance. P Natl Acad Sci USA 100: 9892–9893.
19. Luo YQ (2007) Terrestrial carbon-cycle feedback to climate warming. Annu Rev Ecol Evol S 38: 683–712.
20. Lobell DB, Ortiz-Monasterio JI (2007) Impacts of day versus night temperatures on spring wheat yields: a comparison of empirical and CERES model predictions in three locations. Agron J 99: 469–477.
21. Wan SQ, Xia JY, Liu WX, Niu SL (2009) Photosynthetic overcompensation under nocturnal warming enhances grassland carbon sequestration. Ecology 90: 2700–2710.
22. Bai WM, Xia JY, Wan SQ, Zhang WH, Li LH (2012) Day and night warming have different effect on root lifespan. Biogeosciences 9: 375–384.
23. Wang WQ, Song SQ, Li SH, Gan YY, Wu JH, et al. (2009) Quantitative description of the effect of stratification on dormancy release of grape seeds in response to various temperatures and water contents. J Exp Bot 60 (2): 3397–3406.
24. Trudgill DL, Squire GR, Thompson K (2000) A thermal time basis for comparing the germination requirements of some British herbaceous plants. New Phytol 145: 107–114.
25. Qiu J, Bai Y, Coulman B, Romo JT (2006) Using thermal time models to predict seedling emergence of orchardgrass (*Dactylis glomerata* L.) under alternating temperature regimes. Seed Sci Res 16: 261–271.
26. Khan MA, Ungar IA (1996) Influence of salinity and temperature on the germination of *Haloxylon recurvum*. Ann Bot 78: 547–551.
27. Gul B, Webber DJ (1999) Effect of salinity, light and temperature on germination in *Allenrolfea occidentalis*. Can J Bot 77: 240–246.
28. Khan MA, Gul B, Weber DJ (2000) Germination responses of *Salicornia rubra* to temperature and salinity. J Arid Environ 45: 207–214.
29. Ellis RH, Barrett S (1994) Alternating temperatures and rate of seed germination in Lentil. Ann Bot 74: 519–524.
30. Ekstam B, Johannesson R, Milberg P (1999) The effect of light and number of diurnal temperature fluctuations on germination of *Phragmites australis*. Seed Sci Res 9: 165–170.
31. Markus B (2004) The role of temperature in the regulation of dormancy and germination of two related summer-annual mudflat species. Aquat Bot 79: 15–32.
32. Garcia-Huidobro J, Monteith JL, Squire GR (1982a) Time, temperature and germination of pearl millet (*Pennisetum typhoides* S. & H.) 1. Constant temperature. J Exp Bot 33: 287–295.
33. Garcia-Huidobro J, Monteith JL, Squire GR (1982b) Time, temperature and germination of pearl millet (*Pennisetum typhoides* S. & H.) 2. Alternating temperature. J Exp Bot 33: 297–302.
34. Falster DS, Warton DI, Wright IJ (2006) SMATR: standardised major axis tests and routines, ver 2.0. http://www.bio.mq.edu.au/ecology/SMATR/.
35. Thorpe J, Wolfe SA, Houston B (2008) Potential impacts of climate change on grazing capacity of native grasslands in the Canadian prairies. Can J Soil Sci 88: 595–609.
36. Sage RF, Kubien DS (2003) *Que vadis* C_4? An ecophysiological perspective on global change and the future of C_4 plants. Photosynth Res 77: 209–225.
37. Gordo O, Sanz JJ (2010) Impact of climate change on plant phenology in Mediterranean ecosystems. Glob Change Biol 16: 1082–1106.
38. Zhang H, Zhou D, Wang P, Wang T, Jin Y (2007) Germination responses of four wild species to diurnal increase or decrease in temperature. Seed Sci Technol 35: 291–302.
39. Aronson EL, McNulty SG (2009) Appropriate experimental ecosystem warming methods by ecosystem, objective and practicality. Agr Forest Meteorol 149: 1791–1799.
40. Beier C, Emmett B, Gundersen P, Tietema A, Penuelas J, et al. (2004) Novel approaches to study climate change effects on terrestrial ecosystems in the field: Drought and passive nighttime warming. Ecosystems 7: 583–597.
41. Collins SL, Fargione JE, Crenshaw CL, Nonaka E, Elliott JR, et al. (2010) Rapid plant community responses during the summer monsoon to nighttime warming in a northern Chihuahuan Desert grassland. J Arid Environ 74: 611–617.
42. Phillips CL, Gregg JW, Wilson JK (2011) Reduced diurnal temperature range does not change warming impacts on ecosystem carbon balance of Mediterranean grassland mesocosms. Glob Change Biol 17: 3263–3273.
43. He JS, Wolfe-Bellin KS, Bazzaz FA (2005) Leaf-level physiology, biomass, and reproduction of *Phytolacca americana* under conditions of elevated CO_2 and altered temperature regimes. Int J Plant Sci 166: 615–622.
44. Hattersley PW (1983) The distribution of C_3 and C_4 grasses in Australia in relation to climate. Oecologia 57: 113–128.
45. Sun DL, Pinker R (2014) Factors contributing to the spatial variability of Satellite estimates of diurnal temperature range in the United States. IEEE Geosci Remote S 11: 1524–1528.

21

The Impact of 850,000 Years of Climate Changes on the Structure and Dynamics of Mammal Food Webs

Hedvig K. Nenzén[1]*, Daniel Montoya[2], Sara Varela[3]

1 Département des sciences biologiques, Université du Québec à Montréal, Montréal, Québec, Canada, **2** School of Biological Sciences, Life Sciences Building, University of Bristol, Bristol, United Kingdom, **3** Department of Ecology, Faculty of Science, Charles University, Prague, Czech Republic

Abstract

Most evidence of climate change impacts on food webs comes from modern studies and little is known about how ancient food webs have responded to climate changes in the past. Here, we integrate fossil evidence from 71 fossil sites, body-size relationships and actualism to reconstruct food webs for six large mammal communities that inhabited the Iberian Peninsula at different times during the Quaternary. We quantify the long-term dynamics of these food webs and study how their structure changed across the Quaternary, a period for which fossil data and climate changes are well known. Extinction, immigration and turnover rates were correlated with climate changes in the last 850 kyr. Yet, we find differences in the dynamics and structural properties of Pleistocene *versus* Holocene mammal communities that are not associated with glacial-interglacial cycles. Although all Quaternary mammal food webs were highly nested and robust to secondary extinctions, general food web properties changed in the Holocene. These results highlight the ability of communities to re-organize with the arrival of phylogenetically similar species without major structural changes, and the impact of climate change and super-generalist species (humans) on Iberian Holocene mammal communities.

Editor: Michael Hofreiter, University of York, United Kingdom

Funding: H.N. was supported by: UQAM, NSERC-Discovery, NSERC-Create, Canada Research Chairs. S.V. was supported by 'Support of establishment, development and mobility of quality research teams at the Charles University' (CZ.1.07/2.3.00/30.0022, European Science Foundation and Czech Republic). D.M. was supported by the European Commission (MODELECORESTORATION- FP7 Marie Curie Intra-European Fellowship for Career Development [301124]). The funders had no role in study design, data collection and analysis, decision to publish, or preparation of the manuscript.

Competing Interests: The authors have declared that no competing interests exist.

* Email: hedvig.nenzen@gmail.com

Introduction

Climate change is one of the major drivers affecting the diversity, composition, structure and functioning of ecological communities. Species respond in different ways to climate change which directly affects their persistence within the food web and, consequently, the composition and structure of the community [1,2]. Although evidence of the impacts of climate change mostly comes from studies of modern communities, life on Earth has experienced several climatic perturbations in the past and those changes impacted the composition and structure of ancient communities in similar ways. Understanding how ancient food webs responded to past climate provides information about how communities reorganize across time, and how food webs could respond to contemporary climate change [3,4].

Because the Quaternary fossil record is extensive and the climate changes during this period are relatively well understood, the examination of the Quaternary fossil record could be key to understanding long-term dynamics and structure of ancient biological communities [4]. Climate in the Quaternary is characterized by cyclic climatic changes, oscillating from cold, dry glacial scenarios to warm, wet interglacial scenarios [5]. Changes in annual mean temperature inferred from oxygen isotopes range from an increase of 5°C during the warm scenarios to a decrease of −11°C in the extreme glacial periods [6]. This sequence of successive glacial and interglacial periods had direct and indirect effects on natural communities by forcing species to migrate [4], eliminating species [7], introducing new species, and shaping several broad diversity patterns that we observe today (e.g. [8]).

In the Iberian Peninsula, fossil records show that the composition of species within communities changed during the Quaternary following climate changes (e.g. [9]). Whereas some species went extinct after inhabiting the Iberian Peninsula for long time periods (e.g. *Bos primigenius*, *Crocuta crocuta*), others remained for more than one million years (*Cervus elaphus*, *Sus scrofa*), and certain species appeared for short periods (*Cuon alpinus*, in the Middle Pleistocene; or *Coelodonta antiquitatis*, in the Last Glacial Maximum). Quaternary climate changes co-occurred with extinctions, migrations and the arrival of new species in the Iberian Peninsula that have likely impacted the structure of mammal communities living in this region.

In spite of the observed long-term species changes during the Quaternary, few studies have analysed long series of species turnovers (but see [10]), even fewer have investigated the structure of the mammal communities through time [7,11,12] and none have analyzed food web dynamics across the Quaternary. Analyzing food webs can help us to identify changes in biotic interactions, beyond counting the number of species present in the community [13,14]; these changes can alter the functioning of food webs and ecosystems [15], and affect their future stability and persistence [16,17]. However, food web approaches in paleoecological studies are scarce, and therefore we know little about how past climate changes have affected the long-term dynamics and food web structure of biological communities at regional scales, and to what extent these communities re-organize in response to these impacts (but see [18–21] for temporal food web studies). The reason for this paucity of studies is twofold. First, extending the approaches used in food web theory to ancient communities has been limited by the incompleteness or lack of fossil data. In particular there is little fossil evidence that can be used to establish trophic links, compared to highly-resolved modern networks where changes in feeding interactions can be directly observed (e.g. [22]).

Second, the information from one single fossil site is limited for studies at the community scale where several species are considered. Because of a number of biases that affect the fossil record [23], individual fossil sites often lack species from the regional species pool (the diversity of species at a spatial scale larger than the individual fossil site) and are not an appropriate source of data to construct ancient regional food webs. Here we overcome these limitations by using large-scale cumulative food webs, that are constructed by using fossil information from different sites, and are thus appropriate for comparing food webs over time or space [20]. Moreover, large-scale cumulative webs do not depend on single fossil site records, and minimize problems of undersampling or collection bias [23,24].

In this work, we assemble six large mammal communities in the Iberian Peninsula spanning 850,000 years (Pleistocene and Holocene time periods). To obtain a complete species pool for large Iberian mammals, communities were constructed using published data from 71 fossil sites. Thus, the results presented here describe how large mammal communities at a regional scale changed during the Quaternary climate changes. The aim of this study is to explore the long-term and community structure of a large mammal food web across 850,000 years of climate changes. We specifically address the following questions: (a) how have Quaternary mammal communities changed in terms of dynamics (species extinctions, immigrations, and turnover rates) and food web structure?, and (b) are the changes in the structure and dynamics of these communities associated with Quaternary climate changes?

Methods

Fossil data and food web construction

We assembled six different large mammal food webs across the Quaternary: one for the Early Pleistocene (EP; around 850,000 years before present), one for the Middle Pleistocene (MP; around 450 kyr BP), plus two food webs during the Late Pleistocene, one for the last interglacial maximum, LIM; 120 kyr BP, and one for the last glacial maximum, LGM; 21 kyr BP), and two food webs for the Holocene (H; 10 kyr BP and present, P). For each time, we identified the large mammals (>20 kg), including hominids, that were present using fossil records from several Iberian fossil sites (details about sites and references in Table S1 and figure S1 in File S1). The taxonomy has been revised and unified following [25]. The resulting communities are not observed local communities, but regional food webs constructed using data from 71 fossil sites within the Iberian Peninsula. Large mammals are highly mobile and we assume that their geographic ranges covered the entire Iberian Peninsula. We use the number of glacial cycles between each community, MIS/OIS boundaries (Marine Isotope Stages, or Oxygen Isotope Stages [26]) as a proxy for climatic changes. We assume that a higher number of cycles represent a higher number of recovery and reorganisation periods for the mammal communities, indicating that they experienced greater stress.

Next we determine the trophic links between the species present in each community (figure 1, figure S2 in File S1). This is the most challenging step since diet evidence of extinct species is rare and interactions cannot be observed in the field. Species eaten by Pleistocene carnivores are classically identified from indirect clues, such as cut marks, teeth marks or stone tools marks in the fossil bones [27]. There are methods to distinguish carnivorous vs. herbivorous species [28,29], and to identify ancient species diet requirements using stable isotopes [21] and DNA from coprolites [30]. Here, food web links were designated following three criteria: (i) spatio-temporal co-occurrence as shown by fossil record data (Tables S1, S2 in File S1), (ii) body-size relationships (Table S3 in File S1), and (iii) actualism (applying current species diet to infer past trophic links). The key role of body-size in determining feeding links and structuring food webs is broadly accepted [31]. We complement link information by using actualism, which can be applied to all Quaternary carnivores (e.g. human diet cannot be predicted by body weight alone). Actualism diets follow [32]. The networks analysed here are bipartite and consist of two trophic levels (predators and herbivores/prey).

Community and food web analysis

We explore community long-term dynamics by calculating the number of extinctions, immigrations and species turnover between communities. Extinction and immigration were estimated as the number of species that were lost or gained in each transition. We use Sørensen's dissimilarity index to estimate species turnover rates between the communities [33].

We also explore the possibility that extinction and immigration are non-random by using two plausible hypotheses based on phylogeny and body-size. For immigrating species, we test whether newly immigrated species are phylogenetically similar to the species that went extinct in the preceding time step. Following the phylogenetic conservatism hypothesis, new immigrants are more likely to replace extinct species if they are closely related within the evolutionary tree (because of functional similarity), a phenomenon that seems to be widespread across evolutionary time [34]. To test if species were replaced by closely related species, we count the number of times a species that went extinct was replaced by a species within the same genus (Table S2 in File S1). In the absence of a complete phylogeny this gives a simple measure of phylogenetic relatedness and functional similarity, as closely related species often fulfil the same function in an ecosystem [34–36]. Second, as large animals are likely to go extinct first [37,38], we test whether the body-size distribution of extinct species was random across time. For each food web we randomly remove the same number of species that were observed to go extinct in the next time step. After repeating this 10,000 times, we compare the mean body weight of the randomized and observed food webs.

To examine food web structure we use standard metrics: species richness, number of links and connectance (links/species^2) [39]. Two other food web properties were analysed: vulnerability (mean number of predators per prey), and generality (mean number of prey per predator) [40]. To further characterize generality we sort the number of links per prey and predator and plot the relationship for each food web. Because network structure is related to its stability, we analysed community stability by examining (i) the robustness of ancient food webs to species loss [41–43], and (ii) nestedness [44]. The robustness index measures the topological or structural stability of the food web by simulating how random removal of prey (or predators) induces secondary extinction among the predators (or prey) [43,45]. Robustness is measured as the area under the curve of the number of species

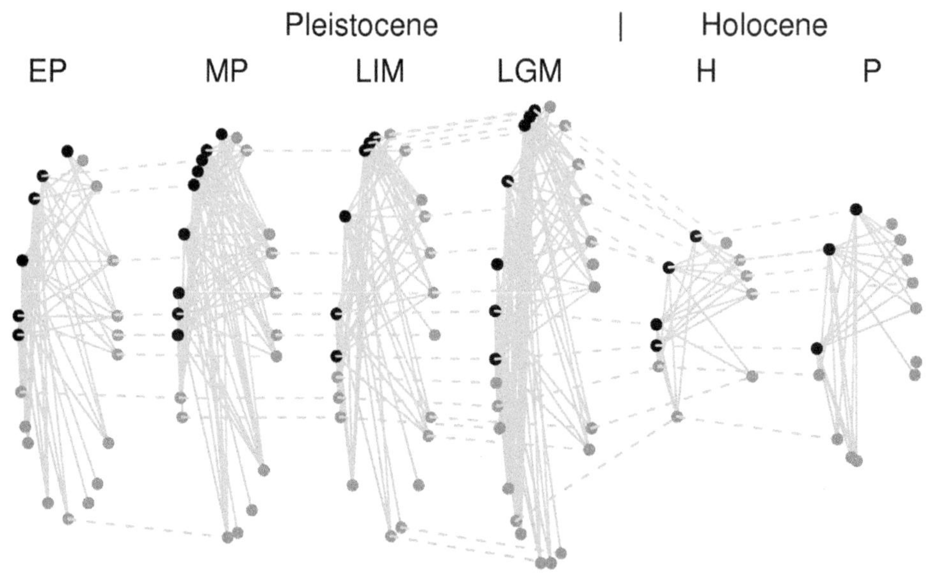

Figure 1. Large mammal food webs of the Iberian Peninsula during the Quaternary. Early Pleistocene (EP; 850,000 years before present, or 850 ky BP), Middle Pleistocene (MP; 450 kyr BP), Last Interglacial Maximum (LIM; 120 kyr BP), Last Glacial Maximum (LGM; 21 kyr BP), Early Holocene (H; 10 kyr BP) and present (P). Each species (black for predator, dark grey for prey) is a node, and each link indicates a trophic interaction. Horizontal links indicate that the species persists between communities.

being removed against the number of secondary extinctions, and ranges from 0 to 1, with high values representing more robust communities (the number of secondary extinctions is lower). We calculate community robustness for removal of both prey and predators individually. Nestedness measures the degree to which the diets of consumers are proper subsets of other, more generalist consumers. The nestedness algorithm used here is based on the nestedness temperature of the interaction matrix, and ranges from 0 which indicates high nestedness, to 100 which indicates no nestedness [46,47].

Finally, we examine the relationships between climate change, species turnover and food web properties. Controlled, replicated experiments cannot be conducted in paleoecological studies, but we can explore whether food web changes in ancient communities are associated to climatic changes in the past. We test if the number of glacial cycles between each community and the observed changes in species composition are related to changes in food web metrics, using the 15 unique pairwise comparisons between all six food webs. Given that non-linear responses of natural communities to climate changes are common (e.g. [48]), we assume that the relationship between the number of glacial cycles, species turnover and food web properties are non-linear, and for that reason we use the Spearman correlation index.

Results

Changes in Quaternary mammal food webs over the last 850,000 years

Pleistocene communities comprise 20–25 species >20 kg (figure 2a), a figure that falls within the range of values reported in comparative analyses of modern [49] and ancient food webs [20,21]. The complexity of these ancient webs is also similar to that observed in modern mammal food webs in Africa where megafauna did not go extinct in the Holocene [50].

During the Pleistocene the rates of extinction were high, (figure 3b, except the LIM-LGM transition), with one species disappearing every 20–36 kyr. Extinction analysis showed no correlation between species' body-size and extinction probability (figure 3e). The number of new species entering Pleistocene communities equalled or exceeded the number of species that went extinct (figure 3c), with one new species entering the Pleistocene food webs every 12–36 kyr. This suggests that extinct species were replaced by new species at similar rates, and that the overall number of species in Pleistocene food webs did not change. Consistent with the phylogenetic conservatism hypothesis, a high proportion of these new species were phylogenetically similar to the extinct species (figure 3f). Despite the high turnover rates (figure 3d), the general diversity and food web properties of the Pleistocene communities remained relatively constant across almost 500 kyr.

The dynamic stability observed in Pleistocene mammal communities changed in the Holocene. Extinction rates dramatically increased (one extinct species every 830 years, figure 3b). These extinctions are non-random (figure 3e), and especially affect the least connected specialist predators (figure S3 in File S1) and species with large body-sizes (figure S4 in File S1). Immigration rate dropped (figure 3c), and, consequently, extinct species were not replaced by new ones. Intriguingly, no predator mammals entered into the food web during the Holocene and the few immigrant species were phylogenetically unrelated (figure 3f). These processes caused a reduction in the number of species at all trophic levels that affected food web structure in two ways: (i) connectance increased, and (ii) generality and vulnerability decreased. The observed differences in other food web properties between Holocene and Pleistocene communities may be a direct consequence of increases in connectance following a reduction in network size and the loss of specialist predators.

Regardless of the changes in community size between Pleistocene and Holocene communities, the robustness of large mammalian communities to secondary extinctions of both prey and predators remained constant across the Quaternary period (figure 2f). We found no correlation between connectance and robustness, nor between species richness and robustness across the Quaternary period (Spearman correlations, $p>0.05$). Nestedness

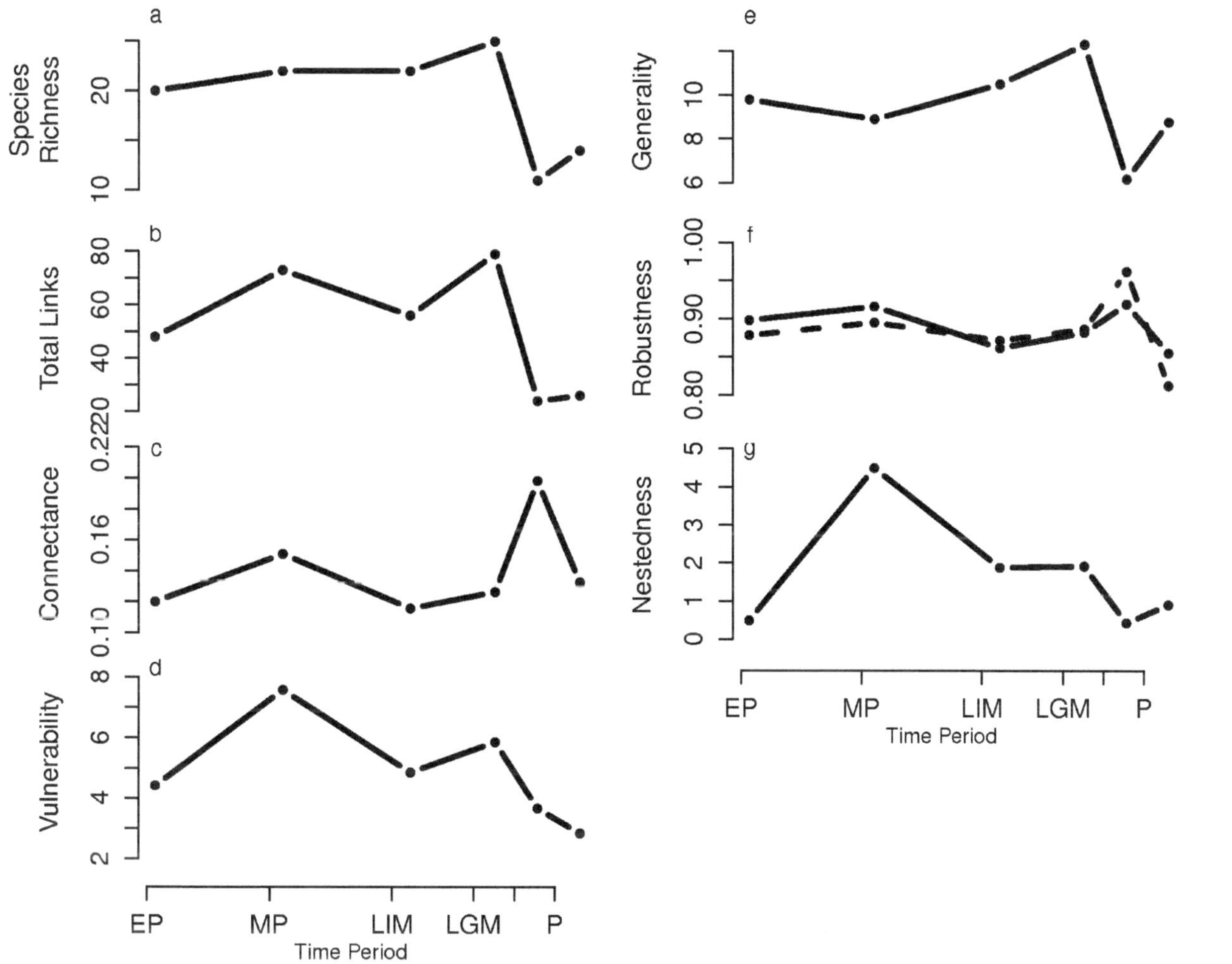

Figure 2. Food web properties at each time step. (a) Species richness (number of species), (b) total links (total number of links or trophic interactions between species), (c) connectance (links/species^2), (d) vulnerability, (e) generality, (f) robustness against removal of prey (solid line) and predators (dotted line), (g) nestedness. The dates are the same as in figure 1. The x-axis is proportional to the time between each community.

patterns, on the other hand, have fluctuated but fall within high nestedness values (figure 2g).

The role of climate change in the dynamics and structure of Quaternary mammal food webs

Overall, turnover rates of Iberian mammal species across the Quaternary are associated with the pattern of glacial cycles (table 1). Quaternary climate changes are positively correlated with species turnover, extinction and immigration rates across all mammal communities. However, glacial-interglacial events show no significant relationship with food web properties, which suggests that climate changes across the Quaternary were not associated with changes in general structure of mammalian communities. We also assessed whether climate change is indirectly associated with food web properties through changes in species composition, yet no significant relationship was observed between species turnover rates and food web metrics (table 2). This suggests that the structure of the Quaternary mammal food webs is independent of observed species turnover rates, and that most of the changes in the food web properties between Pleistocene and Holocene mammal communities are associated with the loss of specialist predators and increases in connectance following a non-random reduction in network size between these two periods of time.

Discussion

Iberian large mammal communities have experienced important dynamical changes across the Quaternary. Extinction, immigration and turnover rates were highly correlated with climate changes in the last 850 kyr, yet food web properties were not significantly altered following these climate changes. Our results suggest that glacial-interglacial cycles are associated with changes in community dynamics, and highlights the ability of communities to re-organize with the arrival of phylogenetically similar species without major changes in food web properties. As a result, ancient mammal food webs were dynamically stable and able to recover from climatic perturbations even though many species went extinct.

Changes in large mammal food web structure and dynamics

Our results show differences in the dynamics and structure of Pleistocene *versus* Holocene mammal communities on the Iberian Peninsula. In the Pleistocene, extinctions were common and affected species irrespective of their body-size. Extinct species in the food web were replaced by newly arriving species that were closely related in the phylogenetic tree, so extinction and immigration rates were balanced. Consequently, food web

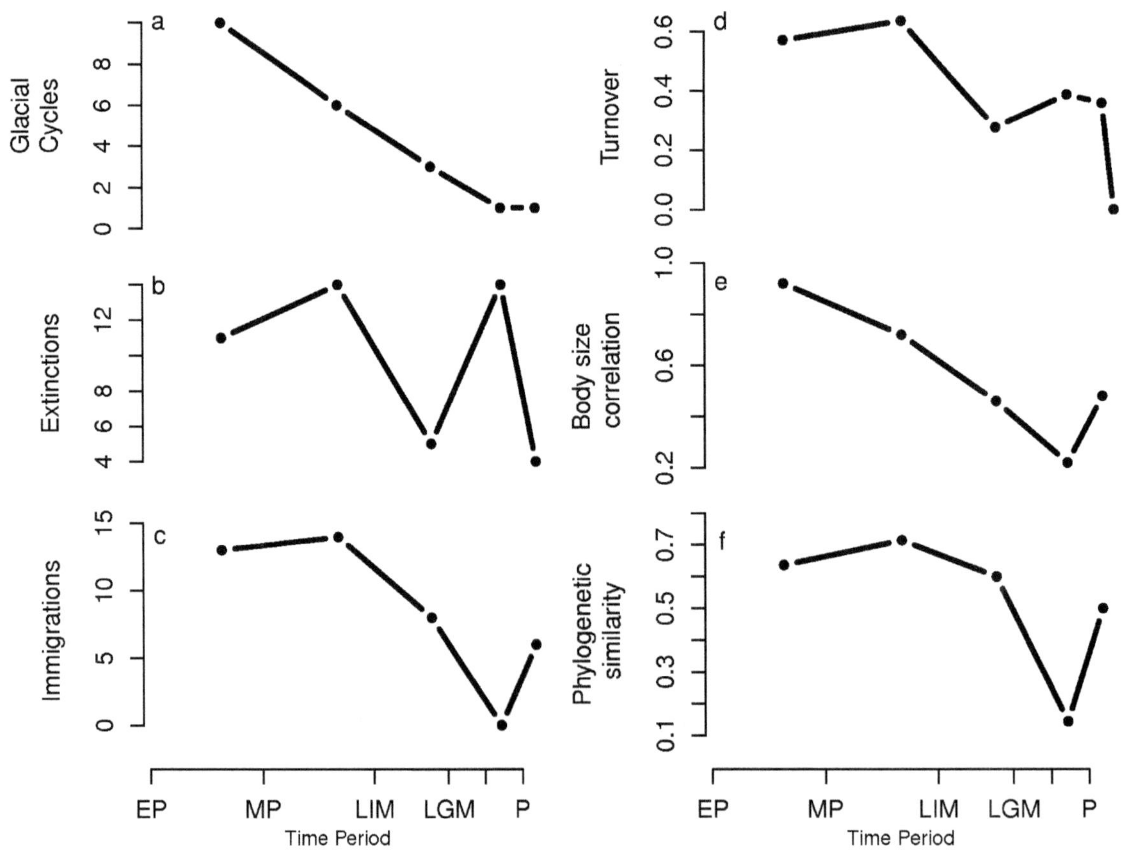

Figure 3. Food web structural changes at each time step. (a) Number of glacial cycles, (b) extinct species and (e) immigrating species, (f) species turnover, (g) correlation between observed and randomly expected body size distribution of extinct species and (f) phylogenetic similarity. The dates are the same as in figure 1.

properties remained unaltered and Pleistocene communities remained structurally stable despite the species identities. These results support the phylogenetic conservatism hypothesis [34], by which new immigrants are more likely to replace extinct species if they are closely-related within the evolutionary tree, as closely related species often fulfil the same function in the ecosystem. However, the extinction-immigration patterns of the Pleistocene are not reproduced in Holocene communities, which contracted as a result of higher extinction rates (56% of species went extinct from the early Pleistocene to Holocene) and low replacement rates. This reduction in community size and loss of specialist species increased connectance, which in turn affected other properties of the food web.

After more than 800,000 years of relative stability, food web properties changed in two steps in the Holocene: species richness and generality first decreased and then increased. The first step was caused by the extinction of large mammals after the Last Glacial Maximum (21–10 kyr BP), when prey such as *Mammuthus primigenius* (woolly mammoth) or *Megaloceros giganteus* (Irish elk) went extinct, followed by large carnivores like *Panthera spelaea* (lion). This is consistent with the observation that modern mammalian communities are remnants of larger Pleistocene

Table 1. Spearman's rank correlation coefficient between climate change (number of glacial-interglacial cycles between scenarios) and species and food web structure parameters for all communities. ρ = Spearman correlation.

Climate change vs.		**ρ**	**S**	**p**
Species	*Species turnover*	0.92	47.54	<0.001
	Extinction	0.74	140.24	0.001
	Immigration	0.65	193.36	0.008
Food web structure	Species Richness	−0.15	654.77	0.585
	Number of food web links	−0.15	646.00	0.584
	Connectance	−0.24	692.00	0.397
	Predator/prey ratio	−0.07	598.00	0.812

Table 2. Spearman's rank correlation coefficient between species and food web structure parameters for all communities.

Species turnover vs.		rho	S	p
Food web structure	Species Richness	−0.06	594.33	0.828
	Number of links	−0.07	600.03	0.800
	Connectance	−0.34	752.17	0.210
	Predator/prey ratio	−0.11	623.05	0.689

mammal communities [51]. This reduction in community size is observed in contemporary mammal communities, and affected food web properties by reducing generality (fewer prey species) and increasing connectance. Theoretical studies suggest that spatially coupled food webs are especially sensitive to the loss of large animals [52], and there is evidence showing that the loss of megaherbivores such as mammoths had large impacts on ecosystem function [53,54]. Consequently, the loss of large mammals may have also affected the ecosystem functions provided by Holocene communities.

In the second step, during the Holocene only human-introduced herbivores, such as *Bos taurus* (cattle) or *Equus caballus* (horse) appeared as new species in the food webs. These introductions slightly increased the number of species and the mean number of prey species per predator, the latter being also a consequence of the existence of generalist predator species, especially the humans. In contrast to Pleistocene communities, the introduced species were not phylogenetically related to the extinct species, and this has likely impacted ecosystem functioning, as certain ecosystem functions may not have been completely replaced by phylogenetically related species.

Robustness and nestedness

Our analysis of robustness reveals across-time similarities in Pleistocene and Holocene communities. Despite the significant extinction and turnover rates across the Quaternary, and the reported changes in community size and food web connectivity, the robustness of mammal communities remained fairly constant for nearly one million years. On the other hand, all our mammal food webs present highly nested structures. Simulations have shown recently that nestedness is a destabilizing force in predator-prey food webs [55], and its absence in other ancient food webs may have promoted stability during the successive climatic changes that occurred in the past [21]. Although the combination of high robustness and high nestedness may have resulted in a neutral effect on stability, nested structures may have also acted as a buffer to secondary extinctions [56], so the interpretation of historical patterns of nestedness of Quaternary mammal communities is not straightforward.

The high nestedness reported in Quaternary mammal communities is explained by the presence of highly generalist predators. Both theoretical and empirical evidence show that super-generalist species become central nodes (most connected species) in the core of the nested community and may increase the overall nestedness [57]. Perturbation events usually favour generalist over specialist species, a pattern observed both in paleoecological [58] and contemporary communities [59], and this contributes to increase the generalist:specialist ratios. Quaternary mammal food webs were no exception to this rule and, as a result of the long-term dynamics following glacial-interglacial cycles, comprised highly generalist species. Specifically, the Holocene period experienced a population growth of the super-generalist anatomically modern humans. Humans not only have a broad diet (figure S2 in File S1), but also introduced domesticated mammals into the food web, intensified the generality and contributed to the high connectivity of Iberian mammal food webs. Anatomically modern humans have been present in Spain since 42 kyr BP, but with a small population size [60]. Other hominids such as *Homo neanderthalensis* were present earlier, but our results suggest that they had a smaller impact on food webs compared to Holocene populations of *Homo sapiens*.

The high impact of omnivore humans on food webs has also been found in contemporary communities. A recent study explored a spatial (rather than temporal) gradient of human impacts in Serengeti food webs, average body mass and species richness also decreased with increasing human impact. Food webs tended to be more generalist following the human-induced extinction of the least-connected, most specialized species [61]. In contrast to Quaternary mammal communities, the Serengeti food web harbours more trophic levels and large predators, which results in a higher species number and lower connectance compared to our food webs.

Conclusion

Paleoecological community studies provide insight into the relationships between biological communities and perturbations such as climate change over long time scales. This work presents a first attempt to track temporal changes in food webs over long time periods. Although the spatial and temporal scales are coarse, the patterns revealed are significant. We found that large mammal communities in the Pleistocene were able to re-organize despite the high extinction rates reported in the last 850,000 years. However, our results indicate that Holocene mammal communities experienced changes that were related to the arrival of new species introduced in the communities by humans (cattle and game species). Future work should aim at investigating the actual changes in temperature at local scales and the duration of those cycles; this information could be used to make cross-comparisons of the species' extinction rates within and among continents, and their effects on ancient food webs, and to understand the mechanisms driving the likely spatial differences in the extinction rates.

Collectively, our results suggest differences between the Pleistocene and Holocene time periods in the structure and dynamics of large mammal food webs. These conclusions have consequences for understanding the current-day sixth extinction event, and to what extent modern communities will be able to re-organize during current climate change. Not only are current rates of change in climate high, but also they interact with other drivers of extinction such as habitat loss and overexploitation, and the feedbacks generated by these processes and biotic interactions may result in higher extinction rates than those expected only from climate change [62,63].

Supporting Information

File S1 Table S1, Reference sources used for determining the presence of species in each of the 71 sites. **Table S2,** The species present (indicated by 1) at each period. Weight categories: 1 = < 45 kg, 2 = 45–90 kg, 3 = 90–360 kg, 4 = 360–1000 kg, 5 = > 1000 kg. Phylogenetic replacement categories: 0 = Not replaced by a species in the same genus in the next time period, 1 = Replaced by a species in the next time period, 2 = Species still present. **Table S3,** Prey weight classes. This information, together with spatio-temporal co-occurrence and actualism (see main text) was used to establish the links between species in the food web. **Figure S1,** Geographic location on the Iberian Peninsula of the Quaternary fossil sites used for constructing the ancient food webs. **Figure S2,** Large mammal food webs of the Iberian Peninsula during the Quaternary. (a) Early Pleistocene (850,000 years before present, or 850 ky BP); (b) Middle Pleistocene (450 ky BP); (c) Last Interglacial Maximum (120 ky BP); (d) Last Glacial Maximum (21 ky BP); (e) Early Holocene (10 ky BP); and (f) Present. Each node (green for prey, red for predator) is a species, and each link indicates a trophic interaction. **Figure S3,** Number of prey per predator in each time period, with species sorted in descending order. The time periods are the same as in figure 1. **Figure S4,** A random extinction experiment demonstrate that the distribution of the number of links, connectance and link density observed Holocene food web is not expected by chance. 1000 food webs have been created extracting randomly 11 species from last glacial maximum food web (the observed number of extinct species in the Holocene).

Acknowledgments

We thank J van der Made, D Storch, T Poisot, M Chevrinais, P Desjardins-Proulx, R Barrientos, A Nenzén and K Orford for their useful comments and advice.

Author Contributions

Conceived and designed the experiments: HN SV DM. Performed the experiments: HN SV DM. Analyzed the data: HN SV DM. Contributed reagents/materials/analysis tools: HN SV DM. Wrote the paper: HN SV DM.

References

1. Petchey OL, McPhearson P, Casey T, Morin P (1999) Environmental warming alters food-web structure and ecosystem function. Nature 402: 69–72. doi:10.1038/47023.
2. Lorenzen ED, Nogues-Bravo D, Orlando L, Weinstock J, Binladen J, et al. (2011) Species-specific responses of Late Quaternary megafauna to climate and humans. Nature 479: 359–364. doi:10.1038/nature10574.
3. Harnik PG, Lotze HK, Anderson SC, Finkel ZV, Finnegan S, et al. (2012) Extinctions in ancient and modern seas. Trends in Ecology & Evolution 27: 608–617. doi:10.1016/j.tree.2012.07.010.
4. Blois JL, Zarnetske PL, Fitzpatrick MC, Finnegan S (2013) Climate Change and the Past, Present, and Future of Biotic Interactions. Science 341: 499–504. doi:10.1126/science.1237184.
5. Kageyama M, Laîné A, Abe-Ouchi A, Braconnot P, Cortijo E, et al. (2006) Last Glacial Maximum temperatures over the North Atlantic, Europe and western Siberia: a comparison between PMIP models, MARGO sea-surface temperatures and pollen-based reconstructions. Quaternary Science Reviews 25: 2082–2102. doi:10.1016/j.quascirev.2006.02.010.
6. Jouzel J, Masson-Delmotte V, Cattani O, Dreyfus G, Falourd S, et al. (2007) Orbital and Millennial Antarctic Climate Variability over the Past 800,000 Years. Science 317: 793–796. doi:10.1126/science.1141038.
7. Ceballos G, Arroyo-Cabrales J, Ponce E (2010) Effects of Pleistocene environmental changes on the distribution and community structure of the mammalian fauna of Mexico. Quaternary Research 73: 464–473. doi:10.1016/j.yqres.2010.02.006.
8. Montoya D, Rodríguez MA, Zavala MA, Hawkins BA (2007) Contemporary richness of holarctic trees and the historical pattern of glacial retreat. Ecography 30: 173–182. doi:10.1111/j.0906-7590.2007.04873.x.
9. Agusti J, Oms O, Remacha E (2001) Long Plio-Pleistocene terrestrial record of climate change and mammal turnover in southern Spain. Quaternary Research 56: 411–418. doi:10.1006/qres.2001.2269.
10. Raia P, Piras P, Kotsakis T (2005) Turnover pulse or Red Queen? Evidence from the large mammal communities during the Plio-Pleistocene of Italy. Palaeogeography, Palaeoclimatology, Palaeoecology 221: 293–312. doi:10.1016/j.palaeo.2005.02.014.
11. Rodríguez J (2004) Stability in Pleistocene Mediterranean mammalian communities. Palaeogeography, Palaeoclimatology, Palaeoecology 207: 1–22. doi:10.1016/j.palaeo.2003.12.016.
12. Palombo MR, Raia P, Giovinazzo C (2005) Early-Middle Pleistocene structural changes in mammalian communities from the Italian peninsula. Geological Society, London, Special Publications 247: 251–262. doi:10.1144/GSL.SP.2005.247.01.14.
13. Jablonski D (2008) Biotic interactions and macroevolution: extensions and mismatches across scales and levels. Evolution 64: 715–739. doi:10.1111/j.1558-5646.2008.00317.x.
14. Bascompte J, Stouffer DB (2009) The assembly and disassembly of ecological networks. Philosophical Transactions of the Royal Society B: Biological Sciences 364: 1781–1787. doi:10.1098/rstb.2008.0226.
15. Thompson R, Brose U, Dunne JA (2012) Food webs: reconciling the structure and function of biodiversity. Trends in Ecology & Evolution 27: 689–697. doi:10.1016/j.tree.2012.08.005.
16. Paine RT (1966) Food Web Complexity and Species Diversity. The American Naturalist 100: 65–75.
17. Gross T, Blasius B (2008) Adaptive coevolutionary networks: a review. Journal of the Royal Society Interface 5: 259–271. doi:10.1098/rsif.2007.1229.
18. Roopnarine PD (2006) Extinction cascades and catastrophe in ancient food webs. Paleobiology 32: 1–19. doi:10.1666/05008.1.
19. Roopnarine PD (2010) Networks, Extinction and Paleocommunity Food Webs. Available: http://precedings.nature.com.ezproxy.its.uu.se/documents/4433/version/2. Accessed 4 June 2011.
20. Dunne JA, Williams RJ, Martinez ND, Wood RA, Erwin DH (2008) Compilation and network analyses of cambrian food webs. PLoS Biology 6: e102. doi:10.1371/journal.pbio.0060102.
21. Yeakel JD, Guimaraes PR, Bocherens H, Koch PL, Guimarães PR (2013) The impact of climate change on the structure of Pleistocene food webs across the mammoth steppe. Proceedings of the Royal Society B: Biological Sciences 280: 20130239. doi:10.1098/rspb.2013.0239.
22. Olesen JM, Stefanescu C, Traveset A (2011) Strong, Long-Term Temporal Dynamics of an Ecological Network. PLoS ONE 6: e26455. doi:10.1371/journal.pone.0026455.
23. Varela S, Lobo JM, Hortal J (2011) Using species distribution models in paleobiogeography: A matter of data, predictors and concepts. Palaeogeography, Palaeoclimatology, Palaeoecology 310: 451–463. doi:10.1016/j.palaeo.2011.07.021.
24. Martinez ND, Hawkins BA, Dawah HA, Feifarek BP (1999) Effects of Sampling Effort on Characterization of Food-Web Structure. Ecology 80: 1044–1055. doi:10.2307/177037.
25. Van der Made J (2005) La fauna del Pleistoceno europeo. In: Carbonell E, editor. Homínidos: las primeras ocupaciones de los continentes. Ariel. pp. 394–432.
26. Lisiecki LE, Raymo ME (2005) A Pliocene-Pleistocene stack of 57 globally distributed benthic δ18O records. Paleoceanography 20: PA1003. doi:10.1029/2004PA001071.
27. Selvaggio M (1994) Carnivore Tooth Marks and Stone Tool Butchery Marks on Scavenged Bones - Archaeological Implications. Journal of Human Evolution 27: 215–228. doi:10.1006/jhev.1994.1043.
28. Boesl C, Grupe G, Peters J (2006) A Late Neolithic vertebrate food web based on stable isotope analyses. International Journal of Osteoarchaeology 16: 296–315. doi:10.1002/oa.834.
29. Newsome SD, Collins PW, Rick TC, Guthrie DA, Erlandson JM, et al. (2010) Pleistocene to historic shifts in bald eagle diets on the Channel Islands, California. Proceedings of the National Academy of Sciences 107: 9246–9251. doi:10.1073/pnas.0913011107.
30. Bon C, Berthonaud V, Maksud F, Labadie K, Poulain J, et al. (2012) Coprolites as a source of information on the genome and diet of the cave hyena. Proceedings of the Royal Society B: Biological Sciences 279: 2825–2830. doi:10.1098/rspb.2012.0358.
31. Woodward G, Ebenman B, Emmerson MC, Montoya JM, Olesen JM, et al. (2005) Body size in ecological networks. Trends in Ecology & Evolution 20: 402–409. doi:10.1016/j.tree.2005.04.005.
32. Owen-Smith N, Mills MGL (2008) Predator-prey size relationships in an African large-mammal food web. The Journal of Animal Ecology 77: 173–183. doi:10.1111/j.1365-2656.2007.01314.x.
33. Baselga A, Orme CDL (2012) betapart: an R package for the study of beta diversity. Methods in Ecology and Evolution 3: 808–812. doi:10.1111/j.2041-210X.2012.00224.x.

34. Gómez JM, Verdú M, Perfectti F (2010) Ecological interactions are evolutionarily conserved across the entire tree of life. Nature 465: 918–921. doi:10.1038/nature09113.
35. Cattin M-F, Bersier L-F, Banasek-Richter C, Baltensperger R, Gabriel J-P (2004) Phylogenetic constraints and adaptation explain food-web structure. Nature 427: 835–839. doi:10.1038/nature02327.
36. Stouffer DB, Sales-Pardo M, Sirer MI, Bascompte J (2012) Evolutionary Conservation of Species' Roles in Food Webs. Science 1018: 1489–1492. doi:10.1126/science.1216556.
37. Voigt W, Perner J, Davis A, Eggers T (2003) Trophic levels are differentially sensitive to climate. Ecology 84: 2444–2453. doi:10.1890/02-0266.
38. Cardillo M, Mace GM, Jones KE, Bielby J, Bininda-Emonds ORP, et al. (2005) Multiple Causes of High Extinction Risk in Large Mammal Species. Science 309: 1239–1241. doi:10.1126/science.1116030.
39. Pascual M, Dunne JA (2005) Ecological Networks: Linking Structure to Dynamics in Food Webs. Oxford University Press.
40. Tylianakis JM, Tscharntke T, Lewis OT (2007) Habitat modification alters the structure of tropical host-parasitoid food webs. Nature 445: 202–205. doi:10.1038/nature05429.
41. Solé R V, Montoya JM (2001) Complexity and fragility in ecological networks. Proceedings of the Royal Society B: Biological Sciences 268: 2039–2045. doi:10.1098/rspb.2001.1767.
42. Dunne JA, Williams RJ, Martinez ND (2002) Food-web structure and network theory: The role of connectance and size. Proceedings of the National Academy of Sciences 99: 12917–12922. doi:10.1073/pnas.192407699.
43. Memmott J, Waser NM, Price MV (2004) Tolerance of pollination networks to species extinctions. Proceedings of the Royal Society B: Biological Sciences 271: 2605–2611. doi:10.1098/rspb.2004.2909.
44. Guimara PR, Ulrich W, Almeida-Neto M, Guimarães PR, Loyola RD, et al. (2008) A consistent metric for nestedness analysis in ecological systems: reconciling concept and measurement. Oikos 117: 1227–1239. doi:10.1111/j.0030-1299.2008.16644.x.
45. Burgos E, Ceva H, Perazzo RPJ, Devoto M, Medan D, et al. (2007) Why nestedness in mutualistic networks? Journal of Theoretical Biology 249: 307–313. doi:10.1016/j.jtbi.2007.07.030.
46. Rodríguez-Gironés MA, Santamaría L (2006) A new algorithm to calculate the nestedness temperature of presence–absence matrices. Journal of Biogeography 33: 924–935. doi:10.1111/j.1365-2699.2006.01444.x.
47. Dormann CF, Gruber B, Fruend J (2008) Introducing the bipartite Package: Analysing Ecological Networks. R news 8: 8–11.
48. Pope KS, Dose V, Da Silva D, Brown PH, Leslie CA, et al. (2013) Detecting nonlinear response of spring phenology to climate change by Bayesian analysis. Global Change Biology 19: 1518–1525. doi:10.1111/gcb.12130.
49. Williams RJ, Martinez ND (2000) Simple rules yield complex food webs. Nature 404: 180–183. doi:10.1038/35004572.
50. Baskerville EB, Dobson AP, Bedford T, Allesina S, Anderson TM, et al. (2011) Spatial Guilds in the Serengeti Food Web Revealed by a Bayesian Group Model. PLoS Computational Biology 7: e1002321. doi:10.1371/journal.pcbi.1002321.
51. Koch PL, Barnosky AD (2006) Late Quaternary Extinctions: State of the Debate. Annual Review of Ecology, Evolution, and Systematics 37: 215–250. doi:10.1146/annurev.ecolsys.34.011802.132415.
52. McCann KS, Rasmussen JB, Umbanhowar J (2005) The dynamics of spatially coupled food webs. Ecology Letters 8: 513–523. doi:10.1111/j.1461-0248.2005.00742.x.
53. Owen-Smith N (1987) Pleistocene Extinctions: The Pivotal Role of Megaherbivores. Paleobiology 13: 351–362.
54. Gill JL, Williams JW, Jackson ST, Lininger KB, Robinson GS (2009) Pleistocene megafaunal collapse, novel plant communities, and enhanced fire regimes in North America. Science 326: 1100–1103. doi:10.1126/science.1179504.
55. Thébault E, Fontaine C (2010) Stability of ecological communities and the architecture of mutualistic and trophic networks. Science 329: 853–856. doi:10.1126/science.1188321.
56. Bastolla U, Fortuna MA, Pascual-García A, Ferrera A, Luque B, et al. (2009) The architecture of mutualistic networks minimizes competition and increases biodiversity. Nature 458: 1018–1020. doi:10.1038/nature07950.
57. Aizen MA, Morales CL, Morales JM (2008) Invasive mutualists erode native pollination webs. PLoS biology 6: e31. doi:10.1371/journal.pbio.0060031.
58. Sahney S, Benton MJ (2008) Recovery from the most profound mass extinction of all time. Proceedings of the Royal Society B: Biological Sciences 275: 759–765. doi:10.1098/rspb.2007.1370.
59. Estes J, Terborgh J, Brashares JS, Power ME, Berger J, et al. (2011) Trophic downgrading of planet Earth. Science 333: 301–306. doi:10.1126/science.1205106.
60. Voight BF, Adams AM, Frisse LA, Qian Y, Hudson RR, et al. (2005) Interrogating multiple aspects of variation in a full resequencing data set to infer human population size changes. Proceedings of the National Academy of Sciences of the United States of America 102: 18508–18513. doi:10.1073/pnas.0507325102.
61. De Visser SN, Freymann BP, Olff H (2011) The Serengeti food web: empirical quantification and analysis of topological changes under increasing human impact. The Journal of Animal Ecology 80: 484–494. doi:10.1111/j.1365-2656.2010.01787.x.
62. Brook BW, Sodhi NS, Bradshaw CJA (2008) Synergies among extinction drivers under global change. Trends in Ecology & Evolution 23: 453–460. doi:10.1016/j.tree.2008.03.011.
63. Zarnetske PL, Skelly DK, Urban MC (2012) Ecology. Biotic multipliers of climate change. Science 336: 1516–1518. doi:10.1126/science.1222732.

22

Incorporating Climate Change and Exotic Species into Forecasts of Riparian Forest Distribution

Dana H. Ikeda[1,2]*, Kevin C. Grady[2,3], Stephen M. Shuster[1,2], Thomas G. Whitham[1,2]

1 Department of Biological Science, Northern Arizona University, Flagstaff, Arizona, United States of America, **2** Merriam-Powell Center for Environmental Research, Northern Arizona University, Flagstaff, Arizona, United States of America, **3** School of Forestry, Northern Arizona University, Flagstaff, Arizona, United States of America

Abstract

We examined the impact climate change (CC) will have on the availability of climatically suitable habitat for three native and one exotic riparian species. Due to its increasing prevalence in arid regions throughout the western US, we predicted that an exotic species, *Tamarix*, would have the greatest increase in suitable habitat relative to native counterparts under CC. We used an ecological niche model to predict range shifts of *Populus fremontii*, *Salix gooddingii*, *Salix exigua* and *Tamarix*, from present day to 2080s, under five general circulation models and one climate change scenario (A1B). Four major findings emerged. 1) Contrary to our original hypothesis, *P. fremontii* is projected to have the greatest increase in suitable habitat under CC, followed closely by *Tamarix*. 2) Of the native species, *S. gooddingii* and *S. exigua* showed the greatest loss in predicted suitable habitat due to CC. 3) Nearly 80 percent of future *P. fremontii* and *Salix* habitat is predicted to be affected by either CC or *Tamarix* by the 2080s. 4) By the 2080s, 20 percent of *S. gooddingii* habitat is projected to be affected by both *Tamarix* and CC concurrently, followed by *S. exigua* (19 percent) and *P. fremontii* (13 percent). In summary, while climate change alone will negatively impact both native willow species, *Tamarix* is likely to affect a larger portion of all three native species' distributions. We discuss these and other results in the context of prioritizing restoration and conservation efforts to optimize future productivity and biodiversity. As we are accounting for only direct effects of CC and *Tamarix* on native habitat, we present a possible hierarchy of effects- from the direct to the indirect- and discuss the potential for the indirect to outweigh the direct effects. Our results highlight the need to account for simultaneous challenges in the face of CC.

Editor: Bazartseren Boldgiv, National University of Mongolia, Mongolia

Funding: This work was supported by the National Science Foundation (NSF) Integrative Graduate Education Research Traineeship (IGERT), Frontiers in Integrative Biological Research (FIBR) grant DEB-0425908, NSF Macrosystems grant DEB-1340852, the Northern Arizona University TRIF program, and Bureau of Land Management Challenge Cost Share grant L12AC20606. The funders had no role in study design, data collection and analysis, decision to publish, or preparation of the manuscript.

Competing Interests: The authors have declared that no competing interests exist.

* Email: Dana.Ikeda@nau.edu

Introduction

Climate change (CC) is predicted to be one of the leading causal agents of future extinctions, impacting biodiversity and ecosystem functioning worldwide [1], [2]. There is considerable evidence that shifts in species distributions both pole-ward and along elevation clines are already occurring [3–7], with regional die-off of species at the trailing edges of their range [8–11]. In the western U.S., recent climatic conditions have become more arid [12]; resulting in high mortality of numerous foundation species at the trailing edge of their distributions, in ecosystems that span from chaparral to alpine forests [10], [13]. Increasing aridity is expected to continue under CC, with models projecting higher temperatures and an increase in drought events in the next 80 years [12], [14], [15].

Projected increases in temperature and drought will negatively impact riparian ecosystems worldwide that have already experienced extensive modifications over the last century [16–18]. Changes in land use, particularly the alteration of flow regimes by damming [19] and extensive cattle grazing over the last century within riparian corridors, have resulted in a 97 percent decline of pre-20th century riparian habitat in the western U.S. [16]. In the southwestern U.S., riparian ecosystems are dominated by *Populus* and *Salix* species [20], which are considered to be foundation species that structure community composition across multiple trophic levels and influence ecosystem processes such as nutrient cycling [21], [22]. Because riparian tree fitness is influenced by a number of different processes (e.g., temperature, soil water availability, flooding regimes), CC will impact riparian species directly by altering growth, phenology and geographic distributions, and indirectly by altering flood regimes, such as the timing of spring runoff and the magnitude of floods (reviewed in [23]).

Rising temperatures, altered precipitation, and novel disturbances associated with land use also make these riparian ecosystems particularly vulnerable to invasive exotic plant species such as those within the *Tamarix* genus [19], [24], [25]. *Tamarix* spp. (*Tamarix chinensis*, *Tamarix ramosissima* and hybrids [26], hereafter referred to as *Tamarix*) have invaded much of the southwestern U.S., making them the third most frequently occurring woody riparian plant in the region [27]. The presence of *Tamarix* can detrimentally impact native riparian trees when combined with other stressors. In areas where the natural flood regime has been altered, such as along dammed river systems, *Tamarix* has invaded habitat once dominated by native *Populus* and *Salix* species [28], [29]. Once established, *Tamarix* has been shown to outcompete native species due to its ability to tolerate a wider range of salinity and soil moisture contents than its native counterparts [29]. In systems where *Tamarix* forms dense monocultures, *Tamarix* has been shown to alter soil salinity [30], [31], hydrology [32], [33], and change the surrounding floral and faunal communities including mycorrhizal mutualists [29], [34], [35]. As such, we also consider species within the *Tamarix* genus to be exotic foundation species [21] due to their role in redefining riparian communities.

An important step toward understanding the impact of CC on riparian ecosystems in the southwest is to assess how CC and a highly competitive species (*Tamarix*) may combine or act synergistically to further extirpate native species. We used the Maxent modeling algorithm [36] to identify areas of expansion and contraction of three native riparian species (*Populus fremontii*, *Salix exigua*, *Salix gooddingii*) and the exotic *Tamarix* under one future climate change scenario and multiple general circulation models (GCMs). Although other studies have created habitat suitability models for *Tamarix* (e.g., [37], [38]), no previous modeling efforts have identified regions where CC is projected to result in 1) native habitat loss and *Tamarix* gain, and 2) potential new range overlaps between natives and *Tamarix*.

Native riparian trees have largely been ignored in research using ecological niche models (ENMs) to assess the impact of CC, especially in respect to altered temperature and precipitation regimes. Projected increases in temperature may exceed the range of tolerance of many native riparian trees [39], [40], (e.g. record temperatures >50°C [39]) resulting in a decline of these stabilizing species, while these same conditions may promote the spread of invasive species [27]. Numerous studies have reported significant effects of temperature and drought on the productivity of riparian species. Grady *et al.* [39] found populations of the riparian trees *P. fremontii* and *S. gooddingii* from warm environments were more productive in a warm common garden than were populations that were from cooler locations, but populations of the riparian shrub *S. exigua* did not perform better based on home site temperature, suggesting local adaptation to temperature in riparian trees but not shrubs. There is conflicting evidence regarding the response of *P. fremontii* and *S. gooddingii* to drought. Some studies predict that *P. fremontii* is more susceptible to drought than *S. gooddingii* [29], [41], whereas others suggest that *S. gooddingii* is less drought-tolerant [42]. Hultine *et al.* [43] found *P. fremontii* to be more sensitive to short-term drought, whereas *S. exigua* had greater sensitivity to long-term drought events.

While it is unclear which native species will be the most susceptible to drought and temperature increases, many studies have found *Tamarix* to be particularly resilient to drought and increased temperature [29], [42], [44] relative to their native counterparts. Together, these studies indicate that climate-induced effects on species' physiological tolerances have the potential to impact their geographic distributions.

In this study we examine effects of CC and exotic species on riparian ecosystems, as quantified by shifts in the geographic distribution of suitable habitat, and test the hypothesis that there will be predictable geographic responses of riparian species to CC. Based on previous studies on the physiological tolerances of native and exotic riparian species to temperature, we predict: 1) *Tamarix* will experience the greatest habitat increase as a result of climate change, with moderate declines of *S. exigua* and extensive habitat reduction of *S. gooddingii*, and *P. fremontii*. This prediction is based on work by Grady et al. [39], which found the productivity of *P. fremontii* to be most sensitive to temperature transfer distance, followed by *S. gooddingii* and *S. exigua*. 2) A greater proportion of future native riparian habitat, relative to current, will be affected by encroachment of *Tamarix* under CC. 3) Cumulative effects of *Tamarix* invasion and habitat loss due to climate change will be most prevalent at the trailing edges of native species distributions where they are already suffering the greatest abiotic stress. By assessing the impact of CC on the distribution of both native and invasive foundation species, we can begin to address the potential combined effects of CC and invasive species on the viability of threatened riparian species distributions.

Methods

Species Data

We obtained *P. fremontii*, *S. exigua*, *S. gooddingii*, and *Tamarix* location points from the Global Biodiversity and Information Facility (GBIF, http://www.gbif.org). Due to the prevalence of hybrids between *T. chinensis* and *T. ramosissima* which are unique to the western United States, we combined occurrences of both species in the U.S. into one *Tamarix* layer [45]. Although there have been questions regarding the accuracy of GBIF data [46], it represents the most comprehensive online database of species location points available. We used an established process to verify locations in the database [47]. Specifically, we removed location points that did not have associated coordinates or were not based on land, and localities where the recorded county did not match with the actual county of collection. A total of 5,758 points were removed in the above process resulting in 739 points for *P. fremontii*, 893 for *S. gooddingii*, 2,092 for *S. exigua* and 1,309 for *Tamarix* that cover their geographic ranges in western North America.

Environmental Variables

Although the species modeled here occupy riparian habitat and are dependent upon environmental variables besides climate, such as flooding, soil salinity, and distance to water [48–50], [37], the focus of this study was to identify areas of climate expansion and contraction across a broad geographic range which could be used to prioritize conservation and restoration efforts. Species within the *Populus* and *Salix* genera are largely dependent upon dynamic river flows and occasional flooding for natural recruitment and survival [51]. However, climate variables typically explain more of the distribution across a large spatial extent than other environmental variables such as soil type and biotic interactions [52]. As such, by including bioclimatic variables to characterize the distribution of these riparian foundation species, we quantify the climatic niche of where *P. fremontii*, *S. exigua* and *S. gooddingii* could occur if flow regimes were suitable.

We used bioclimatic data from the WorldClim database [53] to project the current and future distributions of suitable habitat of all species. To reduce the number of variables included in the final model, we did a preliminary model run for each species and used Maxent's jackknife estimate to examine the permutation importance of the 19 bioclimatic variables (Table 1). Variables with contribution scores <5% were removed from the final model [54]. Variables were averaged for the time periods 1961–1990 and were at a 30 arc-second (~1 km) resolution.

We used a conservative, moderate growth carbon emissions scenario utilized by many scientists [55], A1B [56], and two future time frames, 2050s and 2080s (averaged from 2040 to 2069 and 2070 to 2099, respectively). Because uncertainty in forecasting future climates is mainly attributed to differences in GCMs [57], we used five GCMs that best reflect the current climate of the southwestern US [58]: National Center for Atmospheric Research (NCAR) Community Climate System Model, version 3.0 (CCSM3), Max Planck Institute for Meteorology, Germany (ECHAM5/MPI), Commonwealth Scientific and Industrial Re-

Table 1. The percent contribution and permutation importance (in bold) each bioclimatic variable made to the model building process.

	Species			
Variable	*P. fremontii*	*S. exigua*	*S. gooddingii*	*Tamarix*
BIO1 (Annual Mean Temperature)	**28.98**	**29.36**	-	-
	18.91	17.66		
BIO3 (Isothermality (BIO2/BIO7))	-	-	**18.87**	**10.04**
			17.68	19.98
BIO4 (Temperature Seasonality (Standard Deviation))	**23.80**	**46.39**	**15.83**	**28.91**
	36.71	61.12	5.96	23.22
BIO5 (Max Temperature of Warmest Month)	**16.17**	-	**4.52**	-
	13.47		9.67	
BIO6 (Min Temperature of Coldest Month)	**18.13**	-	-	**6.64**
	12.12			1.34
BIO7 (Temperature Annual Range (BIO5-BIO6))	-	-	**12.86**	-
			6.22	
BIO10 (Mean Temperature of Warmest Quarter)	-	-	-	**13.62**
				18.05
BIO11 (Mean Temperature of Coldest Quarter)	-	-	-	**17.97**
				2.58
BIO12 (Annual Precipitation)	**5.64**	**3.75**	**9.29**	**17.67**
	7.83	2.67	17.67	31.43
BIO13 (Precipitation of Wettest Month)	-	-	**9.19**	-
			4.32	
BIO14 (Precipitation of Driest Month)	-	-	**13.17**	-
			11.86	
BIO15 (Precipitation Seasonality (Coefficient of Variation))	**7.28**	**5.43**	**16.26**	**5.15**
	10.96	2.32	26.62	3.39
BIO17 (Precipitation of Driest Quarter)	-	**15.07**	-	-
		16.23		
Total Temperature	**87.08**	**75.75**	**52.09**	**77.18**
	81.21	78.78	39.54	65.18
Total Precipitation	**12.92**	**24.25**	**47.91**	**22.82**
	18.79	21.22	60.46	34.82

search Organisation CSIRO, Australia (CSIRO-MK3), Centre National de Recherches Meteorologiques, Meteo France, France (CNRM-CM3), and Hadley Centre for Climate Prediction and Research, Met Office United Kingdom, and Hadley Centre Global Environmental Model, version 1 (HadGEM1). These models were statistically downscaled to a 30 arc-second resolution using the Delta Method [59] and downloaded from Global Climate Model data portal (http://www.ccafs-climate.org/data/).

Model Calibration & Evaluation

We used Maxent version 3.3.3e (http://www.cs.princeton.edu/~schapire/maxent) to forecast potential future shifts in suitable habitat of the four species in response to CC. Maximum entropy (Maxent) is a machine learning algorithm which estimates a species ecological niche by finding the distribution with the maximum entropy subject to constraints by environmental variables chosen *a priori* [36], [60]. Studies have shown Maxent to perform well as compared with other presence-background algorithms [49], [61], [62]. Since the selection of background points in Maxent can influence model performance [63], [64], we restricted the selection of background points to the sampling extent using a Minimum Convex Polygon [65]. Additionally, since the species occurrence data did not come from a random sample, there is the potential for sampling bias in the datasets. To remove potential biases, we followed the methods of Elith et al. [66] to create a Kernel Density Estimator (KDE) surface, rescaled from one to 20, from which to draw the 10,000 random background points. To avoid projecting into environments outside those which the models were trained upon, we specified the 'fade-by-clamping' option in Maxent, which removed heavily clamped pixels from the final model predictions [36].

Output format was specified to a logistic form for easy interpretation. We used a 10-fold cross-validation method, which holds 10% of the data for testing and trains the model on the remaining 90% for 10 iterations, in order to test the predictive performance on withheld data [67]. Model performance was evaluated using partial receiver operator characteristics (pROC) following the methods described in Peterson et al. [68]. We utilized the partial ROC software developed by N. Barve [69] in which we specified 1000 iterations with the omission threshold set at ten percent [70]. T-tests were performed to test for statistical significance of pROC values. Linear, quadratic, product, hinge and threshold functions of predictor variables were employed and variable importance was assessed using a jackknife analysis [36]. A binomial probability test was used to assess the accuracy of each predicted distribution compared to that expected by chance [36]. We evaluated variable importance by examining the permutation importance each bioclimatic variable had on the model building process.

To compare suitable and unsuitable habitats, we specified a 10th percentile training presence threshold (T10), which is a fixed threshold where only the lowest 10 percent of predicted values are rejected [70].

Post-hoc Analyses and Calculation of Species Distribution Overlaps

We performed a number of transformations and analyses on the projected habitat models. First, we used "Raster Calculator" in ArcGIS Desktop 9.3 © [71] to identify areas of habitat loss and expansion by subtracting future habitat layers from the previous time period once a threshold had been applied. Second, to identify areas where *Tamarix* would likely interact with *P. fremontii, S. exigua* and *S. gooddingii*, we calculated the projected future overlap between *Tamarix* and the three native species once a threshold had been applied, using "Raster Calculator". Regions where native habitat loss due to CC and future *Tamarix* overlapped were classified as areas where *Tamarix* would have the highest potential to displace *P. fremontii, S. exigua* and *S. gooddingii*. To minimize uncertainty pertaining to differences in GCM projections, all calculations using future projected distributions were based on a liberal threshold (T10) agreed upon by four or more GCMs [72].

Results

Model Results

All four species (*P. fremontii, S. gooddingii, S. exigua*, and *Tamarix*) produced partial ROC values statistically greater than 1.0 (t-test, $p<0.0001$), with values ranging from 1.169 to 1.484 (Table S1 for supporting information). Binomial probabilities were $p<0.0001$, suggesting that all models predicting the distribution of suitable habitat of each species provided significant discriminatory ability as compared with random models (Table S1 for supporting information).

Environmental Variable Importance and Habitat Association

For all three native species and *Tamarix* temperature variables had the greatest contribution to the models, as quantified by permutation importance, with the largest effect for *P. fremontii, Tamarix* and *S. exigua* (Table 1). The distribution of *S. gooddingii* was influenced by both temperature and precipitation (Table 1).

The response curves indicated that suitable habitat generally decreased with excess precipitation, and increased with increasing temperature, up to a threshold (Fig. 1). Overall, these results indicate that for the riparian species examined here, temperature variables have a larger influence in determining suitable habitat than precipitation.

Impact of Climate Change: Range Contraction and Expansion

Under a liberal threshold (T10), four or more GCMs predict that the distribution of suitable habitat for *Tamarix* is expected to shift a total of 71 percent by the end of the century (Fig. 2, Table 2), with a gain of 62 percent and a loss of 9 percent. Of the native species, suitable habitat for *S. exigua* is projected to be the most stable species through time, with only 39 percent of suitable habitat changing (gain of 9 percent and a loss of 30 percent) (Fig. 2, Table 2). While both *P. fremontii* and *S. gooddingii* are expected to gain suitable habitat, over 62 and 38 percent, respectively, up to 40 percent of the suitable habitat for *S. gooddingii* is projected to be lost between now and 2080s (Fig. 2, Table 2). Together, the three native species are expected to have a net gain of nearly 36 percent of current projected suitable habitat by the 2080s (Fig. 2, Table 2).

Impact of Climate Change in the presence of Tamarix

We found that nearly 140 percent of future projected native riparian habitat (defined as containing a minimum of one of the three native species modeled here) in western North America, relative to current projected habitat, is expected to have *Tamarix* present by 2080 (Fig. 3, Table 3). *Tamarix* invasion is expected to occur across 200 percent of *P. fremontii* and 174 percent of *S. gooddingii* by 2080 (Table 3). By 2080, over 102 percent of current *S. exigua* habitat is expected to have *Tamarix*, with the effect spread out relatively evenly between 2050 and 2080 (Table 3). When combined with the effect of CC, nearly 156 percent of native habitat is predicted to be affected by either factor by 2080 (Fig. 3, Table 3). *P. fremontii* is projected to experience the greatest impact of climate change and *Tamarix* invasion, with 205 percent of habitat being affected by 2080, followed by *S. gooddingii* (194 percent) and *S. exigua* (116 percent) (Fig. 3, Table 3).

Climate change and *Tamarix* will also affect native habitat concurrently. We found that 12 percent of current native riparian habitat may be impacted by both CC and the presence of *Tamarix* by 2080 (Fig. 4, Table 3). Nearly 15 percent of *P. fremontii* and *S. exigua* habitat, and 20 percent of *S. gooddingii* habitat is projected to be affected by both factors by 2080 (Fig. 4, Table 3). Contrary to our original hypothesis, concurrent effects were not constrained to the trailing edges of the distribution of native species, although they were more prevalent at the trailing edge for some species, such as *S. exigua* (Fig. 4).

Discussion

This study demonstrates that CC will likely alter the distribution of three native riparian species in western North America. However, there are two important caveats to address in our study. First, despite statistically significant pROC scores (Table S1 for supporting information), the models developed for *P. fremontii* and *S. exigua* showed a relatively high false negative rate (Fig. 1). This is likely a result of over-fitting in the Maxent algorithm [73], and thus projected changes in both *P. fremontii* and *S. exigua* distributions may be especially conservative. Additionally, all future projections for the species modeled here are based upon only five GCMs and one moderate climate change scenario (SRES A1B). Although our rationale for using the five GCMs chosen is based upon previous research of climatic conditions in the southwestern U.S. [58], we did use a moderate emissions scenario, which is conservative in light of current CO_2 emission rates [74]. Thus, model results presented

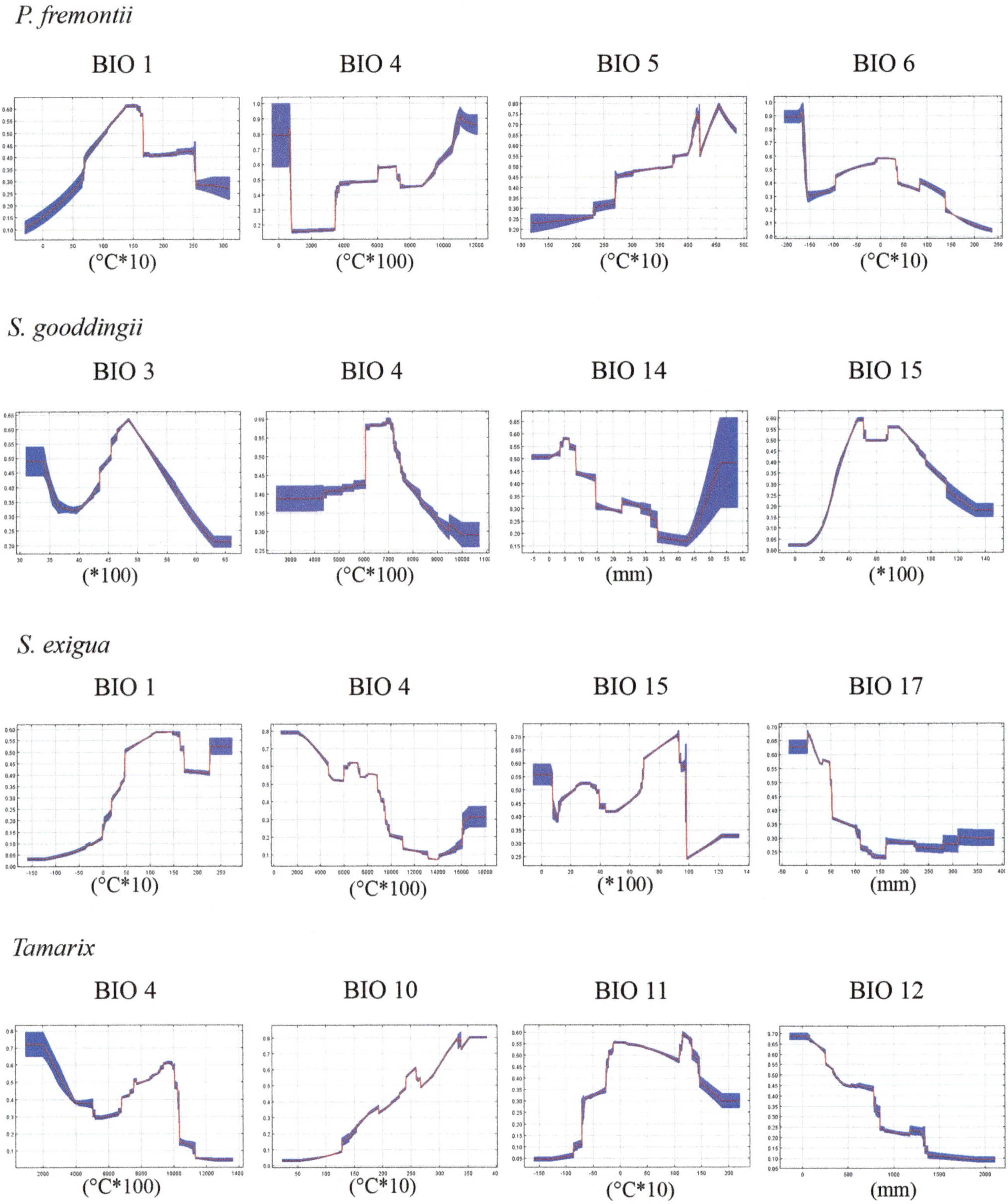

Figure 1. Marginal response curves for each climatic variable. Response curves depicting the probability of suitable habitat related to each climatic variable for *Populus fremontii*, *Salix gooddingii*, *Salix exigua* and *Tamarix*. Units of precipitation variables are measured in millimeters (mm), and temperature variables are degrees Celsius times 10 (°C*10). See Table 1 for full variable names.

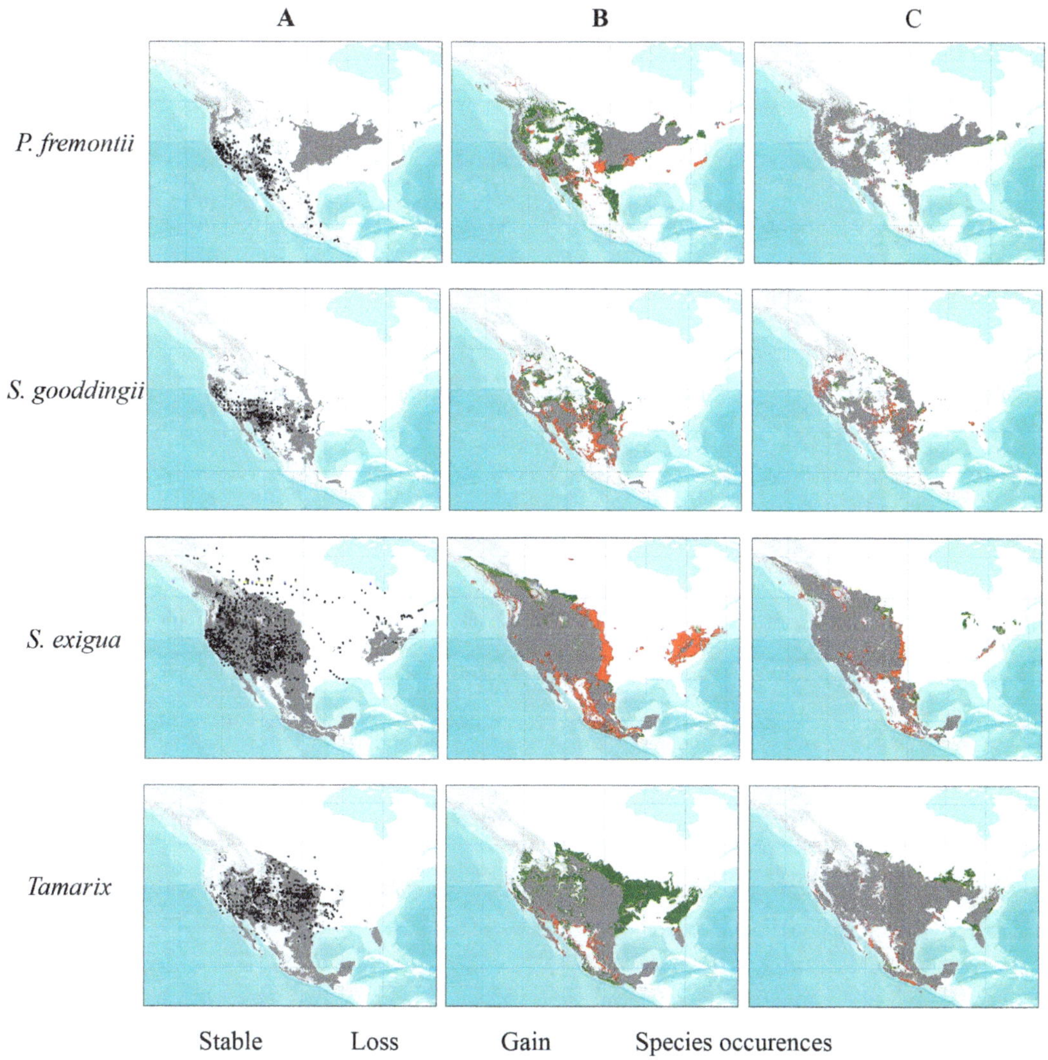

Figure 2. Maxent model outputs showing range expansion and contraction of suitable habitat predicted through time as a result of CC. A) Suitable habitat in current time is shown in grey using a liberal threshold (T10), with the occurrence data overlaid. All occurrence data were downloaded from the Global Biodiversity Information Facility. B-C) Suitable habitat in 2050s and 2080s, predicted by four or more GCMs using a liberal threshold (T10) is also depicted. Areas of loss (red), gain (green), and stable (grey) were calculated by subtracting suitable habitat from the previous time frame.

here are likely conservative estimates of habitat turnover, and actual future climatic conditions may deviate from those used here.

Vulnerability of Native Species to Climate Change

In contrast to our original hypothesis that climate change would have the greatest impact on *P. fremontii*, followed by *S. gooddingii*

Table 2. Predicted percent change (gain and loss) of projected suitable habitat through time using agreement among four or more GCMs and a liberal threshold (T10).

Species	Total % Change Current-2080s	Total % Loss Current-2080s	Total % Gain Current-2080s
P. fremontii	80%	17%	62%
S. exigua	39%	30%	9%
S. gooddingii	77%	40%	38%
Tamarix	70%	9%	62%
Native species	60%	24%	36%

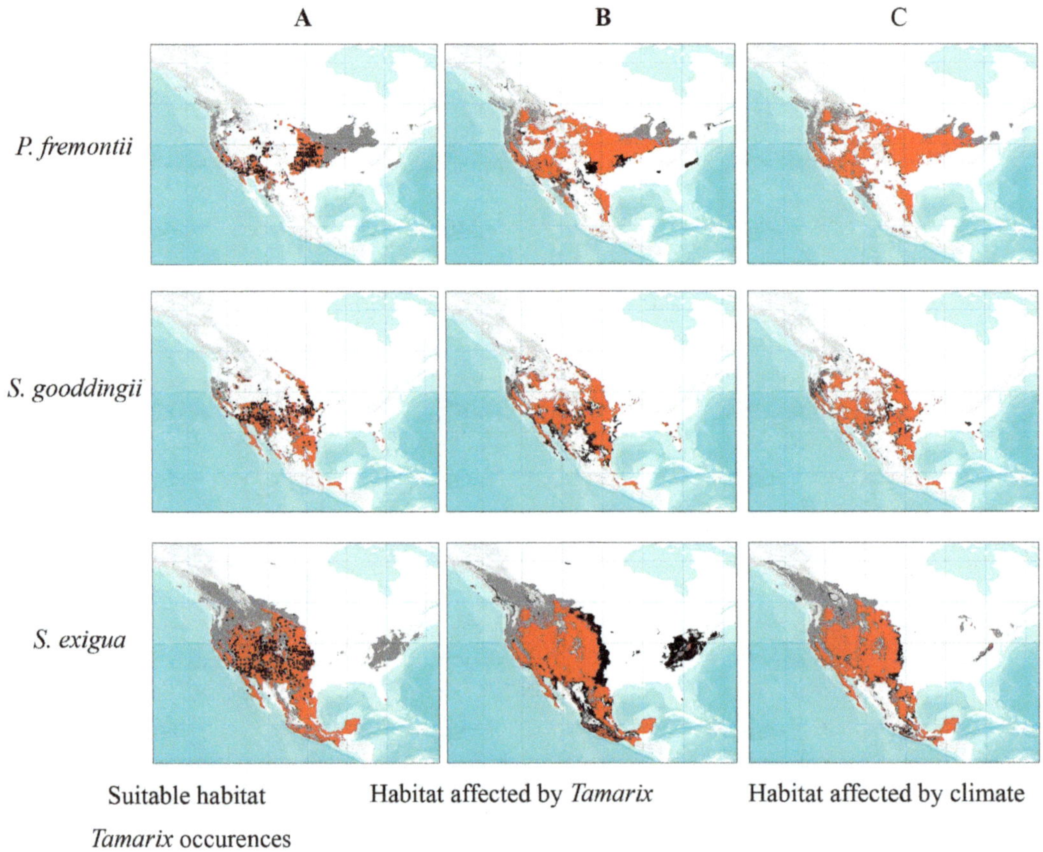

Figure 3. Native species distributions affected by CC or *Tamarix*. Projected distributions of the three native species (grey) showing the range predicted to be affected by either *Tamarix* invasion (red) between current and 2080 (A-C) or CC (black) between 2050 and 2080 (B-C).

and *S. exigua*, we found that *S. gooddingii* is projected to experience the greatest loss of suitable habitat due to CC, followed by *S. exigua* and *P. fremontii* (Table 2). These results are in partial agreement to those reported by Grady *et al.* [39], who experimentally demonstrated greater sensitivity in transfer climate distance of *P. fremontii* and *S. gooddingii* relative to *S. exigua*. In our study, the greatest decline in habitat was predicted in the next thirty years, with moderate declines between 2050 and 2080 (Fig. 2). Although it is unclear why the next thirty years would show greater changes than the following 30 years, it may be that a threshold is reached in the first 30 years that triggers major changes. Following a similar pattern with many other species [1], [75], [76], suitable habitat of the three native species is projected to decline along the trailing edges of the species range with little to moderate increases in habitat at the leading edges (Fig. 2).

Two patterns became apparent when examining the bioclimatic variables which correlated with the distribution of these native riparian species. First, variables related to temperature had the greatest permutation importance on the model building process for *P. fremontii*, and *S. exigua* (Table 1), and for both native species increases in temperature correlated with increased habitat suitability (Fig. 1), with some thresholds present. Second, both precipitation and temperature were important for *S. gooddingii*, with suitable habitat generally increasing with temperature up to a threshold (Table 1). However, increases in precipitation led to a decrease in suitable habitat (Fig. 1). Together, these patterns suggest that while variable contribution differs between the native species, drier, warm climates provide more suitable habitat. While this may be surprising, a number of other studies have documented that water use of riparian species is largely influenced by the availability of groundwater rather than rainfall [77–79].

Tamarix and Climate Change

Current estimates indicate that *Tamarix* occupies an area of 356,241 hectares in the western United States [80]. Our results indicate that the climate niche of *Tamarix* (Fig. 1) is approximately 300 times greater than the realized niche [81], suggesting that there is extensive climatically suitable habitat that this species may yet invade. Additionally, there is substantial overlap in projections of suitable habitat between results generated here and those reported by other studies. Jarnevich *et al.* [37] used 29 climatic, topographic, and geographic variables, including distance to water, to characterize the distribution of *Tamarix*, and found present potential habitat throughout the western US, with distinct patches occurring in Utah and Arizona, along the Pacific crest, regions of Montana and Idaho, and along river systems east of the Rocky Mountain Range. Morisette *et al.* [38] used remote sensing data to examine the current realized niche, and similar to our results, found significant concentration of suitable habitat in the southwestern US, but predicted greater occurrence especially in Texas. The *Tamarix* suitability model developed here predicts suitable habitat in almost all of these regions, despite our use of only five bioclimatic variables, indicating that bioclimatic variables alone can accurately characterize a riparian species niche over broad geographic scales.

The bioclimatic variables describing the niche of *Tamarix* occurrences (Table 1) were consistent with those reported by other

Table 3. The number of cells (km^2) calculated for *P. fremontii*, *S. gooddingii*, *S. exigua*, and all native species together.

	Available habitat	Available habitat	Habitat loss	Habitat Overlap with *Tamarix*	Habitat Loss and *Tamarix*	Available habitat	Habitat loss	Habitat Overlap with *Tamarix*	Habitat Loss and *Tamarix*	Total % impacted by loss	Total % impacted by *Tamarix*	Total % impacted by either factor	Total % impacted by both factors
	current	2050s	2050s	2050s	2050s	2080s	2080s	2080s	2080s	(current-2080s)	(current-2080s)	(current-2080s)	(current-2080s)
	a	b	c	d	e	f	g	h	i	((c+g)/a)*100	((e+h)/a)*100	(((d+e-f)+(g+h-i))/a)*100	((e+i)/a)*100
P. fremontii	3688174	5E+06	535414	3572125	380305	5E+06	108988	3809569	83554	17%	200%	205%	13%
S. exigua	8806324	7E+06	2072957	4662669	1026756	7E+06	578385	4299096	363179	30%	102%	116%	16%
S. gooddingii	2918525	3E+06	736097	2613738	291126	3E+06	424756	2477806	286337	40%	174%	194%	20%
Native habitat	15413023	2E+07	3344468	10848532	1698187	2E+07	1E+06	10586471	148266	29%	139%	156%	12%

Also shown are the percentages of total habitat impacted by; loss of suitable climatic habitat, *Tamarix*, either loss of suitable climatic habitat or *Tamarix*, and both *Tamarix* and loss concurrently.

studies. We found probability of suitable habitat to increase with mean temperature of the warmest quarter, while the remaining bioclimatic variables concerning precipitation had little influence on habitat suitability (Fig. 1). Jarnevich *et al.* [37] also found mean temperature of the warmest quarter to be a significant predictor of suitable habitat while precipitation variables contributed less than 10 percent.

In support of our original hypothesis, we found that *Tamarix* invasion of riparian habitat is expected to increase with CC (Fig. 2, Table 2). Suitable habitat of *Tamarix* is projected to gain approximately 62 percent of the current suitable habitat by 2080 (Table 2). This finding is contrary to that reported by Bradley *et al.* [24], who predicted *Tamarix* distribution to remain relatively stable through time. The discrepancy is likely due to the fact that Bradley *et al.* [24] used a different modeling algorithm, Mahalanobis distance [82], [83], versus the Maxent algorithm used here. Although no studies directly compare the two modeling approaches, there is evidence to suggest that Maxent performs better than Mahalanobis distance [84]. Some of the habitat gain of *Tamarix* predicted here is projected to occur at latitudes higher than where the species is currently found (Fig. 2). Gaskin and Kazmer [85] found *T. chinensis* and *T. ramosissima* introgression to follow strong latitudinal clines, with a higher occurrence of *T. chinensis* alleles at the southern edge and *T. ramosissima* alleles at the northern edge. This pattern is likely driven by greater frost tolerance in *T. ramosissima* than *T. chinensis* [86]. While projected increases in latitude of *Tamarix* (Fig. 2) should increase the prevalence of *T. ramosissima* alleles, warmer climates could decrease the selection pressure for frost tolerance, resulting in *T. chinensis* alleles becoming more widespread.

Implications for Management and Conservation Efforts

In the context of restoration and conservation efforts, our results highlight three main conclusions. First, mitigation efforts should focus on areas where native habitat is projected to be affected by CC and the presence of *Tamarix* (Fig. 4). We found that 20 percent of the distribution of *S. gooddingii* is projected to be affected by both habitat loss and *Tamarix*, followed by 16 percent of *S. exigua* habitat, and 13 percent of *P. fremontii* habitat (Table 3). Given the projected future anthropogenic impacts on riparian habitat (reviewed in [23]), even relatively small projected effects can have disproportionate impacts. This is especially true when the habitat in question is rare- largely due to anthropogenic factors (e.g., cattle grazing, water diversions, farming, dams) [16]. With such rare habitat types, even relatively low mortality may result in bottlenecks and local extinction. For example, in a survey by Gitlin *et al.* [10], less than two percent of the riparian habitat surveyed had *P. fremontii*, but *P. fremontii* experienced greater than 15 percent mortality. Because riparian areas are predicted to be most susceptible to extirpation of native species and endemic species [87], *Tamarix* removal and restored stream flows, which may alleviate loss from predicted CC and increase native species regeneration [88], [51], should be a priority.

Second, rehabilitation of *Tamarix* altered habitat, involving replanting of native riparian species, should begin in areas where *Tamarix* is expected to decline as a result of CC as reflected in Figure 2. Based on the *Tamarix* response function generated here (Fig. 1), we can expect that *Tamarix* will decrease in regions projected to become wetter. Thus, re-vegetation efforts with *Populus* and *Salix* species may be particularly successful [89], [90], especially when planted in proportions which maximize associated community diversity [91].

Lastly, genetics-based approaches offer a method to restore habitat in suboptimal areas (e.g., in regions with a high likelihood

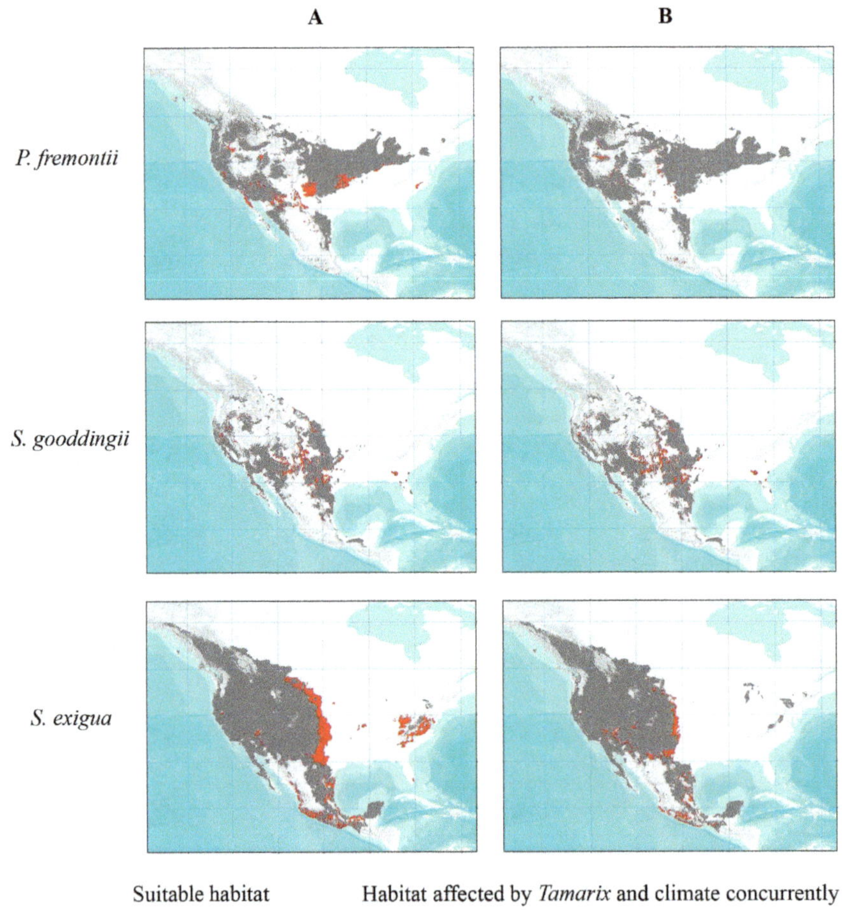

Figure 4. Locations projected to be affected by CC and *Tamarix* concurrently. The projected distributions of three native species depicting the areas with the potential for exotic-by-climate interactions (i.e. the effect of *Tamarix* and habitat loss due to CC) in A) 2050s and B) 2080s.

of excess temperatures, or where *Tamarix* has been present). Planting with source populations known to exhibit increased tolerance to drought and/or higher temperatures, as established through field trials, can increase survival rate and performance [39]. Similarly, managed relocation, based on genetic pre-adaptation, can also be utilized to mitigate the effects of riparian exotics by identifying populations that can better compete with *Tamarix*, as has been shown with other native species and their competing invasives. For example, Goergen *et al.* [92] showed that when native grass species (*Poa secunda* and *Elymus multisetus*) from different source populations were grown with cheatgrass (*Bromus tectorum*), native grasses from source populations of invaded sites better tolerated competition with cheatgrass than native grasses from uninvaded sites. We can take genetics approaches one step further by identifying source populations and genotypes of native species that can simultaneously cope with both CC and invasive species. Such plants may be identified in regions where *Tamarix* occurrences overlap with native species distributions and CC has already resulted in near extirpation of native species [areas with both *Tamarix* occurrences (Fig. 3A) and depicted in black (Fig. 3B)], leaving only genotypes that have persisted in the presence of both *Tamarix* and climate change.

Future Directions: Incorporating Indirect Effects of *Tamarix* and Climate Change on Native Habitat

Although others have reviewed indirect effects of anthropogenic CC on riparian systems in the context of stream flow and human activity (see [23]), there are at least three reasons why *Tamarix* may exert even greater influence on riparian habitats than we predict due to the effects of multiple stressors, community composition and intraspecific variation on the productivity and diversity of native riparian habitat. Thus, our findings may be conservative in light of possible indirect ecological effects between native riparian species, climate and *Tamarix* in the context of CC as follows.

First, *Tamarix* can negatively impact riparian trees already experiencing other stressors. The presence of *Tamarix* seedlings has been shown to be unrelated to *Populus* and *Salix* growth or survival in benign environments [88]; however, *Tamarix* is most abundant along altered streams and rivers where *Populus* recruitment is the lowest [19]. Additionally, Gitlin *et al.* [10] examined the relationship between *Tamarix* density and *P. fremontii* mortality during a severe drought. As *Tamarix* density rose, there was a proportional increase in the mortality of *P. fremontii* ($r^2 = 0.69$) that reached 97 percent when the density of *Tamarix* was approximately 65 percent. Since this mortality relationship was manifested during a record drought, it argues that when *Tamarix* is present, the addition of additional stressor can detrimentally impact a native riparian species. Thus, under CC, we would expect an increase in *Tamarix* due to both an expansion of suitable climatic habitat (as reported here) and projections of escalating human impacts to river systems (reviewed in [23]), and a corresponding rise in mortality of native riparian species due to the presence of *Tamarix* and climate related stress. The combination of these circumstances could act synergistically on *Tamarix* abundance.

Second, *Tamarix* can indirectly impact native riparian habitat by altering the associated community, which can negatively affect both the diversity and productivity of these ecosystems. In *Tamarix* dominated stands, community composition of invertebrates and overall diversity of both invertebrates [93], [94] and birds [95] have been recorded as less than in habitat dominated by native riparian species. Negative effects on biodiversity were also discovered in a stream where aquatic macroinvertebrates were less on *Tamarix* than cottonwood leaf litter [96]. The presence of the tamarisk beetle (*Diorhabda* spp.), which was first released in 2001 as a biocontrol agent for *Tamarix* can also lead to different community assemblages [97]. Furthermore, impacts on the community can occur belowground. Meinhart & Gehring [35] showed that the presence of *Tamarix* altered the arbuscular (AMF) and ectomycorrhizal fungal (EMF) communities of *P. fremontii*, decreasing the colonization rate by half. Since mycorrhizal fungal associations can positively impact the growth of native species [98], and biodiversity promotes ecosystem stability [99], accounting for their presence and structure can further inform how these riparian species and their associated communities might respond to CC.

Lastly, many studies have demonstrated that populations of foundation trees are locally adapted, and intraspecific responses to climate change may be highly variable [77], [100], [101]. Accounting for these population level differences can impact predictions of how species respond to CC. Oney *et al.* [102] compared model results from an ENM (Maxent) and a population-based model (Universal Transfer Functions) for *Pinus contorta* and found that intraspecific genetic variation buffered the species from CC, indicating that ENMs may be under-predicting future suitable habitat. Furthermore, intraspecific genetic variation in productivity will impact associated community members, leading to predictable changes in biodiversity as a result of CC [103]. Although complex, merging the interactions between the indirect effects discussed here and the direct effects reported through our modeling efforts should provide better management strategies to mitigate the effects of both climate change and invasive species on native species and the rich communities they support.

Conclusions

This study is a first step toward integrating multiple environmental challenges (CC and invasive species) to inform conservation and management decisions. Results presented here can help prioritize areas for conservation and restoration by identifying regions where 1) invasive *Tamarix* and CC will jointly impact riparian species in the future, 2) *Tamarix* is expected to increase as a result of climate change, and 3) where the unique evolutionary history of native species is likely to have selected for genotypes that possess high drought tolerance and/or are highly competitive with *Tamarix*. Identifying these superior genotypes may be especially important to utilize in future mitigation efforts. Through provenance trials, we can experimentally identify source populations for use in restoration that are best adapted to future climate, exotics, or climate-by-exotic interactions (e.g., [39], [40]). Together, this information should increase the success of conservation and restoration efforts in the face of multiple stressors.

Supporting Information

Table S1 Partial ROC scores for each of the four species modeled, followed by the standard deviation. An (*) indicates significance at p<0.0001 using a t-test.

Acknowledgments

We thank the Cottonwood Ecology group of Northern Arizona University for their valuable feedback in addition to HM Bothwell, MK Lau, and RR Randall. KR Hultine provided insightful comments on an earlier version of this manuscript. We also thank Dr. Bazartseren Boldgiv and four anonymous reviewers for their valuable comments and suggestions.

Author Contributions

Conceived and designed the experiments: DHI KCG. Performed the experiments: DHI. Analyzed the data: DHI. Contributed reagents/materials/analysis tools: DHI KCG SMS TGW. Wrote the paper: DHI KCG TGW SMS.

References

1. Root TL, Price JT, Hall KR, Schneider SH (2003) Fingerprints of global warming on wild animals and plants. Nature 421: 57–60.
2. Walther G R (2010) Community and ecosystem responses to recent climate change. Philos Trans R Soc Lond B Biol Sci 365: 2019–2024.
3. Parmesan C, Yohe G (2003) A globally coherent fingerprint of climate change impacts across natural systems. Nature 421: 37–42.
4. Walther G-R, Berger S, Sykes MT (2005) An ecological "footprint" of climate change. Proc Biol Sci 272: 1427–1432.
5. Allen CD, Macalady AK, Chenchouni H, Bachelet D, McDowell N, et al. (2010) A global overview of drought and heat-induced tree mortality reveals emerging climate change risks for forests. For Ecol Manage 259: 660–684.
6. Felde VA, Kapfer J, Grytnes J-A (2012) Upward shift in elevational plant species ranges in Sikkilsdalen, central Norway. Ecography 35: 922–932.
7. Bell DM, Bradford JB, Lauenroth WK (2013) Early indicators of change: divergent climate envelopes between tree life stages imply range shifts in the western United States. Glob Ecol Biogeogr 23: 168–180.
8. Williams AP, Allen CD, Macalady AK, Griffin D, Woodhouse CA, et al. (2012) Temperature as a potent driver of regional forest drought stress and tree mortality. Nat Clim Chang 3: 292–297.
9. Breshears DD, Cobb NS, Rich PM, Price KP, Allen CD, et al. (2005) Regional vegetation die-off in response to global-change-type drought. Proc Natl Acad Sci USA 102: 15144–15148.
10. Gitlin AR, Sthultz CM, Bowker MA, Stumpf S, Paxton KL, et al. (2006) Mortality gradients within and among dominant plant populations as barometers of ecosystem change during extreme drought. Conserv Biol 20: 1477–1486.
11. Sánchez Salguero R, Navarro Cerrillo RM, Swetnam TW, Zavala MA (2012) Is drought the main decline factor at the rear edge of Europe? The case of southern Iberian pine plantations. For Ecol Manage 271: 158–169.
12. Garfin G, Jardine A, Merideth R, Black M, LeRoy S, editor (2013) Assessment of Climate Change in the Southwest United States: A Report Prepared for the National Climate Assessment. Washington, DC: Island Press.
13. Brusca RC, Wiens JF, Meyer WM, Eble J, Franklin K, et al. (2013) Dramatic response to climate change in the Southwest: Robert Whittaker's 1963 Arizona Mountain plant transect revisited. Ecol Evol 3: 3307–3319.
14. Seager R, Ting M, Held I, Kushnir Y, Lu J, et al. (2007) Model projections of an imminent transition to a more arid climate in southwestern North America. Science 316: 1181–1184.
15. Dominguez F, Rivera ER, Lettenmaier DP, Castro CL (2012) Changes in winter precipitation extremes for the Western United States under a warmer climate as simulated by regional climate models. Geophys Res Lett 39: 1–7.
16. Noss RF, LaRoe III ET, Scott JM (1995) Endangered ecosystems of the United States: a preliminary assessment of loss and degradation. Biological Report 28. U.S. Department of the Interior, National Biological Service, Washington, D.C., USA.
17. Tockner K, Stanford JA (2002) Riverine flood plains: present state and future trends. Environ Conserv 29: 308–330.

18. Davies PM (2010) Climate change implications for river restoration in global biodiversity hotspots. Restor Ecol 18: 261–268.
19. Merritt DM, Poff NL (2010) Shifting dominance of riparian *Populus* and *Tamarix* along gradients of flow alteration in western North American rivers. Ecol Appl 20: 135–152.
20. Durst SL, Theimer TC, Paxton EH, Sogge MK (2008) Temporal variation in the arthropod community of desert riparian habitats with varying amounts of saltcedar (*Tamarix ramosissima*). J Arid Environ 72: 1644–1653.
21. Ellison AM, Bank MS, Clinton BD, Colburn EA, Elliott K, et al. (2005) Loss of foundation species: consequences for the structure and dynamics of forested ecosystems. Front Ecol Environ 3: 479–486.
22. Whitham TG, Bailey JK, Schweitzer JA, Shuster SM, Bangert RK, et al. (2006) A framework for community and ecosystem genetics: from genes to ecosystems. Nat Rev Genet 7: 510–523.
23. Perry LG, Andersen DC, Reynolds LV, Nelson SM, Shafroth PB (2012) Vulnerability of riparian ecosystems to elevated CO_2 and climate change in arid and semiarid western North America. Glob Chang Biol 18: 821–842.
24. Bradley BA, Oppenheimer M, Wilcove DS (2009) Climate change and plant invasions: restoration opportunities ahead? Glob Chang Biol 15: 1511–1521.
25. Walther G-R, Roques A, Hulme PE, Sykes MT, Pysek P, et al. (2009) Alien species in a warmer world: risks and opportunities. Trends Ecol Evol 24: 686–693.
26. Gaskin JF, Schaal BA (2002) Hybrid *Tamarix* widespread in U.S. invasion and undetected in native Asian range. Proc Natl Acad Sci 99: 11256–11259.
27. Friedman JM, Auble GT, Shafroth PB, Scott ML, Merigliano MF, et al. (2005) Dominance of non-native riparian trees in western USA. Biol Invasions 7: 747–751.
28. Howe WH, Knopf FL (1991) On the imminent decline of Rio Grande cottonwoods in central New Mexico. Southwest Nat 36: 218–224.
29. Busch DE, Smith SD (1995) Mechanisms associated with decline of woody species in riparian ecosystems of the southwestern US. Ecol Monogr 65: 347–370.
30. Taylor JP, Wester DB, Smith LM (1999) Soil disturbance, flood management, and riparian woody plant establishment in the Rio Grande floodplain. Wetlands 19: 372–382.
31. Glenn EP, Morino K, Nagler PL, Murray RS, Pearlstein S, et al. (2012) Roles of saltcedar (*Tamarix* spp.) and capillary rise in salinizing a non-flooding terrace on a flow-regulated desert river. J Arid Environ 79: 56–65.
32. Graf WL (1978) Fluvial adjustments to the spread of tamarisk in the Colorado Plateau region. Geol Soc Am Bull 89: 1491–1501.
33. Hultine KR, Bush SE (2011) Ecohydrological consequences of non-native riparian vegetation in the southwestern United States: A review from an ecophysiological perspective. Water Resour Res W07542, doi:10.1029/2010WR010317.
34. Ellis LM, Crawford CS, Molles MC (1997) Rodent communities in native and exotic riparian vegetation in the middle Rio Grande Valley of central New Mexico. Southwest Nat 42: 13–19.
35. Meinhardt KA, Gehring CA (2012) Disrupting mycorrhizal mutualisms: a potential mechanism by which exotic tamarisk outcompetes native cottonwoods. Ecol Appl 22: 532–549.
36. Phillips S, Anderson R, Schapire R (2006) Maximum entropy modeling of species geographic distributions. Ecol Modell 190: 231–259.
37. Jarnevich CS, Evangelista PH, Stohlgren TJ, Morisette JT (2011) Improving national-scale invasion maps: Tamarisk in the western United States. West North Am Nat 71: 164–175.
38. Morisette JT, Jarnevich CS, Ullah A, Cai W, Pedelty, Jeffrey A, et al. (2006) A tamarisk habitat suitability map for the continental United States. Front Ecol Environ 4: 11–17.
39. Grady KC, Ferrier SM, Kolb TE, Hart SC, Allan GJ, et al. (2011) Genetic variation in productivity of foundation riparian species at the edge of their distribution: implications for restoration and assisted migration in a warming climate. Glob Chang Biol: 1–36.
40. Grady KC, Laughlin DC, Ferrier SM, Kolb TE, Hart SC, et al. (2013) Conservative leaf economic traits correlate with fast growth of genotypes of a foundation riparian species near the thermal maximum extent of its geographic range. Funct Ecol 27: 428–438.
41. Stella JC, Battles JJ (2010) How do riparian woody seedlings survive seasonal drought? Oecologia 164: 579–590.
42. Horton JL, Kolb TE, Hart SC (2001) Physiological response to groundwater depth varies among species and with river flow regulation. Ecol Appl 11: 1046–1059.
43. Hultine KR, Bush SE, Ehleringer JR (2010) Ecophysiology of riparian cottonwood and willow before, during, and after two years of soil water removal. Ecol Appl 20: 347–361.
44. Pockman WT, Sperry JS (2000) Vulnerability to xylem cavitation and the distribution of Sonoran Desert vegetation. Am J Bot 87: 1287–1299.
45. Gaskin JF, Shafroth PB (2005) Hybridization of *Tamarix ramosissima* and *T. chinensis* (Saltcedars) with *T. aphylla* (Athel) (Tamaricaceae) in the southwestern USA determined from DNA sequence data. Madroño 52: 1–10.
46. Yesson C, Brewer PW, Sutton T, Caithness N, Pahwa JS, et al. (2007) How global is the global biodiversity information facility? PLoS One 2: e1124.
47. Ramirez-Villegas J, Jarvis A, Touval J (2012) Analysis of threats to South American flora and its implications for conservation. J Nat Conserv 20: 337–348.
48. Evangelista PH, Kumar S, Stohlgren TJ, Jarnevich CS, Crall AW, et al. (2008) Modelling invasion for a habitat generalist and a specialist plant species. Divers Distrib 14: 808–817.
49. Evangelista PH, Stohlgren TJ, Morisette JT, Kumar S (2009) Mapping invasive Tamarisk (*Tamarix*): a comparison of single-scene and time-series analyses of remotely sensed data. Remote Sens 1: 519–533.
50. Kerns BK, Naylor BJ, Buonopane M, Parks CG, Rogers B (2009) Modeling Tamarisk (*Tamarix* spp.) habitat and climate change effects in the northwestern United States. Invasive Plant Sci Manag 2: 200–215.
51. Rood SB, Samuelson GM, Braatne JH, Gourley CR, Hughes FM, et al. (2005) Managing river flows to restore floodplain forests. Front Ecol Environ 3: 193–201.
52. Pearson RG, Dawson TP (2003) Predicting the impacts of climate change on the distribution of species: are bioclimate envelope models useful? Glob Ecol Biogeogr 12: 361–371.
53. Hijmans RJ, Cameron SE, Parra JL, Jones PG, Jarvis A (2005) Very high resolution interpolated climate surfaces for global land areas. Int J Climatol 25: 1965–1978.
54. Sahlean TC, Gherghel I, Papeş M, Strugariu A, Zamfirescu ŞR (2014) Refining climate change projections for organisms with low dispersal abilities: a case study of the Caspian whip snake. PLoS One 9: e91994.
55. Kim HY, Ko J, Kang S, Tenhunen J (2013) Impacts of climate change on paddy rice yield in a temperate climate. Glob Chang Biol 19: 548–562.
56. IPCC (2007) Climate Change 2007: The physical science basis. Cambridge University Press, Cambridge, UK, p. 996.
57. Buisson L, Thuiller W, Casajus N, Lek S, Grenouillet G (2010) Uncertainty in ensemble forecasting of species distribution. Glob Chang Biol 16: 1145–1157.
58. Garfin GM, Eischeid JK, Lenart MT, Cole KL, Ironside K, et al. (2010) Downscaling climate projections in topographically diverse landscapes of the Colorado Plateau in the arid southwestern United States. The Colorado Plateau IV; Shaping Conservation Through Science and Management. pp. 22–43.
59. Ramirez-Villegas J, Jarvis A (2010) Downscaling global circulation model outputs: the Delta method decision and policy analysis working paper No. 1. Cent Trop Agric Colomb.
60. Phillips SJ, Dudik M (2008) Modeling of species distributions with Maxent: new extensions and a comprehensive evaluation. Ecography 31: 161–175.
61. Elith J, Graham CH, Anderson RP, Dudík M, Ferrier S, et al. (2006) Novel methods improve prediction of species' distributions from occurrence data. Ecography 29: 129–151.
62. Kumar S, Spaulding SA, Stohlgren TJ, Hermann KA, Schmidt TS, et al. (2009) Potential habitat distribution for the freshwater diatom *Didymosphenia geminata* in the continental US. Front Ecol Environ 7: 415–420.
63. Phillips SJ, Dudík M, Elith J, Graham CH, Lehmann A, et al. (2009) Sample selection bias and presence-only distribution models: implications for background and pseudo-absence data. Ecol Appl 19: 181–197.
64. VanDerWal J, Shoo LP, Williams SE (2009) New approaches to understanding late Quaternary climate fluctuations and refugial dynamics in Australian wet tropical rain forests. J Biogeogr 36: 291–301.
65. Flory AR, Kumar S, Stohlgren TJ, Cryan PM (2012) Environmental conditions associated with bat white-nose syndrome mortality in the north-eastern United States. J Appl Ecol 49: 680–689.
66. Elith J, Kearney M, Phillips S (2010) The art of modelling range-shifting species. Methods Ecol Evol 1: 330–342.
67. Elith J, Phillips SJ, Hastie T, Dudík M, Chee YE, et al. (2011) A statistical explanation of MaxEnt for ecologists. Divers Distrib 17: 43–57.
68. Peterson AT, Papeş M, Soberón J (2007) Rethinking receiver operating characteristic analysis applications in ecological niche modeling. Ecol Modell 213: 63–72.
69. Barve N (2008) Tool for Partial-ROC (Biodiversity Institute, Lawrence, KS), ver 1.0.
70. Pearson RG, Thuiller W, Araújo MB, Martinez-Meyer E, Brotons L, et al. (2006) Model-based uncertainty in species range prediction. J Biogeogr 33: 1704–1711.
71. ESRI 2009. ArcGIS Desktop: Release 9.3. Redlands, CA: Environmental Systems Research Institute.
72. Araújo MB, New M (2007) Ensemble forecasting of species distributions. Trends Ecol Evol 22: 42–47.
73. Ochoa-Ochoa L, Urbina-Cardona JN, Vázquez L-B, Flores-Villela O, Bezaury-Creel J (2009) The effects of governmental protected areas and social initiatives for land protection on the conservation of Mexican amphibians. PLoS One 4: e6878.
74. Raupach MR, Marland G, Ciais P, Quere C Le, Canadell JG, et al. (2007) Global and regional drivers of accelerating CO2 emissions. Proc Natl Acad Sci 104: 10288–10293.
75. Thomas CD, Franco AMA, Hill JK (2006) Range retractions and extinction in the face of climate warming. Trends Ecol Evol 21: 415–416.
76. Aitken SN, Yeaman S, Holliday JA, Wang T, Curtis-McLane S (2008) Adaptation, migration or extirpation: climate change outcomes for tree populations. Evol Appl 1: 95–111.
77. Kolb TE, Hart SC, Amundson R (1997) Boxelder water sources and physiology at perennial and ephemeral stream sites in Arizona. Tree Physiol 17: 151–160.

78. Snyder KA, Williams DG (2000) Water sources used by riparian trees varies among stream types on the San Pedro River, Arizona. Agric For Meterology 105: 227–240.
79. Cox G, Fisher DG, Hart SC, Whitham TG (2005) Nonresponse of native cottonwood trees to water additions during summer drought. West North Am Nat 65: 175–185.
80. Sher A, Quigley MF (2013) *Tamarix*: A case study of ecological change in the American west. New York: Oxford University Press. 31 p.
81. Hutchinson GE (1959) Homage to Santa Rosalia or why are there so many kinds of animals? Am Nat 93: 145–159.
82. Farber O, Kadmon R (2003) Assessment of alternative approaches for bioclimatic modeling with special emphasis on the Mahalanobis distance. Ecol Modell 160: 115–130.
83. Tsoar A, Allouche O, Steinitz O, Rotem D, Kadmon R (2007) A comparative evaluation of presence-only methods for modelling species distribution. Divers Distrib 13: 397–405.
84. Jeschke JM, Strayer DL (2008) Usefulness of bioclimatic models for studying climate change and invasive species. New York Acad Sci 24: 1–24.
85. Gaskin JF, Kazmer DJ (2008) Introgression between invasive saltcedars (*Tamarix chinensis* and *T. ramosissima*) in the USA. Biol Invasions 11: 1121–1130.
86. Friedman JM, Roelle JE, Gaskin JF, Pepper AE, Manhart JR (2008) Latitudinal variation in cold hardiness in introduced *Tamarix* and native *Populus*. Evol Appl 1: 598–607.
87. Richardson DM, Holmes PM, Esler KJ, Galatowitsch SM, Stromberg JC, et al. (2007) Riparian vegetation: degradation, alien plant invasions, and restoration prospects. Divers Distrib 13: 126–139.
88. Sher AA, Marshall DL, Taylor JP (2002) Establishment patterns of native *Populus* and *Salix* in the presence of invasive nonnative *Tamarix*. Ecol Appl 12: 760–772.
89. Harms RS, Hiebert RD (2006) Vegetation response following invasive Tamarisk (*Tamarix* spp.) removal and implications for riparian restoration. Restor Ecol 14: 461–472.
90. Bay RF, Sher A (2008) Success of active revegetation after *Tamarix* removal in riparian ecosystems of the southwestern United States: a quantitative assessment of past restoration projects. Restor Ecol 16: 113–128.
91. Bangert R, Ferrier SM, Evans L, Kennedy K, Grady KC, et al. (2013) The proportion of three foundation plant species and their genotypes influence an arthropod community: restoration implications for the endangered Southwestern Willow Flycatcher. Restor Ecol 21: 447–456.
92. Goergen EM, Leger EA, Espeland EK (2011) Native perennial grasses show evolutionary response to *Bromus tectorum* (cheatgrass) invasion. PLoS One 6: e18145.
93. Pendleton RL, Pendleton BK, Finch D (2011) Displacement of native riparian shrubs by woody exotics: effects on arthropod and pollinator community composition. Nat Resour Env Iss 1: 1–12.
94. Anderson B, Russell PE, Ohmart R.D (2004) Riparian revegetation: an account of two decades of experience in the arid southwest. Blythe, California: Avvar Books
95. Brand AL, White GC, Noon BR (2008) Factors influencing species richness and community composition of breeding birds in a desert riparian corridor. Condor 110.
96. Bailey JK, Schweitzer JA, Whitham TG (2001) Salt cedar negatively affects biodiversity of aquatic macroinvertebrates. Wetlands 21: 442–447.
97. Strudley S (2009) Impacts of tamarisk biocontrol (*Diorhabda elongata*) on the trophic dynamics of terrestrial insects in monotypic tamarisk stands. Masters of Science thesis. University of Denver, CO.
98. Smith SE, Read DJ (2008) Mycorrhizal symbiosis. Third edition. Academic Press, New York, New York, USA.
99. Tilman D, Downing JA (1994) Biodiversity and stability in grasslands. Nature 367: 363–365.
100. O'Neill GA, Hamann A, Wang T (2008) Accounting for population variation improves estimates of the impact of climate change on species' growth and distribution. J Appl Ecol 45: 1040–1049.
101. Wang T, O'Neill GA, Aitken SN (2010) Integrating environmental and genetic effects to predict responses of tree populations to climate. Ecol Appl 20: 153–163.
102. Oney B, Reineking B, O'Neill G, Kreyling J (2013) Intraspecific variation buffers projected climate change impacts on *Pinus contorta*. Ecol Evol 3: 437–449.
103. Ikeda DH, Bothwell HM, Lau MK, O'Neill GA, Grady KC, et al. (2013) A genetics-based Universal Community Transfer Function for predicting the impacts of climate change on future communities. Funct Ecol 28: 65–74.

The Invisible Prevalence of Citizen Science in Global Research: Migratory Birds and Climate Change

Caren B. Cooper[1]*, Jennifer Shirk[1], Benjamin Zuckerberg[2]

1 Cornell Lab of Ornithology, Ithaca, New York, United States of America, 2 University of Wisconsin, Madison, Wisconsin, United States of America

Abstract

Citizen science is a research practice that relies on public contributions of data. The strong recognition of its educational value combined with the need for novel methods to handle subsequent large and complex data sets raises the question: Is citizen science effective at science? A quantitative assessment of the contributions of citizen science for its core purpose – *scientific research* – is lacking. We examined the contribution of citizen science to a review paper by ornithologists in which they formulated ten central claims about the impact of climate change on avian migration. Citizen science was never explicitly mentioned in the review article. For each of the claims, these ornithologists scored their opinions about the amount of research effort invested in each claim and how strongly the claim was supported by evidence. This allowed us to also determine whether their trust in claims was, unwittingly or not, related to the degree to which the claims relied primarily on data generated by citizen scientists. We found that papers based on citizen science constituted between 24 and 77% of the references backing each claim, with no evidence of a mistrust of claims that relied heavily on citizen-science data. We reveal that many of these papers may not easily be recognized as drawing upon volunteer contributions, as the search terms "citizen science" and "volunteer" would have overlooked the majority of the studies that back the ten claims about birds and climate change. Our results suggest that the significance of citizen science to global research, an endeavor that is reliant on long-term information at large spatial scales, might be far greater than is readily perceived. To better understand and track the contributions of citizen science in the future, we urge researchers to use the keyword "citizen science" in papers that draw on efforts of non-professionals.

Editor: Robert Guralnick, University of Colorado, United States of America

Funding: These authors have no support or funding to report.

Competing Interests: The authors have declared that no competing interests exist.

* Email: caren.cooper@cornell.edu

Introduction

Citizen science, the practice of involving the public in scientific research, is undergoing a period of rapid growth across numerous disciplines. Effective use of citizen science as a method of data collection relies on techniques from many disciplines, such as science communication, informal science education, and informatics. As an interdisciplinary field of practice, citizen science is experiencing a process of professionalization, which is evident from newly organized societies such as the Citizen Science Association (http://citizenscienceassociation.org), the European Citizen Science Association (http://ecsa.biodiv.naturkundemuseum-berlin.de/), and the Citizen Science Network Australia (http://csna.gaiaresources.com.au). Other hallmarks of professionalization include new educational- and cyber-infrastructure initiatives, international conferences, and numerous advisory boards [1], [2].

Research has documented that citizen science can engage the public through hobbies and games [3], [4], support learning in science, technology, engineering, and math (STEM) [5] provide sources of large and complex datasets that catalyze advances in data visualization [6], [7], advance cyber-infrastructure and new analysis techniques [8], [9], [10], [11], and influence management and policy [12], [13], [14], [15]. But is citizen science effective at science?

Research focused on the scientific value of citizen science has yet to directly quantify the impact of citizen science to any specific area of scientific research, but has instead focused on concerns regarding data quality e.g., [16], [17]. Data from citizen-science efforts can indeed have biases, the largest source of which may originate from the extent to which people self-select to participate, which affect the effort they expend, their level of skill, and the spatiotemporal distribution of the data [18]. Yet, the quality of data collected by volunteers, on a project-by-project basis, has generally been found as reliable as the data collected by professionals in community-based research [19] and contributory projects across a wide variety of subjects, including lady beetles [20], moths [21], wolves [22], trees [23], air pollution [24], light pollution [25], plants [26], pikas [27], invasive plants [28], and bees [29]. Researchers address the biases and variable data quality of volunteered observations through the use of novel online data filters and workflows for detecting erroneous submissions [9], [30], [31], and modified project design and analysis [11], [16], [32]. These advances have been central in the development, vetting, and dissemination of citizen-science datasets that have allowed for new areas of scientific research that would not otherwise be possible; particularly questions requiring data collected over broad spatiotemporal scales.

Many citizen-science programs have operated for decades (some, for more than a century) and frequently span multiple countries and continents [3], [33]. To investigate unanticipated questions arising from global changes, researchers have begun relying on diverse sources of data including the re-purposing of data contributed through citizen science [33]. We expect that research related to climate change may be particularly well suited to draw on citizen science because: (i) the influence of climate change spans broad spatial scales, possibly affecting species throughout their entire ranges, (ii) ecological responses are highly variable across space, that is, not all populations are exposed to similar trends in climate, and (iii) climate-induced impacts occur over long periods of time, from decades to centuries, generally longer than the duration of any scientists' careers or a typical funding cycle.

Given this presumption, we focused on examining citizen science related to research on birds and climate change. Our objectives were to i) quantify the scientific contribution of citizen science to this area of active inquiry, ii) assess whether professionals held volunteer-based research in equal regard to research by professionals, and iii) evaluate the extent to which citizen science was readily visible or noted in the focal studies. Taken together, our goal was to evaluate the use and confidence of citizen science in advancing understanding in an important area of ecology and global change research.

Materials and Methods

One of the most widely cited lines of evidence that species are responding to modern climate change relates to shifts in phenology. Changing spring phenology in migratory birds is a rapidly developing field of study that has been identified as a critical bellwether for assessing the ecological impacts of climate change. We based our study on a review paper by Knudsen et al. [34] to evaluate the influence of citizen science on the study of climate change research. The Knudsen et al. [34] review represents a critical synthesis of the existing scientific support for the patterns, mechanisms, and consequences of phenological changes in bird migration. The authors (all active researchers in the field of migration phenology) reviewed literature to formulate 10 specific scientific claims about migratory birds and global climate change (Table 1). Also, 18 of the 27 authors scored their opinion regarding the amount of research effort so far invested in each claim (hereafter referred to as "knowledge basis") and whether each claim held in general (hereafter referred to as "support"), on a continuous scale from 0 (least) to 10 (most) (Table 1). They report the mean for each value and the associated standard deviation.

There were 205 papers referenced in the 10 claims, 15–36 references per claim, in Knudsen et al. [34]. We excluded 42 references that were reviews (including books, chapters, and meta-analyses) from our analysis. We were able to classify 171 of the 173 research papers according to the sources of data: either as including observations collected by volunteers (citizen science) or not (professionals). We noted sources of museum collections (n = 4) and found two of these papers also used data from a citizen science source; we classified the remaining two papers using museum collections as professional. Some (n = 45) papers were referenced in more than one claim, thirteen of which appear in 2–5 claims. We computed the Pearson's correlation between the mean knowledge basis and percent citizen science per claim, and between the mean support and proportion of citizen science per claim. We repeated these analyses using the reported standard deviation associated with knowledge basis and support in relation to proportion of citizen science per claim.

For papers that used citizen-science data, we classified the type of project as: large-scale coordinated scheme (dispersed network of volunteers following a protocol sharing centralized objectives and data management), local volunteers following a protocol but with multiple sampling objectives (e.g., ringers at an observatory), local groups (e.g., local bird clubs and societies), and other (e.g., journals and diaries of amateur naturalists). We also noted the terminology we relied on to classify each citizen-science paper, including explicit mention of volunteers, and/or a specific program, other term, or through contact with the author.

Results

We examined 173 original research papers that were used by Knudsen and colleagues to formulate 10 central claims about the impacts of climate change on avian migration [34]. We found that 85 of the 171 papers that we could classify were based on citizen science, constituting 5 to 20 papers per claim (Appendix S1). Citizen science heavily informed claims related to ecological patterns and consequences and was less frequently cited for claims about mechanisms (Table 1).

Data from a wide range of citizen-science efforts were included in these papers, including observations from large-scale coordinated programs (n = 35 papers), local volunteers following protocols (n = 40), ad hoc counts from local bird clubs (n = 12 papers), other volunteer sources (n = 4), and some papers (n = 6) drew on several of these citizen-science efforts.

Of the 84 papers that were based on citizen science, 74 were published in, or prior to, 1995, which was when "citizen science" was coined in the context of bird research [35]. Despite the importance of citizen science in substantiating the claims in Knudsen et al. [34], the term "citizen science" *never* appeared in any referenced publications (Appendix S1). The term "volunteer" was used in 37 of the citizen-science papers, typically only in the acknowledgements (Appendix S1). We viewed 37 citizen-science papers as "invisible" because our ability to identify these as citizen science was based on i) the name of specific programs (n = 35; e.g., BTO Common Bird Census), ii) the mention of general programs or efforts (n = 7; e.g., local ornithological societies), iii) terms "bird watchers," "ringers," "bander," "public," "naturalist," and "people" (n = 42), iv) by contacting authors and data-providing organizations for confirmation (n = 11 authors contacted and 9 responded), or v) a combination of these identifiers (n = 22). The papers that were difficult to classify were evenly distributed across claims, comprising roughly half of the citizen-science papers per claim.

Knudsen et al. [34] provided the mean values of the opinions of 18 of their authors on the strength of the knowledge basis and support for each of the claims they reviewed. We found that the mean values of the expert opinions were not correlated to the proportion of citizen science supporting each claim ($r = -0.02$, $p = 0.9$ for knowledge basis, Figure 1; $r = 0.07$, $p = 0.8$ for support). Similarly, the standard deviation associated with the mean values of the expert opinions were not correlated to the proportion of citizen science supporting each claim ($r = -0.38$, $p = 0.3$ for knowledge basis; $r = 0.17$, $p = 0.6$ for support).

Discussion

Our findings are strongly indicative of the usefulness and credibility of citizen science in the field of global change research. We found that more than half the central claims about the impacts

Table 1. Ten claims with mean values for support and knowledge basis from Knudsen et al. [34].

	#	Statement	Expert Opinion			Citizen Science	
			Support	Knowledge basis	# papers*	#	%
Pattern	1	Spring migration advances due to global climate change	8.52	8.28	32	17	53
	2	Phenological response depends on migration distance	5.68	5.99	22	11	50
	3	Climate change affects migratory distance and routes	4.86	3.83	20	15	77
Mechanism	4	Controlling mechanisms are hardwired	4.06	3.79	21	5	24
	5	Changes are mainly due to phenotypic plasticity	4.84	3.79	15	5	33
	6	Phenotypic variability is mainly due to weather *en route*	4.87	4.42	29*	14	48
	7	Responses are constrained by the annual cycle	5.54	3.27	24	11	46
Consequence	8	Increased trophic mismatch on breeding grounds	4.97	4.00	21	5	24
	9	Climate change causes population declines	4.53	3.69	35	20	57
	10	Climate change affects community composition	4.10	2.97	21*	14	67

*excluded papers that we were unable to classify (1 in claim 6 and 1 in claim 10).
The number of papers used for assessment of each claim, and the number and percent of papers that used citizen science.

of climate change on avian migration were based on studies that depended on data from citizen scientists. In addition, we did not find any statistical relationship between the knowledge basis or scientific support for each claim and the proportion of citizen science. The use of citizen science data in an active field of ecological research, such as migration phenology, is strong evidence that any stigma associated with the use of data collected by volunteers is unwarranted. Yet, the contributions of citizen science were not readily detectable in most cases. Thus, the stigma may persist unless researchers begin to draw attention to the citizen-science elements in their research papers.

Citizen science was more critical in supporting claims related to ecological patterns and consequences when compared to claims about mechanisms (Table 1). The reason for this finding highlights a simultaneous strength and limitation of citizen science. One of the primary motivations of developing and deploying a citizen science project is to collect data over spatiotemporal scales that would be difficult (if not impossible) with professional scientists. As a result, these programs have been particularly useful for documenting broad-scale patterns (e.g., macroecology) or long-term consequences (e.g., population trends). Many of these programs, in many ecological disciplines in addition to ornithology, were initiated for purposes other than documenting the ecological responses to climate change, but as our findings emphasize, the scale of data collection has proven essential for analyzing patterns and consequences. Cost savings are another advantage of re-purposing existing data. Climate change studies focused on mechanisms tend to involve experimental investigation (e.g., active or passive warming devices) or more intensive field studies (e.g., use of geolocators). That being said, data generated from long-term citizen science programs have also been repurposed to focus on mechanistic hypotheses and objectives. As an example, several of the papers classified as supporting mechanisms of migration focused on explicit hypotheses such as the geographical variation in phenological mismatch [36], [37] and buffer effects in population dynamics [38]; these hypotheses could not be tested without data from coordinated banding studies involving volunteers in different regions.

Given the invisible prevalence of citizen science in advancing this one area of global change research, we suspect it also common in many other areas of inquiry such as studies of land-use change, invasive species, and environmental pollutants, to name a few. We urge future use of consistent terminology and acknowledgement to facilitate tracking the impact of citizen science across numerous disciplines. Specifically, we urge use of the keyword phrase "citizen science" in papers that rely on scientific contributions from the public. Continued assessment of the value of citizen science in other areas of research could help increase overall public participation as well as identify new frontiers in multiple research fields and improve the interdisciplinary practice of citizen science.

An additional consequence of the invisibility of the scientific impact of citizen science is that projects may miss the broader social impacts of their work. Unique positive societal impacts, such as increased scientific literacy, depend on participants being not merely engaged as instruments or human sensors, but upon being informed and engaged with research progress and outcomes [39]. Yet, many long-running volunteer efforts did not originate with the specific purpose of understanding the consequences of global climate change, and as a result, most of these projects were not designed to foster communication of scientific findings back to project participants; this is particularly true for studies using data from online repositories. Explicit recognition of citizen science in published papers could promote the communication linkages necessary for broader impacts by helping shift public discourse

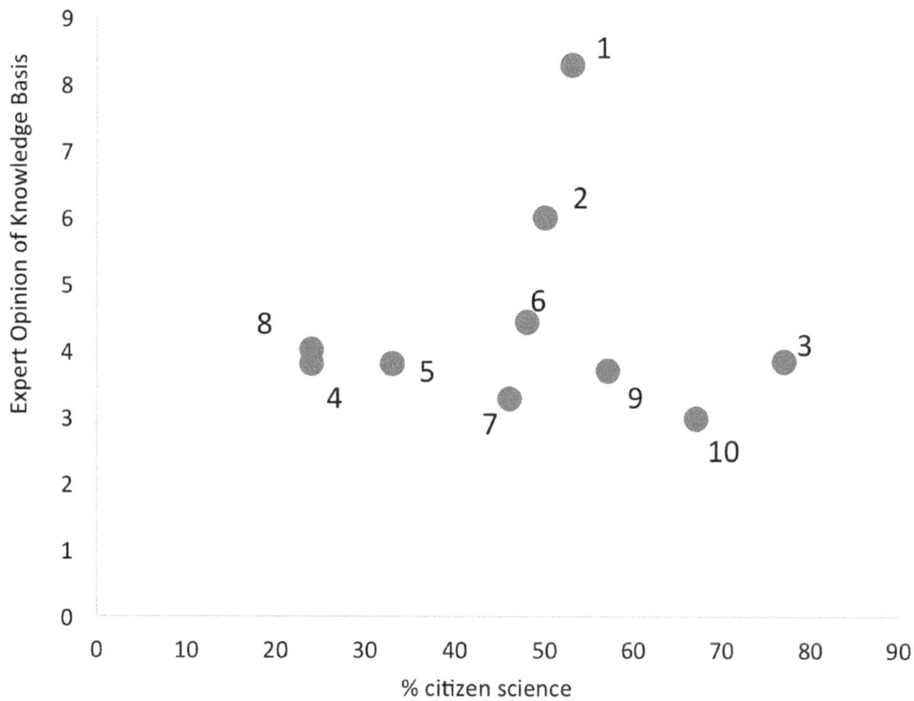

Figure 1. No relationship between the mean values of the expert opinions on the knowledge base of each claim in Knudsen et al. [34] were not correlated to the proportion of citizen science supporting each claim.

associated with modern climate change from controversy to acceptance. Our findings demonstrate the exceptional value of the efforts of thousands of participants whose data informed the 10 claims, and point to the potential of the millions of global participants whose "invisible" efforts may be contributing to new discoveries.

Supporting Information

Appendix S1 Research papers in Knudsen et al. [34], classified as involving data from citizen science (yes = 1, no = 0).

Acknowledgments

We thank R Bonney, A Shwartz, and an anonymous reviewer for comments and insights that improved this manuscript.

Author Contributions

Conceived and designed the experiments: CC BZ. Analyzed the data: CC. Contributed reagents/materials/analysis tools: CC JS. Contributed to the writing of the manuscript: CC JS BZ.

References

1. Miller-Rushing A, Benz S (2013) Workshop 1: Conference on Public Participation in Scientific Research 2012: An International, Interdisciplinary Conference. Bull Ecol Soc Am 94: 112–117.
2. Silvertown JA (2009) A new dawn for Citizen Science. Trends Ecol Evol 24: 467–471.
3. Roy HE, Pocock MJ, Preston CD, Roy DB, Savage J (2012)Understanding Citizen Science and Environmental Monitoring. Final Report on behalf of UK-EOF. NERC Centre for Ecology & Hydrology and Natural History Museum. 179 p.
4. Dickinson JL, Zuckerberg B, Bonter DN (2010) Citizen Science as an Ecological Research Tool: Challenges and Benefits. Annual Rev Ecol Evol Syst 41: 149–172.
5. Bonney R, Cooper CB, Dickinson J, Kelling S, Phillips T, et al. (2009) Citizen Science: a developing tool for expanding science knowledge and scientific literacy. BioSci 59: 977–984.
6. Hochachka WM, Fink D, Hutchinson RA, Sheldon D, Wong W, et al. (2012) Data-intensive science applied to broad-scale citizen science. Trends Ecol Evol 27: 130–137.
7. Kelling S, Hochachka WM, Fink D, Riedewald M, Caruana R, et al. (2009) Data-intensive Science: A New Paradigm for Biodiversity Studies. BioSci 59: 613–620.
8. Newman G, Graham J, Crall AW, Laituri M (2011) The art and science of multi-scale citizen science support. Ecol Infor 6: 217–227.
9. Sullivan BL, Wood CL, Illiff MJ, Bonney RE, Fink D, et al. (2014) The eBird enterprise: An integrated approach to development and application of citizen science. Biol Cons 169: 31–40.
10. Switzer A, Schwille K, Russell E, Edelson D (2012) National Geographic FieldScope: a platform for community geography. Front Ecol Envir 10. 334–335.
11. Fink D, Hochachka WM, Zuckerberg B, Winkler DW, Shaby B, et al. (2010) Spatiotemporal exploratory models for broad-scale survey data. Front Ecol Envir 20: 2131–2147.
12. Danielsen F, Burgess ND, Jensen PM, Pirhofer-Walzl K (2010) Environmental monitoring: the scale and speed of implementation varies according to the degree of people's involvement. J Appl Ecol 47: 1166–1168.
13. McCormick S (2012) After the Cap: Risk Assessment, Citizen Science and Disaster Recovery. Ecol Soc 17: 31.
14. Cornwell ML, Campbell LM (2011) Co-producing conservation and knowledge: Citizen-based sea turtle monitoring in North Carolina, USA. Soc Stud Sci 42: 101–120.
15. Bonilla NO, Scholl J, Armstrong M, Pieri D, Otero B, et al. (2012) Ecological Science and Public Policy: an Intersection of Action Ecology. Front Ecol Envir 93: 340–345.
16. Engel SR, Voshell JR Jr (2002) Volunteer biological monitoring: can it accurately assess the ecological condition of streams? Am Entomol 48: 164–177.
17. Genet KS, Sargent LG (2003) Evaluation of methods and data quality from a volunteer-based amphibian call survey. Wildl Soc Bull 31: 703–714.
18. Cooper CB, Hochachka WH, Dhondt AA (2012) The opportunities and challenges of Citizen Science as a tool for ecological research. In: Dickinson JL, Bonney B, editors.Citizen Science: Public Collaboration in Environmental Research.Ithaca: Cornell University Press. pp. 99–113.

19. Danielsen F, Jensen PM, Burgess ND, Altamirano R, Alviola PA, et al. (2014) A multicountry assessment of tropical resource monitoring by local communities. BioSci 64: 236–251.
20. Gardiner MM, Allee LL, Brown PMJ, Losey JE, Roy HE, et al. (2012) Lessons from lady beetles: accuracy of monitoring data from US and UK citizen-science programs. Front Ecol Envir 10: 471–476.
21. Bates AJ, Sadler JP, Everett G, Grundy D, Lowe N, et al. (2013) Assessing the value of the Garden Moth Scheme citizen science dataset: how does light trap type affect catch? Entom Expt Appl 146: 386–397.
22. Miller DAW, Weir LA, McClintock BT, Grant EHC, Bailey LL, et al. (2013) Experimental investigation of false positive errors in auditory species occurrence surveys. Eco Appl 22: 1665–1674.
23. Galloway AWE, Tudor MT, Vander Haegen WM (2006) The reliability of citizen science: A case study of Oregon white oak stand surveys. Wildl Soc Bull 34: 1425–1429.
24. Tregidgo DJ, West SE, Ashmore MR (2013) Can citizen science produce good science? Testing the OPAL Air Survey methodology, using lichens as indicators of nitrogenous pollution. Envir Pollut 182: 448–451.
25. Kyba CCM, Wagner JM, Kuechly HU, Walker CE, Elvidge CD, et al. (2013) Citizen Science Provides Valuable Data for Monitoring Global Night Sky Luminance. Sci Rep 3, 10.1038/srep01835
26. Gollan J, de Bruyn LL, Reid N, Wilkie L (2012) Can Volunteers Collect Data that are Comparable to Professional Scientists? A Study of Variables Used in Monitoring the Outcomes of Ecosystem Rehabilitation. Envir Manage 50: 969–978.
27. Moyer-Horner L, Smith MM, Belt J (2012) Citizen science and observer variability during American pika surveys. J Wildl Manage 76: 1472–1479.
28. Jordan RC, Brooks WR, Howe DV, Ehrenfeld JG (2012) Evaluating the Performance of Volunteers in Mapping Invasive Plants in Public Conservation Lands. Envir Manage 49: 425–434.
29. Kremen C, Ullmann KS, Thorp RW (2011) Evaluating the Quality of Citizen-Scientist Data on Pollinator Communities. Con Bio 25: 607–617.
30. Bonter DN, Cooper CB (2012). A process for improving data quality and a strategy for ensuring sustainability in a citizen science project. Front Ecol Envir 10: 305–307.
31. Parsons J, Lukyanenko R, Wiersma Y (2011). Easier citizen science is better. Nature 471: 37–37.
32. Cooper CB (2014) Is there weekend bias in clutch-initiation dates from citizen science? Implications for studies of avian breeding phenology. Intl J Biometeor in press.
33. Devictor V, Whittaker RJ, Beltrame C (2010) Beyond scarcity: citizen science programmes as useful tools for conservation biogeography. Div Distrib 16: 354–362.
34. Knudsen E, Lindén A, Both C, Jonzén N, Pulido F, et al. (2011) Challenging claims in the study of migratory birds and climate change. Biol Rev 86: 928–946.
35. Bonney R, Ballard H, Jordan R, McCallie E, Phillips T, et al. (2009) Public Participation in Scientific Research: Defining the Field and Assessing Its Potential for Informal Science Education. A CAISE Inquiry Group Report. Washington, D.C.: Center for Advancement of Informal Science Education (CAISE).
36. Both C (2010) Flexibility of timing of avian migration to climate change masked by environmental constraints en route. Current Biol 20: 243–248.
37. Both C, te Marvelde L (2007) Climate change and timing of avian breeding and migration throughout Europe. Climate Res 35: 93–105.
38. Gill JA, Norris K, Potts PM, Gunnarsson TG, Atkinson PW, et al. (2001) The buffer effect and large scale regulation in migratory birds. Nature 412: 436–438.
39. Lawrence A (2009) The first cuckoo in winter: phenology, recording, credibility, and meaning in Britain. Global Envir Change 19: 173–179.

Climate Change Sensitivity Index for Pacific Salmon Habitat in Southeast Alaska

Colin S. Shanley*, David M. Albert

The Nature Conservancy, Juneau, Alaska, United States of America

Abstract

Global climate change may become one of the most pressing challenges to Pacific Salmon conservation and management for southeast Alaska in the 21st Century. Predicted hydrologic change associated with climate change will likely challenge the ability of specific stocks to adapt to new flow regimes and resulting shifts in spawning and rearing habitats. Current research suggests egg-to-fry survival may be one of the most important freshwater limiting factors in Pacific Salmon's northern range due to more frequent flooding events predicted to scour eggs from mobile spawning substrates. A watershed-scale hydroclimatic sensitivity index was developed to map this hypothesis with an historical stream gauge station dataset and monthly multiple regression-based discharge models. The relative change from present to future watershed conditions predicted for the spawning and incubation period (September to March) was quantified using an ensemble global climate model average (ECHAM5, HadCM3, and CGCM3.1) and three global greenhouse gas emission scenarios (B1, A1B, and A2) projected to the year 2080. The models showed the region's diverse physiography and climatology resulted in a relatively predictable pattern of change: northern mainland and steeper, snow-fed mountainous watersheds exhibited the greatest increases in discharge, an earlier spring melt, and a transition into rain-fed hydrologic patterns. Predicted streamflow increases for all watersheds ranged from approximately 1-fold to 3-fold for the spawning and incubation period, with increased peak flows in the spring and fall. The hydroclimatic sensitivity index was then combined with an index of currently mapped salmon habitat and species diversity to develop a research and conservation priority matrix, highlighting potentially vulnerable to resilient high-value watersheds. The resulting matrix and observed trends are put forth as a framework to prioritize long-term monitoring plans, mitigation experiments, and finer-scale climate impact and adaptation studies.

Editor: Inés Álvarez, University of Vigo, Spain

Funding: This project was funded by the Alaska Sustainable Salmon Fund (www.akssf.org; project #44550) and the Gordon and Betty Moore Foundation (www.moore.org). The funders had no role in study design, data collection and analysis, decision to publish, or preparation of the manuscript.

Competing Interests: The authors have declared that no competing interests exist.

* Email: cshanley@tnc.org

Introduction

Pacific Salmon (*Oncorhynchus* spp.) are a key cultural [1], ecological [2], and economic [3] driver in southeast Alaska with global socio-ecological value [4,5]. In a recent study by the U.S. Forest Service, 96% of Alaskans said that salmon are essential to the Alaskan way of life [6]. More than 50 species of animals feed on spawning salmon each year, and 1 in 10 jobs in southeast Alaska is supported by salmon [6]. Yearly salmon productivity has varied considerably by species and watershed—the regional productivity has remained relatively resilient [3], although the influence of hatchery enhancements on wild salmon stocks is of concern [7–9]. Sustained production and harvest opportunities are likely the result of a portfolio effect [10] whereby thousands of watersheds and tens-of-thousands of streams provide a diversity of freshwater habitats, promoting phenotypic diversity, which buffers regional salmon returns against variability in ocean [11], near-shore [12], and freshwater conditions [13]. Looking to the future, predicted hydrologic change associated with climate change may very well present the biggest challenge to Pacific Salmon conservation and management in the 21st Century [13–20].

Southeast Alaska is predicted to experience the largest change in winter days above freezing in all of North America due to climate change [21]. While boreal and arctic Alaska is expected to see the largest changes in absolute temperature [22], a small increase in temperature in southeast Alaska could have transformative ecological effects (e.g., [23–27]). The mean winter temperature is currently near freezing at −4°C [28]; therefore, with a relatively small increase in temperature, many watersheds and portions of sub-basins will no longer receive winter precipitation as snow, and will transition into rain-fed systems. Many watersheds will likely see decreased total snowpack in headwaters that have provided water storage and moderated flows for salmon streams throughout the summer, and maintained cooler summer stream temperatures [17,25,26].

Watersheds of southeast Alaska can be generally categorized into three hydrologic types: (1) rain-fed, (2) snow-fed, and (3) glacial [29]. Each hydrologic type is expected to exhibit a spectrum of change associated with climate change [17,18,29] and provide unique challenges to resident salmon population adaptation [16,30]. In general, rain-fed systems will likely see increased winter flows, reduced summer flows, and higher stream temperatures all year [16,17,19,20]. Snow-fed systems will likely see more variable discharge patterns with increased rain-on-snow events that can cause flooding, an earlier spring melt, and some may transition into rain-fed hydrologic types due to loss of adequate headwater snowpack [17,29,31]. Glacial systems will likely see increased discharge year-round (until some systems lose glaciers altogether), generally colder summer stream temperatures while glaciers are still present, and warmer winter water temperatures [24,32–35].

Pacific Salmon are affected by hydroclimatic factors at every stage of their lifecycle. Starting with egg incubation in freshwater habitats, water temperatures have the greatest effect on development rates [36]. In addition, winter flow extremes can scour salmon eggs from mobile spawning substrates (i.e., smaller gravels) and cause increased sedimentation that eliminates or degrades habitat by reducing oxygenation [19]. Emergent fry and juvenile salmonids require adequate stream flows to maintain high quality, protected off-channel habitats in both summer and winter to support growth and survival [36]. Smolt migration requires adequate stream flows and appropriate temperature cues to reach estuarine rearing habitats in synchrony with high food availability in the near-shore marine environment [37,38]. Development and spawning success during the adult life-phase is in-turn dependent on global ocean circulation patterns, feed availability, and adequate stream flows and suitable temperatures for return migration and spawning [36].

A review of the current literature suggests egg-to-fry lifestage survival under projected climate change may be one of the most important limiting factors for salmon productivity and conservation in southeast Alaska freshwater habitats [15,18,19,27,29]. Recent studies in Washington State have detailed information on air-to-stream water temperature relationships [19], extreme flow events, and survival estimates [15]; population modeling studies thus far suggest egg-to-fry survival may be a key limiting factor under a range of climate change scenarios for the Pacific Northwest [15]—in addition to a suite of temperature related factors that may be less of a concern to salmon in northern parts of their range [17]. However, summer habitat conditions warrant further research in southeast Alaska [17,38]. Washington State also has generally larger river systems that are more affected by climate model predictions for drier, hotter summers for interior parts of the state [19]. Southeast Alaska's mostly small and steep watersheds may behave more like the Olympic Peninsula watersheds where the scouring of eggs due to winter flooding events is currently thought to be more of concern than summer habitat conditions (e.g., [27]). Warmer winter water temperatures in southeast Alaska may accelerate salmon development rates, increase off-channel rearing habitats, and improve productivity in some watersheds (e.g., moderately glacial systems) in northern parts of their range [17,36,38]; this has been evidenced by increased salmon catch records during the warm-phase of the Pacific Decadal Oscillation in Alaska (PDO; [26]). However, depending on specific watershed physiographic characteristics (e.g., mean elevation, slope, and floodplain viability), stream channel geomorphology, genetic diversity, and capacity for phenotypic plasticity [16,20,30,39], directionally changing flow regimes will likely transition into less favorable habitats due to more frequent scouring events in segments of some watersheds [15].

Two primary questions were asked in this paper: (1) effects of future projection trends in temperature and precipitation on seasonal discharge patterns across southeast Alaska, and (2) vulnerability or resilience of watersheds to hydrologic change in relation to the current distribution of high-value salmon habitat. The project was conducted in five phases: (1) development of a comprehensive historical stream gauge station database for the region; (2) development of a transboundary geospatial database of watershed physiographic and climatic characteristics for historical and projected temperature and precipitation; (3) building and testing multiple regression-based monthly discharge models using AIC model selection; (4) mapping the regional discharge model results for projected change during the spawning and incubation period (September to March), and; (5) combining a climate change sensitivity index with an index of current salmon habitat and species diversity to develop a research and conservation priority matrix.

Methods

Historical discharge patterns

A database of historical stream gauge stations was developed for southeast Alaska using USGS mean monthly gauge station records and USGS gauge station catchment polygons. For gauge stations where a catchment polygon was not available, watersheds were delineated using the Shuttle Radar Topography Mission (SRTM) digital elevation model (DEM; 25 m) and the ArcGIS 10.1 Watershed delineation tool (ESRI, Redlands, CA). An inventory of all gauge stations ever recorded via the USGS National Water Inventory System was conducted; only those gauge stations without human alteration (e.g., dams, hydro plants, etc.) were obtained for analysis. When more than one gauge station existed in a watershed, the station with the longest period of record was selected. In order to achieve a spatial distribution necessary for landscape modeling, ≥5 years of monthly discharge data was used as a cut-off for modeling purposes. Furthermore, data post-1976 Pacific Decadal Oscillation (PDO) was also used as a cut-off so as not to confound the analysis with PDO cycles [13,40]. Monthly means (cubic feet per second) for each gauge station across the period of record available were calculated for model development.

Watershed physiography and climatology

A transboundary geospatial database of physical watershed characteristics, and historical and projected climatologies were developed to model present and future patterns of stream discharge for all watersheds. The SRTM DEM was used for the basis of analysis in all U.S. watersheds, and U.S. portions of transboundary rivers. The best available DEMs were used for Canadian reaches (Table 1). The climate modeling software ClimateWNA 4.62 [41] was used at a 1 km spatial resolution to map gridded estimates of monthly temperature and precipitation from the PRISM climate model [42–44]. The three top performing global climate models for the region (ECHAM5, HadCM3, and CGCM3.1; [22,45]) from the Intergovernmental Panel on Climate Change (IPCC) Fourth Assessment were averaged into a single ensemble model for analysis. This ensemble model was used to project temperature and precipitation for the year 2080 using three global greenhouse gas emission scenarios to identify regional trends [41,46]: B1 (low growth), A1B (moderate growth), A2 (high growth). The National Hydrography Dataset (NHD) was used to delineate lake coverage within the U.S., and the best available datasets were used for Canada (Table 1). Glacier coverage for all watersheds was delineated with the Alaska Department of Natural Resources 1:2,000,000 glacier coverage. The transboundary watershed polygons (n = 1784) developed by the U.S. Forest Service were used to create a regional watershed database. Mean values were calculated for temperature, precipitation, and elevation. Percent coverage was calculated for lakes and glaciers. Values for the historical gauge station catchment polygons and the complete regional watershed polygons were calculated separately.

Table 1. Climatic and physiographic geospatial data sources used to develop multiple regression-based monthly discharge models for southeast, Alaska, USA.

Variable	Description
Basin Area	USGS gauge station catchment polygons were used for the historical gauge station analysis. The USFS transboundary watershed layer, derived from primarily USGS HUC10 polygons, was used for the regional analysis.
Precipitation	ClimateWNA version 4.62 [41] 1 km downscaling of PRISM climate model [42] with monthly means generated from the period 1961–1990.
Temperature	ClimateWNA version 4.62 [41] 1 km downscaling of PRISM climate model [42] with monthly means generated from the period 1961–1990.
Elevation	Alaska: Shuttle Radar Topography Mission DEM (25 m), British Columbia: Terrain Resource Information Management Program DEM (25 m), Yukon: Department of Environment DEM (70 m). These DEMs were combined and resampled to 70 m.
Lakes	Alaska: National Hydrography Dataset, British Columbia: British Columbia Watershed Altus, Yukon: North American Water Polygons by ESRI. These polygons were combined into one lake coverage.
Glaciers	Alaska Department of Natural Resources 1:2,000,000 glacier coverage.

Discharge models and sensitivity index

A hierarchy of multiple regression models aimed at explaining monthly discharge using six potential explanatory variables were tested (Table 2). These models were tested using multiple regression and AIC model selection criterion [47]. Seven *a priori* models were developed, starting with the simplest model where basin area is the only explanatory variable, to more complex models incorporating climatology and physiographic setting. The same models were tested for each month for consistency and comparability. The most parsimonious, best-fit models with the lowest ΔAIC scores and highest weights were selected for an accuracy evaluation and the final hydroclimatic sensitivity index.

The eight most regionally relevant discharge model variables used in Wiley and Curran [48] for the state of Alaska and conterminous basins in Canada were considered for modeling: basin area, main channel length, mean channel slope, mean basin elevation, % lakes, % forest, % glaciers, mean precipitation and mean temperature, as well as an estimate of precipitation as snow [44]. A Pearson's pairwise correlation analysis was conducted on the data matrix for each month to identify any collinearities ($|r| \geq 0.7$). The resulting dataset used for analysis was paired down to six key variables: basin area, mean basin elevation, % lakes, % glaciers, mean monthly precipitation, and mean monthly temperature. The physiographic variables (i.e., elevation, lakes, and glaciers) and monthly discharge values were normalized with log transformations [49].

The best-fit multiple regression equations for each month were first run on the historical gauge station database. A yearly hydrograph was plotted for a cross-section of rain-fed, snow fed, and glacial systems for observed, predicted, and future projections for visual interpretation. The model accuracy was evaluated by comparing the observed monthly means with the predicted monthly means in a percent error matrix. The best-fit monthly regression equations were then run on the regional database to calculate monthly means for present and future conditions across all watersheds.

A hydroclimatic sensitivity index was calculated for all watersheds by averaging the percent change in predicted discharge (A1B emission scenario) from September to March when salmon eggs are in the gravel, and broken into a relative rank index by standard deviations from the mean. A salmon habitat and species diversity index was calculated using the 2012 Alaska Department of Fish and Game (ADF&G) Anadromous Waters Catalog (AWC). The AWC mapped presence of salmon species by stream reach was converted to the kilometers of salmon stream in each watershed, weighted by the number of species (≤6 including Pacific salmon and steelhead, *O. mykiss*), and scaled by the percent of total salmon streams in the region. The final priority matrix was the hydroclimatic sensitivity index and the salmon habitat and species diversity index split into four simple risk-value categories with median value cut-points.

Results

The historical gauge station database (n = 41) showed a relatively well distributed spatial pattern with station locations spanning the latitudinal gradient of southeast Alaska from Ketchikan (55° latitude) in the south to Yakutat (59° latitude) in the north (Figure 1, Table S1). The gauge stations catchments were also distributed among islands (n = 18), mainland (n = 23), transboundary/interior (n = 3), and glacial systems (n = 13).

The AIC model selection process showed a characteristic seasonal pattern in variable selection. The simplest three variable models were in April and October with basin area, temperature and precipitation as the most parsimonious, best-fit model (Table 2). Elevation was selected as an additional variable for the early summer months of May, June, July, and again in the fall for the month of November (Table 2). Glaciers were added as a variable in August and September, and then again in December through March (Table 2). The most complex 6-variable model with the addition of lakes was selected in late winter months, February and March (Table 2). The regression coefficients matched seasonal patterns in discharge. Precipitation had a positive association in all models (Table 3). Temperature had a positive association in all months except for mid-summer, June, July and August (Table 3). Elevation had a positive association in summer months, May to September, and a negative association in winter months, November to March (Table 3). Glaciers had a positive association in summer months, August to September, and negative association in winter and spring months, November to March (Table 3). Lakes had a positive coefficient in all models in which it was included (Table 3). All models had high correlation values (Adjusted $R^2 = 0.964–0.980$) with the best models in summer months (Table 3).

Climate model projection trends for gauge station catchments in year 2080 show an increase in temperature, precipitation, and discharge, with the exception of July and August discharge decreases (Figure 2, Table S2, Table S3, Table S4, and Table S5). The mean monthly temperature for gauge station catchments is projected to increase from 3.4°C to an emission scenario range of 5.7–7.1°C, with the largest increases in late fall through spring (Figure 2). The mean monthly precipitation increased from 259.7

Table 2. Differences in AIC scores (ΔAIC), weights (AIC*w*), and number of model parameters (*k*) used to develop monthly multiple regression-based discharge models for southeast, Alaska, USA.

Models		Month					
		January		February		March	
Predictors	*k*	Δ AIC	AIC*w*	Δ AIC	AIC*w*	Δ AIC	AIC*w*
Area	1	69	0.000	79	0.000	63	0.000
Area + Precip	2	55	0.000	71	0.000	53	0.000
Area + Precip + Temp	3	6	0.024	13	0.001	9	0.006
Area + Precip + Temp + Elev	4	2	0.214	8	0.010	6	0.021
Area + Precip + Temp + Elev + Lakes	5	3	0.106	2	0.219	3	0.136
Area + Precip + Temp + Elev + Glac	6	0	0.477	3	0.159	1	0.338
Area + Precip + Temp + Elev + Glac + Lakes	7	2	0.180	0	0.611	0	0.499
		April		May		June	
Predictors	*k*	Δ AIC	AIC*w*	Δ AIC	AIC*w*	Δ AIC	AIC*w*
Area	1	62	0.000	39	0.000	56	0.000
Area + Precip	2	44	0.000	23	0.000	45	0.000
Area + Precip + Temp	3	0	0.529	24	0.000	36	0.000
Area + Precip + Temp + Elev	4	2	0.207	0	0.484	0	0.501
Area + Precip + Temp + Elev + Lakes	5	4	0.076	2	0.184	2	0.229
Area + Precip + Temp + Elev + Glac	6	3	0.136	1	0.243	2	0.185
Area + Precip + Temp + Elev + Glac + Lakes	7	5	0.052	3	0.089	4	0.086
		July		August		September	
Predictors	*k*	Δ AIC	AIC*w*	Δ AIC	AIC*w*	Δ AIC	AIC*w*
Area	1	60	0.000	53	0.000	36	0.000
Area + Precip	2	43	0.000	29	0.000	8	0.010
Area + Precip + Temp	3	27	0.000	13	0.001	10	0.004
Area + Precip + Temp + Elev	4	0	0.483	0	0.372	3	0.105
Area + Precip + Temp + Elev + Lakes	5	2	0.220	7	0.013	5	0.040
Area + Precip + Temp + Elev + Glac	6	2	0.193	0	0.343	0	0.467
Area + Precip + Temp + Elev + Glac + Lakes	7	3	0.104	1	0.272	0	0.375
		October		November		December	
Predictors	*k*	Δ AIC	AIC*w*	Δ AIC	AIC*w*	Δ AIC	AIC*w*
Area	1	47	0.000	60	0.000	60	0.000
Area + Precip	2	22	0.000	44	0.000	45	0.000
Area + Precip + Temp	3	0	0.406	0	0.325	8	0.009
Area + Precip + Temp + Elev	4	1	0.309	0	0.407	3	0.096
Area + Precip + Temp + Elev + Lakes	5	2	0.124	0	0.323	2	0.153
Area + Precip + Temp + Elev + Glac	6	3	0.115	0	0.330	0	0.431
Area + Precip + Temp + Elev + Glac + Lakes	7	4	0.046	2	0.186	1	0.311

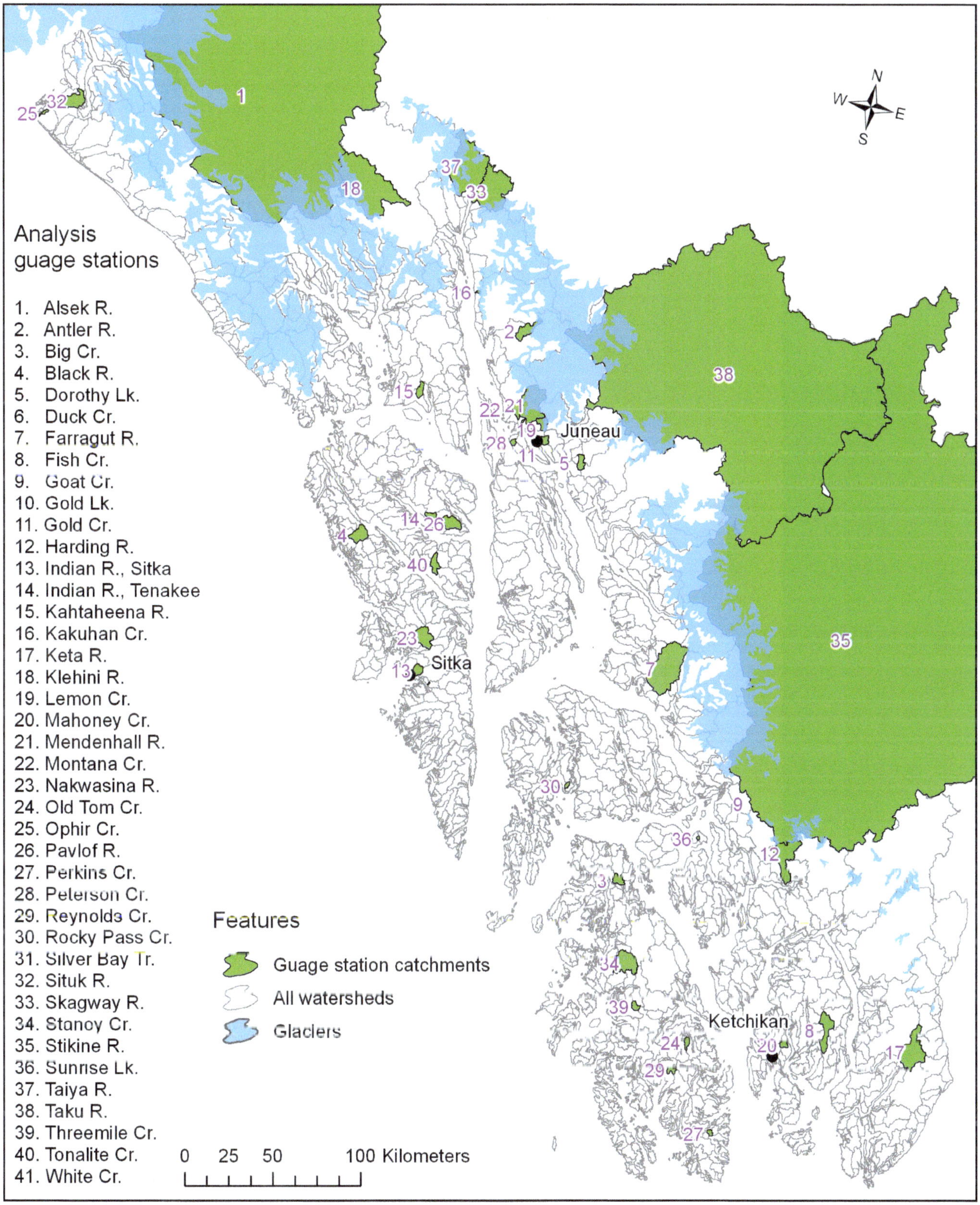

Figure 1. The watersheds of southeast, Alaska, USA, with the 41 gauge station catchments used for the development of regional multiple regression-based monthly discharge models.

Table 3. Coefficients (β), standard error (SE), and adjusted R^2 values for the most parsimonious multiple regression-based monthly discharge models for southeast, Alaska, USA.

	Constant	Area	Precip	Temp	Elevation	Glaciers	Lakes	Adj. R^2
January	−5.410**	0.986**	0.002**	0.059**	−0.167**	−0.911*		0.967
	(0.269)	(0.032)	(0.0002)	(0.009)	(0.077)	(0.508)		
February	−6.327**	1.097**	0.002**	0.093**	−0.192**	−0.981*	2.765*	0.973
	(0.280)	(0.033)	(0.0003)	(0.011)	(0.073)	(0.522)	(1.361)	
March	−6.192**	1.036**	0.002**	0.094**	−0.154*	−1.146*	2.318	0.964
	(0.336)	(0.037)	(0.0004)	(0.015)	(0.079)	(0.571)	(1.501)	
April	−7.090**	1.054**	0.002**	0.134**				0.973
	(0.298)	(0.032)	(0.0003)	(0.015)				
May	−7.339**	0.979**	0.003**	0.068**	0.409**			0.980
	(0.432)	(0.030)	(0.0004)	(0.020)	(0.072)			
June	−6.923	0.946**	0.003**	−0.011	0.609**			0.980
	(0.576)	(0.031)	(0.0007)	(0.026)	(0.083)			
July	−6.210**	0.965**	0.002**	−0.068**	0.535**			0.978
	(0.668)	(0.033)	0.0007	(0.029)	(0.088)			
August	−6.299**	0.989**	0.002**	−0.039	0.322**	0.780		0.979
	(0.595)	(0.033)	(0.0005)	(0.025)	(0.081)	(0.615)		
September	−6.806**	0.991**	0.002**	0.043**	0.219**	1.129**		0.979
	(0.430)	(0.030)	0.0002	(0.018)	(0.073)	(0.530)		
October	−6.430**	1.029**	0.001**	0.070**				0.975
	(0.311)	(0.032)	(0.0001)	(0.013)				
November	−5.935**	1.031**	0.001**	0.068**	−0.106			0.972
	(0.291)	(0.032)	(0.0002)	(0.010)	(0.071)			
December	−5.492**	0.989**	0.001**	0.056**	−0.158**	−1.025**		0.967
	(0.295)	(0.033)	(0.0002)	(0.010)	(0.075)	(0.480)		

Standard errors are reported in parentheses.
*$p<0.10$,
**$p<0.05$.

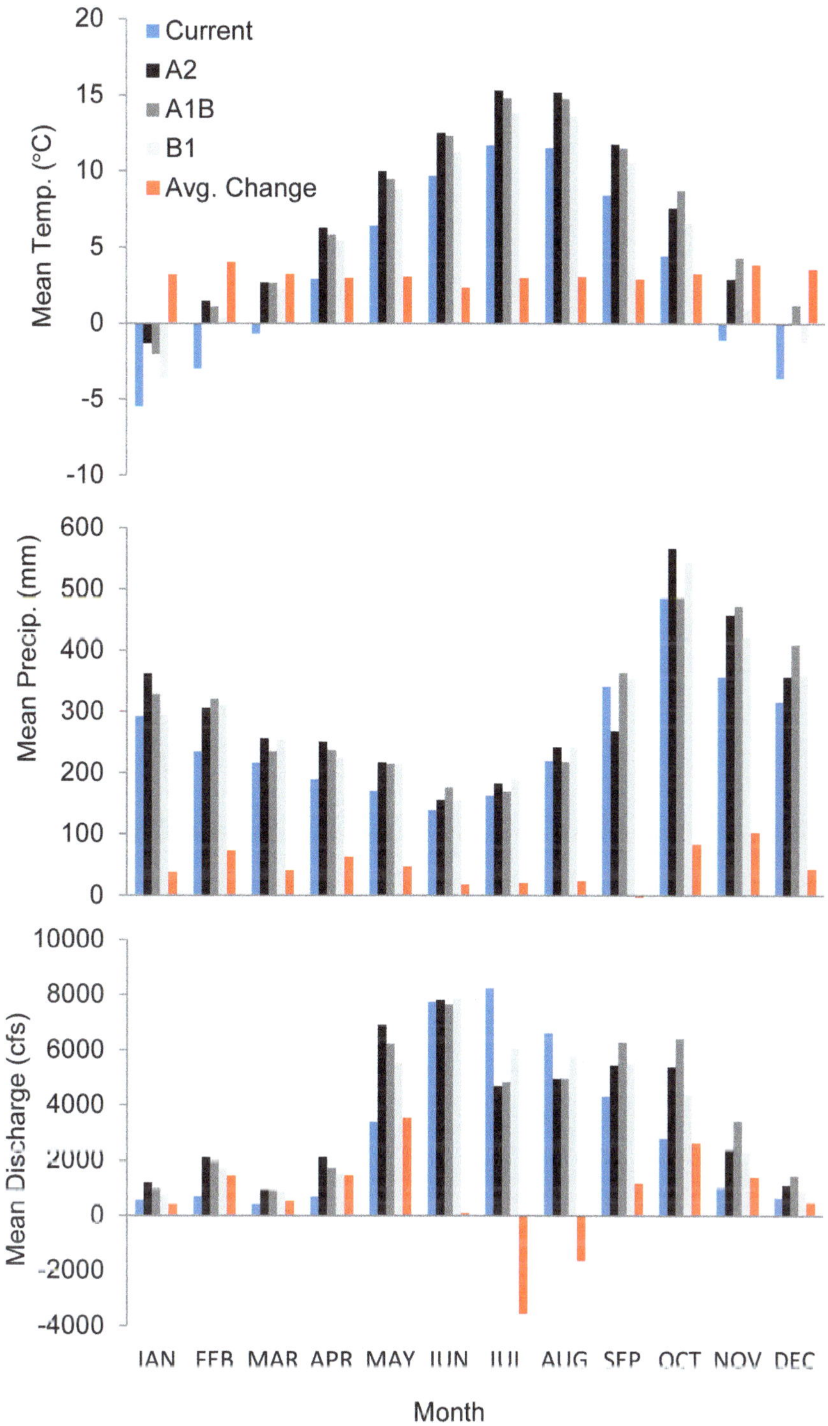

Figure 2. The historical mean (1961–1990) temperature (°C), precipitation (mm), and modeled discharge (cubic feet per second) for the 41 analysis gauge station catchments in southeast, Alaska, USA. These are shown next to projected changes for the year 2080 using an ensemble global climate model average (ECHAM5, HadCM3, and CGCM3.1) and three global greenhouse gas emission scenarios (B1, A1B, and A2) using multiple regression-based monthly discharge models.

to 297.0–302.4 mm (14.4–16.4% change), with the largest increases in late fall (Figure 2). The mean monthly modeled discharge increased from 3088.3 to 3599.6–3905.0 cfs (16.6–26.4% change), with the greatest increases in early fall and spring (Figure 2).

The yearly hydrograph plots for a cross-section of rain-fed, snow-fed, and glacial systems showed characteristic seasonal discharge patterns (Figure 3). Threemile Creek near Klawock, on southern Prince of Wales Island, is a small rain-fed system that generally followed seasonal precipitation patterns with peak flows in the fall with little water storage (Figure 3). The projected

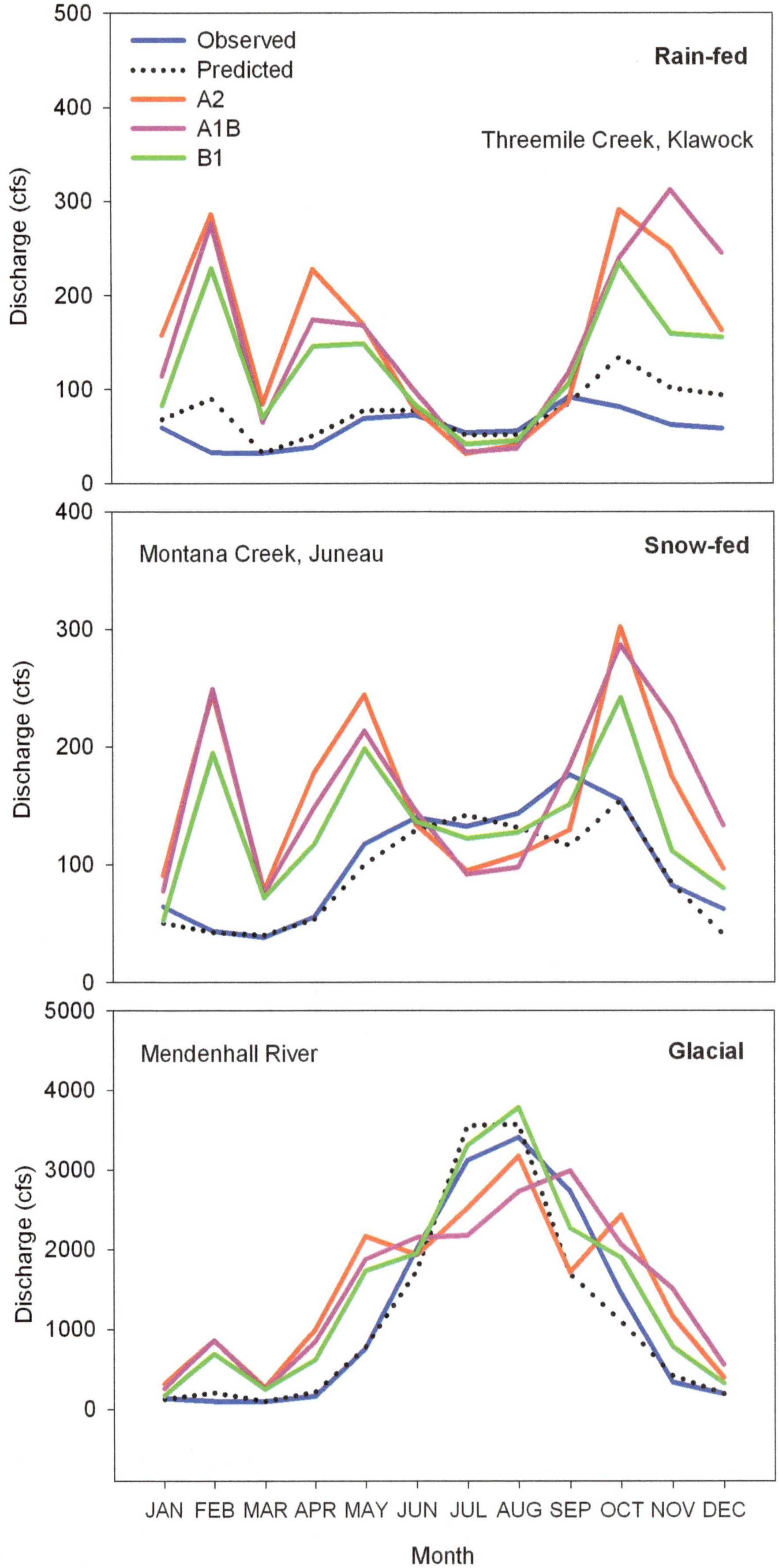
Observed
Predicted
A2
A1B
B1
Rain-fed
Threemile Creek, Klawock
Discharge (cfs)
500
400
300
200
100
0
400
Montana Creek, Juneau
Snow-fed
Discharge (cfs)
300
200
100
0
5000
Mendenhall River
Glacial
Discharge (cfs)
4000
3000
2000
1000
0
JAN FEB MAR APR MAY JUN JUL AUG SEP OCT NOV DEC
Month

Figure 3. Yearly hydrographs of monthly means for a cross section of rain-fed, snow-fed, and glacial watersheds in southeast, Alaska, USA. Recorded observations, modeled predictions, and future projections for the year 2080 are plotted for comparison using multiple regression-based monthly discharge models and an ensemble global climate model average (ECHAM5, HadCM3, and CGCM3.1) run for three global greenhouse gas emission scenarios (B1, A1B, and A2).

hydrograph for Threemile Creek showed increased peak flows in the spring and fall, and similar or lower discharge in summer months (Figure 3). Montana Creek near Juneau is a medium-sized snow-fed system with discharge patterns that showed a spring melt and fall peak (Figure 3). The projected hydrograph for Montana Creek showed increased spring and fall peak flows, and lower discharge in summer months (Figure 3). The Mendenhall River near Juneau is a large glacial system with characteristic peak flows in mid-summer with a bell-shaped yearly hydrograph (Figure 3). The projected hydrograph for the Mendenhall River showed generally increased discharge year-round with a more spread out summer peak (Figure 3).

The historical gauge station model evaluation showed that monthly discharge was on average over predicted (Table S6). The mean monthly discharge absolute error for all models was 36.4%. The most accurate models were in summer months with a mean absolute error of 28.5%. The mean absolute error during the analysis period of September to March was 41.9%.

The regional hydroclimatic sensitivity index showed northern mainland and steeper, snow-fed mountainous watersheds with current winter temperatures close to freezing exhibiting the greatest change (Figure 4). The highest ranking group (>1.5 SD; 184–280%) were distributed along the northern mainland, including some glacial systems, and others were scattered throughout the region. Next, there was a group of watersheds with high spatial variability and moderate ranking (0.50–1.5 SD; 156–184%) with the highest concentration in the southeast corner of the region. Watersheds with average percent change (−0.50–0.50; 128–156%) occurred along the mainland, and included many glacial systems and mountainous areas of the major island groups. Fair to low ranking sensitivity (−0.50–<−1.5 SD; 93–128%) watersheds were located throughout the region in low elevation areas, and included several of the larger transboundary watersheds. The combined salmon habitat and species diversity hydroclimatic sensitivity index showed highest priority in systems that exhibited a combination of high salmon habitat and species diversity with changing hydrology (Figure 5).

Discussion

The most striking result of the analysis was the transition in mean winter temperatures across the freezing threshold for many watersheds, and how it translated into changes in regional discharge patterns that are important for the reproduction and survival of Pacific Salmon in southeast Alaska. Combined with predicted increases in precipitation across all months, mean monthly discharge was forecasted to increase by approximately 1-fold to 3-fold during September to March when salmon eggs are in the gravel and exposed to more frequent scouring events. Plots of yearly hydrographs showed substantially increased peak flows in rain-fed and snow-fed systems in early spring and late fall across emission scenarios. The hydroclimatic sensitivity index showed the interaction between projected climate trends and the region's diverse physiography—a relatively predictable pattern that could be used as a preliminary framework to develop targeted long-term monitoring and potential mitigation strategies. The combined salmon habitat and species diversity sensitivity index showed clusters of high-value watersheds that could be prioritized for: (1) conservation of available genetic and life history diversity (e.g., run timing; [30]); (2) evaluation of local, reach-scale geomorphic sediment mobility and susceptibility to scour; (3) restoration to improve natural variability and general ecological resilience [15,50]; (4) monitoring harvest pressures and escapement goals in light of new environmental factors, and; (5) developing finer-scale climate impact and salmonid adaptation studies (e.g., [19,20,27,34,38,51]).

The ability of the discharge models to accurately reflect seasonal flow patterns among watersheds and watershed types is a positive indicator for the models' ability to predict regional trends in ungauged watersheds across the landscape. An evaluation of model selection, coefficient direction, and yearly hydrographs agreed with the current general understanding of regional hydrologic patterns [13,24,32,52]. The models general over-prediction of discharge is common among discharge models [48]. The percent error scores (both positive and negative) are likely attributed to a combination of sources: resolution of the climate data; older and more remote faulty gauge station records, especially during winter months when stations are more prone to freezing issues, and; water storage ecological processes (e.g., wetlands) simply not captured in the generalized linear regional model. This regional modeling effort is an important step forward in the development of future predictive models for southeast Alaska. Uncertainty associated with the climate modeling and potential effects could be further reduced with higher spatial and temporal resolution datasets that can capture ecological processes such as geomorphologic change and generate flood frequency statistics. Future forecasting studies would also benefit from a broader stream gauge network that represents the spectrum of potentially vulnerable to resilient salmon producing watersheds, with an emphasis on monitoring extreme events such as the frequency of flooding, coupled with experimental studies on egg-to-fry survival and salmonid adaptive capacity.

Nested within the hydroclimatic sensitivity framework, river channel types will change with flow rates and different species of salmon will likely be affected differently based on their body size, freshwater life-history and residence time, which warrants further research. The body size of salmon species directly relates to the burial depth of eggs and therefore exposure to scouring events in mobile spawning substrates [53,54]. The spawning location in watersheds directly relates to the rate of flow and sedimentation exposure [19]. The highest flow and sedimentation rates are generally expected in the central parts of the mainstem and significant tributaries, as opposed to the headwaters or floodplains [19]. Steelhead (*O. mykiss*) spawn in the spring within the highest reaches of the watershed [36] and could, for example, be less prone to winter souring events. Coho salmon (*O. kisutch*), sockeye salmon (*O. nerka*), and Chinook salmon (*O. tshawytscha*) spawn in the mid-portions of the watersheds, have longer residence times [36], and may therefore be the most exposed species to flooding events; however, their relative body sizes will help egg burial depths and exposure to scour [53]. Pink salmon (*O. gorbuscha*) and chum salmon (*O. keta*) spawn and rear in the lowest reaches of watersheds and migrate a few weeks after they emerge from the gravel (April–May) [36], and may therefore be less vulnerable to scouring where floodplains are intact. However, potential sea level

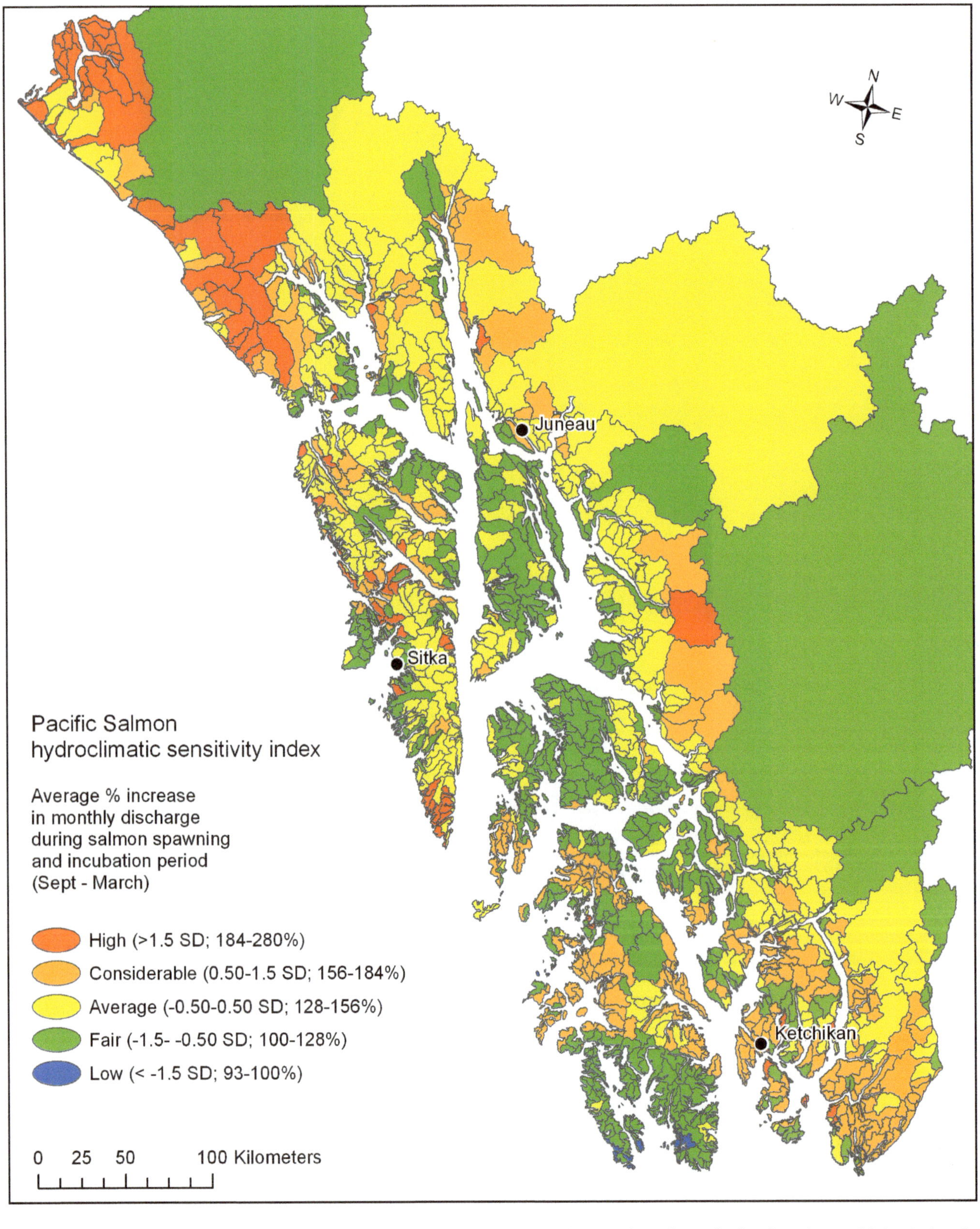

Figure 4. The watersheds of southeast, Alaska, USA (n = 1784) showing the Pacific Salmon hydroclimatic sensitivity index for predicted hydrologic change. This was derived using multiple regression-based monthly discharge models, an ensemble global climate model average (ECHAM5, HadCM3, and CGCM3.1) for temperature and precipitation, and the A1B global greenhouse gas emission scenario projections for the year 2080 to illustrate regional trends.

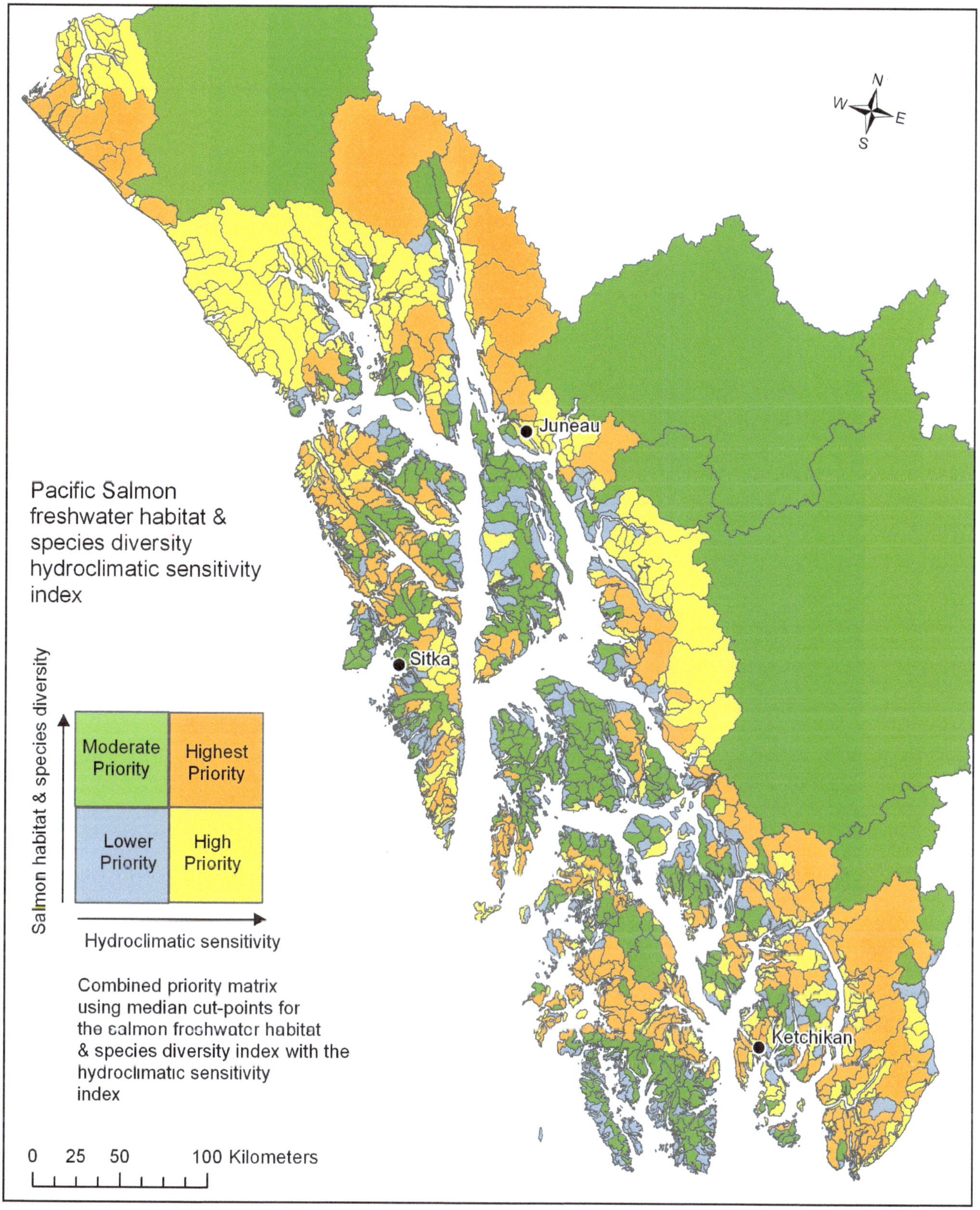

Figure 5. The watersheds of southeast, Alaska, USA (n = 1784) showing the Pacific Salmon freshwater habitat and species diversity index combined with the hydroclimatic sensitivity index developed in this study.

rise is a factor for low-elevation pink salmon and chum salmon spawning habitat in southern southeast Alaska [17,55].

Hydrologic projections for Washington State predict a complete loss of snow-fed hydrologic systems by the year 2080 under the same A1B emission scenario, with only a few watersheds in the North Cascade mountains retaining transitional rain/snow characteristics [18,56]. The snow-fed watersheds are predicted to transition into rain-fed hydrologic patterns with low flows in the summer and more intense flooding events during the winter months [18]. Increases in temperature appear to have a stronger influence on flows than changes in precipitation in Washington State [19]; this also appears to be the case with the largest changes in projected flows for southeast Alaska occurring during months when many watersheds cross the freezing threshold. Thus far, simulation studies suggest low-elevation floodplain and wetland connectivity restoration efforts to slow increased flow rates and improve summer water storage are the most effective climate change mitigation strategies in population viability models for Washington State [15]. These ecosystem engineering techniques could be further tested in historically impacted watersheds in southeast Alaska [57].

The results of this analysis should be treated as a hypothesis of potential change, and a framework for finer-scale experimental studies that investigate the long-term effects of changing hydrologic regimes, inter-annual variability, extreme events, and salmonid adaptive capacity in southeast Alaska. As global climate models, hydrologic forecasting and downscaling techniques improve, the absolute values for projected temperature, precipitation, and stream discharge will change and model predictions will improve. The results of this study provide a compelling case for how climate change trends could interact with the regions diverse physiography in a relatively predictable pattern where research and mitigation experiments should be prioritized for effective salmon conservation investments.

Supporting Information

Table S1 Recorded mean monthly discharge (cubic feet per second) for 41 southeast, Alaska, USA gauge stations with ≥5 year period of record.

Table S2 Multiple regression-based monthly discharge model predictions for 41 southeast, Alaska, USA gauge station catchments.

Table S3 Multiple regression-based monthly discharge models for projected future discharge (cubic feet per second) for 41 gauged catchments in southeast, Alaska, USA using an ensemble global climate model average (ECHAM5, HadCM3, and CGCM3.1) and the B1 global greenhouse gas emission scenario for the year 2080.

Table S4 Multiple regression-based monthly discharge models for projected future discharge (cubic feet per second) for 41 gauged catchments in southeast, Alaska, USA using an ensemble global climate model average (ECHAM5, HadCM3, and CGCM3.1) and the A1B global greenhouse gas emission scenario for the year 2080.

Table S5 Multiple regression-based monthly discharge models for projected future discharge (cubic feet per second) for 41 gauged catchments in southeast, Alaska, USA using an ensemble global climate model average (ECHAM5, HadCM3, CGCM3.1) and the A2 global greenhouse gas emission scenario for the year 2080.

Table S6 Multiple regression-based monthly discharge model percent error ((predicted-observed)/observed) for 41 southeast, Alaska, USA gauged catchments.

Acknowledgments

C.L. Woll, M.V. McPhee, and two anonymous reviewers made helpful comments on the developing manuscript.

Author Contributions

Conceived and designed the experiments: CS DA. Performed the experiments: CS. Analyzed the data: CS. Contributed reagents/materials/analysis tools: CS. Wrote the paper: CS DA.

References

1. Bottom DL, Jones KK, Simenstad CA, Smith CL (2009) Reconnecting social and ecological resilience in salmon ecosystems. Ecology and Society 14: 5.
2. Gende SM, Edwards RT, Willson MF, Wipfli MS (2002) Pacific salmon in aquatic and terrestrial ecosystems. Bioscience 52: 917–928.
3. Clarke JH, McGregor A, Mecum RD, Krasnowski P, Carroll A (2006) The commercial salmon fishery in Alaska. Alaska Fishery Research Bulletin 12: 1–146.
4. Gustafson RG, Waples RS, Myers JM, Weitkamp LA, Bryant GJ, et al. (2006) Pacific salmon extinctions: quantifying lost and remaining diversity. Conservation Biology 21: 1009–1020.
5. Pinsky ML, Springmeyer DB, Goslin MN, Augerot X (2009) Range-wide selection of catchments for Pacific Salmon conservation. Conservation Biology 23: 680–691.
6. USFS (2012) Salmon in the Tongass National Forest. R10-PR-028. Juneau, Alaska: USDA Forest Service.
7. Araki H, Cooper B, Blouin MS (2007) Genetic effects of captive breeding cause a rapid, cumulative fitness decline in the wild. Science 318: 100–103.
8. Buhle ER, Holsman KK, Scheuerell MD, Albaugh A (2009) Using an unplanned experiment to evaluate the effects of hatcheries and environmental variation on threatened populations of wild salmon. Biological Conservation 142: 2449–2455.
9. Ruggerone GT, Peterman RM, Dorner B, Myers KW (2010) Magnitude and trends in abundance of hatchery and wild Pink salmon, Chum salmon, and Sockeye salmon in the North Pacific Ocean. Marine and Coastal Fisheries: Dynamics, Management, and Ecosystem Science 2: 306–328.
10. Schindler DE, Hilborn R, Chasco B, Boatright CP, Quinn TP, et al. (2010) Population diversity and the portfolio effect in an exploited species. Nature 465: 609–612.
11. Mantua NJ, Hare SR, Zhang Y, Wallace JM, Francis RC (1997) A Pacific interdecadal climate oscillation with impacts on salmon production. Bulletin of the American Meteorological Society 78: 1069–1079.
12. Koski KV (2009) The fate of coho salmon nomads: The story of an estuarine-rearing strategy promoting resilience. Ecology and Society 14: 4. http://www.ecologyandsociety.org/vol14/iss11/art14/.
13. Neal EG, Walter MT, Coffeen C (2002) Linking the pacific decadal oscillation to seasonal stream discharge patterns in southeast Alaska. Journal of Hydrology 263: 188–197.
14. Adkison MD, Finney BP (2003) The long-term outlook for salmon returns to Alaska. Alaska Fishery Research Bulletin 10: 83–94.
15. Battin J, Wiley MW, Ruckelshaus MH, Palmer RN, Korb E, et al. (2007) Projected impacts of climate change on salmon habitat restoration. PNAS 104: 6720–6725.
16. Beechie T, Buhle E, Ruckelshaus M, Fullerton A, Holsinger L (2006) Hydrologic regime and the conservation of salmon life history diversity. Biological Conservation 130: 560–572.
17. Bryant MD (2009) Global climate change and potential effects on Pacific salmonids in freshwater ecosystems of southeast Alaska. Climatic Change 95: 169–193.
18. Elsner MM, Cuo L, Voisin N, Deems JS, Hamlet AF, et al. (2010) Implications of 21st century climate change for the hydrology of Washington State. Climatic Change: DOI: 10.1007/s10584-10010-19855-10580.

19. Mantua N, Tohver I, Hamlet A (2010) Climate change impacts on streamflow extremes and summertime stream temperature and their possible consequences for freshwater salmon habitat in Washington State. Climatic Change: DOI 10.1007/s10584-10010-19845-10582.
20. Taylor SG (2008) Climate warming causes phenological shift in Pink Salmon, Oncorhynchus gorbuscha, behavior at Auke Creek, Alaska. Global Change Biology 14: 229–235.
21. Meehl GA, Tebaldi C, Nychka D (2004) Changes in frost days in simulations of twentyfirst century climate. Climate Dynamics 23: 495–511.
22. Walsh JE, Chapman WL, Romanovsky V, Christensen JH, Stendel M (2008) Global climate model performance over Alaska and Greenland. Journal of Climate 21: 6156–6174.
23. Hennon PE, D'Amore DV, Schaberg PG, Wittwer DT, Shanley CS (2012) Shifting climate, altered niche, and a dynamic conservation strategy for Yellow-cedar in the North Pacific coastal rainforest. Bioscience 62: 147–158.
24. Hood E, Berner L (2009) Effects of changing glacial coverage on the physical and biogeochemical properties of coastal streams in southeastern Alaska. Journal of Geophysical Research 114: G03001, doi:03010.01029/02009JG000971.
25. Hodgson S, Quinn TP (2002) The timing of adult sockeye salmon migration into freshwater: adaptations by populations to prevailing thermal regimes. Canadian Journal of Zoology 80: 542–555.
26. Mote PW, Parson EA, Hamlet AF, Keeton WS, Lettenmaier D, et al. (2003) Preparing for climatic change: The water, salmon, and forests of the Pacific Northwest. Climatic Change 61: 45–88.
27. Milner AM, Robertson AL, McDermott MJ, Klarr MJ, Brown LE (2013) Major flood disturbance alters river ecosystem evolution. Nature Climate Change 3: 137–141.
28. Simpson JJ, Hufford GL, Daly C, Berg JS, Fleming MD (2005) Comparing maps of mean monthly surface temperature and precipitation for Alaska and adjacent areas of Canada produced by two different methods. Arctic 58: 137–161.
29. Edwards RT, D'Amore DV, Norberg E, Biles F (2013) Riparian ecology, climate change, and management in North Pacific Coastal Rainforests. In: G. H. Orians and J. Schoen, editors. North Pacific temperate rainforests: ecology and conservation. Seattle, WA: University of Washington Press. pp. 43–72.
30. Halupka KC, Willson M, Bryant MD, Everest FH, Gharrett AJ (2003) Conserving of population diversity of Pacific salmon in southeast Alaska. North American Journal of Fisheries Management 23: 1057–1086.
31. Stewart IT, Cayan DR, Dettinger MD (2004) Changes in snowmelt runoff timing in western North America under a 'business as usual' climate change scenario. Climatic Change 62: 217–232.
32. Neal EG, Hood E, Smikrud K (2010) Contribution of glacier runoff to freshwater discharge into the Gulf of Alaska. Geophysical Research Letters 37: L06404, doi:06410.01029/02010GL042385.
33. Nolin AW, Phillippe J, Jefferson A, Lewis SL (2010) Present-day and future contributions of glacier runoff to summertime flows in a Pacific Northwest watershed: Implications for water resources. Water Resources Research 46: W12509 DOI:12510.11029/12009WR008968.
34. Fellman JB, Nagorski S, Pyare S, Vermilyea AW, Scott D, et al. (2013) Stream temperature response to variable glacier cover in coastal watersheds of southeast Alaska. Hydrologic Processes doi: 10.1002/hyp.9742
35. Milner AM, Brown LE, Hannah DM (2009) Hydroecological response of river systems to shrinking glaciers. Hydrologic Process 23: 62–77.
36. Quinn TP (2005) The behavior and ecology of Pacific Salmon and trout. Seattle, WA.: University of Washington Press. 378 p.
37. Wipfli MS, Baxter CV (2010) Linking ecosystems, food webs, and fish production: subsidies in salmonid watersheds. Fisheries 35: 373–387.
38. Kovach RP, Joyce JE, Echave JD, Lindberg MS, Tallmon DA (2013) Earlier migration timing and decreasing phenotypic variation for multiple salmonid species. PLOS ONE 8: e53807.
39. Kovach RP, Gharrett AJ, Tallmon DA (2012) Genetic change for earlier migration timing and loss of biocomplexity in a pink salmon population. Proceeding of the Royal Society Biology 279: 3861–3869.
40. Schwarz TC (2010) Analysis of select stream discharge models in southeast Alaska. Masters Thesis, Cornell University, Ithaca, New York.
41. Wang T, Hamann A, Spittlehouse DL, Murdock TQ (2012) ClimateWNA–High-resolution spatial climate data for western North America. Journal of Applied Meteorology 51: 16–29.
42. Daly C, Gibson WP, Taylor GH, Johnson GL, Pasteris P (2002) A knowledge-based approach to the statistical mapping of climate. Climate Research 22: 99–113.
43. Hamann A, Wang GA (2005) Models of climatic normals for genecology and climate change studies in British Columbia. Agriculture and Forest Meteorology 128: 211–221.
44. Wang T, Hamann A, Spittlehouse D, Aitken SN (2006) Development of scale-free climate data for western Canada for use in resource management. International Journal of Climatology 26: 383–397.
45. Radic V, Clarke GKC (2011) Evaluation of IPCC models' performance in simulating late-twentieth-century climatologies and weather patterns over North America. Journal of Climate 24: 5257–5274.
46. Fussel H (2009) An updated assessment of the risks from climate change based on research published since the IPCC fourth assessment report. Climatic Change 97: 469–482.
47. Anderson DR (2008) Model based inference in the life sciences: a primer on evidence. New York, New York: Springer. 184 p.
48. Wiley JB, Curran JH (2003) Estimating annual high-flow statistics and monthly and seasonal low-flow statistics for ungauged sites on streams in Alaska and conterminous basins in Canada. Water-Resources Investigations Report 03-4114. Anchorage, Alaska: U.S. Geological Survey.
49. McGarigal K, Cushman S, Stafford S (2000) Multivariate statistics for wildlife ecology and research. New York: Springer.
50. Bisson PA, Dunham JB, Reeves GH (2009) Freshwater ecosystems and resilience of Pacific salmon: Habitat management based on natural variability. Ecology and Society 14: 45. http://www.ecologyandsociety.org/vol14/iss41/art45/.
51. Dorava JM, Milner AM (2000) Role of lake regulation on glacier-fed rivers in enhancing salmon productivity: the Cook Inlet watershed, south-central Alaska, USA. Hydrologic Processes 14: 3149–3159.
52. Schoch GM, Albert DM, Shanley CS (2013) An estuarine habitat classification for a complex fjordal island archipelago. Estuaries and Coasts: doi 10.1007/s12237-12013-19622-12233.
53. Montgomery DR, Beamer EM, Pess GR, Quinn TP (1999) Channel type and salmonid spawning distribution and abundance. Canadian Journal or Fisheries and Aquatic Sciences 56: 377–387.
54. Jonsson B, Finstad AG, Jonnson N (2012) Winter temperature and food quality affect age at maturity: an experimental test with Atlantic salmon (*Salmo salar*). Canadian Journal or Fisheries and Aquatic Sciences 69: 1817–1826.
55. Larsen CF, Motyka RJ, Freymueller JT, Echelmeyer KA, Ivins ER (2005) Rapid viscoelastic uplift in southeast Alaska caused by post-Little Ice Age glacial retreat. Earth and Planetary Science Letters 237: 548–560.
56. Mote PW, Salathe EP (2010) Future climate in the Pacific Northwest. Climatic Change DOI 10.1007/s10584-010-9848-z.
57. Albert DM, Schoen JW (2013) Use of historical logging patterns to identify disproportionately logged ecosystems within temperate rainforests of southeastern Alaska. Conservation Biology 27: 774–784.

25

Global Agricultural Land Resources – A High Resolution Suitability Evaluation and Its Perspectives until 2100 under Climate Change Conditions

Florian Zabel*, Birgitta Putzenlechner, Wolfram Mauser

Department of Geography, Ludwig Maximilians University, Munich, Germany

Abstract

Changing natural conditions determine the land's suitability for agriculture. The growing demand for food, feed, fiber and bioenergy increases pressure on land and causes trade-offs between different uses of land and ecosystem services. Accordingly, an inventory is required on the changing potentially suitable areas for agriculture under changing climate conditions. We applied a fuzzy logic approach to compute global agricultural suitability to grow the 16 most important food and energy crops according to the climatic, soil and topographic conditions at a spatial resolution of 30 arc seconds. We present our results for current climate conditions (1981–2010), considering today's irrigated areas and separately investigate the suitability of densely forested as well as protected areas, in order to investigate their potentials for agriculture. The impact of climate change under SRES A1B conditions, as simulated by the global climate model ECHAM5, on agricultural suitability is shown by comparing the time-period 2071–2100 with 1981–2010. Our results show that climate change will expand suitable cropland by additionally 5.6 million km^2, particularly in the Northern high latitudes (mainly in Canada, China and Russia). Most sensitive regions with decreasing suitability are found in the Global South, mainly in tropical regions, where also the suitability for multiple cropping decreases.

Editor: Juergen P. Kropp, Potsdam Institute for Climate Impact Research, Germany

Funding: This research was carried out within the framework of the GLUES (Global Assessment of Land Use Dynamics, Greenhouse Gas Emissions and Ecosystem Services) Project, which has been supported by the German Ministry of Education and Research (BMBF) program on sustainable land management (FKZ 01LL0901E). (http://modul-a.nachhaltiges-landmanagement.de/en/). The funders had no role in study design, data collection and analysis, decision to publish, or preparation of the manuscript.

Competing Interests: The authors have declared that no competing interests exist.

* Email: f.zabel@lmu.de

Introduction

Natural constraints are limiting the land's suitability for agriculture and cultivation practices. They consist of prevailing local climatic, soil and topographic conditions determining the available energy, water and nutrient supply for agricultural crops. Besides natural conditions, complex interactions of social, economic, political, and cultural aspects determine whether and how land is used for agriculture. Agricultural land has become one of the largest terrestrial biomes on the planet, occupying approx. 40% of the land surface [1]. Thereby, a variety of different land use types and intensities determine heterogeneously distributed patterns, including e.g. the choice of crop varieties, irrigation practices, fertilization, terracing and the level of technological input [2]. Thus, natural constraints are to a limited extent suspended by human actions [3].

The demand for agricultural products is expected to increase by 70–110% by 2050, driven by a projected world population of 9 billion people, increasing meat consumption and a growing use for bio-based materials and biofuel [4–15].

An increase in agricultural production can be accomplished by agricultural intensification and expansion, while considering social and environmental externalities and changing climate conditions [5,16]. Bruinsma [16] concluded that additionally 1.2 million km^2 of converted land are projected to be necessary until 2030 and another 5% up to 2050 with most land expected to be transformed in South America and Sub Saharan Africa, while latest studies project an increase of cropland between 10-25% by 2050 compared to 2005 for different socio-economic and climate scenarios [17]. Nonetheless, the expansion of agricultural land into forested or protected areas must be viewed critically, in order to conserve valuable ecosystem services e.g. for regulating climate or conserving biodiversity [5–8].

Changing patterns of temperature and precipitation and man-made degradation affect the suitability of land for agricultural use. For example, 19-23 ha of suitable land are lost per minute due to soil erosion and desertification [18,19]. Additionally, the area of suitable land is decreasing due to urbanization, with an estimate of 1.5 million km^2 until 2030 [20,21].

When focusing on the natural potentials of land for agricultural use, suitability analyses give local evidence on todays and future availability and quality. Thus, they help answering questions for managing a transition towards a more environmentally efficient and sustainable land use and involve better information on the global scale impacts of land use decisions [1].

Table 1. List of investigated food, feed and energy crops.

Crop name
Barley (*hordeum vulgare*)
Cassava (*manihot esculenta*)
Groundnut (*arachis hypogaea*)
Maize (*zea mays*)
Millet (*pennisetum americanum*)
Oil palm (*elaeis guineensis*)
Potato (*solanum tuberosum*)
Rapeseed (*brassica napus*)
Paddy rice (*oryza sativa*)
Rye (*secale cereale*)
Sorghum (*sorghum bicolor*)
Soy (*glycine maximum*)
Sugarcane (*saccharum officinarum*)
Sunflower (*helianthus annus*)
Summer wheat (*triticum aestivum*)
Winter wheat (*triticum aestivum*)

The relationship between climate, soil, topography and agricultural suitability has long been recognized. As such, suitability analysis combine heterogeneous soil, terrain and climate information and determine whether specific crop requirements are fulfilled under the given local conditions and assumptions. A variety of regional suitability studies for specific crops exist [22–28], while only a few exist on a global scale and for a broad variety of crops [3,29–31].

In the meantime, global soil and topography data are available at high spatial resolution and global climate models have improved their capabilities and spatial resolution. Previous analysis showed that questions of scale play a major role in suitability analysis as coarse data affect the validity of results [32]. In this context, we present our results in modelling global crop-suitability using a fuzzy logic approach at a spatial resolution of 30 arc seconds. The results of this approach include the potentially suitable area for agriculture differentiated for 16 crops for rainfed and irrigated conditions, the start of the growing cycles and the number of crop cycles. We analyze global distribution of agricultural suitability and changes until 2100 considering the numbers of crop cycles.

Thereby, we identify changes, opportunities and challenges in global agriculture related to the expansion of agricultural land competing with protected and forested areas as ecosystem services.

Material and Methods

Local climate, soil and topography determine the natural suitability of land for agricultural use. Thereby, the climatic, soil and topographic requirements may vary over a wide range of different agricultural crops. This analysis investigates the suitability for the following 16 crops that are most important for the global economy, food security and biofuel issues (see Table 1).

We aggregated the world into 23 regions in order to regionally analyse the results (see Fig. 1). We applied a fuzzy-logic approach [33,34] in order to calculate the crops' suitability on the globe at a spatial resolution of 30 arc seconds (0.00833°, approx. 1 km^2 at the equator). The length of the growing cycle (*lgc*) and the '*membership functions*' that describe the crop-specific requirements for each of the crops during the growing period (Fig. 2) are derived from [35].

The membership functions representing climate constraints describe the degree of membership of each selected crop with regard to mean temperature and total precipitation during its

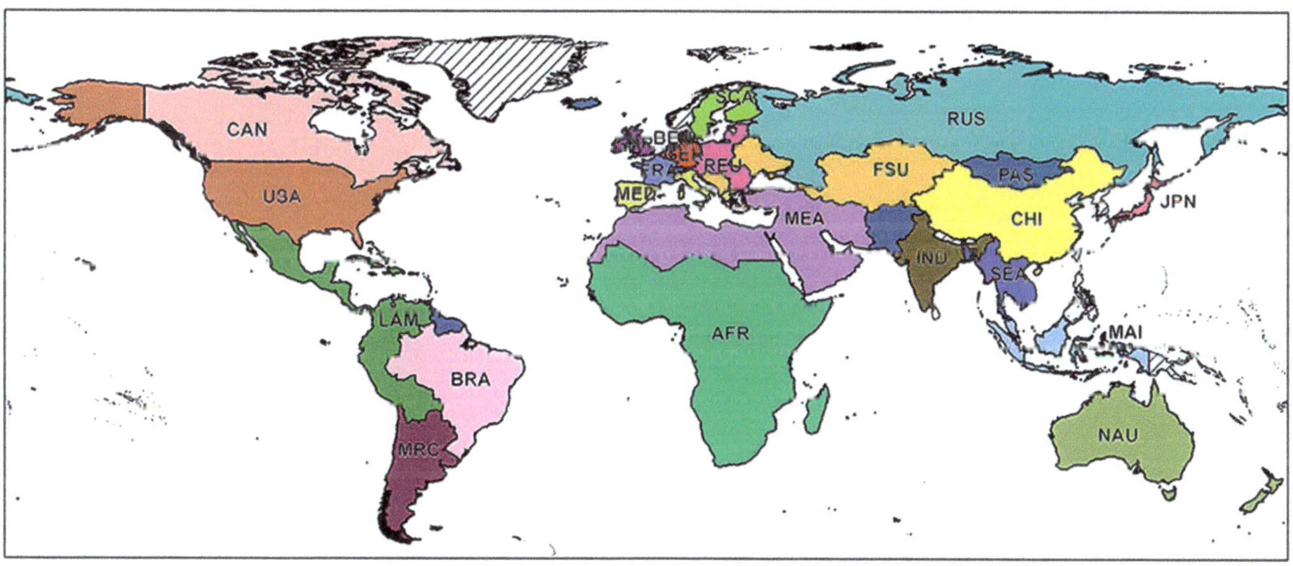

Figure 1. Map of the 23 world regions: AFR (Sub Saharan Africa), ANZ (Australia, New Zealand), BEN (Belgium, Netherlands, Luxemburg), BRA (Brazil), CAN (Canada), CHN (China), FRA (France), FSU (Rest of Former Soviet Union and Rest of Europe), GBR (Great Britain), GER (Germany), IND (India), JPN (Japan), LAM (Rest of Latin America), MAI (Malaysia, Indonesia), MEA (Middle East, North Africa), MED (Italy, Spain, Portugal, Greece, Malta, Cyprus), PAC (Paraguay, Argentina, Chile, Uruguay), ROW (Rest of the World), REU (Austria, Estonia, Latvia, Lithuania, Poland, Hungary, Slovakia, Slovenia, Czech Republic, Romania, Bulgaria), RUS (Russia), SCA (Finland, Denmark, Sweden), SEA (Cambodia, Laos, Thailand, Vietnam, Myanmar, Bangladesh), USA (United States of America).

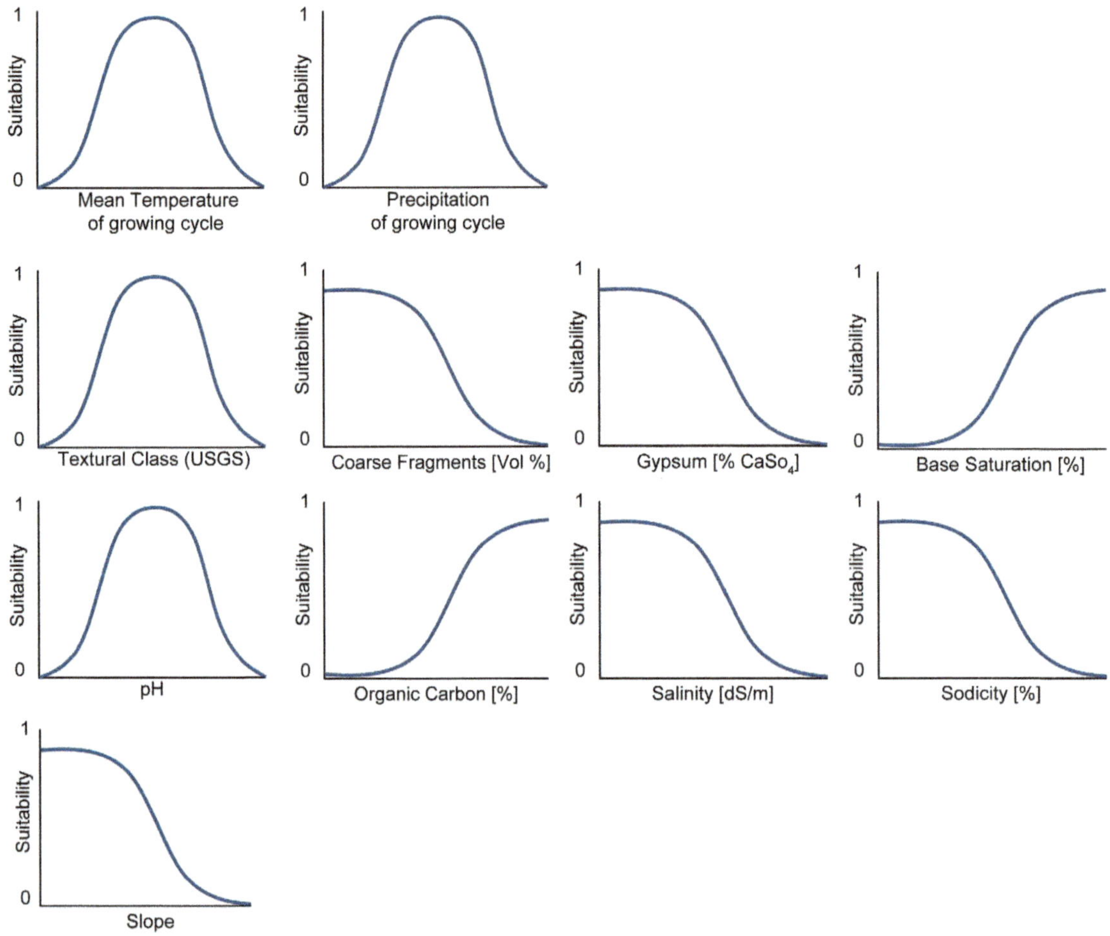

Figure 2. Membership functions for climatic, soil and topographic conditions.

respective growing cycle. Depending on the crop, membership functions have different curves according to [35]. Three shapes are in principle possible: 'more is better', 'less is better' and 'optimum'. For temperature e.g., the suitability is increasing from a minimum towards an optimal temperature and again decreasing until a maximum temperature is reached (Fig. 2). Eight soil parameters are considered: texture, proportion of coarse fragments and gypsum, base saturation, pH content, organic carbon content, salinity, and sodicity. Terrain is considered by the slope. The fuzzy-logic approach calculates *fuzzy values* based on the ecological rules (between 0 and 1), which determine the crops' suitability on a specific location by the lowest membership value of all parameters.

An overview of the applied global datasets is given in Table 2. The climate data applied in this study are outputs from the global circulation model ECHAM5 of the Max-Planck Institute for Meteorology (MPI-M) [36,37]. It uses radiative forcing, sea surface temperature and sea ice concentrations from a 20th century/ SRES A1B scenario simulation. The 6-hourly dataset (temperature, precipitation) are converted to daily values for the climate period of 1981–2010 and 2071–2100. The daily data is spatially downscaled from its original resolution of 0.56° to 0.00833° (30

Table 2. Applied global datasets.

Parameter	Source	Detailed Description
Climate	ECHAM5	[37]
Soil	Harmonized World Soil Database (HWSD)	[41]
Topography	Space Shuttle Topography Mission (SRTM)	[39]
Crop-requirements	FAO Land Evaluation Part III: Crop Requirements	[35]
Irrigation	Global Map of Irrigation Areas (GMIA) v5.0	[44]
Protected Areas	International Union for Conservation of Nature (IUCN) Protected Areas	[45]
Forested Areas	GlobCover 2009	[46]

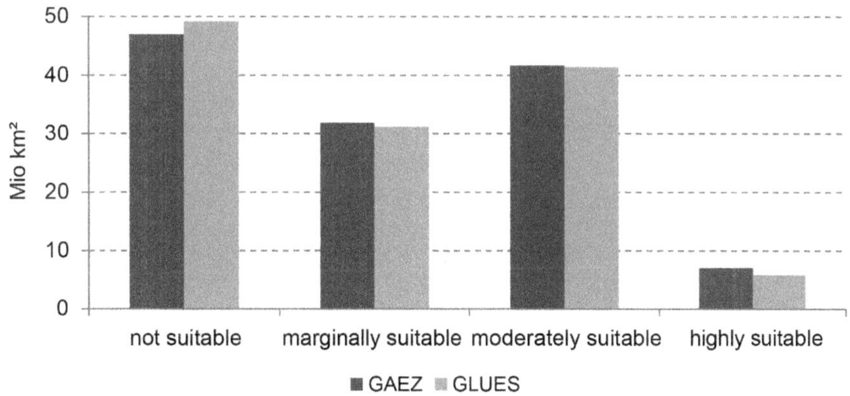

Figure 3. Global comparison of agriculturally suitable area between GAEZ (Baseline period 1961–1990) and GLUES (1961–1990).

arc seconds), based on an approach by [38], using sub-grid terrain information provided by the SRTM-dataset [39]. A bias-correction is executed during the downscaling procedure for temperature and precipitation based on monthly derived factors from the WorldClim dataset [40].

Mean temperature ($\bar{T}$) and total precipitation ($\bar{P}$) are calculated over the length of the growing cycle for each day of the year (*doy*) (see eq. 1 and eq. 2). Starting on the 1st of January (*doy* = 1), the growing cycle is shifted day by day until the 31st of December (*doy* = 365). The suitability value (*S*) is calculated for each *doy* as in eq. 3 for $\bar{T}$ and $\bar{P}$ according to the membership function (*mf*).

$$\bar{T}_{doy} = \overline{T_{doy}, \ldots, T_{doy+lgc}} \qquad \text{(eq.1)}$$

$$\bar{P}_{doy} = \sum_{doy}^{doy+lgc} P \qquad \text{(eq.2)}$$

$$S(\bar{T}_{doy}) = mf(\bar{T}_{doy}) \;;\; S(\bar{P}_{doy}) = mf(\bar{P}_{doy}) \qquad \text{(eq.3)}$$

Since the natural suitability of crop growth is limited by the minimum value, the smaller value of the temperature and precipitation fuzzy value determines the climate suitability $S(C)$ which is calculated for each *doy* (eq. 4).

$$S(C_{doy}) = min\ \{S(\bar{T}_{doy}), S(\bar{P}_{doy})\} \qquad \text{(eq.4)}$$

Among all daily fuzzy values of $S(C)$ within the year, the maximum of $S(C)$ determines the climate suitability over the growing cycle and thus, the optimal start of the growing cycle (eq. 5) for cultivation of a single crop within the entire growing season.

$$S(C_{start\ of\ the\ growing\ cycle}) = max\ \{S(C_{doy\ 1}), \ldots, S(C_{doy\ 365})\} \qquad \text{(eq.5)}$$

In order to allow for the calculation of multiple cropping, the fuzzy values for each possible combination of days for the start of the growing cycle are tested as to how often they would fit within one year. The number of multiple cropping is selected that generates the highest accumulated value. Multiple cropping and the start of the growing cycle(s) are obtained for single, double and triple cropping. Hereby, the start of the growing cycle(s) in the context of this paper describes an optimal time for cultivation of a crop to reach the maximum suitability within a year. Crop mixing is not considered. Regarding temporal demands for technical field work, we assume a break of two weeks between crop cycles.

Moreover, the following assumptions are made: At least 20 mm of precipitation are required within the first two weeks of the growing season in order to provide enough soil moisture for germination. No day within the growing period must be below 5°C and below 1°C for winter crops. Vernalisation requirements are considered separately from the growing period for winter

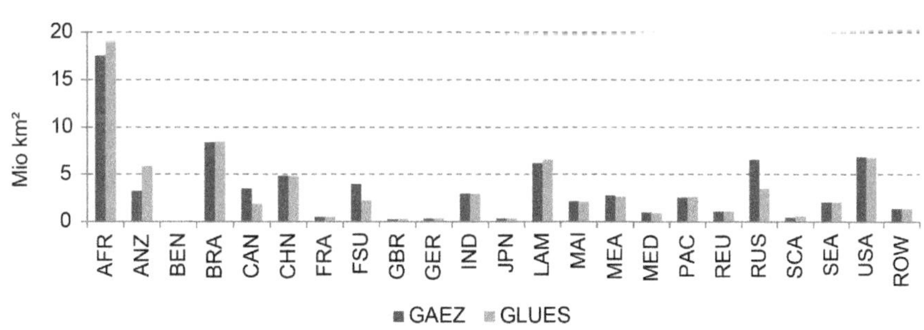

Figure 4. Comparison of total agriculturally suitable area of GAEZ (Baseline period 1961–1990) and GLUES (1961–1990) for different regions.

Table 3. Classified suitability considering rainfed conditions (1981–2010).

Not Suitable	Marginally Suitable	Moderately Suitable	Highly Suitable
49.8 million km^2	30.6 million km^2	41.3 million km^2	5.8 million km^2

crops: Vernalisation period starts 150 days before the start of the growing period. At least 20 days below 5°C must exist during the vernalisation period and there must not exist more than 3 days below −30°C. In order to consider permafrost conditions that exclude agricultural use, mean annual temperature must not be below 0°C. Mean daily incoming solar radiation must exceed 60 W/m^2 to provide enough energy for crop growth.

Thus, suitability values, number of crop cycles and the start of the growing cycle are calculated on each land surface pixel for both rainfed and irrigated conditions. For irrigated conditions, fuzzy values for precipitation are neglected during the calculation process. Due to a lack of global information on irrigation practices, we assume perennial irrigation on irrigated areas.

Besides climatic constraints, soil properties are limiting agricultural suitability. According to the membership functions (Fig. 2), the fuzzy values representing each of the soil properties are calculated. The minimum of the eight values represents the value of the soil suitability. Soil information was taken from the Harmonized World Soil Database (HWSD) [41], considering the topsoil (0–30 cm) of the dominant and all (up to 8) component soils at a spatial resolution of 30 arc seconds [42]. Within the calculation of soil suitability, fuzzy values of each of the component soils are calculated and weighted according to their share.

The suitability for crops to be cultivated is decreasing with increasing slope (see Fig. 2). The slope must not exceed 16% for the considered crops, except for oil palm and paddy rice. The slope was calculated and resampled to 30 arc seconds from Shuttle Radar Topography Mission (SRTM) data [39].

Across all climate, soil and topography fuzzy values, the lowest fuzzy value quantifies the crops' suitability at a certain location. The highest value across all crops determines the suitability for agriculture at a certain location.

This methodology does not allow for yield estimations, in which socio-economic and bio-physical aspects, which our approach does not consider, play an important role. However, this approach is well suited to draw conclusions about where areas are agriculturally suitable and how these areas may change with future climate conditions.

Results

The Earth surface consists of 510 million km^2 of which 149 million km^2 are land surface. Up to 60°S, excluding Antarctica, and considering a lack of input data, in total 127.5 million km^2 of land surface remain to be analyzed regarding their suitability for agriculture. We classified the results of the suitability analysis into four categories: not suitable (0), marginally suitable (>0.0), moderately suitable (>0.33) and highly suitable (>0.75).

Comparison

Our results (further named GLUES in the Figures) highly correlate with existing studies, such as the GAEZ approach [29], when comparing the area of each of the four classified categories in each of the 23 World Regions ($R^2 = 0.99$).

The global aggregation of the classified areas and the regional distribution of not suitable and suitable areas show a high level of agreement (Fig. 3 and 4). Compared to the distribution of global cropland in the year 2000 [43], our approach identifies 95.5% of current cropland as suitable.

Rainfed

For the period 1981–2010, our suitability analysis shows that in total 77.7 million km^2 are potentially suitable for purely rainfed agricultural cultivation, while 49.8 million km^2 are not suitable for rainfed conditions (Table 3). Further, 30.6 million km^2 are marginally suitable, 41.3 million km^2 are moderately suitable and 5.8 million km^2 are highly suitable (Table 3).

Irrigation

Irrigated agriculture produces 40% of the world's food (FAO) on 3.1 million km^2 [44]. When considering irrigation, suitability is area weighted according to the fraction of rainfed and irrigated agricultural area (given by GMIA Version 5.0 [44]). Thereby, irrigation increases suitability on irrigated areas in global average by 0.13, adds 1.8 million km^2 of suitable land (Table 4) and allows for multiple cropping on 1.2 million km^2 (assuming sufficient water available for irrigation). Accordingly, huge areas e.g. in the Nile and Ganges delta are only becoming suitable due to irrigation. Overall, 79.6 million km^2 are suitable with spatially varying patterns (Fig. 5).

Figure 5 represents the global distribution of agricultural suitability as a result of local climate, soil and terrain conditions. In boreal regions, the growing season over all stages of phenology usually is too short for cultivation. The temperate zones seasonally have adequate temperatures and enough precipitation and often sufficient soil, while in subtropical regions, the annual distribution of precipitation strongly determines crop growth and soils often are alkaline. In inner tropics have adequate temperature and moisture throughout the year, but soil quality often restricts cultivation due to low organic content and acidity [3].

Table 4. Classified suitability considering rainfed and irrigated conditions (1981–2010).

Not Suitable	Marginally Suitable	Moderately Suitable	Highly Suitable
48.0 million km^2	31.8 million km^2	41.8 million km^2	5.9 million km^2

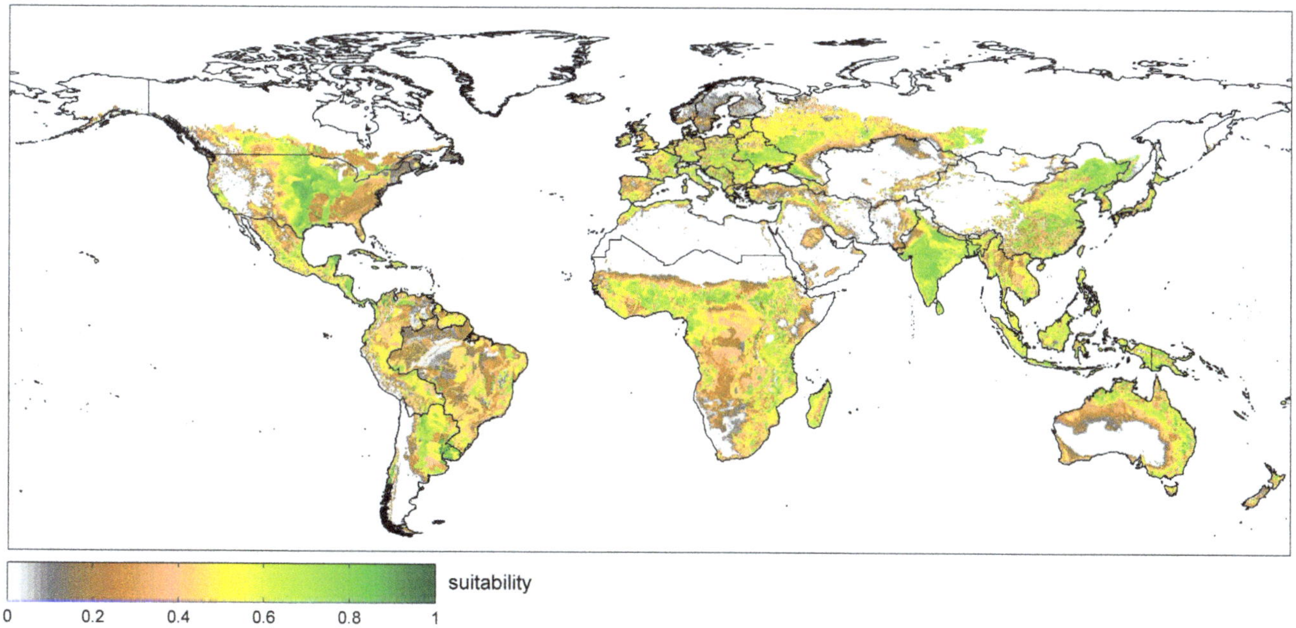

Figure 5. Agricultural suitability considering rainfed conditions and irrigated areas (1981–2010).

Protected Areas

Protected areas globally account for 8.3 million km^2. Information on actual protected areas is gathered from IUCN [45]. When excluding protected areas from the suitability calculation, 74.8 million km^2 remain suitable for cultivation. Thereby, protected areas are mainly situated in not suitable or marginally suitable areas (Table 5). Only 2% (0.2 million km^2) of the global protected area are located on land highly suitable for agriculture, 25% (2.1 million km^2) are on moderately suitable land, 30% (2.4 million km^2) on marginally suitable land while 43% (3.6 million km^2) are situated on unsuitable land. Overall, only 57% of global protected areas are suitable for agriculture.

Forested Areas

Dense forests are highly important to provide numerous ecosystem services. Densely forested areas account for 23.3 million km^2 according to GlobCover [46] and 23.5 million km^2 according to [47]. GlobCover defines forests as being dense when 75% of the pixel is forest [46]. Only 1.5 million km^2 or 6.2% of the global densely forested areas are currently protected.

4.9% (1.1 million km^2) of the densely forested areas (excluding forests within protected areas) are located in highly suitable land, 49.4% (11.1 million km^2) in moderately suitable land, 37.5% (8.4 million km^2) in marginally suitable land and only 8.2% (1.9 million km^2) are situated on unsuitable land. Overall, 92% of densely forested areas are potentially suitable for agriculture which indicates that global forests are subject to increasing societal stress.

Current Use of Suitable Land and Trade-Offs

When excluding both, protected areas and dense forests from the suitability calculation, 54.1 million km^2 remain suitable (Table 6). In comparison, currently used agricultural land (including pasture) today covers 49.1 million km^2, of which 15.5 million km^2 (status for 2011) are arable land (land under temporary and permanent crops; double-cropped areas are counted only once) [48]. Accordingly, 91% of all suitable land is already occupied by agriculture when today's protected and densely forested areas are preserved in the future. This illustrates that agricultural expansion is only possible by substituting other uses/covers of land which causes high social and ecological externalities. Figure 6 gives an overview of the current use/cover of suitable areas in the different regions of the world.

Figure 6 shows, that the current fraction of suitable area, which is not protected or dense forest is highly variable across regions. The most efficient use of current agriculturally suitable land is obvious in the USA, where only 2% of currently suitable land is not yet used or protected/dense forest.

In Africa, about 20% of the agriculturally suitable area is currently not used for agriculture or is statistically not recorded in the data of currently used agricultural land (Ramankutty *et al.*, 2008). This shows the extraordinary potentials of Africa for future expansion of agricultural land. However, agricultural expansion would always take place at ecological costs (e.g. conversion of tropical rainforest, grassland and savannah). In Latin America large suitable areas are protected or covered with dense forest and

Table 5. Classified suitability for 1981–2010 considering rainfed and irrigated conditions, excluding protected areas.

Not Suitable	Marginally Suitable	Moderately Suitable	Highly Suitable
44.4 million km^2	29.4 million km^2	39.7 million km^2	5.7 million km^2

Table 6. Classified suitability for 1981–2010 considering rainfed and irrigated conditions, excluding protected and densely forested areas.

Not Suitable	Marginally Suitable	Moderately Suitable	Highly Suitable
42.6 million km^2	21.0 million km^2	28.6 million km^2	4.6 million km^2

Table 7. Classified suitability for 2071–2100 considering rainfed and irrigated conditions, excluding protected and densely forested areas.

Not Suitable	Marginally Suitable	Moderately Suitable	Highly Suitable
37.8 million km^2	24.8 million km^2	30.2 million km^2	3.9 million km^2

the current fraction of remaining suitable area is smaller than in Africa, India is the prototype of a country, which is already using very large parts of its suitable agricultural land - and by for using the largest proportion (58%) of current cropland. Australia and larger parts of Asia still have reasonable land resources left for future expansion (Fig 6).

Future Change

For the investigation of future agricultural suitability for the time-period 2071–2100 as determined by the simulated climate effects of the SRES A1B emission scenario, we assume no changes in irrigated areas, soil properties, terrain or any adaptations, such as crop breeding. As result, when again excluding protected and densely forested areas, the global area being highly suitable for agriculture decreases from 4.6 to 3.9 million km^2, while marginally and moderately suitable areas increase (Table 7). In total, agriculturally suitable areas increase by 4.8 million km^2 due to the selected climate change scenario. However, most of the additional area is only marginally suitable for agricultural use.

Without excluding any areas, the impact of climate change increases the potentially suitable areas on the globe by 5.6 million km^2. Marginally suitable areas increase by 4.2 million km^2, moderately suitable areas increase by 2.3 million km^2, while highly suitable areas decrease by 0.8 million km^2 (Fig. 7).

A more regional analysis shows that the world is divided into regions that receive additional suitable land and regions where land that used to be suitable turns into not suitable land (Fig. 8). Regions in the northern hemisphere, such as Canada (+2.1 million km^2 of suitable land), Russia (+3.1 million km^2) and China (+0.9 million km^2), benefit most.

On the global scale, suitability improves on 18.7 million km^2 and worsens on 22.2 million km^2. In total, the area with decreasing suitability is 3.5 million km^2 more than the area with increasing suitability (Fig. 9). The highest absolute net loss of suitable areas is found in Sub-Saharan Africa.

Thereby, the globally averaged suitability value (averaged over all suitable areas), decreases from 0.41 to 0.39. The greatest losses of suitability are simulated for France and the Mediterranean (Fig. 10). The changing suitability is mapped in Fig. 11.

Growing Cycle and Multiple Cropping

The seasonal development of temperature and precipitation determines the length of the growing season, the start of the growing cycle and the potential number of annual cropping. Thus, the option of multiple cropping represents an important measure for farmers to increase production. Figure 12 shows the spatial distribution of the start of the growing cycle for the time period 1981–2010, exemplarily for maize.

Changing climate does not only affect the suitability of land, but also the start and length of the growing cycle. As an example, the start of the growing cycle for maize in Germany shifts in average 23 days earlier in time, when comparing the period of 2071–2100 with 1981–2010. The shift of growing cycles again influences the possibility for multiple cropping. Today's maximal achievable multiple cropping according to the course of temperature and precipitation is shown in Fig. 13.

Our results suggest that climate change has huge impacts on the areas suitable for multiple cropping under the assumed climate scenario. Until 2100, 6.0 million km^2 are globally lost for triple cropping until 2100, while the area which is suitable for double cropping increases by 2.3 million km^2. Multiplying the area with

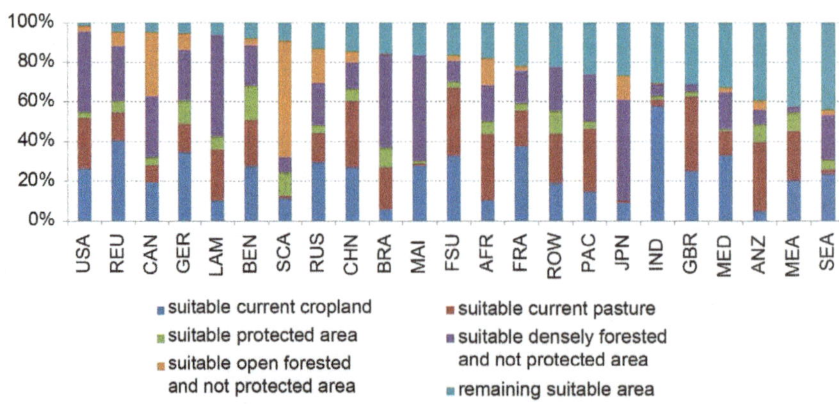

Figure 6. Current use of suitable areas (1981–2010), considering forest cover [46], **protected areas** [45] **and current pasture and cropland (Ramankutty *et al.*, 2008).** If forested areas are agriculturally suitable and protected, they are attributed to 'suitable protected area'.

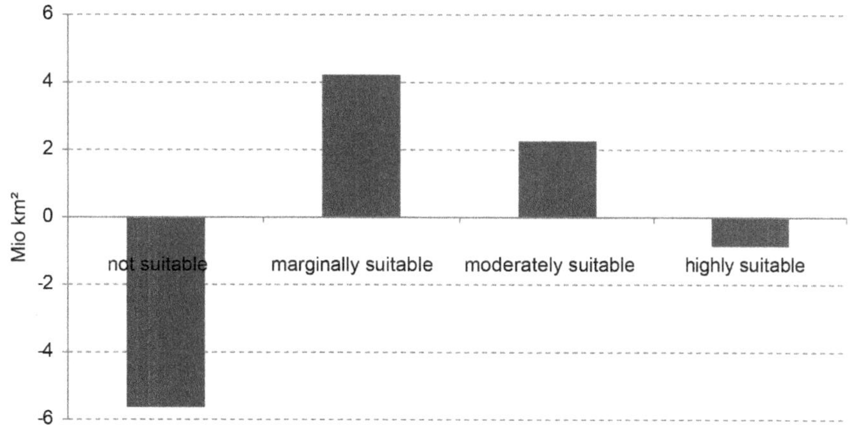

Figure 7. Global changes in agricultural suitability categories (million km^2) between 1981–2010 and 2071–2100.

the number of cycles, this means a global decrease of 13.4 million km^2. Most of the increase in double cropping areas results from the transformation from triple to double cropping. Again, no change in irrigation is assumed in this calculation.

The largest decrease in multiple copping area can be found in Brazil (BRA) and in Sub-Saharan Africa (AFR), where areas suitable for triple cropping decrease by 1.7 (AFR) and 2.9 million km^2 (BRA) (Fig. 14), while the area for double cropping increases by 0.2 and 1.3 million km^2, respectively. In total, this means a decrease of multiple cropping area by 1.5 (AFR) and 1.6 million km^2 (BRA). This is equivalent to the amount of 4.7 and 6.1 million km^2 respectively, which are lost for agriculture, when multiplying the area with the number of possible crop cycles. This corresponds to 20.2 (AFR) and 28.8% (BRA) of today's potentially suitable area for multiple cropping. In the same manner, France (FRA) and the Mediterranean (MED) lose 24.1 (FRA) and 13.2% (MED) of their total equivalent area when considering the change of multiple cropping, which means a decrease by 93 (FRA) and 55% (MED) according to the multiple cropping area of 1981–2010. Regions where areas that potentially allow for more than one crop cycle increase due to climate change are CHI, IND, JPN, MEA, REU, RUS and USA, while the total area considerably increases mainly in the USA for both, double (0.35 million km^2) and triple (0.12 million km^2) cropping (Fig. 14).

Conclusions

The analyses of the present situation demonstrats that there is extraordinary potential e.g. for Sub Saharan Africa for future expansion of agricultural land without expanding into protected or forested areas. Further research is necessary to identify the environmental and social costs and consequences of agricultural expansion in these regions. Also further investigation is needed to give answers on how this land could be managed sustainable with benefit to local food systems and socio-economy.

Our results show at high spatial resolution how agricultural suitability may change until 2100 due to changing climate under the chosen scenario (SRES A1B), assuming no adaptation measurements by farmers. First, suitable areas increase especially in the northern regions such as Canada, China and Russia, where new land will be available for agricultural use. The increase in suitable areas mainly takes place in sparsely populated areas, which could imply a lack of labor for open up new agricultural land and prepare soils. Certainly, it will be related with high investment costs and it will take a long time to extend agriculture here. Secondly, global average suitability decreases under the chosen climate scenario. Especially the extend of highly suitable areas is reduced by the effect of climate change. Finally, suitable areas indirectly are reduced due to a substantial global reduction of the suitability for multiple cropping, especially in Sub Saharan Africa, and Brazil.

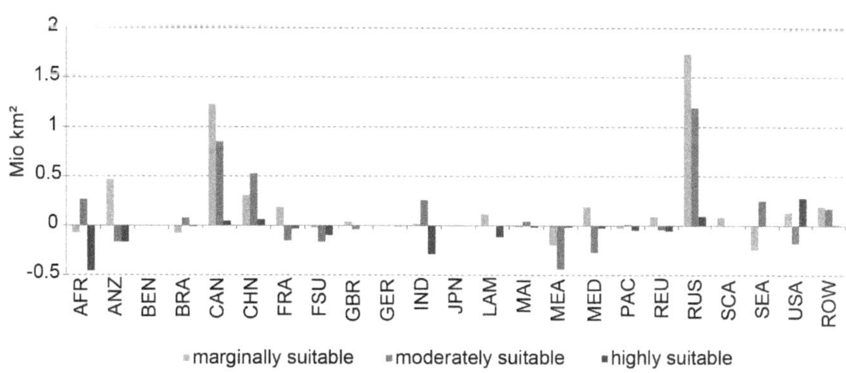

Figure 8. Regional change in agricultural suitability categories (million km^2) between 1981–2010 and 2071–2100.

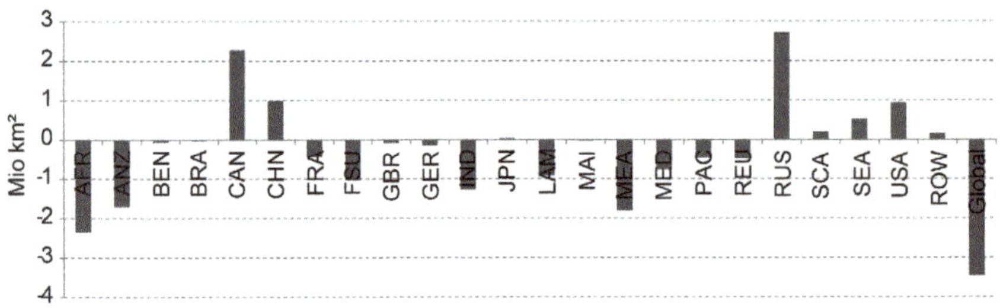

Figure 9. Regional change of agriculturally suitable area due to A1B climate change scenario between 1981–2010 and 2071–2100.

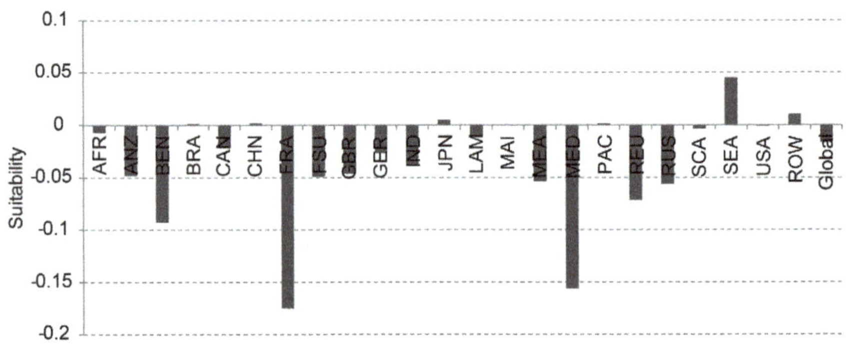

Figure 10. Regional changes in the average suitability between 1981–2010 and 2071–2100.

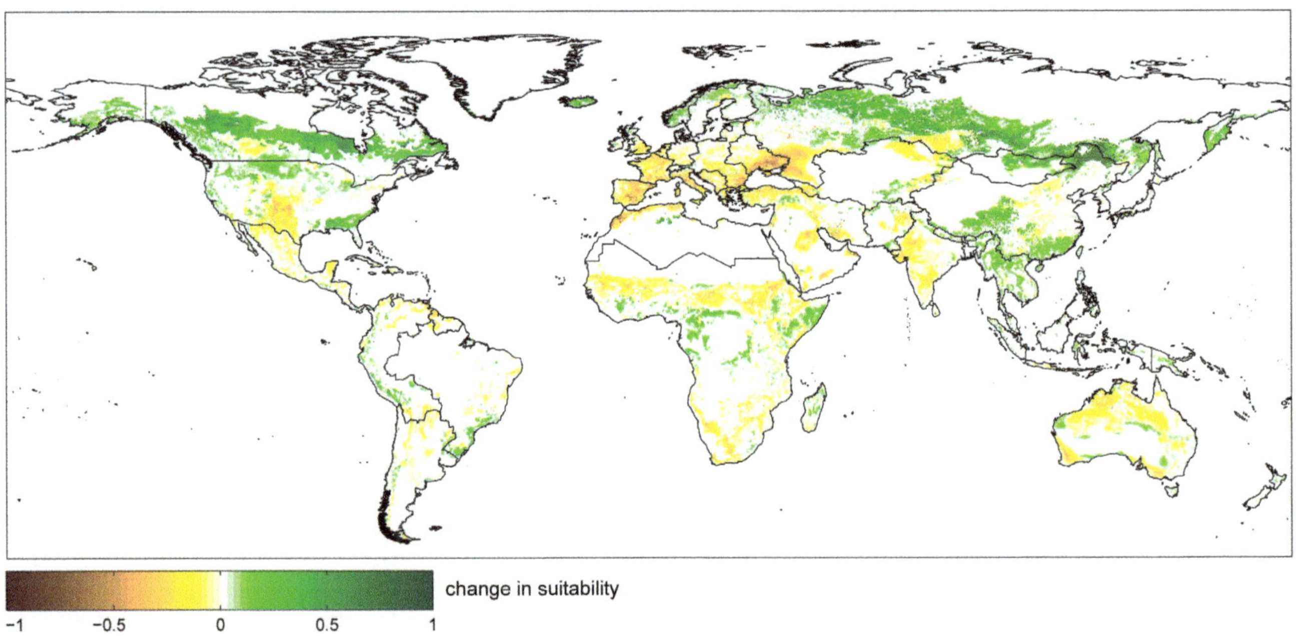

Figure 11. Change in agricultural suitability between 1981–2010 and 2071–2100. Green areas indicate an increase in suitability while brown areas show a decreasing suitability.

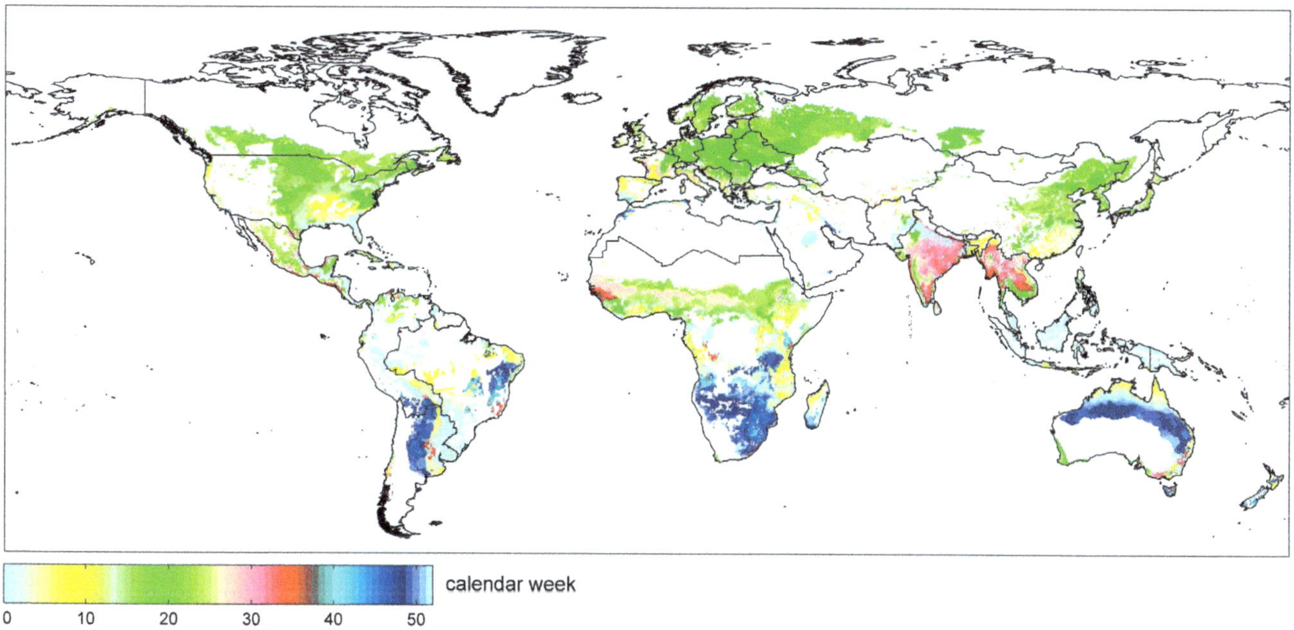

Figure 12. Start of the growing cycle for maize (1981–2010). The start of the growing cycle is illustrated for rainfed conditions and for irrigated conditions on predominantly irrigated areas (irrigated area > 50%). In case of multiple cropping, the map shows the start of the first growing cycle.

Overall, the Global North regionally increases suitability and the number of crop cycles, while the Global South and the Mediterranean area lose agriculturally suitable land without adaptations. This will decisively affect smallholder farmers as their options for adaptations through e.g. irrigation are limited.

Scientific knowledge on the geographical distribution has decisively being increased with the availability of global data sets, also based on remote sensing. The tensions between both limits of land expansion and intensification within the context of sustainable agricultural intensification stresses the ongoing debate on

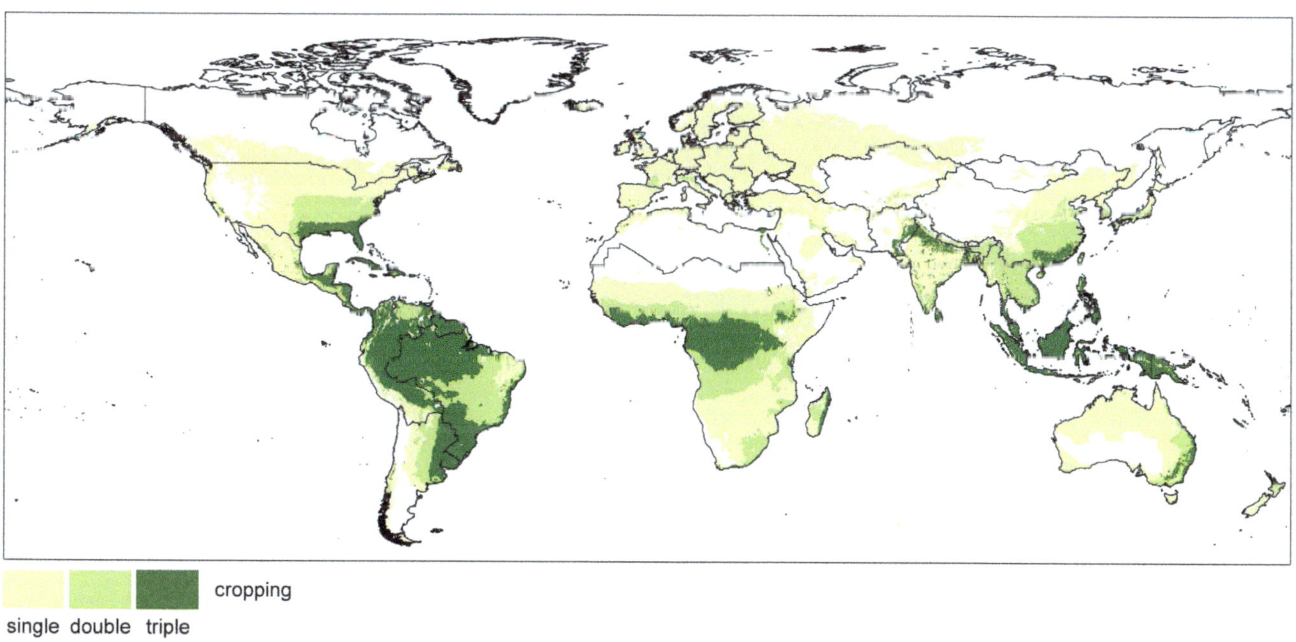

Figure 13. Suitable areas for single, double and triple cropping (1981–2010). Multiple cropping is illustrated for rainfed conditions and for irrigated conditions on predominantly irrigated areas (irrigated area > 50%).

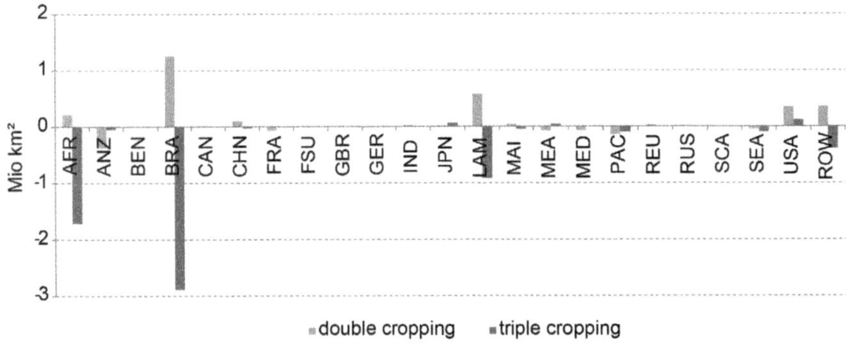

Figure 14. Change in area suitable for double and triple cropping (million km^2) between 1981–2010 and 2071–2100.

global land management, considering the complex interplay and trade-offs between different uses of land and ecosystem services.

Acknowledgments

This research was carried out within the framework of the GLUES (Global Assessment of Land Use Dynamics, Greenhouse Gas Emissions and Ecosystem Services) Project. Thanks to all project members and to Jonas Maier who contributed to this study as a student assistant.

Author Contributions

Conceived and designed the experiments: FZ BP WM. Performed the experiments: FZ. Analyzed the data: FZ. Contributed reagents/materials/ analysis tools: FZ. Contributed to the writing of the manuscript: FZ.

References

1. Foley JA, DeFries R, Asner GP, Barford C, Bonan G, et al. (2005) Global Consequences of Land Use. Science 309: 570–574.
2. Václavík T, Lautenbach S, Kuemmerle T and Seppelt R (2013) Mapping global land system archetypes. Global Environmental Change 23: 1637–1647.
3. Ramankutty N, Foley JA, Norman J, McSweeney K (2002) The global distribution of cultivable lands: current patterns and sensitivity to possible climate change. Global Ecology and Biogeography 11: 377–392.
4. Alexandratos N, Bruinsma J (2012) World agriculture towards 2030/2050: the 2012 revision. ESA Working Paper No 12-03. Rome: FAO.
5. Tilman D, Balzer C, Hill J, Befort BL (2011) Global food demand and the sustainable intensification of agriculture. Proceedings of the National Academy of Sciences 108: 20260–20264.
6. Foley JA, Ramankutty N, Brauman KA, Cassidy ES, Gerber JS, et al. (2011) Solutions for a cultivated planet. Nature 478: 337–342.
7. Godfray HCJ, Beddington JR, Crute IR, Haddad L, Lawrence D, et al. (2010) Food Security: The Challenge of Feeding 9 Billion People. Science 327: 812–818.
8. Pretty J, Sutherland WJ, Ashby J, Auburn J, Baulcombe D, et al. (2010) The top 100 questions of importance to the future of global agriculture. International Journal of Agricultural Sustainability 8: 219–236.
9. Gregory PJ and George TS (2011) Feeding nine billion: the challenge to sustainable crop production. Journal of Experimental Botany.
10. Ray DK, Mueller ND, West PC, Foley JA (2013) Yield Trends Are Insufficient to Double Global Crop Production by 2050. PLoS ONE 8: e66428.
11. Vermeulen S, Zougmoré R, Wollenberg E, Thornton P, Nelson G, et al. (2012) Climate change, agriculture and food security: a global partnership to link research and action for low-income agricultural producers and consumers. Current Opinion in Environmental Sustainability 4: 128–133.
12. Kastner T, Rivas MJI, Koch W, Nonhebel S (2012) Global changes in diets and the consequences for land requirements for food. Proceedings of the National Academy of Sciences 109: 6868–6872.
13. Cassidy ES, West PC, Gerber JS, Foley JA (2013) Redefining agricultural yields: from tonnes to people nourished per hectare. Environmental Research Letters 8: 034015.
14. Erb K-H, Haberl H, Krausmann F, Lauk C, Plutzar C, et al. (2009) Eating the Planet: Feeding and fuelling the world sustainably, fairly and humanely – a scoping study. Social Ecology Working Paper. Alpen-Adria Universität Klagenfurt.
15. Spiertz JHJ, Ewert F (2009) Crop production and resource use to meet the growing demand for food, feed and fuel: opportunities and constraints. NJAS - Wageningen Journal of Life Sciences 56: 281–300.
16. Bruinsma J (2011) The resources outlook: by how much do land, water and crop yields need to increase by 2050? In: P. Conforti , editor. Looking ahead in world food and agriculture: Perspectives to 2050. Rome: FAO.
17. Schmitz C, van Meijl H, Kyle P, Nelson GC, Fujimori S, et al. (2014) Land-use change trajectories up to 2050: insights from a global agro-economic model comparison. Agricultural Economics 45: 69–84.
18. Pimentel D, Harvey C, Resosudarmo P, Sinclair K, Kurz D, et al. (1995) Environmental and Economic Costs of Soil Erosion and Conservation Benefits. Science 267: 1117–1123.
19. UNCCD (2014) Desertification - The Invisible Frontline. Bonn, Germany: United Nations Convention to Combat Desertification.
20. Seto KC, Fragkias M, Güneralp B, Reilly MK (2011) A Meta-Analysis of Global Urban Land Expansion. PLoS ONE 6: e23777.
21. Avellan T, Meier J, Mauser W (2012) Are urban areas endangering the availability of rainfed crop suitable land? Remote Sensing Letters 3: 631–638.
22. Teka K, Haftu M (2012) Land Suitability Characterization for Crop and Fruit Production in Midlands of Tigray, Ethiopia. Momona Ethiopian Journal of Science 4: 12.
23. Kalogirou S (2002) Expert systems and GIS: an application of land suitability evaluation. Computers, Environment and Urban Systems 26: 89–112.
24. Baja S, Chapman DM, Dragovich D (2002) A Conceptual Model for Defining and Assessing Land Management Units Using a Fuzzy Modeling Approach in GIS Environment. Environmental Management 29: 647–661.
25. Braimoh AK, Vlek PLG, Stein A (2004) Land Evaluation for Maize Based on Fuzzy Set and Interpolation. Environmental Management 33: 226–238.
26. Kurtener D, Torbert HA, Krueger E (2008) Evaluation of Agricultural Land Suitability: Application of Fuzzy Indicators. In: O . Gervasi, B . Murgante, A . Laganà, D . Taniar, Y . Mun and M . Gavrilova, editors. Computational Science and Its Applications – ICCSA 2008. Springer Berlin Heidelberg. pp. 475–490.
27. Nisar Ahamed TR, Gopal Rao K, Murthy JSR (2000) GIS-based fuzzy membership model for crop-land suitability analysis. Agricultural Systems 63: 75–95.
28. Van Ranst E, Tang H, Groenemam R, Sinthurahat S (1996) Application of fuzzy logic to land suitability for rubber production in peninsular Thailand. Geoderma 70: 1–19.
29. IIASA/FAO (2012) Global Agro-ecological Zones (GAEZ v3.0) - Model Documentation. IIASA, Laxenburg, Austria and FAO, Rome, Italy: IIASA, FAO.
30. Fischer G, Hizsnyik E, Prieler S, Wiberg D (2011) Scarcity and abundance of land resources: competing uses and the shrinking land resource base. SOLAW Background Thematic Report. FAO.
31. Lane A, Jarvis A (2007) Changes in Climate will modify the Geography of Crop Suitability: Agricultural Biodiversity can help with Adaptation. Journal of SAT Agricultural Research 4: 12.
32. Avellan T, Zabel F, Mauser W (2012) The influence of input data quality in determining areas suitable for crop growth at the global scale – a comparative analysis of two soil and climate datasets. Soil Use and Management 28: 249–265.
33. Burrough PA (1989) Fuzzy mathematical methods for soil survey and land evaluation. Journal of Soil Science 40: 477–492.
34. Burrough PA, Macmillan RA, van Deursen W (1992) Fuzzy classification methods for determining land suitability from soil profile observations and topography. Journal of Soil Science 43: 193–210.
35. Sys CO, van Ranst E, Debaveye J, Beernaert F (1993) Land evaluation: Part III Crop requirements. Agricultural Publications. Brussels: G.A.D.C.
36. Bengtsson L, Hodges KI, Keenlyside N (2009) Will Extratropical Storms Intensify in a Warmer Climate? Journal of Climate 22: 2276–2301.

37. Jungclaus JH, Keenlyside N, Botzet M, Haak H, Luo JJ, et al. (2006) Ocean Circulation and Tropical Variability in the Coupled Model ECHAM5/MPI-OM. Journal of Climate 19: 3952–3972.
38. Marke T, Mauser W, Pfeiffer A, Zängl G, Jacob D, et al. (2013) Application of a hydrometeorological model chain to investigate the effect of global boundaries and downscaling on simulated river discharge. Environ Earth Sci: 1–20.
39. Farr TG, Rosen PA, Caro E, Crippen R, Duren R, et al. (2007) The Shuttle Radar Topography Mission. Reviews of Geophysics 45: RG2004.
40. Hijmans RJ, Cameron SE, Parra JL, Jones PG, Jarvis A (2005) Very high resolution interpolated climate surfaces for global land areas. International Journal of Climatology 25: 1965–1978.
41. FAO/IIASA/ISRIC/ISSCAS/JRC (2012) Harmonized World Soil Database (version 1.2). FAO, Rome, Italy and IIASA, Laxenburg, Austria.
42. Avellan T, Zabel F, Putzenlechner B, Mauser W (2013) A Comparison of Using Dominant Soil and Weighted Average of the Component Soils in Determining Global Crop Growth Suitability. Environment and Pollution 2: 11.
43. Ramankutty N, Evan AT, Monfreda C, Foley JA (2008) Farming the planet: 1. Geographic distribution of global agricultural lands in the year 2000. Global Biogeochemical Cycles 22: GB1003.
44. Siebert S, Henrich V, Frenken K, Burke J (2013) Global Map of Irrigation Areas version 5. Rheinische Friedrich-Wilhelms-University, Bonn, Germany/Food and Agriculture Organization of the United Nations, Rome, Italy.
45. IUCN (2008) Guidelines for Applying Protected Area Management Categories In: N. Dudley, editor. Gland, Switzerland: IUCN.
46. Bontemps S, Defourny P, Bogaert Ev, Arino O, Kalogirou V (2009) GLOBCOVER 2009, Products Description and Validation Report. ESA, Univ. catholique de Louvain.
47. Hansen MC, Potapov PV, Moore R, Hancher M, Turubanova SA, et al. (2013) High-Resolution Global Maps of 21st-Century Forest Cover Change. Science 342: 850–853.
48. FAOSTAT (2014) FAOSTAT Land USE module. Retrieved 24 February 2014. Available at: http://faostat.fao.org/site/377/DesktopDefault.aspx?PageID=377#ancor.

Permissions

All chapters in this book were first published in PLOS ONE, by The Public Library of Science; hereby published with permission under the Creative Commons Attribution License or equivalent. Every chapter published in this book has been scrutinized by our experts. Their significance has been extensively debated. The topics covered herein carry significant findings which will fuel the growth of the discipline. They may even be implemented as practical applications or may be referred to as a beginning point for another development.

The contributors of this book come from diverse backgrounds, making this book a truly international effort. This book will bring forth new frontiers with its revolutionizing research information and detailed analysis of the nascent developments around the world.

We would like to thank all the contributing authors for lending their expertise to make the book truly unique. They have played a crucial role in the development of this book. Without their invaluable contributions this book wouldn't have been possible. They have made vital efforts to compile up to date information on the varied aspects of this subject to make this book a valuable addition to the collection of many professionals and students.

This book was conceptualized with the vision of imparting up-to-date information and advanced data in this field. To ensure the same, a matchless editorial board was set up. Every individual on the board went through rigorous rounds of assessment to prove their worth. After which they invested a large part of their time researching and compiling the most relevant data for our readers.

The editorial board has been involved in producing this book since its inception. They have spent rigorous hours researching and exploring the diverse topics which have resulted in the successful publishing of this book. They have passed on their knowledge of decades through this book. To expedite this challenging task, the publisher supported the team at every step. A small team of assistant editors was also appointed to further simplify the editing procedure and attain best results for the readers.

Apart from the editorial board, the designing team has also invested a significant amount of their time in understanding the subject and creating the most relevant covers. They scrutinized every image to scout for the most suitable representation of the subject and create an appropriate cover for the book.

The publishing team has been an ardent support to the editorial, designing and production team. Their endless efforts to recruit the best for this project, has resulted in the accomplishment of this book. They are a veteran in the field of academics and their pool of knowledge is as vast as their experience in printing. Their expertise and guidance has proved useful at every step. Their uncompromising quality standards have made this book an exceptional effort. Their encouragement from time to time has been an inspiration for everyone.

The publisher and the editorial board hope that this book will prove to be a valuable piece of knowledge for researchers, students, practitioners and scholars across the globe.

List of Contributors

Andreas Huth
Department of Ecological Modeling, Helmholtz Centre for Environmental Research (UFZ), Leipzig, Germany

Alvaro G. Gutiérrez
Department of Ecological Modeling, Helmholtz Centre for Environmental Research (UFZ), Leipzig, Germany
Forest Ecology Group, Institute of Terrestrial Ecosystems Department of Environmental Sciences, Swiss Federal Institute of Technology (ETH Zürich), Zürich, Switzerland

Juan J. Armesto
Instituto de Ecología y Biodiversidad (IEB), Santiago, Chile
Departamento de Ecología, Facultad de Ciencias Biológicas, Universidad Catolica de Chile, Santiago, Chile

M. Francisca Díaz
Departamento de Ciencias Biológicas, Facultad de Ciencias Biolo´ gicas, Universidad Andrés Bello, Santiago, Chile

Eli D. Lazarus
Environmental Dynamics Lab, Earth Surface Processes Group, School of Earth and Ocean Sciences, Cardiff University, Cardiff, Wales, United Kingdom

Brian J. McGill
School of Biology and Ecology, University of Maine, Orono, Maine, United States of America

Viviane G. Ferro, Adriano S. Melo and Rafael Loyola
Departamento de Ecologia, Universidade Federal de Goiás, Goiânia, Goiás, Brazil

Priscila Lemes
Programa de Pós-Graduação em Ecologia e Evolução, Universidade Federal de Goiás, Goiânia, Goiás, Brazil

Kristjan Zobel
Department of Botany, Institute of Ecology and Earth Sciences, University of Tartu, Tartu, Estonia

Qiaoying Zhang
Department of Botany, Institute of Ecology and Earth Sciences, University of Tartu, Tartu, Estonia
Department of Environmental Science and Engineering, Qilu University of Technology, Jinan, Shandong, China

Yunchun Zhang
Department of Environmental Science and Engineering, Qilu University of Technology, Jinan, Shandong, China

Shaolin Peng
State Key Laboratory of Biocontrol, School of Life Sciences, Sun Yat-sen University, Guangzhou, Guangdong, China

Abbey Rosso and Peter Neitlich
National Park Service, Winthrop, Washington, United States of America

Robert J. Smith
Department of Botany and Plant Pathology, Oregon State University, Corvallis, Oregon, United States of America

Danielle Fraser and Natalia Rybczynski
Department of Biology, Carleton University, Ottawa, Ontario, Canada
Palaeobiology, Canadian Museum of Nature, Ottawa, Ontario, Canada

Christopher Hassall
Department of Biology, Carleton University, Ottawa, Ontario, Canada
School of Biology, University of Leeds, Leeds, United Kingdom

Root Gorelick
Department of Biology, Carleton University, Ottawa, Ontario, Canada
Department of Mathematics and Statistics, Carleton University, Ottawa, Ontario, Canada
Institute of Interdisciplinary Studies, Carleton University, Ottawa, Ontario Canada

Uttam Babu Shrestha
Institute for Agriculture and the Environment, University of Southern Queensland, Toowoomba, Queensland, Australia

Kamaljit S. Bawa
Department of Biology, University of Massachusetts, Boston, Massachusetts, United States of America
Ashoka Trust for Research in Ecology and Environment (ATREE), Bangalore, India

Fang-Zhou Ma, Zhi-Chuang Lü and Ren Wang
State Key Laboratory for Biology of Plant Diseases and Insect Pests, Institute of Plant Protection, Chinese Academy of Agricultural Sciences, Beijing, P. R. China

Fang-Hao Wan
State Key Laboratory for Biology of Plant Diseases and Insect Pests, Institute of Plant Protection, Chinese Academy of Agricultural Sciences, Beijing, P. R. China
Center for Management of Invasive Alien Species, Ministry of Agriculture, Beijing, P. R. China

Maria João Santos
Center for Spatial and Textual Analysis, Spatial History Project, and Bill Lane Center for the American West, Stanford University, Stanford, California, United States of America
Terry Watt
Terry Watt Planning Associates, San Francisco, California, United States of America

Stephanie Pincetl
California Center for Sustainable Communities, Institute of the Environment University of California Los Angeles, Los Angeles, California, United States of America

Jean-Franc¸ois Silvain
Institut de Recherche pour le Développement (IRD), UR 072, Laboratoire Evolution, Génomes et Spéciation, UPR 9034, Centre National de la Recherche Scientifique (CNRS), Gif sur Yvette, France et Université Paris-Sud 11, Orsay, France

Emile Faye
Institut de Recherche pour le Développement (IRD), UR 072, Laboratoire Evolution, Génomes et Spéciation, UPR 9034, Centre National de la Recherche Scientifique (CNRS), Gif sur Yvette, France et Université Paris-Sud 11, Orsay, France
UPMC Univ Paris06, Sorbonne Universités, Paris, France
Facultad de Ciencias Exactas y Naturales, Pontificia Universidad Católica del Ecuador, Quito, Ecuador

Olivier Dangles
Institut de Recherche pour le Développement (IRD), UR 072, Laboratoire Evolution, Génomes et Spéciation, UPR 9034, Centre National de la Recherche Scientifique (CNRS), Gif sur Yvette, France et Université Paris-Sud 11, Orsay, France
Facultad de Ciencias Exactas y Naturales, Pontificia Universidad Católica del Ecuador, Quito, Ecuador
Instituto de Ecología, Universidad Mayor San Andrés, Cotacota, La Paz, Bolivia

Mario Herrera
Facultad de Ciencias Exactas y Naturales, Pontificia Universidad Católica del Ecuador, Quito, Ecuador
Lucio Bellomo
Mediterranean Institute of Oceanography (MIO) CNRS/INSU, IRD, UM 110, Université de Toulon, La Garde, France

Heather J. Lynch
Department of Ecology and Evolution, Stony Brook University, Stony Brook, New York, United States of America

Mathew R. Schwaller
Mesoscale Atmospheric Processes Laboratory, NASA/ Goddard Space Science Center, Greenbelt, Maryland, United States of America

Camerron M. Crowder and Virginia M. Weis
Department of Integrative Biology, Oregon State University, Corvallis, Oregon, United States of America

Wei-Lo Liang
Institute of Marine Biology, National Dong Hwa University, Pingtung, Taiwan, R.O.C.

Tung-Yung Fan
Institute of Marine Biology, National Dong Hwa University, Pingtung, Taiwan, R.O.C. National Museum of Marine Biology and Aquarium, Pingtung, Taiwan, R.O.C.

Simón B. Castillo and Francisco Bozinovic
Departamento de Ecología and Center of Applied Ecology & Sustainability (CAPES), Facultad de Ciencias Biológicas, P. Universidad Católica de Chile, Santiago, Chile

José M. Rojas
Departamento de Ecología and Center of Applied Ecology & Sustainability (CAPES), Facultad de Ciencias Biológicas, P. Universidad Católica de Chile, Santiago, Chile
Centro de Investigación e Innovación para el Cambio Climático, Universidad Santo Tomás, Santiago, Chile

Sebastián Abades
Departamento de Ecología and Center of Applied Ecology & Sustainability (CAPES), Facultad de Ciencias Biológicas, P. Universidad Católica de Chile, Santiago, Chile
Instituto de Ecología y Biodiversidad, Santiago, Chile

Guillermo Folguera
Filosofía de la Biología Group, Facultad de Ciencias Exactas y Naturales & Facultad de Filosofía y Letras, Universidad de Buenos Aires, Buenos Aires, Argentina

Johanna Varner and M. Denise Dearing
Department of Biology, University of Utah, Salt Lake City, Utah, United States of America

José Manuel Ortega-Egea , Nieves García-de-Frutos and Raquel Antolín-Ló pez
Department of Economics and Business, University of Almería (ceiA3), Almeńa, Spain

Tianxing Wang, Jiancheng Shi, Yingying Jing, Tianjie Zhao, Dabin Ji and Chuan Xiong
State Key Laboratory of Remote Sensing Science, Institute of Remote Sensing and Digital Earth, Chinese Academy of Sciences. Beijing, China

Emily M. York and Wayne D. Lord
W. Roger Webb Forensic Science Institute, University of Central Oklahoma, Edmond, Oklahoma, United States of America
Department of Biology, University of Central Oklahoma, Edmond, Oklahoma, United States of America

Christopher J. Butler
Department of Biology, University of Central Oklahoma, Edmond, Oklahoma, United States of America

Wei-wei Zhou, Hong-man Chen, Jie-qiong Jin, Jun-xiao Yang, Yun-yu Wang, Jing Che and Ke Jiang
State Key Laboratory of Genetic Resources and Evolution, and Yunnan Laboratory of Molecular Biology of Domestic Animals, Kunming Institute of Zoology, Chinese Academy of Sciences, Kunming, China

Bao-lin Zhang
State Key Laboratory of Genetic Resources and Evolution, and Yunnan Laboratory of Molecular Biology of Domestic Animals, Kunming Institute of Zoology, Chinese Academy of Sciences, Kunming, China
Laboratory for Conservation and Utilization of Bio-resources, Yunnan University, Kunming, China

Robert W. Murphy
State Key Laboratory of Genetic Resources and Evolution, and Yunnan Laboratory of Molecular Biology of Domestic Animals, Kunming Institute of Zoology, Chinese Academy of Sciences, Kunming, China
Centre for Biodiversity and Conservation Biology, Royal Ontario Museum, Toronto, Ontario, Canada

Ya-ping Zhang
Laboratory for Conservation and Utilization of Bio-resources, Yunnan University, Kunming, China

Yuanchun Zou, Guoping Wang, Xiaonan Lou, Xiaofei Yu and Xianguo Lu
Key Lab of Wetland Ecology and Environment, Northeast Institute of Geography and Agroecology, Chinese Academy of Sciences, Changchun, China

Michael Grace
Water Studies Centre and School of Chemistry, Monash University, Clayton, Australia

Hongxiang Zhang, Yingxin Huang and Daowei Zhou
Northeast Institute of Geography and Agroecology, Chinese Academy of Sciences, Changchun, China

Guangdi Li
Northeast Institute of Geography and Agroecology, Chinese Academy of Sciences, Changchun, China

Qiang Yu
State Key Laboratory of Forest and Soil Ecology, Institute of Applied Ecology, Chinese Academy of Sciences, Shenyang, China

Wei Zheng
Graham Centre for Agricultural Innovation, Wagga Wagga, New South Wales, Australia

Yu Tian
Animal
Science and Technology College, Jilin Agricultural University, Changchun, China

Yantao Song
College of Environment and Resources, Dalian Nationalities University, Dalian, China

Hedvig K. Nenzén
Département des sciences biologiques, Universitédu Québec áMontréal, Montréal, Québec, Canada

Daniel Montoya
School of Biological Sciences, Life Sciences Building, University of Bristol, Bristol, United Kingdom

Sara Varela
Department of Ecology, Faculty of Science, Charles University, Prague, Czech Republic

Dana H. Ikeda, Stephen M. Shuster and Thomas G. Whitham
Department of Biological Science, Northern Arizona University, Flagstaff, Arizona, United States of America
Merriam-Powell Center for Environmental Research, Northern Arizona University, Flagstaff, Arizona, United States of America

Kevin C. Grady
Merriam-Powell Center for Environmental Research, Northern Arizona University, Flagstaff, Arizona, United States of America
School of Forestry, Northern Arizona University, Flagstaff, Arizona, United States of America

Caren B. Cooper and Jennifer Shirk
Cornell Lab of Ornithology, Ithaca, New York, United States of America

Benjamin Zuckerberg
University of Wisconsin, Madison, Wisconsin, United States of America

Colin S. Shanley and David M. Albert
The Nature Conservancy, Juneau, Alaska, United States of America

Florian Zabel, Birgitta Putzenlechner and Wolfram Mauser
Department of Geography, Ludwig Maximilians University, Munich, Germany

Index